WITHDRAWN

LARAMIDE FOLDING ASSOCIATED WITH BASEMENT BLOCK FAULTING IN THE WESTERN UNITED STATES

The Geological Society of America, Inc.
Memoir 151

Laramide Folding Associated with Basement Block Faulting in the Western United States

Edited by
VINCENT MATTHEWS III

1978

Copyright 1978 by The Geological Society of America, Inc.
Copyright is not claimed on any material prepared by
U.S. Government employees within the scope of their employment.
Library of Congress Catalog Card Number 78-54346
ISBN 0-8137-1151-7

Published by
THE GEOLOGICAL SOCIETY OF AMERICA, INC.
3300 Penrose Place
Boulder, Colorado 80301

Printed in the United States of America

Contents

Preface	vi
Faulting and forced folding in the Rocky Mountains foreland . . . .*David W. Stearns*	1
Seismic interpretation of basement block faults and associated deformation . . . . *W. R. Sacrison*	39
Some two-dimensional kinematic analyses of the drape-fold concept . . . . . *David M. Weinberg*	51
Experimental folding of rocks under confining pressure: Part VI. Further studies of faulted drape folds . . . . *John M. Logan, M. Friedman, and M. T. Stearns*	79
Laramide folding associated with basement block faulting along the northeastern flank of the Front Range, Colorado . . . .*Vincent Matthews III and David F. Work*	101
Laramide structures and basement block faulting: Two examples from the Big Horn Mountains, Wyoming . . . .*John C. Palmquist*	125
Geometric analysis of multiple drape folds along the northwest Big Horn Mountains front, Wyoming . . . .*Martha Tirey Stearns and David W. Stearns*	139
Origin of Elk Mountain anticline, Wyoming . . . . . .*James E. McClurg and Vincent Matthews III*	157
Laramide structure of the Black Hills uplift, South Dakota–Wyoming–Montana . *Alvis L. Lisenbee*	165
A relationship between strike-slip faults and the process of drape folding of layered rocks . . . . *Robert A. Cook*	197
Monocline fold pattern of the Colorado Plateau . . . .*George H. Davis*	215
Development of monoclines: Part I. Structure of the Palisades Creek branch of the East Kaibab monocline, Grand Canyon, Arizona . . . . *Ze'ev Reches*	235
Development of monoclines: Part II. Theoretical analysis of monoclines . . . . *Ze'ev Reches and Arvid M. Johnson*	273
Analytical solutions applied to structures of the Rocky Mountains foreland on local and regional scales . . . . *Gary Couples and David W. Stearns*	313
Comments on applications of boundary-value analyses of structures of the Rocky Mountains foreland . . . . *Gary Couples*	337
Plate tectonics of the Laramide orogeny . . . .*William R. Dickinson and Walter S. Snyder*	355
Index	367

Preface

Papers in this volume describe various aspects of Laramide deformation in the Rocky Mountains foreland and the Colorado Plateau. This collection of papers is not the result of a formal symposium, rather it is the product of coincidence. Three years ago I realized that several geologists were in the final stages of investigations dealing with Laramide deformation and that the results of their investigations were being reported at regional and national meetings around the country. I explored the possibility of collecting these recent reports into one volume, because much of the older literature on Laramide deformation in the Rocky Mountains is scattered in journals and guidebooks of local and regional societies. This Memoir is the result; the quality and scope of the included papers far exceed my initial hopes.

Most of the structures produced by Laramide deformation in the Rocky Mountains are interpreted by the authors to be drape folds (forced folds), that is, the geometry of the folds in the ductile sedimentary strata is determined by the differential uplift of underlying brittle blocks of structural basement. The widespread occurrence of drape folds in the Rockies was first pointed out in 1965 by J. J. Prucha, J. A. Graham, and R. P. Nikelsen in an article titled, "Basement Controlled Deformation in Wyoming Province of Rocky Mountain Foreland." Since then, many workers have concluded that most of the Laramide structures in the Rocky Mountains foreland are drape folds. However, most structural geology textbooks lack information on the concept of drape folding. The AGI *Glossary of Geology* gives an inadequate and misleading definition for the term "drape fold." It is hoped that this collection of papers will focus attention on the significant advances that have been made toward understanding foreland-type deformation in the Western United States.

The first paper in this volume is by D. W. Stearns and is a state-of-the-art review of research on the concept of forced folding. Specific examples of forced folds in Arizona, Utah, Colorado, Wyoming, and South Dakota are found in the papers by G. H. Davis, R. A. Cook, V. Matthews and D. F. Work, M. T. Stearns and D. W. Stearns, J. C. Palmquist, J. E. McClurg and V. Matthews, and A. L. Lisenbee. Successful efforts to create drape folds in the laboratory are reported in papers by G. H. Davis and by J. M. Logan, M. Friedman, and M. T. Stearns. Discussions of new theoretical solutions are found in the papers by Z. Reches and A. M. Johnson, G. Couples, and G. Couples and D. W. Stearns. D. M. Weinberg discusses and contrasts the kinematics of the Paleozoic and Mesozoic strata in forced folds. W. R. Sacrison presents seismic evidence for forced folds in the subsurface of Wyoming. Z. Reches argues that a monocline in the Grand Canyon is a buckle fold, rather than a drape fold. In the final paper, W. R. Dickinson and W. S. Snyder discuss how these foreland structures may fit into the framework of plate tectonics.

VINCENT MATTHEWS III
Amoco Production Company
Security Life Building
Denver, Colorado 80202

Faulting and forced folding in the Rocky Mountains foreland

David W. Stearns
Department of Geology and Center for Tectonophysics
Texas A&M University
College Station, Texas 77843

ABSTRACT

The structural basement in the Rocky Mountains foreland deforms by faulting and rigid-body rotations. The faults at the interface between the sedimentary and crystalline rocks can be anything from low-angle reverse to normal faults. Despite the range of geometries, the total assemblage of faults and rotations is best explained by a movement system that is dominated by vertical motions along faults, many of which are curved in cross section. The first causes deep within the crust are not sufficiently well documented in the geophysical record to justify a firm interpretation. However, there are certain conditions recorded in the geologic history of the region and in the surface structures that place constraints even on speculations. With these constraints, vertical movements seem more likely to dominate than horizontal movements. The sedimentary layers are deformed primarily by forced folding. Their final geometry is a product of several parameters such as welding and stratigraphic make-up. Measurements on natural folds demand that the section either thinned or detached. Detachment without appreciable thinning further requires that (1) large displacements occur within the sedimentary section and (2) at the termination of these folds, the movement must be in several directions. Geologists' intuition as to how the layered rocks achieved their shapes is not always correct, but field data combined with experimental and theoretical data provide a basis for understanding these folds. It is concluded that the structural style in the Rocky Mountains foreland is not unique, but rather it is only an excellent example of a more universal class of deformation, namely, forced folding.

INTRODUCTION

The structures within the Rocky Mountains foreland have long attracted the attention of the geological fraternity. Well they should have because they are young enough to have many of their features preserved, they are well exposed both vertically and horizontally, they present perplexing problems to challenge the intellect, and they occur in magnificent terrains that excite geologists. The features have been mapped, modeled, drilled, theorized about, and shot seismically. They have created joyous careers for some and lifelong enemies and frustration for others. Therefore, to try to combine all of the science and emotion into one unprejudiced account would only be possible by a writer outside the geological fraternity. I am, thankfully, not outside of this fraternity, so my admission from the beginning is that I am writing from 20 years of active, but prejudiced, interest in the subject.

This paper will be based primarily on the work of the past 20 years, and there are two major objectives. The first is to describe the general tectonic class to which most of the structures in the Rocky Mountains foreland belong. The second is to review what is known about these structures and what can be concluded from them.

Prucha and others (1965) referred to the Rocky Mountains foreland as the Wyoming province because the styles of deformation in the entire region are so characteristic of the state of Wyoming with the exception of the western Wyoming thrust belt. In their use of "Wyoming province" they emphasized that the area characterized by this structural style extends far beyond the boundaries of Wyoming into parts of Montana, Colorado, Utah, New Mexico, and Arizona. That the layered rocks in this province are capable of considerable folding has been emphasized by Berg (1962), Prucha and others (1965), Stearns (1971), Stearns and others (1975), and Cook and Stearns (1975), to mention only a few. How the layered rocks fold, however, is a subject more debated. In this paper I will argue that within this region, the fold style is a manifestation of a universal tectonic type: forced folding.[1]

FORCED FOLDS

A forced fold is one in which the final overall shape and trend are dominated by the shape of some forcing member below. Imagine a series of wooden blocks of irregular size and shape on a table top; the blocks are covered by a layer of some pliable material, such as a rubber sheet. If the blocks were then differentially tilted and rotated and the sheet was forced to conform to the irregular surface, the resulting anticlines and synclines would be forced folds. This is opposed to end loading the rubber sheet where the size, shape, and location of the resulting folds would be controlled by the geometry and physical properties of the rubber sheet (free folds). That the fold style in the Wyoming province is primarily one of forced folding will be justified below, but some support for this idea already exists in the literature (Stearns, 1971; Stearns and others, 1975; Stearns and Weinberg, 1975). This fold style has also been correctly referred to as "drape folding," but a drape fold is only a specific type of the more general class.

In general, the forcing member for a forced fold can be anything from an intrusive sill (Johnson, 1970; Savage, 1974), to faulted basement (Stearns, 1971), to a faulted massive sedimentary unit such as the Ellenberger Dolomite in the Permian basin (Elam, 1969). Another way to describe a forced fold is to consider it as a fold type that allows rocks to go from a discrete discontinuity in the displacement field (fault) to a more widespread or integrated total displacement. In Figure 1a, the discrete discontinuity in the displacement field that is implied by the fault

---

[1]The term "forced folding" was proposed, but not published, by the late George M. Sowers (personal commun. about 1971).

in line AA' at point X has been absorbed by the folded layer BB' and distributed over region Y as illustrated in Figure 1b.

Because forced folding is dependent on the shape of the forcing member, originally horizontal sedimentary layers are generally loaded at high angles to layering. When loaded at low angles to layering, free folding usually occurs even though in such a loading system, forced folds can develop as a secondary result.

The large strength anisotropy imparted to most sedimentary rocks because of their depositional layering is very important to the forced-folding process. In order to accomplish any folding, but especially forced folding, a great deal of internal slip (along bedding planes) is required. The cohesive strength across natural depositional layering is usually much less in sedimentary rocks than across any other original plane through the intact material (Donath, 1961). Bedding planes become prime candidates for slip planes because the shear stress needed to cause slip along them essentially must only exceed the coefficient of sliding friction. Furthermore, natural bedding enables slip to be distributed over large volumes of rocks in small increments without having to overcome the bulk cohesive strength for every new slip plane. On other planes, not only must the coefficient of friction be exceeded before sliding can occur, but the cohesive strength must be overcome as well.

The ease with which forced folding occurs and the specific geometries that result are the combined effect of several physical parameters, but none is probably more important than depth of burial during deformation. That is, the low cohesive strength due to bedding becomes less and less an advantage with increasing depth. As burial increases, the normal stresses across the flat-lying bedding planes increase (assuming normal pore pressure), and therefore, the amount of shear stress needed for slip along them correspondingly increases. In other words, in beds loaded at high angles to their layering, increasing burial tends to make faulting an easier mechanism of deformation than folding until burial becomes so great that the entire rock mass becomes ductile (metamorphism). It is likely that deeply buried sections (about 7,000 m deep) would fault and behave as part of the faulted forcing member provided that they remained brittle. The sharp discontinuity between faulting and folding illustrated in Figure 1 is only possible when dealing with a brittle and statistically homogeneous forcing member beneath a shallowly buried sequence of layered rocks. Such a system produces a lower region that is fault-dominated, (the lower forcing member) and a section above that is fold-dominated. If, however, the layered sequence is thick (in excess of 5,000 or 6,000 m), there very well may develop three distinct vertical zones in which different mechanisms predominate. In the forcing member, faulting is the dominant mechanism; in the very shallow layered rocks, folding is the dominant mechanism; but between areas of dominant faulting and dominant folding, there may be sequences in which mixed faulting and folding occur. One excellent exposure of this type of behavior crops out

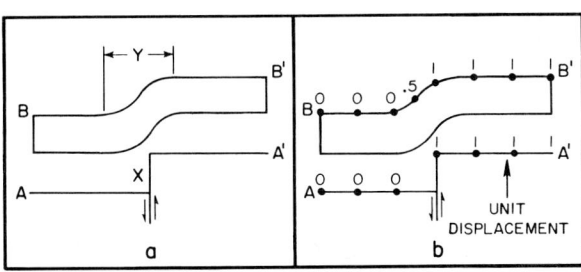

Figure 1. Schematic illustration of displacement in a forced fold compared to displacement in the faulted forcing member. The discontinuity in line AA' is spread out over region Y in the fold.

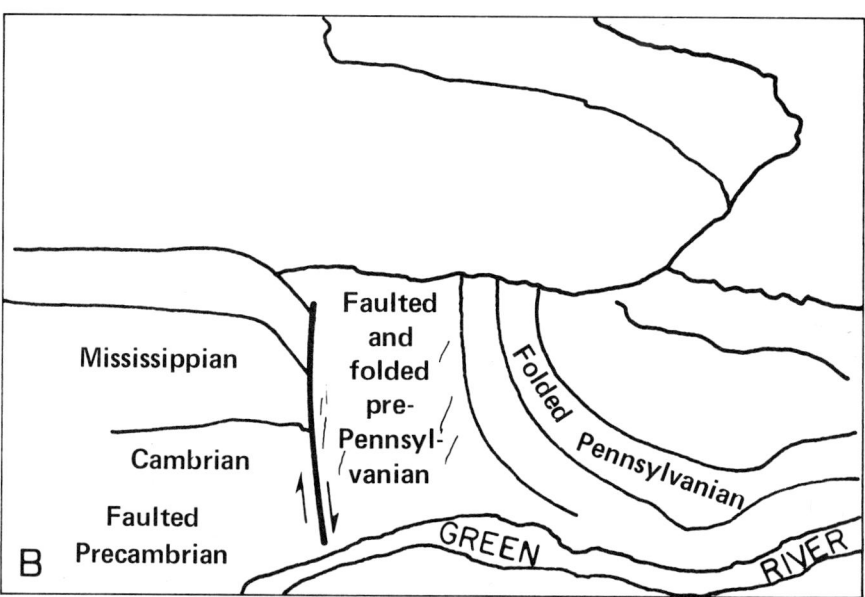

Figure 2. (A) Photograph of section of rock that is fault-dominated at the bottom and fold-dominated at the top but with mixed faulting and folding in the middle. (B) Tracing of photograph in Figure 2A; ages of rock units are shown.

along the Green River in the eastern Uinta Mountains and was mapped by Untermann and Untermann (1965). Here the layered rock sequence is thicker than 5,000 m, and the upper part of the Precambrian and the lower part of the Cambrian section are dominated by faulting (Fig. 2). The Upper Cambrian through Lower Pennsylvanian rocks are deformed by a combination of faulting and folding, but in the younger rocks, faulting is totally absent and folding is the dominant mechanism.

A second condition that greatly affects the style of forced folding is how strongly

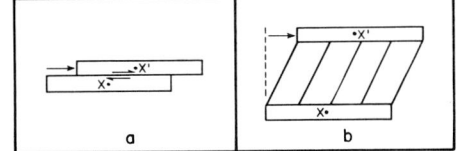

Figure 3. Schematic illustration of two types of offsets. (a) Slip of one rock layer past the one immediately below. (b) Slip of one layer relative to another below by flow within an intervening layer.

the layered sequence is welded to the forcing member. If the layered rocks are not easily decoupled from the forcing member, so that interstratal slip cannot become effective, the layered rocks tend to fault rather than develop folds. How strongly the layered rocks are welded to the forcing member can be a function of burial alone, because the higher the normal stress across the base of the layered sequence, the more difficult it is to decouple these rocks from the forcing member. However, burial is not the only condition that affects the welding. The ductility of the unit immediately above the forcing member also affects the degree of welding. At least two types of offset can occur (Figs. 3a and 3b). In the first case, slip occurs on a single plane within the layered rocks, and there is discrete separation between previously adjacent points. The physical parameters that affect such an offset, often called a detachment, are the normal stress across the plane and the coefficient of sliding friction between the materials of the two layers. (The cohesive strength across the contact is ignored here because it would only vary slightly with changes in rock type.) Considerable offset of one layer relative to another also can be accomplished across intervening thick ductile units without ever involving the frictional characteristics of discrete continuous planes (Fig. 3b). Separation of points X and X' is the same in each case, but in Figure 3b it has occurred owing to the ductility of the intervening unit. Because this is primarily a ductility control, it would be enhanced by higher confining pressures. The higher pressures, however, would make the mechanism illustrated in Figure 3a more difficult. Any weak, ductile stratigraphic unit is a good candidate for accomplishing the sort of offset illustrated in Figure 3b. Obvious examples are thick clay-shale or salt units. Field experience has shown that intuition is not always good in determining what rock types will allow offsetting in a section because of their ductility. For example, in the Freezeout Mountains, Precambrian granitic material serves as the forcing member. It is immediately overlain by a thin limestone (Mississippian) and a thick sequence of bedded sandstones and limestones of the Casper Formation (Pennsylvanian). Such a section might not seem to be ductile, but in this case the Pennsylvanian sandstones served as an excellent decoupling unit because of their ability to flow cataclastically. Here, owing to the bulk ductility of the sandstone, drape folds of at least 1,000-m (3,000-ft) displacement have formed without faulting of the sedimentary section.

Two other factors that affect the eventual geometries in forced folding are the physical make-up of the forcing member and that of the folded sequence. If the forcing member is, for example, shallowly buried granitic material, not only is faulting the dominant mechanism, but folding is completely excluded (for reasons, see Stearns, 1975). If the forcing member is a massive dolomite unit such as the Ellenberger Dolomite of West Texas, then because of both its brittleness and its lack of closely spaced bedding, faulting may still be the dominant mechanism. However, there may be small amounts of folding that accompany the faulting in the forcing member.

The make-up of the folded part of the system also has a great deal of control on the resulting geometry. If the stratigraphic layers above the forcing member have physical properties so that thinning occurs at lower stress differentials than are required for massive interstratal slip, the units fold over the forcing member, but are attenuated or thickened as is demanded by the kinematics of the fold. If the physical make-up of the layered rocks is such that interstratal slip occurs at lower stress differentials than does appreciable thinning, individual layers experience lateral mass transport into the fold, and little or no thinning accompanies the folding process. The tendency for any given stratigraphic sequence to behave according to either of the endpoints may be more a product of their bulk behavior than the individual characteristics of any specific layer (Stearns, 1969). Examples of thinned sections have been reported for Mesozoic sandstone and shale sequences at Casper Mountain, Wyoming (Vaugh, 1976), Hamilton dome, Wyoming (Berg, 1976), in Pennsylvanian sandstone sections in the eastern Uinta Mountains (Cook and Stearns, 1975), and in Triassic sandstone units in the Uncompahgre uplift (Stearns and Jamison, 1977). Examples of nonthinned sections have also been reported (Stearns, 1971; Stearns and others, 1975; Stearns and Stearns, this volume). It is important to note that although the end-product geometry is different for rocks that thin as opposed to those that simply translate without thinning, this is only a difference in mechanistic response within the layered rocks to the same causative loading conditions, not a difference in the fold type.

The final parameter that seemingly affects the behavior of the layered rocks is the angle at which the fault leaves the forcing member. Because forced folding (as defined in this paper) usually is produced by loads at high angles to the bedding, faults that leave the forcing member as high-angle reverse faults (60° or steeper) or as very high angle normal faults (75° or greater) are more effective in producing the folds. If the layered rocks in the immediate vicinity of the top of the forcing member are put into too much extension (normal faults that dip less than 75°), the layered sequences tend to fault rather than fold. Although precise numbers for the exact amount of extension that causes this phenomenon in natural situations are not available, the general principle is in accord with the laboratory findings of Heard (1960). He found that rocks are considerably more brittle in extension than when put into even slight compression. The requirement for compression, as opposed to extension, by high-angle reverse faulting should not, however, be confused with low-angle reverse faulting (overthrusting), which tends to produce faulting and/or free folding in the layered rocks. The precise dip angles for fault control from natural examples are not available, but it should be noted that there are no known examples of drape folds over proved normal faults that dip 60° or less. Drape folding can occur over very high angle normal faults. For example, the frontal normal fault at Rattlesnake Mountain in Wyoming dips about 85° where exposed, and there is considerable drape folding above this displacement (Stearns, 1971). In the Owl Creek Mountains, however, the Boyson fault leaves the basement as a 60°-dipping normal fault. Over this fault the same sequence of layered rocks, under essentially the same deformation conditions as existed at Rattlesnake Mountain, are faulted through with little or no folding (Fanshawe, 1939). If the fault leaves the basement at low angles (less than 45°) it also continues into the layered rocks (Stearns and others, 1975).

FORCED FOLDS IN THE ROCKY MOUNTAINS FORELAND

Statement of the Problem

Even casual study of such a simplified map as is shown in Figure 4 indicates several of the major tenants that must be included in a discussion of this region.

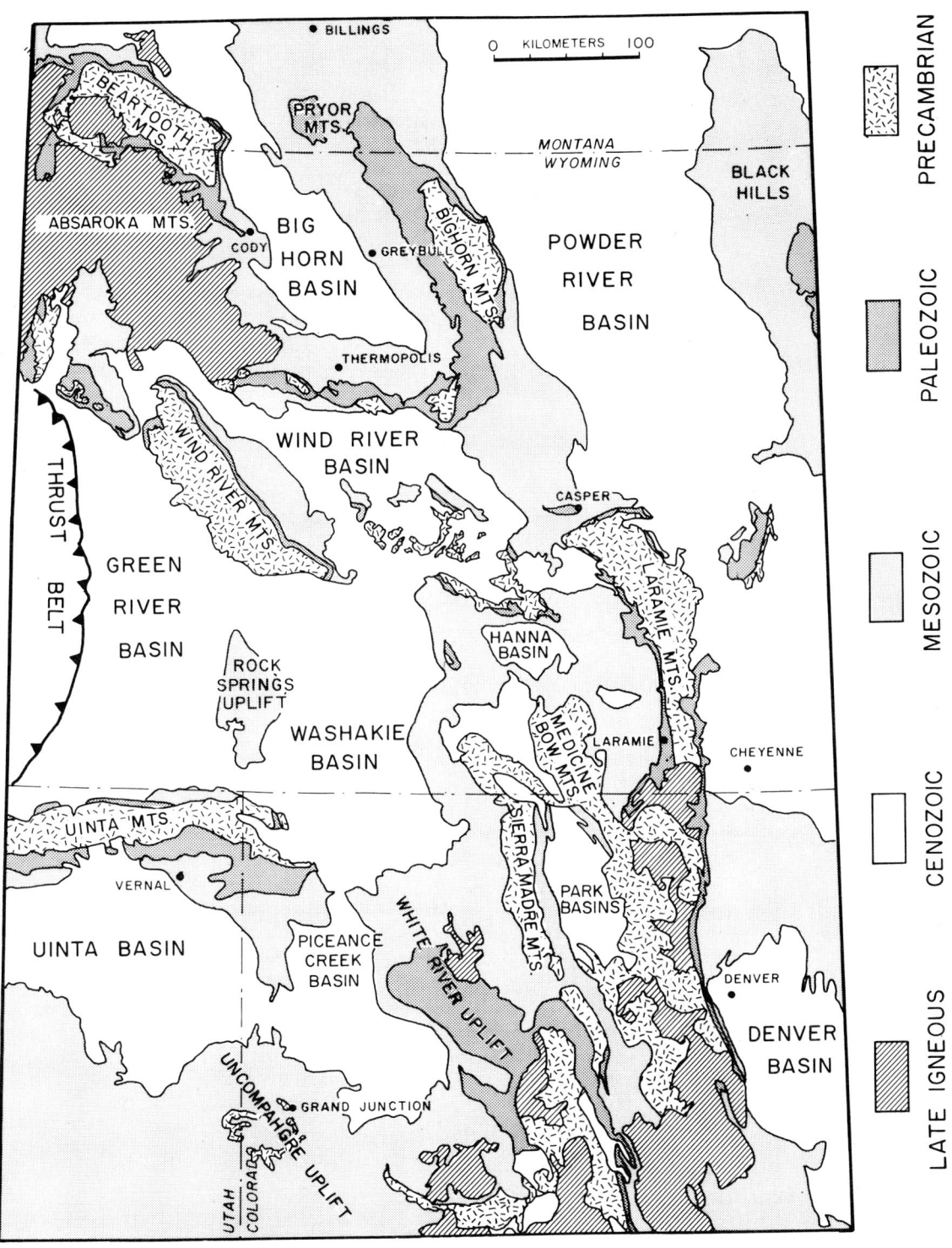

Figure 4. Generalized geologic map of part of the Rocky Mountains foreland.

Because most of the major mountain ranges are cored with Precambrian basement rock, it is evident that the basement is involved in the structural style. It is also evident that although there are statistically dominant trends to the structural features, the fabric of the structural trends is not as strong as it is in thrust belts. From the distribution of mountains, uplifts, and basins, it is apparent that whatever the deformation style, large-scale differential vertical movements occurred. The basins, like the mountains, have several different trends and shapes. Some of them are nearly oval (Bighorn Basin), others triangular (Wind River Basin), and others almost round (Hanna Basin).

There are several features concerning the overall style that are not observable from such a simplified map, but are definitely facts in the geologic record. Throughout the region, the Precambrian basement surface was peneplaned before the transgression of Cambrian seas. This peneplaned Precambrian surface is for the most part a crystalline basement in which planar anisotropies play a very small mechanical role. The two areas of exception are in and around the Uinta Mountains and in the southwestern corner of the Wind River Mountains, where the upper Precambrian section does contain rocks with sufficient layering to impart a strength anisotropy. This peneplaned surface was deformed during the Laramide orogeny by a series of rigid-body rotations that resulted in absolute upward and absolute downward motions of the original planar surface. The peneplaned surface at the beginning of the Laramide orogeny, as attested to by sedimentary thicknesses, was a relatively flat-lying surface except in the westernmost sections of Wyoming. This surface at the beginning of deformation was between 3,000 and 3,500 m (10,000 and 12,000 ft) below sea level. Today, mountain peaks with eroded Precambrian rocks range up to 4,800 m (15,500 ft) above sea level, and the original planar Precambrian surface in the deeper parts of basins is between 9,000 and 9,500 m (30,000 and 40,000 ft) below sea level. Most of the basins are strongly asymmetric. Most of the mountain blocks are rotated so that dips on the previously horizontal Precambrian surface range between 10° and 16°. However, a few of the mountain blocks, notably the Uncompahgre uplift and the Beartooth Mountains, are plateau-type uplifts where the elevated Precambrian surface is nearly horizontal. Some of the rotated blocks are rotated toward, and some away from, an adjoining basin. Some basin edges are formed by a single, uniformly rotated mountain block (for example, the south side of the Wind River Basin, which is formed by the north flank of the Wind River Mountains). Other basin edges, such as the eastern flank of the Bighorn Basin, are formed by separate blocks rotated in different directions. Along the northeastern flank of the Bighorn Basin, the mountain blocks are rotated away from the Bighorn Basin and toward the Powder River Basin; in the southern part of the Bighorn Basin, however, the mountain blocks are rotated toward the Bighorn Basin and away from the Powder River Basin. One final observational fact is that within Wyoming, some mountain blocks are bounded by faults that are very high angle, whereas other mountain blocks are bounded by much lower angle reverse faults. In general there are more high-angle major faults in the northern part of Wyoming and more low-angle major faults in the southern part. It is not impossible to find a low-angle fault feature in the northern part of the province, nor is it impossible to find a high-angle fault feature in the southern part of the province; however, the occurrence of high-angle faults is more common in the north than in the south.

These observations are neither profound nor new, but they all must be included in any attempt at explaining the structural style in a general way. They cannot be taken one at a time, or conveniently ignored. If a system is going to explain the structural behavior of this region, it must be able to account for all of these facts within the framework of the explanation offered.

The structural style under consideration has been variously categorized in the geologic literature as reviewed by Berg (1962). Terms such as "block faulting" or "Rocky Mountains foreland faulting" have been applied. These descriptive terms serve well to designate the geographic area or to characterize the particular geometric form. However, they do little to specify the actual structural problem under attack. To specify properly the structural problem of the region it should be stated both mechanically and geologically. In the most general mechanical terms the problem is that of the large-scale behavior of layered, inhomogeneous, anisotropic sequences of varied lithic type (sedimentary rocks) as they are deformed over rotated blocks of shallowly buried, statistically isotropic, homogeneous, continuous basement. The geologic problem is that of the response of layered sedimentary rocks as they are deformed over blocks of Precambrian crystalline basement that are bounded by faults with various dip angles. Stearns and others (1975) argued that if the faults are high angle (from 75° normal to 60° reverse), the layered sequences in the Rocky Mountains foreland are folded over the tops of the basement blocks (forced folding). Both the mechanical and the geologic statements of the problem emphasize the behavior of the forcing member and the behavior of folded layered rocks. It is only when the layered rocks are faulted through that there is any remote continuity of structural styles between the basement and the overlying sedimentary veneer. For these reasons, then, the response of the basement will be considered separately from the response of the overlying sedimentary rocks. In some cases, the two responses will be similar, but in most, they will be distinctly different.

Structural Response of the Basement

The term "basement" as used here has only a mechanical connotation. In the Rocky Mountains foreland the basement is all of Precambrian age, but this is only an accident of the particular rocks that occur within the region, and there is no necessity to place an age restriction on the structural basement. The upper surface of any structural basement should be that level below which there is no reasonable expectation of the occurrence of significant mechanical layering. The basement is, therefore, that mass of rock which is statistically homogeneous, isotropic, and continuous. Throughout the foreland the basement is what the field geologist would refer to casually as Precambrian "granite." Although in detail the rock is not always mineralogically a granite, it is a crystalline material in which layering plays no role. Precambrian rocks having either sedimentary layering or closely spaced metamorphic foliation should not be considered basement. Regions in the foreland where Precambrian rocks with such layering occur are the extreme southern end of the Wind River Mountains, the southern Front Range below the Canon City embayment, and the Uinta Mountains.

The material properties of the basement in the foreland are such that the rock behaves brittlely up to the point of rupture unless it is subjected to very high confining pressures and/or temperatures (Moho conditions or very near intrusions). Certainly the burial conditions during Laramide deformation in the Rocky Mountains foreland (less than 4,500 m) require that the upper several thousand metres of basement behave as a brittle material. This fact, which is so clearly demonstrated in the laboratory (Borg and Handin, 1966), has not been easily accepted by many geologists, who still insist upon significant folding of the basement. If one accepts that during Laramide deformation, for some reason unknown to physics, these basement rocks changed their mechanical properties and folded, one must also accept an extremely coincident event of post-Laramide erosion—that is, the contact between Precambrian granite and Cambrian sedimentary rocks is nowhere significantly arched or folded, although this contact is well exposed in the Wind River, Big Horn, Owl Creek, and Gros Ventre Mountains as well as over vast regions in the Uncompahgre Plateau. The contact certainly is rotated, but the dips in the rotated blocks are uniform and arching is absent. Folding of the basement has been *interpreted* by connecting two linear rotations in opposite directions with a curved surface, or else strictly by interpretation of nonexposed basement. If the basement is truly folded on a significant scale, nowhere is this folding exposed for direct observation. As will be discussed below, at the sharp edges and especially corners of brittle basement blocks, there can be a certain amount of closely spaced breakage of the brittle block that results in a small, highly broken, curved contact, but this behavior is on a scale that is trivial relative to the total deformation. With regard to this problem of folded basement, considerable attention has been drawn to a small fold at the Cambrian-Precambrian contact above Manitou Springs, Colorado. (This outcrop has since been removed during highway construction.) Even though the fold was less than 30 m (100 ft) in amplitude, it had been used as a demonstration of folded basement. However, both Hudson (1955) and Stearns (1971) showed that the upper Precambrian surface is weathered into a gravel, and it is this material that is folded with the Cambrian rocks. In this locality, continuous basement only exists below the weathered zone in the Precambrian rock. For all of these reasons, then, it will be assumed for the remainder of this paper that the basement, as defined here, did not fold significantly under the physical environment to which it was subjected during Laramide deformation.

The basement does fault, and it faults by a myriad of different fault types and inclinations. As Wisser (1957) pointed out, major mountain blocks in the Rocky Mountains foreland are bounded by virtually every type of fault that is geometrically classified. Prucha and others (1965) argued that many of the geometries they observed were best rationalized by a fault that was curved and steepened with depth (upthrust fault). I have argued that in order to avoid volume problems, all major faults in the basement (with the exception of certain normal faults) must be curved (Stearns, 1975). It is, therefore, a matter of direct observation that many different fault types coexist in the foreland, and from the material properties of the basement, it is a requirement that these faults, at least in the upper several thousand metres of the basement, must be curved in order to produce the observed rotations without deforming the basement beyond its ductile limits.

What is not agreed upon is the general tectonic system that produced these faults. Are they produced solely by (1) horizontal compression involving large-scale underthrusting or overthrusting of the basement, or are they produced primarily by (2) differential vertical movements without large-scale overthrusting or underthrusting. Deep geophysical properties of the Earth and crust-mantle relations are not sufficiently well understood in the foreland to conclusively answer these questions, so the arguments persist. There are, however, certain geologic arguments that can be presented for or against either case. They are based upon surface considerations and cannot be used conclusively, but at present they form the only basis for decision. Each of the two above-listed possibilities will be taken up separately.

Arguments Concerning Horizontal Motions

There are two principle reasons that make large-scale horizontal motions in the basement of the Rocky Mountains foreland an appealing movement scheme. The

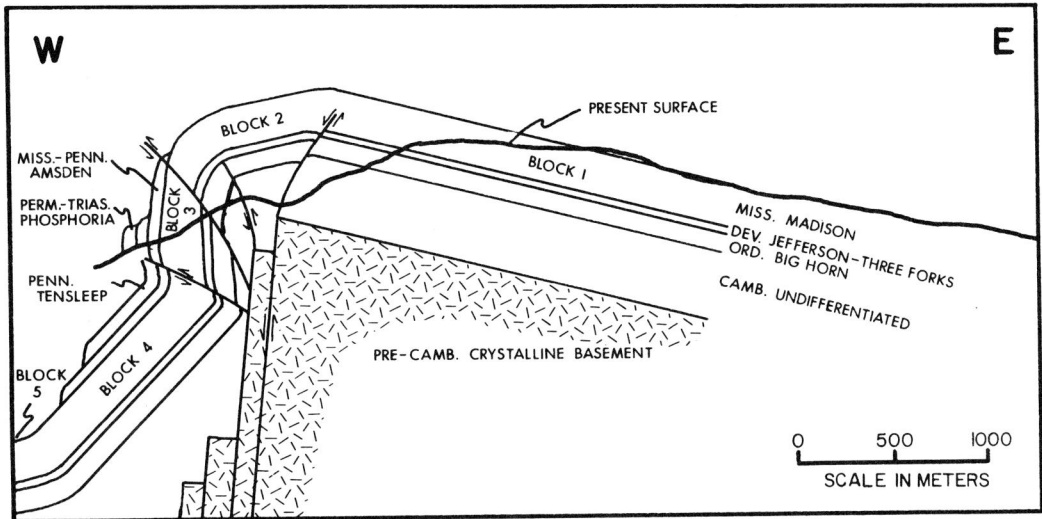

Figure 5. Cross section of Rattlesnake Mountain west of Cody, Wyoming. Block designations after Stearns (1971).

first stems directly from the observation that there are areas in the province where the basement rides out and over the sedimentary veneer along low-angle fault zones. Examples of this are particularly well displayed along the southwestern front of the Wind River Mountains. Seismic studies and drilling have established that such features also exist along the southern boundary of the Owl Creek Mountains and along the southwestern front of the Granite Mountains. There are other regions, such as along the front of the Beartooth Mountains and the north and south flank of the Uinta Mountains, in which sedimentary rocks are repeated across low-angle faults that must be associated with the basement deformation. The second reason is that with large-scale horizontal motions, certain folding problems in the overlying sedimentary layers are more easily rationalized. That is, superincumbent folding in the layered rocks would be more easily explained if the basement had moved horizontally. I have pointed out (Stearns, 1971) that many of the layered rocks in the Wyoming province form drape folds which are accompanied by little or no thinning of the Paleozoic carbonate sections. Rattlesnake Mountain is one of these structures (Fig. 5). If only this cross section is considered, it is much easier to explain the combination of continuous folding and nonthinning by shortening the basement. There are, however, several observations that make large-scale horizontal motions in the basement less appealing, and furthermore, neither of the reasons that make horizontal motions attractive is incompatible with differential vertical uplift. If the long, rotated blocks, such as the dip slope of the Wind River Mountains into the Wind River Basin, are assumed to be underlain by low-angle thrust faults, a series of geometric near-impossibilities arise. These geometric difficulties do not manifest themselves when only two-dimensional cross sections are considered. However, if broader regions and three dimensions are considered at the same time as cross sections, then horizontal motions become less appealing. For example, Figure 6 is a simplified map of the mountains and basins that appear in Figure 4 with the movement directions that are necessary if each of the major mountain blocks is underlain by an overthrust (solid arrows) or an underthrust (open arrows). Each of the solid arrows is drawn assuming that major basement

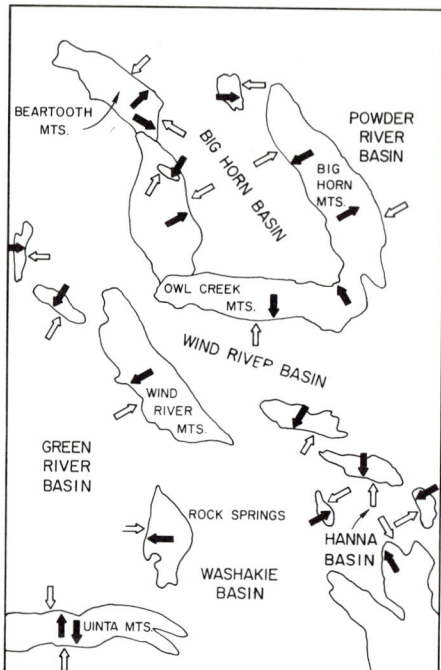

Figure 6. Generalized map of mountains and basins showing movement pattern required for underthrusting (open arrows) and overthrusting (solid arrows).

blocks (like those shown in Fig. 5) are underlain by a fault that flattens with depth ("sled runner") and that the upper plate is the active-movement plate. The open arrows are drawn as if each of the major basement blocks are underlain by a shallow-dipping fault in which the basinward block moves underneath an uplifted mountain block. Figure 6 indicates that although single cross sections through any of the blocks may appeal to low-angle horizontal motions, when the entire region is considered, the horizontal motions must have been in many different directions. Multiple sources, one for each horizontal motion, then need to be found. It is difficult, if not impossible, to imagine a driving system that would produce horizontal motions in so many different directions. Even this is not the most damaging argument to such a scheme. The patterns shown in Figure 6 are even more paradoxical when the behavior of the upper basement surface in the Rocky Mountains foreland is considered. This behavior pattern is one of rupture and rigid-body rotation. Consider, for example, the difficulty in explaining the near-perpendicular horizontal movements that would be required in the Beartooth Mountains. Such motions in brittle materials would produce unacceptable volume discrepancies that would seemingly result in large holes in the basement. The problem in and around the Beartooth Mountains (Fig. 6) is expressed again and again within the foreland. For example, in the north end of the Bighorn Basin, if underthrusting is accepted, it is necessary to underthrust in two directions away from the center of the basin. The Big Horn Mountains present a different, but equally perplexing problem. The major blocks north of Greybull, Wyoming, are tilted toward the Powder River Basin—which would indicate that the northern part of the mountain range moved to the southwest if we invoke low-angle faults. The block south of Greybull, Wyoming, is tilted toward the Bighorn Basin; this would imply a large-scale horizontal motion to the northeast. Again, this seemingly leads to an uncompensated-volume system. Small basins such as the Hanna Basin present even more complexing problems. The Hanna Basin is about 30 to 40 km (20 to 25 mi) across, and yet it is 9 km (30,000 ft) deep. If horizontal motions are accepted as the dominant pattern of basement movement (either overthrusting or underthrusting) in the Hanna Basin, again an extreme paradox results. Underthrusting would produce a large void in the basement in the center of the basin, and overthrusting would require a system that can drive thrusts in an almost circular pattern. The large change in the horizontal-motion direction between the Wind River Mountains and the Owl Creek Mountains that flank the Wind River Basin would be equally difficult to explain. Between the Gros Ventre Mountains and Teton Mountains, a severe movement shift would be required. Likewise, the Green River Basin presents a problem to accepting this movement pattern. If the Wyoming thrust belt is due to underthrusting (as suggested by Royse and others, 1975) and the Wind River Mountains are also underthrust, again there is a severe volume problem in the middle of the Green River Basin. To explain the Uinta Mountains with overthrusting (black arrows in Fig. 6) is virtually an impossibility. Underthrusting in this case, for just the single mountain range, is a more acceptable movement scheme, but the underthrusting that would be required on the north flank of the Uinta Mountains is somewhat in conflict with the underthrusting that would be required for the Rock Springs uplift or the Wind River Mountains.

Another reason for doubting large-scale horizontal motions in the upper several thousand metres of basement in the Rocky Mountain foreland arises from examining the corners of basement blocks in three dimensions as opposed to concentrating on two-dimensional cross sections near the middle of the blocks. The geometry near the termination of a block, if horizontal motions in the basement were responsible for the observed structures, is illustrated in Figures 7a through 7d. Figure 7a is a schematic drawing of a rotated basement block relative to a horizontal plane where the arrows indicate the required motions of the block, if the fault flattens with depth as indicated by the dashed line. Figure 7b is a map view of the resulting Precambrian surface. Figure 7c illustrates what would be expected in the overlying sedimentary rocks if the motions in Figure 7a were correct. Arrow 1 across the front of the block in Figure 7c represents a simple fold. At the termination of the block, wrench motion is necessary; arrow 2 in Figure 7c schematically illustrates what the required motions would be. Figure 7d is a map projection of Figure 7c. The configurations represented in Figures 7c and 7d simply are not seen at the terminations of any of the blocks in the Wyoming province. Rather, Figures 7e and 7f illustrate reality. Over the uplifted and rotated blocks, the sedimentary veneer continuously drapes in two directions as illustrated by arrows 1, 2, and 3 in Figure 7e. The fold dies out along the end of the block as illustrated by arrows 2 and 3, and the map pattern illustrated in Figure 7f is the resulting fold configuration in layered sedimentary rocks. Figure 8 is a photograph at the end of Rattlesnake Mountain in a position that is approximated by arrow 2 in Figure 7e. It can be seen in the photograph of Figure 8 that the sedimentary units drape continuously over the end of the block, the fold dies out toward the right-hand side of the photograph, and there is no indication of any lateral motion on the end of the blocks. Rattlesnake Mountain is not a singular example of this occurrence. Areas in which I have observed behavior similar to that illustrated in Figure 7f are Rattlesnake Mountain, southeastern corner of the Beartooth Mountains, the Prior Mountains, numerous occurrences north of the Five Springs area in the Big Horn Mountains, the Elk basin structure, the Dry Fork Ridge structure in the northeastern Big Horn Mountains, the front of the Big Horn Mountains north of Greybull, structures in and around and including Circle Ridge anticline in the western Owl Creek Mountains, the northern Tetons, the central and southern Gros

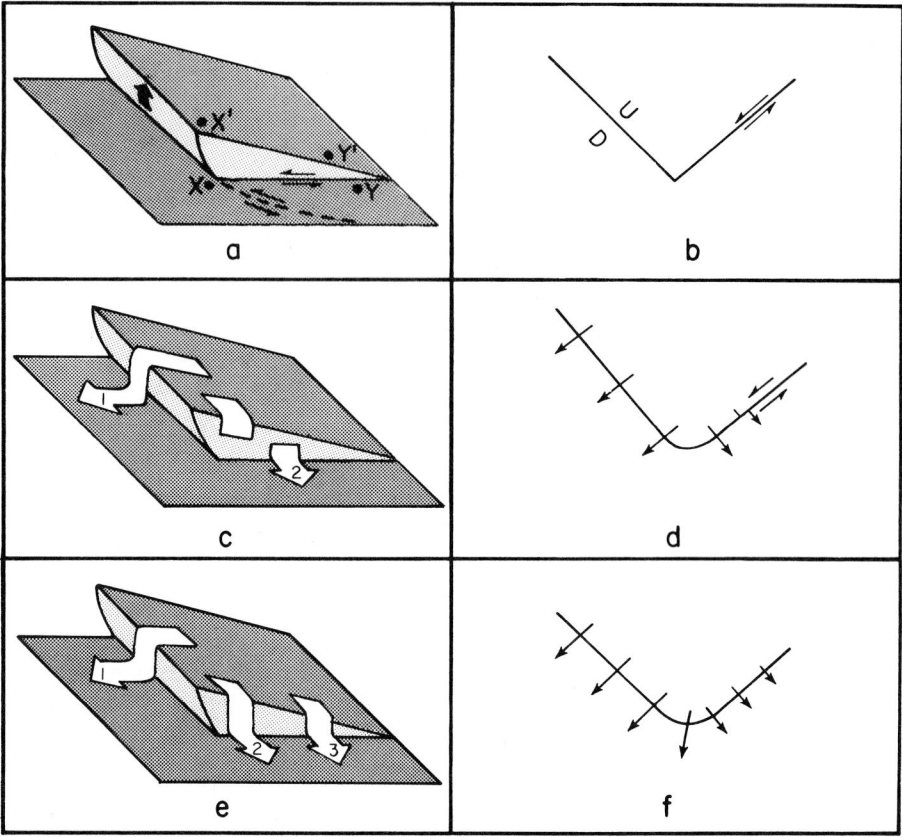

Figure 7. Schematic diagrams showing block configurations if they are underlain by a low-angle thrust fault (7a–7d) opposed to what actually occurs (7e and 7f); 7b, 7d, and 7f are simplified maps of the block configurations shown in 7a, 7c, and 7e, respectively.

Figure 8. Photograph of the southeastern end of the Rattlesnake Mountain block showing continuous forced folds with no lateral offset. Continuity of fold over end of block can be seen on the steep hillside (at arrow).

Ventre Mountains, Casper Mountain, Seminoe Mountains, the Freezeout Mountains, Elk Mountain in the northern Medicine Bow Mountains, several places in the Front Range between Fort Collins and Denver, the eastern Uinta Mountains, and several places along the northeastern front of the Uncompaghre uplift. These occurrences, coupled with the fact that no place in the entire Rocky Mountains foreland have I ever seen a wrench fault large enough to accommodate implied horizontal motion, leads me to believe that the patterns illustrated in Figures 7e and 7f are the common behavioral patterns for the area and must be considered in any explanation of the region.

A third observational fact that casts doubt on large-scale horizontal motions is the plateau uplifts that occur within the Rocky Mountains foreland. Although the usual style is one in which the basement block is uplifted and rotated, there are areas in which the uplift results in nearly flat-topped blocks such as the Beartooth Mountains (Foose and others, 1961), the Cottonwood Canyon region in the northern Big Horn Mountains (Stearns and Stearns, this volume), and the Uncompahgre uplift (Lowman, 1963; Stearns, 1971). A rotated basement block such as that illustrated in Figure 5 could at least be geometrically produced by low-angle overthrust or underthrust faults in the basement. However, a large flat-topped block bounded on several sides by steep faults such as the Beartooth Mountains would be exceedingly difficult to produce by horizontal motions on low-dipping fault planes.

For all of these reasons it is difficult to accept horizontal motions on low-dipping planes in the upper several thousand metres in the basement. That is not to say that motions in the deep crust or upper mantle may not contain a horizontal component, but for that portion of the basement, at least the upper 6,000 m (20,000 ft), where brittle behavior must dominate, the observable geometries are difficult to rationalize with large-scale horizontal motions. If, however, large horizontal motions are interpreted, they must apply to all of the regional observations together, not just single cross sections through the centers of large blocks.

Arguments Concerning Vertical Movements

The opposite extreme of large-scale horizontal motions in the basement is a scheme in which most of the movements are vertical. Is there evidence that such motions have occurred within the Rocky Mountains foreland? As stated above, the peneplaned Precambrian surface at the beginning of Laramide deformation was approximately 3,000 to 3,500 m (10,000 to 12,000 ft) below sea level. Examination of the map in Figure 4 shows that Laramide deformation resulted in uplifted mountain blocks and downdropped basin blocks. Because the Precambrian surface in all of the basins is now much lower than 3,500 m (12,000 ft) below sea level and in all of the mountains it is considerably higher than that elevation, the inescapable conclusion is that the deformation produced large absolute up and absolute down vertical motions. Furthermore, the different trends of mountain systems and basins within the same general region would be allowed by vertical motions, whereas horizontal motions would normally result in uniform trends (Fig. 6). A pattern of vertical movement in the western North American continent is not unique to Laramide deformation. The evidence for this statement arises from what geologists know best of all: the stratigraphy of the layered Phanerozoic rocks. From Cambrian through Mississippian time, the western part of the continent underwent a long-wavelength, high-amplitude, low-average-displacement rate, differential vertical movement. Such a motion is verified by simple observation of the stratigraphy. Less than 1,500 m (5,000 ft) of Cambrian through Mississippian rocks in eastern Colorado are represented by more than 21,000 m (70,000 ft) of rocks of the same age in California (Gilluly, 1963). During Cambrian time there was transgression of the western continent (Haun and Kent, 1965) that moved from west to east and placed all of the upper Precambrian surface from California to Colorado at sea level (beach deposits) at some time during the Cambrian (Haun and Kent, 1965). By the end of Mississippian time, less than 1,500 m of rock had accumulated in western Colorado, but in excess of 21,000 m of rock had accumulated during the same period in California; this implies differential vertical motion of at least 19,500 m (65,000 ft) across the continent. Furthermore, this long-wavelength, high-amplitude, slow movement was relatively uniform along strike for thousands of kilometres. There is ample stratigraphic evidence to support the idea that the region from eastern Colorado to eastern Nevada was subjected to shorter-wavelength, intermediate-amplitude, rapid differential vertical motions during Pennsylvanian time. Source areas such as the Ancestral Rockies and Uncompahgre uplift became emergent and were not completely covered again until Cretaceous time. Other source areas such as the Emery arch probably remained subaqueous. Many basins were downdropped and became local depositional areas. Examples are the Bird Springs basin of southern Nevada, the Oquirrh basin in Utah, the Paradox basin in Utah and Colorado, the Maroon basin in Colorado, and the depositional area mostly in Wyoming that received the sediments that compose the Weber, Casper, and Tensleep Formations, all of Pennsylvanian age. Unlike the earlier downwarping that was continuous all along the western part of the continent, the Pennsylvanian motions produced a series of local source areas and depositional centers that in a sense fragmented the earlier more orderly system. Some of these Pennsylvanian differential vertical movements occurred over very short lateral distances. The zero isopach for the Cutler Formation (Pennsylvanian and Permian) on the northeastern side of the Paradox basin is only 8 km (5 mi) away from a section of rock 4,900 m (16,000 ft) thick. Most of the lower part of this formation is a sequence of coarse conglomeratic alluvial fans coming off from the newly created faulted front of the Uncompahgre uplift. Not until Cretaceous time were the irregularities caused by the Pennsylvanian differential vertical motions even approximately smoothed out. During Cretaceous time the western part of the continent again experienced a long-wavelength, relatively high-amplitude, slow differential vertical motion. Cretaceous stratigraphy clearly reveals that the central part of the area became emergent at least by Early Cretaceous time, and two troughs formed on either side of the uplift (Gilluly, 1963). The eastern downwarp, the Rocky Mountains trough, by the end of Cretaceous time was sufficiently depressed to accumulate in excess of 6,000 m (20,000 ft) of marine sediments (Haun and Kent, 1965). The exact elevation of the positive area is not known, but even the fact that it was emergent coupled with the depth of the adjoining trough indicates the magnitude of the differential vertical motions. Furthermore, these motions were very regular because all of the basin fill is shallow-water sediment. Finally, beginning in Paleocene time the eastern part of this area once again underwent a series of short-wavelength, intermediate-amplitude, differential vertical motions that produced the Rocky Mountains foreland, complete with its mountains and intermountain basins.

The cumulative post-Cambrian stratigraphy, therefore, demonstrates that the western part of the continent has long been subjected to differential vertical motions. There is at least a hint of cyclicity to these motions that begin with long-wavelength, widespread, slow movements and culminate in short-wavelength, irregular, rapid movements. The evidence for these movements comes from stratigraphy. This evidence, however, does not speak to the cause, but only to the result. Anyone attempting to explain causes must incorporate all of the stratigraphic information and not use just preselected parts of the total.

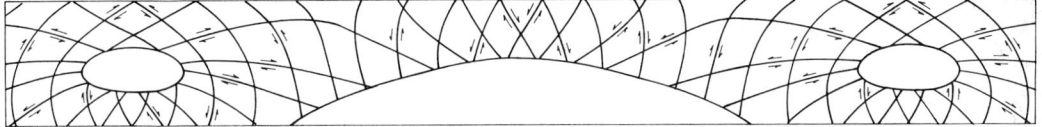

Figure 9. Diagram of potential faults that would be produced in a continuous, elastic block by differential vertical loads at the bottom (after Hafner, 1951). The lower boundary loads are in the form of a smooth sinusoid, the end loads are linear increases in the horizontal component of stress (normal burial), and the upper boundary represents the air-rock interface. Arrows indicate sense of shear on potential faults.

There are certain theories of deformation that help to explain the vertical movements during just the Laramide orogeny. Hafner (1951) dealt with the potential fault planes that might form in a homogeneous, isotropic, continuous, elastic segment of the Earth's crust due to differential vertical motions at the base of the unit. Hafner's paper not only addressed potential fault patterns in differential vertical uplift, but it is a hallmark paper in the geologic understanding of faulting in general. Up until 1951, and unfortunately for many years after, the geological fraternity tended to think of areas as if each were dominated by a single fault type—such as areas of normal faulting or areas of thrust faulting. Hafner's most important contribution was showing that from geologically realistic boundary conditions, a multiplicity of fault types can form as a result of a single loading condition in one area. His assumptions for material behavior were that the body is homogeneous, continuous, and isotropic and that it will behave up to the point of rupture as a linearly elastic body. These assumptions may not apply to all rocks, such as layered rocks of varying ductility or rocks near the base of the crust or the upper mantle. However, the material assumptions are not in contradiction to either field-observed or laboratory-measured behavioral characteristics of the upper several thousand metres of Precambrian crystalline basement in the Rocky Mountains foreland. Hafner's solution is a static solution that automatically limits its applicability to the formation of the potential shear fractures along which later displacement can occur. It does not address itself to large displacements that occur after the shear fractures form. His two-dimensional solution restricts any application to the centers of uniform regions that are long with respect to their width. This assumption, plane strain, could be applicable through the centers of basins or long mountain blocks, but definitely could not be applied near the terminations of blocks nor the ends of the basins. Figure 9 illustrates the solution to one set of boundary values that Hafner solved. In reality this solution is that of a thick beam bend occurring in elastic materials. In nonmathematical terms, it states that bending moments due to the uplift and downdrop at the base of the block must be added to standard-state conditions due to burial in order to produce the true stress field in the block. Because the beam is so thick, stress differences away from the center due to bending become large enough to cause rupture in rock materials (such as granite) before any observable deflection occurs at the upper surface.

There is only one way to test any theory that is based on so many assumptions. That way is simply to test a fit between what the theory predicts and what reality has shown. I have attempted to fit breaks predicted by Hafner to a region of the crust across the northern Bighorn Basin (Stearns, 1975). The apparent fit between the reality of this region and the theory was remarkably good, if the complexities of the real problem are considered. Through application of the Hafner shear fracture trajectories (potential faults), I was able to match the following features: (1) proper amounts of rotation for large basement blocks, (2) parallel or opposing rotations

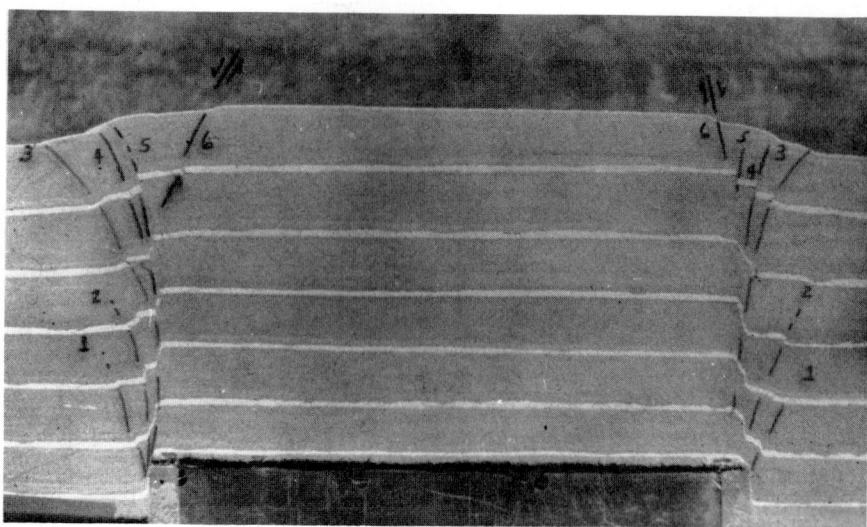

Figure 10. Photograph of the front view of a sandbox experiment. The lower piston has been displaced about 3 cm. Faults are traced on front; numbers refer to the order of their formation.

in adjoining blocks, (3) the proper position for known large-scale faults across the basin, and (4) the proper amount of throw on these faults. Furthermore, the composite movement on all of the faults selected from the Hafner diagram produced a basement configuration that would allow an asymmetrical basin the size and shape of the Bighorn Basin to form, and most importantly, this could be achieved without either creating space problems in the basement or violating the tenants of brittle behavior and rigid-body rotations for basement materials.

I concluded that not all of the problems of basement deformation in the Bighorn Basin could be solved by the theory of linear elasticity (Stearns, 1975). On the other hand, the remarkable agreement between the predictions of theory and the observations of the major structures in the northern Bighorn Basin cannot be ignored. I therefore concluded that "the correlation between prediction and observation is good enough for acceptance of the theory as the basis of the genesis of these regional structures. Furthermore, the geometrical relations of this area are explained by faults that are mechanically compatible."

Sanford (1959) did a study similar to that of Hafner. The major difference between the two studies, with the exception of techniques involving displacement rather than stress analysis, occurs in the lower boundary condition. Sanford studied a discontinuous step function at the base of his model that would serve as an analogue computer to the analytical solution. In his now-famous model studies he reproduced the lower boundary condition by means of a movable hydraulic piston placed at the bottom and in the center of an aquarium filled with sand. Activation of this lower piston in an upward direction produces a series of faults that start out as vertical faults and flatten toward the surface to become first high-angle reverse faults and finally thrust faults near the sand-air interface. Behind these curved reverse faults a series of normal faults develop as a consequence of the displacements that occur along the reverse breaks (Fig. 10). It should be noted that when fault 1 in Figure 10 formed, the piston was not yet fully displaced. These experiments are very reproduceable, and the sequence of events is as follows: a fault takes off from the corner of the piston as a high-angle fault, begins to curve, then dies out; a second fault forms near the vertical portion of the first fault and continues higher into the section with the same pattern. In the experiment illustrated in Figure 10, fault 3 finally broke through to the surface, and as it was forming, fault 6 (a normal fault) formed behind the sweep of curved reverse faults 1 through 5. The precise number of faults that are formed varies, but the pattern of small faults dying out and then new faults forming as the displacement propagates upward is the common pattern. It must be remembered, however, that the horizontal parallel markers shown in Figure 10 do not constitute layering; they are merely original horizontal markers to keep track of the displacement field, and the sand in the box is a very homogeneous, isotropic material with no effects of layering. Therefore, any applicability that such experiments might have to reality will be restricted to rocks with those properties. For the foreland, this means that application is restricted to the Precambrian crystalline basement and must exclude the heterogeneous, anisotropic, inelastic layered sedimentary rocks above the basement.

The differences between the results shown in Figures 9 and 10 are striking and important. The main difference stems from the lower boundary condition. The condition that Hafner (1951) used (Fig. 9) is one in which the load differential is distributed over a broad region in the form of a sinusoid. This is in contrast to the lower boundary condition used by Sanford (1959). In Sanford's experiments, all of the differential displacement is distributed over a small region (the discontinuous step at the base of the block in Fig. 10). It can be seen that Hafner's lower boundary condition produces shear fractures throughout a broad region, and if activated, these fractures produce a series of rotated blocks across the entire area. However, Sanford's solution suggests that faults would be formed only at the edges of a plateau uplift. His solution should, then, best fit regions of plateau uplifts that are fault-bounded, such as the Beartooth Mountains (Foose and others, 1961). Indeed, this solution shows a remarkably good fit to the deformation along the front of the Beartooth Mountains, where drilling has established that a number of reverse faults, some of which are low angle, produce repeated sections of rock. Back in the Beartooth Mountains, where sedimentary rocks are in contact with Precambrian crystalline basement, the major faults are normal faults. This is precisely what would be expected if the basement in the Beartooth Mountains was being deformed by a system similar to that modeled by Sanford. As will be discussed below, faults that leave the basement at a low angle tend to propagate up through the sedimentary veneer as low-angle reverse faults. Therefore, faults 2 and 3 in Figure 10 would produce repeated sections in the sedimentary veneer, and faults 4 and 5 would break through and allow a plateau uplift of the crystalline basement. Such a movement plan does not result in the mechanical difficulties illustrated for the Beartooth Mountains in Figure 6 in which the mechanism of overthrusting or underthrusting is called upon. A slightly different problem, but still requiring differential vertical movements, is presented by Couples and Stearns (this volume) and fits the Beartooth deformation even better. Nevertheless, the work of Sanford (1959) remains a good model for this type of uplift.

Figures 9 and 10 represent two theoretical models for the types of faults that would be produced in thick brittle materials due to differential vertical loads deep within the crust. As valuable as they are for helping to understand certain regions, they represent but two loading conditions. Many other ways of loading the lower boundary can easily be imagined for a region as vast as the Rocky Mountains foreland. Despite this fact the models of Hafner and Sanford published during the 1950s remained until recently the only theoretical models with which to work. Couples (1977) introduced solutions to several different boundary-value problems that may be applicable in regions where the conditions of Hafner's or Sanford's models do not fit the observed geology. Some of the practical applications of these new models are presented by Couples and Stearns (this volume). Two extremely important implications come out of the new models. As is pointed out in detail elsewhere in this volume (Couples and Stearns), rotations of large blocks such as the Wind River Mountains in a direction opposite the motion on curved reverse faults no longer present a paradox. A slight change in the boundary conditions produces, in near proximity to each other, reverse faults and normal faults that can produce the previously difficult-to-explain geometries such as the Wind River Mountains and the Owl Creek Mountains (Couples and Stearns, this volume). The second important feature to come out of the new models is that Couples (1977, and this volume) demonstrates that horizontal end loads on a block, when superimposed on deep-seated differential vertical loads, alter the predicted fault pattern in detail but not in character. That is, the superposition of a horizontal load does not produce exclusively low-dipping overthrust-type faults. The dominant movements from such superimposed loads can still be vertical. The applicability of these new solutions clearly demonstrate a need for more boundary-value solutions with differing, but realistic, boundary loads.

The Uinta Mountains present a perplexing geometry that is relatively unusual in the Rocky Mountains foreland. The Uinta Mountains are flanked on both sides by large, curved reverse faults and seem to have been produced by a "mushroomlike" uplift. It is difficult to conceive of such a configuration resulting from a single loading condition. However, sandbox models such as those run by Sanford (1959) with slightly different loading conditions show that similar geometries are not incompatible with differential vertical movements. Sandbox experiments were run in which the rigid piston that produces step displacements was replaced by a partially inflated weather balloon. This partially inflated balloon, when covered with sand, forms an initial boundary that is somewhat elliptical in shape (Fig. 11a). With the sand covering the balloon to the top of the sandbox, continued inflation of the balloon produces a loading condition (Fig. 11b) that is significantly different from those used by Hafner (1951), Sanford (1959), or Couples (1977, and this volume). The overall geometry of the "faults" so produced is shown in Figure 12 and is remarkably similar to that observed in the Uinta Mountains. The geometry includes a series of curved reverse faults on either side of the uplift and a large extension zone in the middle. Although this similarity to the geometry of the Uinta Mountains does not prove that they formed this way, it does demonstrate that their geometry is consistent with differential vertical movement in the total absence of horizontal motions.

The Colorado Front Range north of Denver, like the Uinta Mountains, present a complicated geometric form whose origin has long been argued. This range in all probability represents still a different loading condition that to date has not been examined in either a model or theory. However, Matthews (1976) and Matthews and Work (this volume) clearly show that differential vertical movement of discrete rigid basement blocks is the only system that is compatible with all of the geologic observations. Prucha and others (1965) showed that vertical uplift was compatible with the observations around Milner Mountain near Fort Collins, Colorado. Matthews and Work (this volume) have extended the principle in a rational fashion from north of the Wyoming state line to Denver. They clearly show along this continuous mountain front that there are three distinct styles that operate in adjacent areas. The styles range from curved reverse faulting (Denver) to gentle block motions in the north. None of these distinctive styles, however, is incompatible with differential vertical motions of rigid blocks within the basement. Palmquist (this volume) likewise shows that geologic observations made along the eastern front and southern part of the Big Horn Mountains are best reconciled by differential

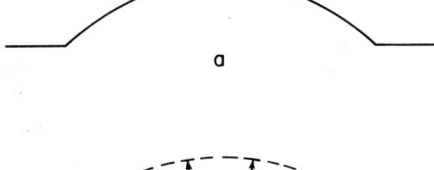
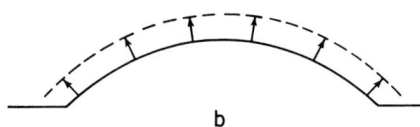

Figure 11. Illustration of the loading condition produced in homogeneous sand by starting with a partially inflated balloon (a) and further inflating it (b).

vertical block motions in the basement. He further points out that many structures that appear to be compressive in nature are secondary features formed by primary movements along high-angle faults.

It is concluded, therefore, that although large-scale basement shortening by underthrusting or overthrusting makes rationalization of the sedimentary veneer an easier task in cross section, total motions in the basement (Fig. 6) as well as block terminations (Fig. 7) make its acceptance as a unifying theory unappealing. However, the application of models for differential vertical uplift (Hafner, 1951; Sanford, 1959; Couples, 1977, and this volume) to specific mountain ranges produces a relatively good fit between prediction and actuality. Such theories have been applied to the northern Bighorn Basin (Stearns, 1975), the Beartooth Mountains, the Wind River Mountains, and the Owl Creek Mountains (Couples and Stearns, this volume) in a direct sense. Other regions such as the Front Range and the Big Horn Mountains as well as small features in the Beartooth Mountains and Seminoe Mountains are best explained by this style. In other cases, such as the Uinta Mountains, it has been shown that the total geometry is compatible with differential vertical motion (Fig. 12). In summary, then, it seems that most direct information concerning the real structures that involve basement material supports differential vertical motion in the upper part of the basement as the major component of the displacement field. Furthermore, these motions are accomplished by rigid-body rotations in the brittle basement.

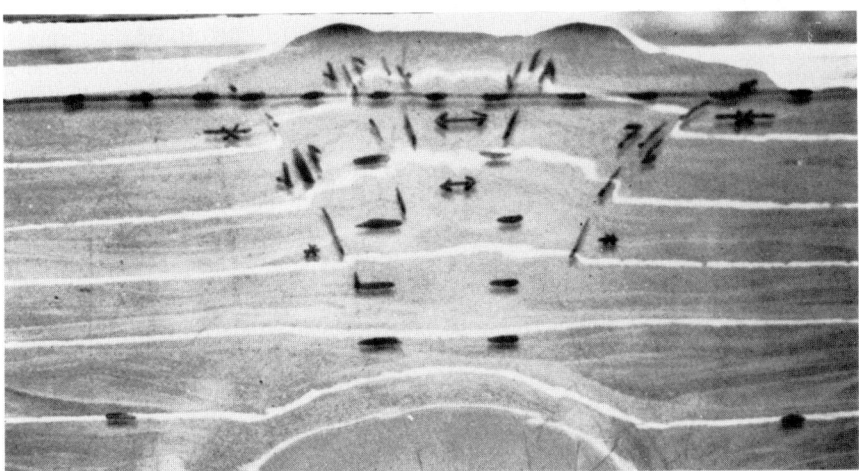

Figure 12. Photograph of faults produced in a sandbox by inflating a balloon (lower center) beneath the sand.

BEHAVIOR OF LAYERED ROCKS

The make-up of the stratigraphic section influences the final fold geometry; however, our knowledge of forced folding is too meager to completely specify the details of that geometry at this time. That the final shape is at least partially controlled by the stratigraphy is an inescapable fact, and even now certain characteristics of the sedimentary section that do play a role in fold shape can be specified. Perhaps the most important factor is whether the folded strata are welded to the forcing member. The largest unfaulted folds occur where the layers above the forcing member are the freest to slip independently of the forcing member. A second factor that greatly affects the ultimate shape of the fold is the presence or absence of some controlling stratigraphic package that behaves in a relatively rigid fashion—that is, some stratigraphic unit that is thick enough to control the shape of the fold, but under the conditions of deformation is unable to thin or attenuate. If the sedimentary section contains a series of units that are unable to thin, even though other parts of the section are capable of thinning, it is comparable to placing a strut or reinforcing member in building materials. The behavior of the rigid unit then controls the shape of the fold.

There are at least three different general classes of sedimentary sections that will be considered here. First is a nonwelded section that contains a stiff or nonthinning stratigraphic unit that controls the shape of the fold. The second generalized section contains a welded, stiff, nonthinning controlling member. The third class of section is welded to the forcing member but is ductile and capable of thinning during the folding process. Between the first and third class is an almost continuous spectrum of behavior, and sharp division lines between the classes are nonexistent. However, there is enough difference in the behavior to speak about typical sections of each class to provide a standard of comparison. These three general classes are represented in various parts of the Rocky Mountains foreland.

Nonwelded, Nonthinning Sections

The first class of section exists throughout most of Wyoming, except in the extreme southern parts. The Precambrian crystalline basement which is the forcing member is immediately overlain by 275 to 400 m (900 to 1,300 ft) of Cambrian section that is dominantly shale. The shale behaves in bulk as a ductile material and allows nonwelded offset of the overlying layers in the manner illustrated in Figure 3b. From a structural standpoint the rest of the Paleozoic section behaves more or less as a single unit. The Ordovician, Devonian, and Mississippian rocks are almost entirely carbonates. The lower part of this middle Paleozoic section is composed of thick, bedded dolomites and limestones. The Pennsylvanian Amsden is a shaley unit, but too thin (less than 65 m) to have much control. This shale is overlain by a sandstone (Tensleep Formation) and Permian carbonates. The entire unit in the northern part of the Rocky Mountains foreland serves as a nonthinning strut and is about 600 m (2,000 ft) thick. This strut in turn is overlain by a series of Mesozoic clastic rocks that behave in a much different manner than the Paleozoic carbonates, but whose shape is more or less controlled by the behavior of the carbonate strut (Weinberg, this volume).

The behavior of this carbonate strut is best seen by observing its shape in a mature, well-developed forced fold. Perhaps the best exposure of such a fold both laterally and vertically occurs at Rattlesnake Mountain west of Cody, Wyoming. The overall structure at Rattlesnake Mountain (discussed by Stearns, 1971) is shown in cross section in Figure 5. Above the Cambrian and below the Triassic rocks, faulting plays only a minor role in the deformation and is restricted to the lower part of the post-Cambrian sedimentary section. Upward from the middle of the Paleozoic section, significant faulting is totally absent. To help clarify the discussion, certain blocks have been numbered 1 through 5 as indicated in Figure 5. These numbers refer only to the geometry of the post-Cambrian–pre-Triassic rocks. Block 1 is the gentle flank of the structure, and its dip conforms to the rotation of the upper basement surface. There is a reversal of a few degrees between blocks 1 and 2, so that block 2 dips as much as 15° in a direction opposite to that of block 1. Block 3 is everywhere steep; in most sections, its dips are nearly vertical, and it is connected to block 2 by a sharp hinge zone. Block 4 dips about 45° in the same direction as does block 2. The nature of the lowest block (5) on Rattlesnake Mountain is completely conjectural, but corresponding blocks are well exposed in several similar drape folds in the northwest Big Horn Mountains. These block designations apply only to this specific well-developed structure, but such designations will be useful in talking about deviations from the well-developed fold.

It can been seen in Figure 5 that the large basement fault with approximately 2,300 m (7,000 ft) of throw dies out upward quickly in the sedimentary section. The localized displacement along the fault in the basement is accommodated by folding in the sedimentary rocks as is schematically illustrated in Figure 1. The normal fault that marks the contact between blocks 1 and 2 in Figure 5 displaces layers higher in the section than most of the other subsidiary faults in the folded sedimentary rocks. In general as uplift continues, displacement by folding must reach a limit where folding can no longer keep pace with basement faulting. The layers would then fault through, probably along the normal fault shown between blocks 1 and 2. Examination of analogous, but slightly larger, structures in the nearby Beartooth Mountains leads to the suspicion that the sedimentary veneer on Rattlesnake Mountain has just about reached this point of separation. In the Beartooth Mountains, faults with 3,000 m (10,000 ft) of throw in the basement also cut the folds in the sedimentary rocks.

There are three important aspects of this geometry that should be emphasized. The first is that the hinges between adjoining blocks are fixed early in the deformation and remain hinge lines as the fold develops. That is, the hinge does not migrate through the beds during the deformation. The field evidence for this stems from the fact that the hinge regions are shattered by fracturing, and the segments between hinges are not only unshattered but, in general, they are linear noncurved segments (see Stearns, 1971). The second fact of nature that must be contended with is that there is no appreciable thinning within the carbonate strut across the fold (Stearns, 1971; Stearns and Stearns, this volume). These two observational facts, when combined, necessitate decoupling of the carbonate strut at least near the base of the section. Evidence for such decoupling is found at Rattlesnake Mountain (Stearns, 1971); it has occurred in both of the ways illustrated in Figure 3. The Cambrian shales behaved in a ductile fashion and responded quite differently from the overlying carbonate section, especially in the region between the basement and block 4 (see Fig. 5). In this region the Cambrian shales contain numerous large internal structures that are not present in the overlying sedimentary veneer, and there is an overall angular disconformity between Cambrian beds and the strata above. This type of behavior results in decoupling like that shown in Figure 3b. Discrete displacement across a single bedding plane (Fig. 3a) is found at the top of the first dolomite layer in the carbonate section. Such slip occurring between dolomite layers is understandable in light of experimental work by Logan and others (1972) that demonstrated that a brittle material sliding on a brittle material (for example, dolomite) has a lower coefficient of sliding friction than that for brittle materials sliding on ductile ones. There may be other detachments within the Paleozoic carbonate section that contribute to the lateral transport into the fold area, but as of now they remain unidentified. The movement path during the fold history of the Paleozoic carbonate package is not intuitively obvious. Detailed work on this kinematic pattern is fully discussed by Weinberg (this volume). His analysis, based on the model just presented, studies displacement into the fold of the carbonates as a function of vertical displacement and rotation of the basement block.

The sequence of development of the blocks (Stearns, 1971) empirically is determined from observing folds in the foreland at various stages of basement-fault displacement. Block 3 rotates to a near-vertical position before blocks 4 or 5 are activated (Fig. 13). In Figure 5 notice that the Ordovician Bighorn Dolomite (base of the Paleozoic strut) is nearly in contact with the basement block because the entire Cambrian section has flowed into what otherwise would have been a void created by the faulting. Once the hinge between blocks 3 and 4 has been laterally transported into this position, it can no longer move horizontally, and the development of block 4 is necessitated. Incipient formation of such a block 4 is observed in the northern Big Horn Mountains, and in these cases block 3 has already migrated to a vertical position. The existence of block 4 on all large drape folds is more difficult to prove than blocks 1, 2, and 3 because of depth of erosion required

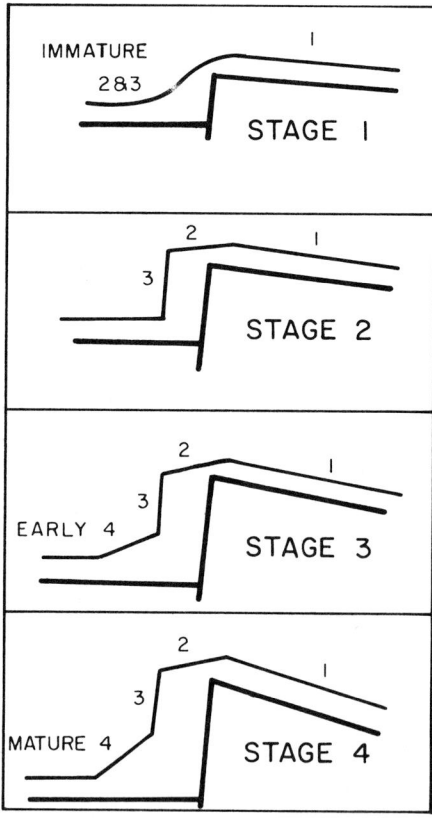

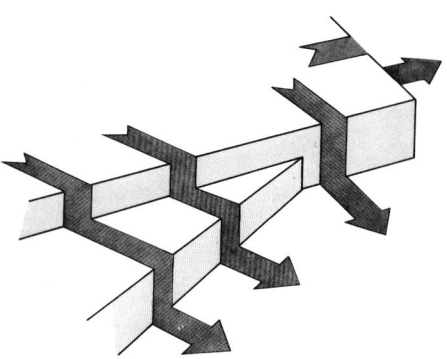

Figure 13. Postulated order of block development in the Paleozoic carbonates during the growth of a forced fold in the northern Rocky Mountains foreland. Numbers refer to blocks defined in the text.

Figure 14. Schematic illustration of typical block configuration along mountain fronts in the northern Rocky Mountains foreland. Arrows represent folded sedimentary layers.

to expose block 4. A block 4 is well-developed, however, on other deeply dissected structures, but the universal presence of block 4 in such situations is still a matter of conjecture.

To this point, the discussion has been restricted to cross-sectional geometries near the centers of single rotated blocks. However, one of the characteristics of forced folding in the Rocky Mountains foreland is that the folds do not form in parallel fold trains. The shapes, trends, and sizes of the forcing member, which in the Rocky Mountains foreland is Precambrian crystalline basement, determine the ultimate geometry of the folds in the layered rocks. The basement blocks and, therefore, the folds terminate abruptly along strike. This type of termination is illustrated in Figure 7e. The interior angles at block corners usually are between 70° and 120°. If the block is rotated, the folding dies out along the terminating fault in the downdip direction (Fig. 7e). Therefore, unlike free folds, some of which tend to die out in long plunges along strike, forced folds die out by turning an abrupt corner and losing throw in a direction at high angles to the average fold strike. Furthermore, when several basement blocks with different strikes and different rotations adjoin one another, the resulting fold geometry can become very complex and unpredictable (Fig. 14). Although Figure 14 is idealized, it is taken from actual cases and is representative of the types of complex geometries that can result from multiple block rotations in the same area. Even though on the state geological maps the mountain fronts in northern Wyoming appear to be straight, they are in fact composed of multiple blocks that abruptly change strike along the front. That is, the front is rarely formed by a single fault the length of the mountain system, but rather by a complex of rotated blocks. This situation gives rise to abrupt changes in the strike of the frontal fold. This is particularly well illustrated along the western front of the Big Horn Mountains between Lovell and Greybull, Wyoming, along the steep western front of the Gros Ventre Mountains, along the steep south flank of the Seminoe Mountains, along the south flank of the San Juan Mountains, and all along the east side of the Front Range. That the Paleozoic rocks are able to conform to such complicated forced shapes without thinning or faulting appreciably is a poorly understood fact. However, lack of understanding should not be confused with the fact that the continuous folds do exist. Such geometries are discussed in detail for an area in the northern Big Horn Mountains by Stearns and Stearns (this volume). They discuss, primarily from field observations, the facts of these geometries and rule out certain obvious explanations. Although their purpose is primarily to define the problem, they conclude that whatever the total mechanisms are for achieving these complicated shapes, bedding-plane detachments and movements in three directions are necessary.

Perhaps the most perplexing, and difficult to understand, feature of the forced folds in the Rocky Mountains foreland is the behavior of the stiff carbonate strut at the termination or corners of the forcing block. The features of such corner areas are discussed by Stearns and Stearns (this volume). That the layered carbonates can conform to the shape of the forcing member without thinning, faulting, or the creation of subsidiary folds is an observational fact. This leads to the inevitable conclusion that the carbonate strut must be detached from the underlying materials and free to translate in virtually any direction required by the folding. Although the forcing member may be more highly broken at such corners, as will be discussed below, layered carbonates accomplish their new shape with a smoothness and continuity that is difficult to accept. Such corners are, however, observable in folds too numerous to list within northern Wyoming and in such sufficient numbers that the observational fact must be accepted as part of reality. That the carbonate strut must detach and slide into the fold in many directions is further attested to by the work of Vaughn (1976). In her work at Casper Mountain she studied a series of corner configurations within the more-ductile Mesozoic part of the section. She found that at the corners, these more-ductile rocks show a considerable amount of local thickening and thinning in order to accomplish the fold process. Such thickness changes require that a great deal of material must either move into or away from the corner area. If, on the other hand, the geometry is accomplished with no thinning or rupture, as in the carbonate strut, it must mean that differential motions in many directions occur along bedding-plane detachments. The fact that these movement patterns cannot be reconciled into a rational framework at this point should not be confused with the evidence for their existence.

The behavior and shape of the Mesozoic rocks during forced folding in the northern Rocky Mountains foreland are not as well defined nor as well studied as for the Paleozoic section. The primary reason for this lack of observational control in the Mesozoic rocks is erosion. Most of the mountain ranges, from which the Paleozoic information is derived, are eroded either completely through the Mesozoic strata or at least to the lower sequences. Furthermore, even on the low mountain flanks where Mesozoic rocks are preserved, they tend to develop soil cover and vegetation much more rapidly than do the Paleozoic carbonates. As a consequence there simply is not as much exposure of Mesozoic rocks on these forced folds compared to the exposure of Paleozoic rocks. This is particularly true in the critical hinge areas where outcrop is sparse in the Paleozoic part but nearly absent in the Mesozoic part of the section. However, certain generalities at least can be made. The most striking observation is that within the Mesozoic section the distinct block shape of the deformed Paleozoic strata gives way to a more uniform, continuous fold that is best represented by arcs of large circles as opposed to straight linear segments (see Weinberg, this volume). This may indicate that the folded Paleozoic rocks serve as a loading condition for a new forced fold in the Mesozoic rocks. Exactly where within the Mesozoic strata this transition occurs is not known. The smoother folding is usually in existence by the middle of the Mesozoic section, and quite frequently the Triassic strata (Chugwater) still show good conformity to the blocks of Paleozoic rocks. However, even in the Chugwater where there is some hinge exposure, the flexing is not as sharp as in the Paleozoic rocks. Therefore, it seems that the transition from sharp, linear segmented folds to smooth rounded folds occurs gradually. The material properties of the sandstones and shales that make up the Mesozoic section are such that thickness changes by ductile flow are more likely in Mesozoic than in Paleozoic strata. That this generality is true is substantiated on many folds where thinning and thickening, particularly in the shale units, are noted. However, the order or pattern, if it exists, has not yet been delineated. Certainly, the best understanding of the overall Mesozoic section results from the work of Weinberg (this volume) in which he considers the kinematics demanded of the Mesozoic section by the folding of the Paleozoic rocks. His studies show that under very reasonable conditions and at certain stages of the folding, an excess of Mesozoic rock material may be expected. The manifestations of such a movement pattern are certainly recorded in many areas. In drilling it is not uncommon to find many more repeated sections of Mesozoic rocks on the flanks of forced folds than of Paleozoic rocks. In addition, Weinberg (this volume) describes a class of secondary folds (drape subsidiary folds) that are frequently present in the Mesozoic section on forced folds, but are absent in the Paleozoic rocks. The precise mechanisms within the Mesozoic section, nonetheless, remain an unsolved problem. Particularly important for future studies is the determination as to the relative role of the two types of bedding-parallel offset illustrated in Figure 3. It is not known at this time whether there are favored slip horizons that produce displacement such as in Figure 3a or whether thick sequences behave as shown in Figure 3b with no sharp bedding discontinuities.

Welded, Nonthinning Sections

During Paleozoic time the area of south-central Wyoming was intermittently high relative to the shelf region of northern Wyoming. The most important stratigraphic change (from a structural point of view) that resulted is that the thick Cambrian shale section, present in northern Wyoming, was completely replaced by a thin transgressive sandstone of Late Cambrian age. In addition, the carbonate strut that is well defined in northern Wyoming was reduced in thickness so that it consists essentially of just the Mississippian Madison Limestone. The Pennsylvanian sandstones that were present in northern Wyoming are still well developed in southern Wyoming, but the Permian section lost many of the limestone units, which were replaced by red beds and limey siltstones. The net result is that not only has the strut in the package been reduced in thickness, but by the disappearance of the Cambrian shale section, the layered rocks have become more welded to the basement than in northern Wyoming.

Although this region has not been studied as thoroughly as the areas farther north and south, even a reconnaissance trip through the region indicates that forced folding is still a dominant structural style. Drape folds in the Paleozoic rocks can be observed at Casper Mountain and throughout all of the mountain systems that flank the Hanna Basin. The main differences seem to be that although block 1 and block 3 are well developed, the hinges between blocks 1 and 3 are not as distinct and as sharp as they are farther north. That is to say there is more broad arching between blocks 1 and 3 in southern Wyoming structures than in those farther north. Furthermore, there is definitely a tendency for the folds to fault through at lesser displacements than there was farther north. In the area of Rattlesnake Mountain, the layered rocks are able to fold without rupturing over basement faults with 2,500 m (8,000 ft) of throw. More work needs to be done in the south-central region of the Wyoming province to pin the limiting displacement down more precisely. The displacements range from about 1,250 to 1,500 m (4,000 to 5,000 ft) before separation occurs and perhaps in some cases as little as 1,000 m (3,000 ft). Another difference that seems to occur is that the basement frontal fault system contains more small splinter faults than occur in the northern part of the province where detachment or offset of the layered rocks is more easily accomplished. This splintering probably forms in order to accommodate the space requirements near the interface between basement and layered rocks. Figure 5 shows a triangular region formed between the base of the Paleozoic carbonates, the frontal fault in the basement, and the downthrown basement surface. Where thick shale sections exist, this triangle is filled by ductile flow of the shales themselves. In southern Wyoming where the ductile shales are absent, the basement seems to fragment into small frontal splinters that partially satisfy the space filling of this triangle. Vaughn (1976) reported such behavior along the Casper Mountain front (Fig. 15), and it can also be observed in the deep canyons in the Seminoe Mountains. This development of multiple splinters along the main fault front has also been reproduced experimentally (Fig. 16). Beneath the folded, layered material and in front of the main fault there are a series of small curved reverse faults in the brittle material that essentially accommodate the space requirements between the faulted, brittle layers and the folded, more-ductile layers.

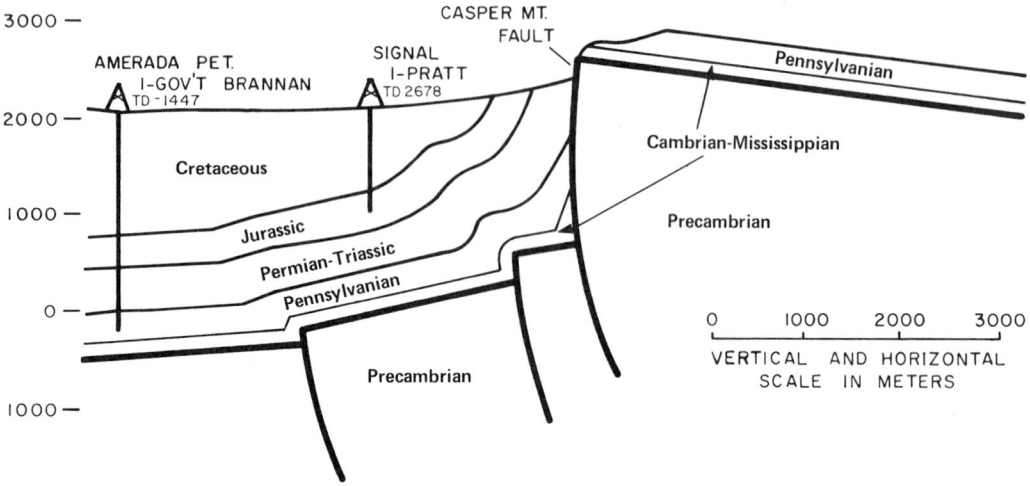

Figure 15. Cross section of Casper Mountain (after Vaughn, 1976).

Nowhere in southern Wyoming can a block 4 be observed along the drape-fold fronts. This could be simply a coincidence of erosional depth. However, there are arguments indicating that block 4 may not form under these circumstances. The kinematic studies of Weinberg (this volume) show that block 4 does not begin to form until there is about 1,500 m (5,000 ft) of displacement on the fold. Should rupture of the folded layers by faulting occur at this point, there would be no need for block 4 to form in the fold. Therefore, it is reasonable to speculate that in the regions where the layered rocks are more welded to the basement and faulting of the sedimentary section occurs at lesser displacements, a block 4 may never develop.

Welded, Ductile Sections

The third type of section to be considered is that of a welded, but ductile structural unit lying immediately above the forcing member. Such a situation is well exposed in the Colorado National Monument on the northeast flank of the Uncompahgre uplift. Here, Precambrian crystalline basement blocks have been differentially uplifted; this forced Mesozoic rocks, primarily eolian sandstones, to be folded over the blocks.

The folds formed during Laramide deformation and the unconformity between the Precambrian and the overlying Triassic sedimentary rocks reflect Ancestral Rocky Mountain movements (Pennsylvanian and Permian). The Triassic Chinle Formation (20 to 25 m thick), which lies directly on the Precambrian basement, is composed of the typical red continental sandstones, siltstone, and shale common to the Colorado Plateau. Overlying the Chinle is the Upper Triassic Wingate Formation; a cross-bedded eolian sandstone approximately 100 m thick. The Lower Jurassic Kayenta Formation is a cross-bedded, highly lenticular, medium- to coarse-grained sandstone with a thickness of about 25 m. The overlying Jurassic Entrada Formation is similar to the Wingate Formation in that it is a massive, cross-bedded sandstone. Where it is flat-lying, the Entrada Formation is about 40 m thick. All of these rocks were once overlain by the Jurassic Summerville and Morrison Formations and a thick section of Cretaceous rocks. Because rocks

Figure 16. Thin section of experimentally deformed sequence of sandstone and limestone. The lowest sandstone member has been displaced along a 60°-dipping precut fault; this produced the deformation in the overlying material. Under the conditions of the experiment, the sandstone is brittle relative to the limestone (from Friedman and others, 1976).

younger than the Entrada Formation are not preserved within the drape folds, they will not be considered here.

The folds form over a series of steep faults in the Precambrian basement rocks (northeast flank of the Uncompahgre uplift). The zone of faulting and flexing is about 1.6 km wide with the maximum fault throws on the order of 350 m (1,200 ft). On the downthrown side of the zone, older beds are covered by alluvium from the Colorado River. On the upthrown side, the beds are flat-lying and form the Uncompahgre Plateau.

In the Colorado National Monument, at least within the area of exposure, the rock column consists of only two mechanical units: the Precambrian crystalline basement and the overlying clastic rocks. This is in contrast to Rattlesnake Mountain where in excess of 300 m of ductile shale separates the basement from the folded layered rocks. The basement faults die out upward within the first 30 m of the Triassic sedimentary rocks.

Several canyons cut through the sedimentary rocks into the basement and trend normal to faults so that they provide excellent exposures in the vertical plane. Most of the data for the controlled cross section in Figure 17 are subject to direct observation. This cross section at first glance is similar to those across Rattlesnake Mountain: the hinges between the blocks are unfaulted, and the blocks are well defined. The principal difference between the folds in the two areas is the lack of thinning at Rattlesnake Mountain and the extreme thinning at the Uncompahgre uplift. This means that there is no need for detachment at the Uncompahgre uplift because volume remains constant owing to thinning across the fold. The Wingate sandstone has, for example, attenuated from 110 m in block 1 to 59 m in the upper part of block 3. There are other canyons along this front where the Wingate is thinned to less than 30 m. This thinning is accomplished by cataclastic flow (Stearns, 1969). Individual beds within the Wingate Formation are capable of large flow by cataclasis. A bed of 2-m thickness can be reduced to less than a few centimetres by flow that is accomplished by both internal fracturing of the grains and macrofracturing within the formation. This thinning attests to the very macroscopically ductile behavior of the sandstone units in this area.

It is concluded that if as a whole the layered section immediately above the forcing member is very ductile, the layers thin, and there is no need for detachment. This, then, represents the same mechanical system as is represented in the Northern Rocky Mountains, but the mechanism of fold formation differs because the material response of the overlying rocks is different.

Another example of welded, ductile sections occurs along the southeastern flank of the Hanna Basin, in the vicinity of the Freezeout Mountains. Here the basement is overlain by a very thin section of Mississippian limestone that in turn is overlain by thick clastic sequences of the Casper Formation (Pennsylvanian). Even this section is capable of involvement in forced folds with up to at least 1,000 m (3,000 ft) of displacement without rupturing, as can be seen along the fronts of the Freezeout Mountains. However, the Casper Formation along these folds is thinned by cataclastic flow. Whether the thinning completely accommodates the necessary geometry or whether a combination of detachment plus thinning is responsible is not known. Detailed measurements along these mountain fronts need to be made in order to answer this question more precisely.

An area that is transitional between welded, nonthinning and welded, ductile sections is along the Front Range of Colorado. Here the entire sub-Pennsylvanian carbonate section was removed by erosion following uplift of the Ancestral Rocky Mountains. The Fountain Formation (Pennsylvanian) lies immediately on top of Precambrian granite. That the lower part of the stratigraphic section is more faulted in this area than in northern Wyoming is demonstrated by Matthews and Work (this volume). However, even this section is capable of considerable drape folding over the uplifted basement blocks, as is discussed by Matthews and Work. The most remarkable aspect of these sections is that they can fold as much as they do without faulting. One of the better-exposed folds of this type occurs at Elk Mountain west of Laramie, Wyoming (McClurg and Matthews, this volume). Elk Mountain has a well-developed block 1. There is sufficient relief along its steep front to show that forced folding in the layered rocks is continuous for at least 1,000 m (3,000 ft) over the basement fault.

Behavior at the Corners of Basement Blocks

As discussed by Stearns and Weinberg (1975) and shown experimentally by Friedman and others (1976), the sharp corners of the basement blocks are frequently

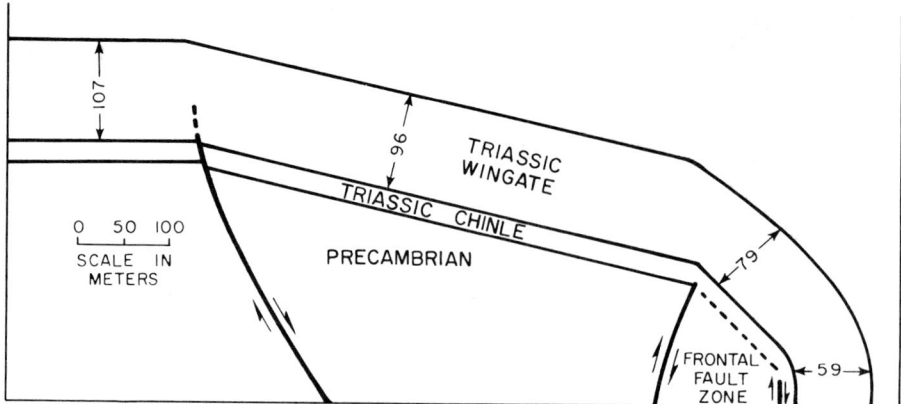

Figure 17. Controlled cross section along North Entrance Canyon in the Uncompahgre uplift showing the thinning in the Wingate Formation as it passes through a forced fold over rigid basement blocks.

sights of anomalous behavior in the forcing member. As the sedimentary rocks drape around the sharp corners, stress concentrations are created in the brittle basement. As a result, large-scale cataclastic flow can develop in these corner areas. Pieces of the sharp brittle corners are literally torn off by the folding of the overlying continuous layers, and pseudofolding of the basement results. In these regions, instead of a uniformly dipping, upper basement surface, large blocks of granite are crushed and rotated relative to one another so that a smooth, rounded fold can develop in the layered rocks. The scale at which this pseudofolding of the forcing member occurs is trivial with respect to the scale of the larger feature of which it is a part. This sort of behavior at corners should not be confused with the overall behavior of the basement block, which remains brittle and rigid.

There is another type of fold that results from block rotations but is not a forced fold. This fold results when blocks are rotated in the same general direction, but along different axes of rotation, as illustrated in Figure 18. This sort of fold results from an excess of material in the layered rocks where the two blocks join one another. As the blocks continue to rotate and the fold grows, the hinge area migrates through the beds. That is, this type of fold, unlike the drape fold, does not have fixed hinges in the Paleozoic strata. A particularly well-exposed example of this sort of folding is the Pat O'Hara structure that lies between the rotated Rattlesnake and Dead Indian blocks near Cody, Wyoming. The Rattlesnake Mountain block strikes northwest and is rotated toward the Bighorn Basin. It adjoins the Dead Indian Hill block, which is also rotated toward the Bighorn Basin, but which strikes more northward than the Rattlesnake Mountain block. Between the two blocks, the Pat O'Hara structure occurs. It terminates where the two blocks adjoin, and it broadens rapidly basinward. In such folds because the hinges do migrate through the beds, the Paleozoic rocks are completely shattered throughout the fold, not just at hinge lines.

FIRST MOVEMENTS OR ULTIMATE CAUSES

In the past few years a great deal has been learned about deep-crustal or upper-mantle movements that are the ultimate cause for surficial mountain terrains. We owe much of our current thinking on ultimate causes to solid-earth geophysicists, thermodynamicists, and oceanographers. However, most of the data that have led to a better understanding of the deeper Earth have been accumulated from the ocean or near continental edges. There is, perhaps, at least a tendency on the parts of some writers to make an unsubstantiated extrapolation or extension of these data to continental interiors. Ultimately, there may be a justification for

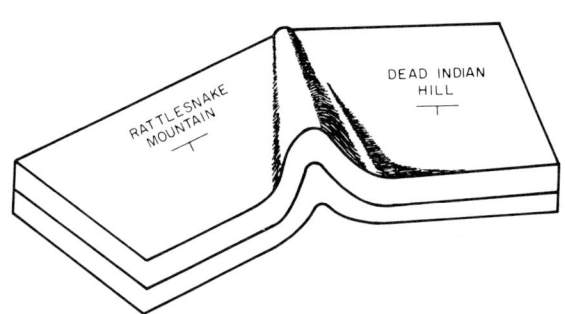

Figure 18. Sketch of the type of fold that can form when two basement blocks are rotated toward one another. This is not a forced fold.

such extrapolations. However, it seems that with the current state of knowledge, there is more speculation than there is substantiation when it comes to dealing with first causes for continental-interior movements. Wholesale extrapolations of proved continental-edge systems into the interior of continents, with no geophysical justification, may be unwarranted. Nonetheless, certain observations regarding the surficial structures can be made so that at least the problem can be specified if not solved. These specifications can at least place some restraints on an ultimate solution once the thermodynamics of continental interiors is more fully understood.

The first of these observations is that within the Wyoming province throughout Phanerozoic time and especially during Cretaceous and Tertiary time, differential vertical movements at the surface have dominated over horizontal movements. The very fact that there is a systematic development of deep intermountain basins surrounded by large uplifted mountains makes this conclusion unavoidable even though the whole continent may have been translated horizontally by plate motion. Furthermore, whatever the deep-seated motions are that produced the surficial motions, a fault *system* that arises from broad-scale uplift and downdropping seems to fit best with the observational facts. Although in detail the theory may be incorrect and the necessary assumptions may be geologically naive, solutions similar to those of Hafner (1951), Sanford (1959), and Couples (1977, and this volume) explain too many features of the region to be ignored. These features include an intermixture of fault types, rotations of large blocks, position of faults, and curvature of faults. Therefore, it would seem that whatever is postulated within the deeper portions of the Earth as a first cause, it must be able to produce broad-scale, absolute upward motions as well as broad-scale downward motions.

As argued earlier, localized horizontal motions as an explanation for the various orientations of mountain fronts seems implausible. However, that there is some deep-seated (lower-crust or upper-mantle), east-to-west transport of material would at least seem to be substantiated on the largest scale. By the end of Triassic time the upper part of the crust dipped gently westward throughout the Rocky Mountains foreland. However, by mid-Cretaceous time, conditions had drastically

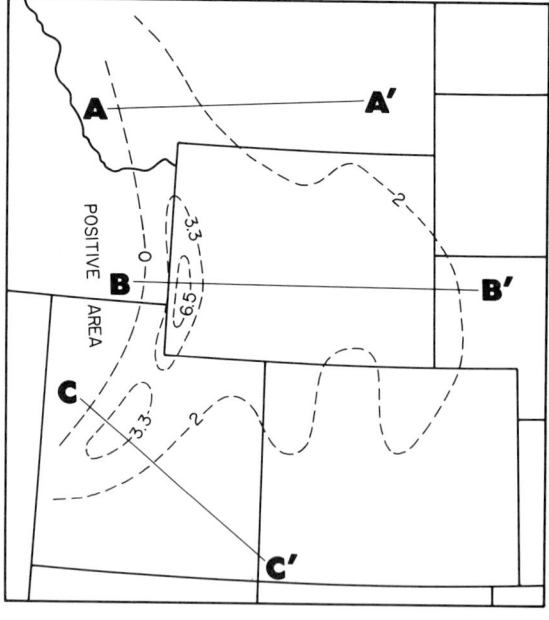

Figure 19. Isopachs (in kilometres) for Upper Cretaceous sedimentary rocks in the Rocky Mountains foreland (modified from Haun and Kent, 1965).

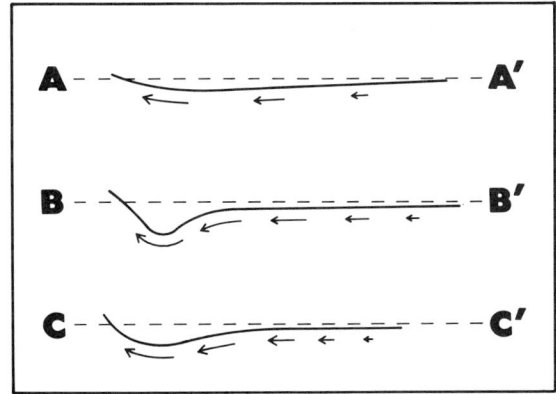

Figure 20. Cross sections in deep crust taken from data in Figure 19. Lines located in Figure 19. The arrow lengths are proportional to the amount of transfer required to produce the deviation from horizontal "dashed lines."

changed with the development of a positive area in the west and an asymmetrical trough immediately to the east (Fig. 19). Not only was the trough asymmetrical, but in it along its axis, lows of different depths were developed. The positive area changed strike considerably along trend. In order for this to happen in the shallow crust, there must have been material transferred from deep beneath the trough (near the Moho) into the newly created positive area. Certainly, along the positive area there are many instances of regenerated rock intruded into the shallow crust. Such transfer could have produced east-to-west transport in the lower crust or upper mantle. Furthermore, the amount of material withdrawn from the trough region would have been somewhat in proportion to the depth of the trough (Fig. 20). Because of the irregular nature of the trough, the net results would be differential lateral motion from north to south that died out to the east (Fig. 21).

The main justification for even considering differential lateral motions within the deep material comes from the broad-scale observations made by Sales (1968). The similarity he achieved in his barite-mud model when compared to the overall

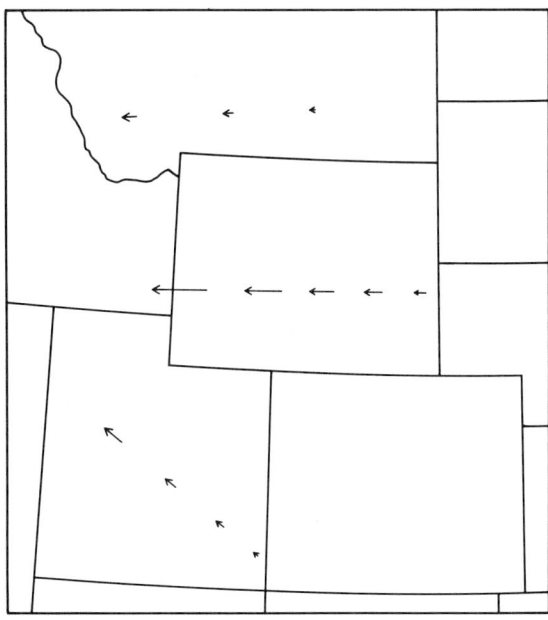

Figure 21. Schematic illustration of the relative lateral motions deep within the crust to produce the upper-crust configuration shown in Figure 19. The arrow positions and lengths are taken from Figure 20.

distribution of mountain systems and basins within the foreland is a remarkable geometric comparison. To produce a geometric similarity to any single mountain system or basin within such a modeling system could very well be ascribed to coincidence. However, in Sales's models, which are produced by a widespread mechanical couple in ductile material, there is geometric correspondence to virtually all of the major features in the state of Wyoming east of the thrust belt. Applying the uplifts and depressions that Sales produced in a very ductile material to the upper part of the brittle Precambrian basement may be mechanically naive. However, the upper surface of the Sales model could be considered to be in the lower crust or upper mantle where owing to pressure and temperature conditions, the rock materials can behave in a highly ductile fashion. As such it might serve as a model to the loading conditions that produce the discrete faults in the upper, brittle part of the basement. This, of course, is rank speculation, but the simultaneous creation of so many different geometries compared to what is actually seen in the field makes it an attractive speculation at least. It is also interesting to note that the widest, and best developed, part of the foreland (that is, most of Wyoming) occurs just to the east of the deepest part of the Cretaceous trough. Evidence for an eastward decrease in the amount of horizontal compression is presented in Couples and Stearns (this volume).

Even if there is some correspondence at depth (such as postulated by Sales, 1968) that is produced by differential lateral motions, it should be pointed out that such a system would need to be decoupled from the upper crust (both upper basement and layered sedimentary rocks). The justification for such a statement lies in the fact that lateral motions (that is, along wrench faults) play virtually no role in the displacement patterns in the upper basement or layered sedimentary rocks of the foreland. As was pointed out above, differential rotations of blocks can give a pattern similar to that which would be produced by lateral faulting (Fig. 7). However, examination in the field of mountain front after mountain front clearly demonstrates that there are no large-scale lateral motions in the surface rocks. Of all of the classical fault types, wrench faulting plays the least role in the formation of surface structures in the Rocky Mountains foreland. Nowhere in the surface rocks are there large-scale offsets that could be considered controlling features. Therefore, if lateral motions play a large role in the formation of the mountain structures in the Wyoming province, they must be deep within the crust and essentially decoupled from the upper-crustal materials. That is, their role can only be setting up broad-scale upwarps and downwarps in the ductile materials that in turn produce the lower loading condition for the upper, much more brittle materials.

CONCLUSIONS

There is an entire class of folds that can develop when loads are at high angles to planes of anisotropy within rock sections. These folds are forced folds. In most cases this anisotropy is sedimentary layering. Within this general class there are particular adjectives that apply to certain folds, such as drape folds or diapiric folds. Furthermore, in the case of drape folding, the kinematics and dynamics that determine the ultimate form depend upon many physical parameters such as rock type, depth of burial, degree of welding, and whether the rocks are in layer-parallel extension or compression.

The Wyoming province serves as an excellent example of some of the types of forced folding as demonstrated by field exposures over the entire province.

Virtually all mountain ranges within the Wyoming province exhibit some form of forced folding in the layered sedimentary rocks. In addition to field evidence for the existence of this type of structure, such folds have also been created experimentally. When layered rock materials in the laboratory are subjected to differential vertical movements of a homogeneous, brittle member from below (forcing member), many of the features seen in the field can be reproduced in the laboratory (Logan and others, this volume).

If first causes in the Rocky Mountains foreland are ever to be understood, more geophysical studies are needed. It is unlikely that the true causes at depth will be unraveled from scattered surface data alone. In addition, it may be misleading to make wholesale extrapolations from the thermodynamics of continental edges to continental interiors. Even if such extrapolations turn out to be justified, they are not more than lucky guesses at this time without geophysical studies to back them.

There are numerous ways in which layered rocks can be loaded at high angles to their planes of anisotropy, and more of these loading systems should be investigated. The smooth sinusoids of Hafner (1951), the sharp steps of Sanford (1959), and the sawtooth configuration of Couples (this volume) are but a few of the conditions that lead to faults that load the layered rocks, and further work is needed to delineate other possible loading conditions.

Bedding-plane detachments play a very major role, at least in the Wyoming province, in the resulting structures. Therefore, lateral motion of layered rocks during the folding process deserves considerably more attention than it has received in the past. Much data (see Stearns, 1975; Vaughn, 1976; Cook and Stearns, 1975; Weinberg, this volume; Stearns and Stearns, this volume) indicate that some of these lateral displacements may be large. Furthermore, in order to produce these continuous folds, lateral displacements must occur in three dimensions. Therefore, if understanding of this folding process is to be enhanced, the three-dimensional kinematics on a large scale must be investigated. From a geologic standpoint this would seem to be the most necessary area of new work. That is, cross sections of regions near the centers of the blocks are relatively well delineated in terms of the final configuration of the fold. However, the internal displacements that are required to produce such large-scale folding are but remotely understood, and the total displacement field required within large blocks with thousands of square metres of surface area has never been properly investigated.

REFERENCES CITED

Berg, Robert R., 1962, Mountain flank thrusting in Rocky Mountain foreland, Wyoming and Colorado: Am. Assoc. Petroleum Geologists Bull., v. 46, p. 2019–2032.

——1976, Deformation of Mesozoic shales at Hamilton Dome, Bighorn Basin, Wyoming: Am. Assoc. Petroleum Geologists Bull., v. 60, p. 1425–1433.

Borg, I., and Handin, J., 1966, Experimental deformation of crystalline rocks: Tectonophysics, v. 3, p. 249–368.

Cook, Robert A., and Stearns, David W., 1975, Mechanisms of sandstone deformation; a study of the drape folded Weber sandstone in Dinosaur National Monument, Colorado and Utah, *in* Bolyard, D. W., ed., Symposium on deep drilling frontiers of the central Rocky Mountains: Denver, Rocky Mtn. Assoc. Geologists, p. 21–32.

Couples, Gary, 1977, Stress and shear fracture (fault) patterns resulting from a suit of complicated boundary conditions: Pure and Applied Geophysics, v. 115, p. 113–133.

——1978, Comments on applications of boundary-value analyses of structures of the Rocky Mountains foreland, *in* Matthews, V., III, ed., Laramide folding associated with basement block faulting in the Western United States: Geol. Soc. America Mem. 151 (this volume).

Couples, Gary, and Stearns, D. W., 1978, Analytical solutions applied to structures of the Rocky Mountains foreland on local and regional scales, *in* Matthews, V., III, ed., Laramide folding associated with basement block faulting in the Western United States: Geol. Soc. America 151 (this volume).

Donath, F. A., 1961, Experimental study of shear failure in anisotropic rocks: Geol. Soc. America Bull., v. 72, p. 985–990.

Elam, J. G., 1969, Tectonic style in the Permian Basin and its relationship to cyclicity *in* Chuber, S., and Elam, J., eds., Cyclic sedimentation in the Permian Basin: West Texas Geol. Soc. Pub., p. 56–69.

Fanshawe, J. R., 1939, Structural geology of Wind River Canyon area, Wyoming: Am. Assoc. Petroleum Geologists Bull., v. 23, p. 1439–1492.

Foose, R. M., Wise, D. V., and Garbarini, G. S., 1961, Structural geology of the Beartooth Mountains, Montana and Wyoming: Geol. Soc. America Bull., v. 72, p. 1143–1172.

Friedman, M., Handin, J., Logan, J. M., Min, K. D., and Stearns, D. W., 1976, Experimental folding of rocks under confining pressure: Pt. III, Faulted drape folds in multilithologic layered specimens: Geol. Soc. America Bull., v. 87, p. 1049–1066.

Gilluly, James, 1963, The tectonic evolution of the western United States: Geol. Soc. London Quart. Jour., v. 119, p. 133–174.

Hafner, W., 1951, Stress distribution and faulting: Geol. Soc. America Bull., v. 62, p. 373–398.

Haun, John D., and Kent, Harry C., 1965, Geologic history of Rocky Mountain region: Am. Assoc. Petroleum Geologists Bull., v. 49, p. 1781–1800.

Heard, Hugh C., 1960, Transition from brittle fracture to ductile flow in Solenhofen limestone as a function of temperature, confining pressure, and interstitial fluid pressure, *in* Griggs, D. T., and Handin, J., eds., Rock deformation: Geol. Soc. America Mem. 79, p. 193–226.

Hudson, F. S., 1955, Folding of unmetamorphosed strata superjacent to massive basement rocks: Am. Assoc. Petroleum Geologists Bull., v. 39, p. 2038–2052.

Johnson, Arvid M., 1970, Physical processes in geology: San Francisco, Freeman, Cooper & Co., 577 p.

Logan, J. M., Iwasaki, T., Friedman, M., and Kling, S. A., 1972, Experimental investigation of sliding friction in multilithologic specimens, *in* Pincus, H., ed., Geological factors in rapid excavation: Geol. Soc. America Eng. Geology Case History no. 9, p. 55–67.

Logan, J. M., Friedman, M., and Stearns, M. T., 1978, Experimental folding of rocks under confining pressure: Pt. VI, Further studies of faulted drape folds, *in* Matthews, V., III, ed., Laramide folding associated with basement block faulting in the Western United States: Geol. Soc. America Mem. 151 (this volume).

Lowman, S. W., 1963, Geologic map of the Grand Junction area, Colorado: U.S. Geol. Survey Misc. Inv. Map I-404, c scale 1:31,680.

Matthews, Vincent, III, 1976, Mechanisms of deformation during the Laramide orogeny in the Front Range, Colorado, *in* Epis, R. C., and Weimer, R. J., eds., Studies in Colorado field geology: Colorado School Mines Prof. Contr., no. 8, p. 398–402.

Matthews, Vincent, III, and Work, David F., 1978, Laramide folding associated with basement block faulting along the northeastern flank of the Front Range, Colorado, *in* Matthews, V., III, ed., Laramide folding associated with basement block faulting in the Western United States: Geol. Soc. America Mem. 151 (this volume).

McClurg, J. E., and Matthews, Vincent, III, 1978, Origin of Elk Mountain anticline, *in* Matthews, V., III, ed., Laramide folding associated with basement block faulting in the Western United States: Geol. Soc. America Mem. 151 (this volume).

Palmquist, John C., 1978, Laramide structures and basement block faulting: Two examples from the Big Horn Mountains, Wyoming, *in* Matthews, V., III, ed., Laramide folding associated with basement block faulting in the Western United States: Geol. Soc. America Mem. 151 (this volume).

Prucha, John J., Graham, John A., and Nickelson, Richard P., 1965, Basement controlled deformation in Wyoming province of Rocky Mountains foreland: Am. Assoc. Petroleum Geologists Bull., v. 49, p. 966–992.

Royse, F., Jr., Warner, M. A., and Reese, D. L., 1975, Thrust belt structural geometry and related stratigraphic problems, Wyoming–Idaho–northern Utah, *in* Bolyard, D. W., ed., Symposium on deep drilling frontiers of the central Rocky Mountains: Denver, Rocky Mtn. Assoc. Geologists, p. 41–54.

Sales, John K., 1968, Crustal mechanics of Cordilleran foreland deformation: A regional and scale-model approach: Am. Assoc. Petroleum Geologists Bull., v. 52, p. 2016–2044.

Sanford, A. R., 1959, Analytical and experimental study of simple geologic structures: Geol. Soc. America Bull., v. 70, p. 19–52.

Savage, William Z., 1974, Stress and displacement fields in stably folded rock layers [Ph.D. dissert.]: College Station, Texas A&M Univ., 193 p.

Stearns, David W., 1969, Fracture as a mechanism of flow in naturally deformed layered rocks, *in* Baer, A. J., and Norris, D. K., eds., Conference on research in tectonics, kink bands and brittle deformation: Canada Geol. Survey Paper 68–52, p. 79–95.

——1971, Mechanisms of drape folding in the Wyoming province: Wyoming Geol. Assoc., 23rd Ann. Field Conf., Wyoming Tectonics Symp., Guidebook, p. 125–143.

——1975, Laramide basement deformation in the Bighorn Basin—The controlling factor for structures in the layered rocks: Wyoming Geol. Assoc., 27th Ann. Field Conf., Geology and mineral resources of the Bighorn Basin, Guidebook, p. 82–106.

Stearns, David W., and Jamison, W. R., 1977, Deformation of sandstones over basement uplifts, Colorado National Monument, *in* Veal, H. D., ed., Exploration frontiers of the Central and Southern Rockies: Rocky Mtn. Assoc. Geologists Guidebook, 1977, p. 31–39.

Stearns, David W., and Weinberg, David M., 1975, A comparison of experimentally created and naturally formed drape folds: Wyoming Geol. Assoc., 27th Ann. Field Conf., Geology and mineral resources of the Bighorn Basin, Guidebook, p. 159–166.

Stearns, David W., Sacrison, W. R., and Hanson, R. C., 1975, Structural history of southwestern Wyoming as evidenced from outcrop and seismic, *in* Bolyard, D. W., ed., Symposium on deep drilling frontiers of the central Rocky Mountains: Denver, Rocky Mtn. Assoc. Geologists, p. 9–20.

Stearns, Martha T., and Stearns, David W., 1978, Geometric analysis of multiple drape folds along the northwest Big Horn Mountains front, Wyoming, *in* Matthews, V., III, ed., Laramide folding associated with basement block faulting in the Western United States: Geol. Soc. America Mem. 151 (this volume).

Untermann, G. E., and Untermann, B. R., 1965, Geologic map of the Dinosaur National Monument, Colorado-Utah: Utah Geol. and Mineralog. Survey.

Vaughn, Patty H., 1976, Mesozoic sedimentary rock features resulting from volume measurements required in drape folds at corners of basement blocks—Casper Mountain area, Wyoming [M.S. thesis]: College Station, Texas A&M Univ., 93 p.

Weinberg, D. M., 1978, Some two-dimensional kinematic analyses of the drape-fold concept, *in* Matthews, V., III, ed., Laramide folding associated with basement block faulting in the Western United States: Geol. Soc. America Mem. 151 (this volume).

Wisser, Edward, 1957, Deformation in structural geology: Deformation in the Cordilleran region of the Western United States: Colorado School Mines Quart., v. 52.

MANUSCRIPT RECEIVED BY THE SOCIETY JUNE 27, 1977
MANUSCRIPT ACCEPTED AUGUST 25, 1977

Printed in U.S.A.

Geological Society of America
Memoir 151

Seismic interpretation of basement block faults and associated deformation

W. R. SACRISON
Amoco Production Company
Security Life Building
Denver, Colorado 80202

ABSTRACT

The use of reflection seismic data is necessary for the proper interpretation of buried block fault, drape-fold structural features. By the application of certain basic principles, it is possible to make reasonable structural interpretations in areas where seismic information is of poor quality or where complete coverage cannot be acquired. A seismic-model study and a number of seismic sections from various Wyoming basins illustrate the capabilities and limitations of reflection seismic techniques for defining block fault, drape-fold structures.

INTRODUCTION

Recent advances in seismic-data acquisition and processing enable geophysicists to make structural interpretations that would have been difficult or impossible ten or fifteen years ago. The most important data-gathering advancement was the implementation of CDP or multifold recording. Digital processing of seismic data has provided the means for geophysicists to greatly improve resolution of data, migration techniques, the conversion of time sections to depth sections, and seismic modeling procedures. The seismic lines shown in this paper utilize part or all of these techniques to illustrate how reflection seismic methods can be used to define drape-fold structures.

SEISMIC MODEL STUDY

Figure 1 is a geologic model of the stratigraphic units in a typical block fault, drape-fold structure in the northern Bighorn Basin of Wyoming. Sufficient well control is available in the area to make a reasonably accurate model of the true structural configuration. Note that no faults are present. Note also that the stratigraphic units dip evenly west from the right side of the model to the base of the asymmetric structure located at the left side of the model.

Figure 2 is a ray-path diagram that shows the seismic rays that have been generated at the surface, reflected from the Dakota horizon, and that have returned to receiver positions on the surface. The segment of the Dakota Formation indicated by the brace or bracket shows that part of the subsurface from which no seismic energy would be reflected back to the receivers along a conventional seismic spread. The segment of no energy return would represent blocks 3 and 4 of Stearns's drape-fold model that is discussed in another paper of this memoir. How much, if any, of blocks 3 and 4 would be seen on seismic lines across drape folds depends on the steepness of the beds and the configuration of the seismic field technique.

Figure 3 is a synthetic seismic section that was generated from the geologic model of Figure 1. Most interpreters would probably place a fault near the left center of the synthetic section. We know from the geologic model, however, that the section is not faulted. A velocity anomaly has also been created just west of the center part of the section. The velocity anomaly was caused by the lateral velocity change created when the faster-velocity lower Mesozoic and Paleozoic rocks were uplifted and the slower velocity upper Mesozoic and Tertiary rocks were eroded from the uplifted area. The resulting lateral-velocity change from the top of the structure to the area east of the structure causes the velocity anomaly. The velocity anomaly is one of the classic pitfalls in seismic interpretation.

Figure 4 is an actual seismic section across the area from which the geologic model was made. Note the similarity between the synthetic section and the actual section.

Although the model study demonstrates that pseudo-structures may exist near drape-fold structures, interpreters should certainly not assume that all time anomalies

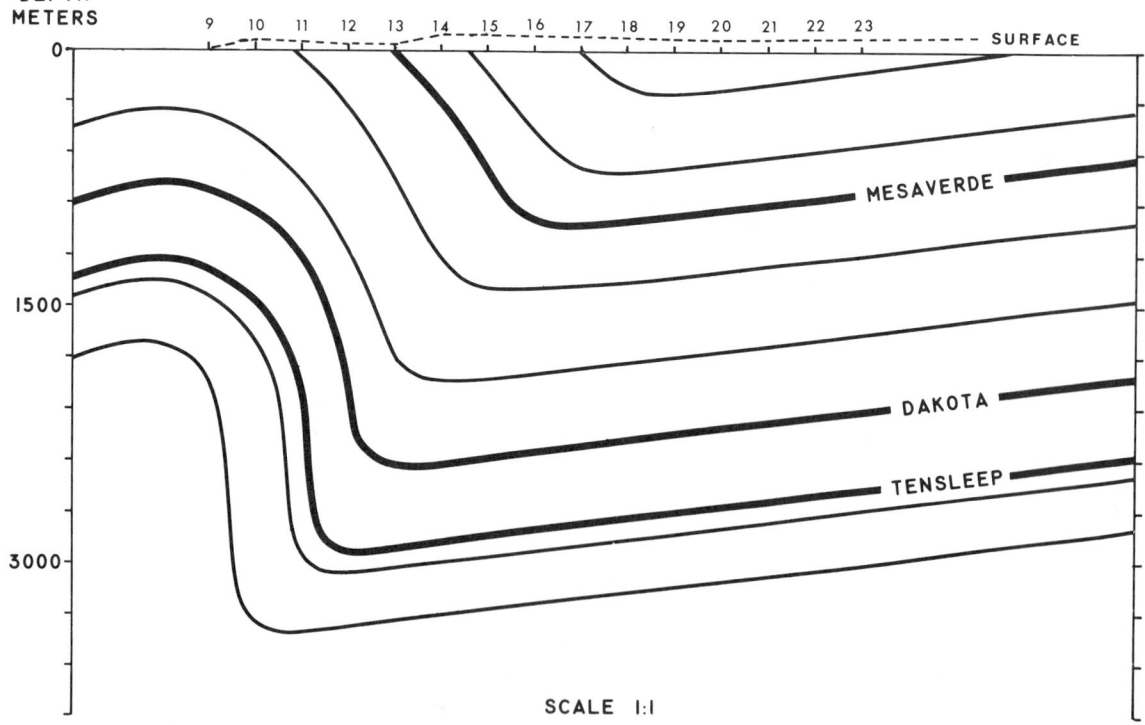

Figure 1. Geologic model of drape fold in northern Bighorn Basin, Wyoming.

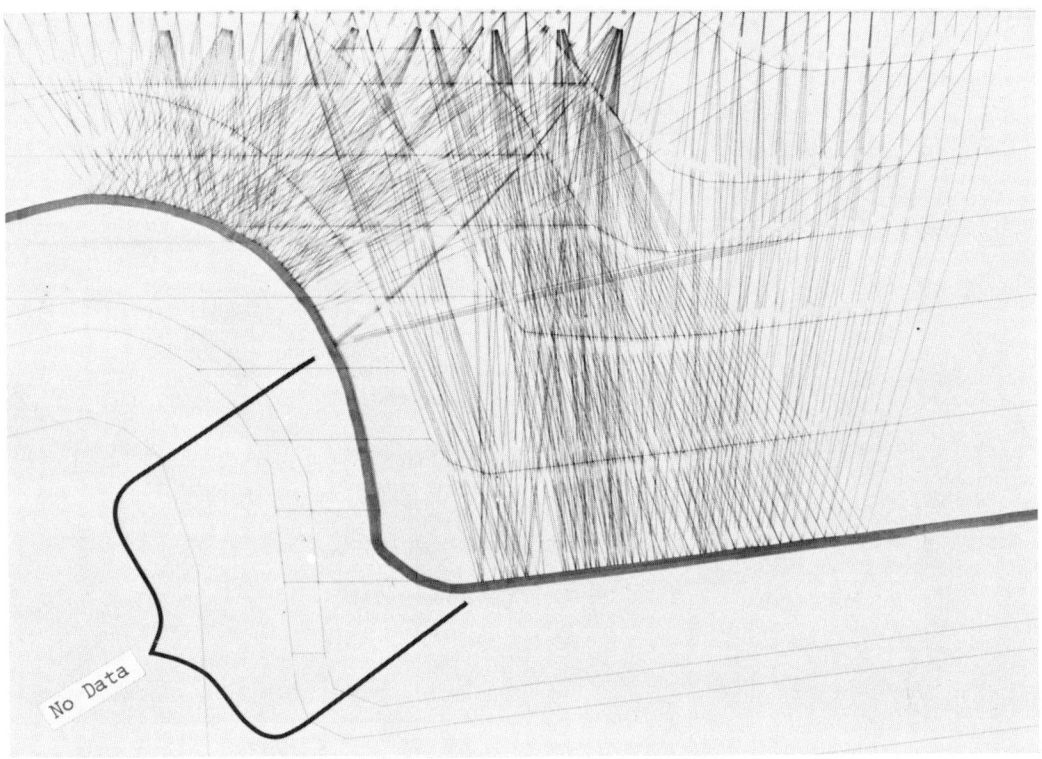

Figure 2. Ray-path diagram. Heavy line = Dakota horizon.

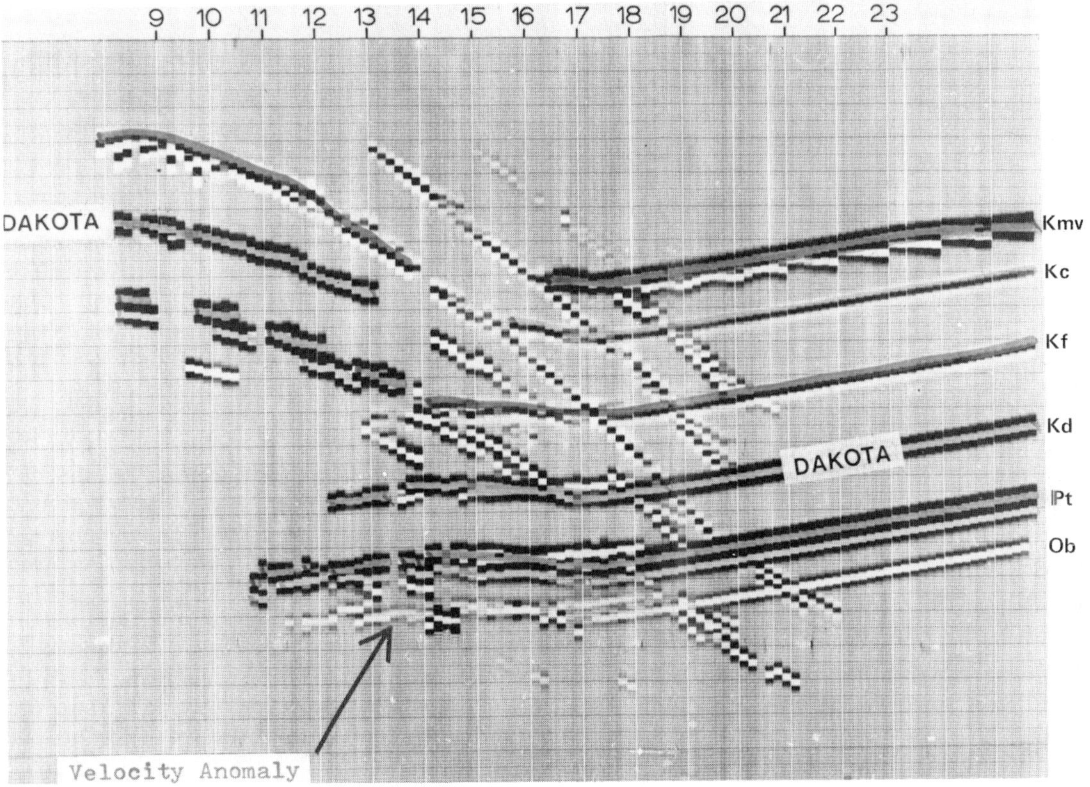

Figure 3. Synthetic seismic section.

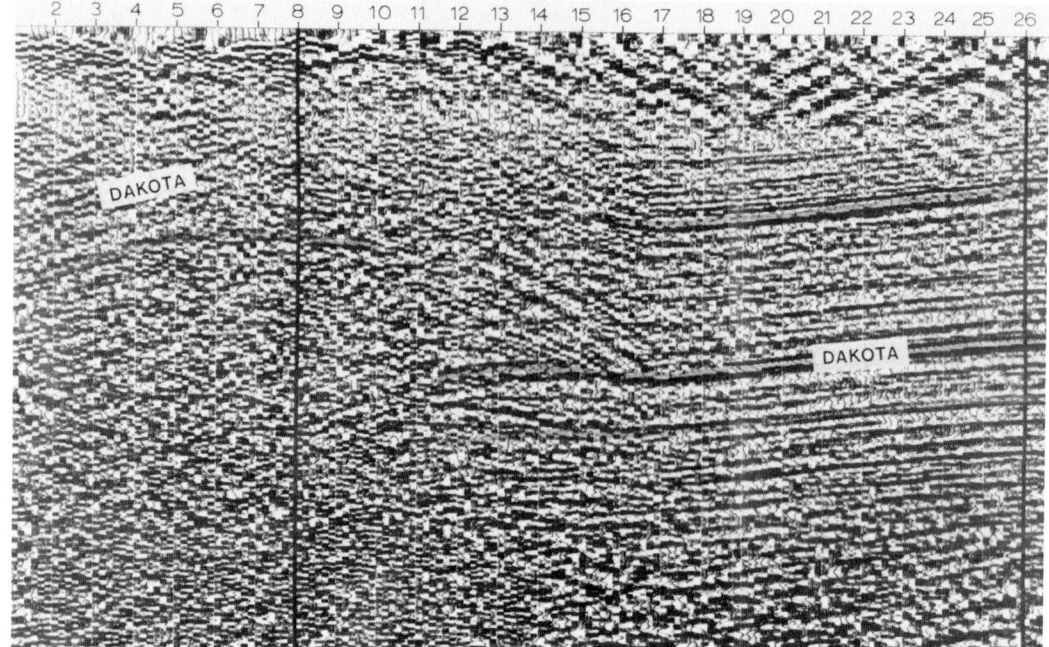

Figure 4. Actual seismic section.

in similar positions are false. A careful study of the velocities should be made to either confirm or condemn the feature. If results are inconclusive, it may be less costly to drill a dry hole on a velocity anomaly than to miss an oil field.

SEISMIC SECTIONS

The seismic sections, unless otherwise indicated, are nonmigrated time sections. Migration is the procedure whereby reflections from dipping horizons are moved to their proper position on seismic sections. Figures 5 and 6 illustrate the effect of migration on an irregular reflection surface. Figure 5 is a nonmigrated time section, and Figure 6 is a migrated time section of the same seismic line. Notice that migration improves reflection continuity, steepens the flanks and decreases the apparent size of structural highs, and expands the synclines. Migrated sections more nearly represent the true structural configuration than do nonmigrated sections.

Most of the sections have a vertical exaggeration of approximately 2:1 to 3:1, as indicated by the triangle at the side of the sections. Consequently, the dips appear steeper than they really are. All but one of the seismic lines have only one triangle shown. A series of triangles would more accurately depict the horizontal to vertical scale comparison, because rock velocities usually increase with increasing depth. Therefore, the vertical exaggeration on time sections would become less with increasing time. The single triangle represents an average vertical exaggeration.

Figure 7 is a nonmigrated seismic line recorded in the northern Bighorn Basin. It shows several prominent drape-fold structures. The Precambrian, Madison, Dakota, and Frontier horizons are indicated on the section. Although the exact location of the basement surface below the drape folds is difficult to determine, it can be established that the basement faults are either high-angle normal or high-angle reverse. The dark lines extending upward from the steep face of the basement

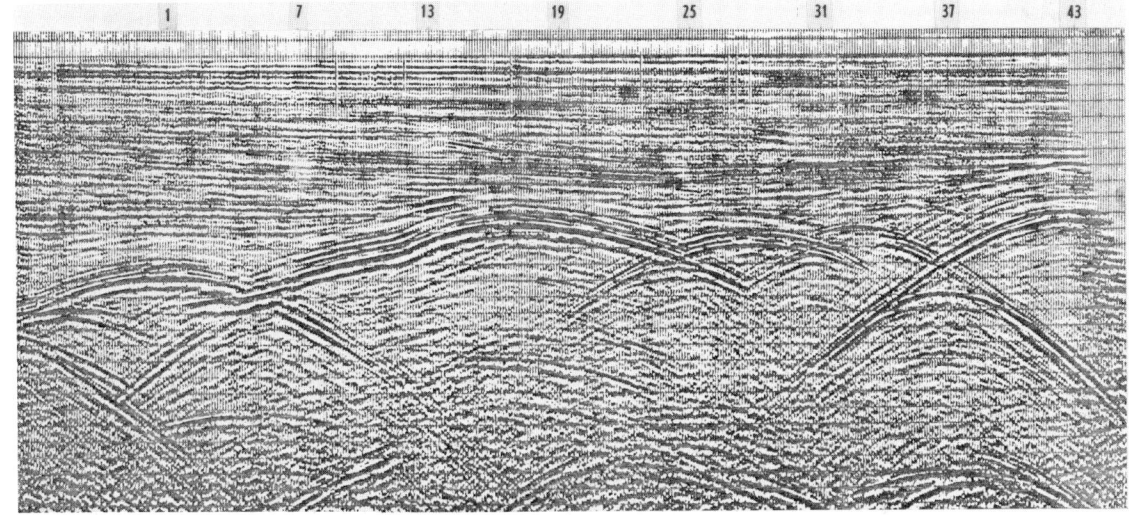

Figure 5. Nonmigrated seismic section.

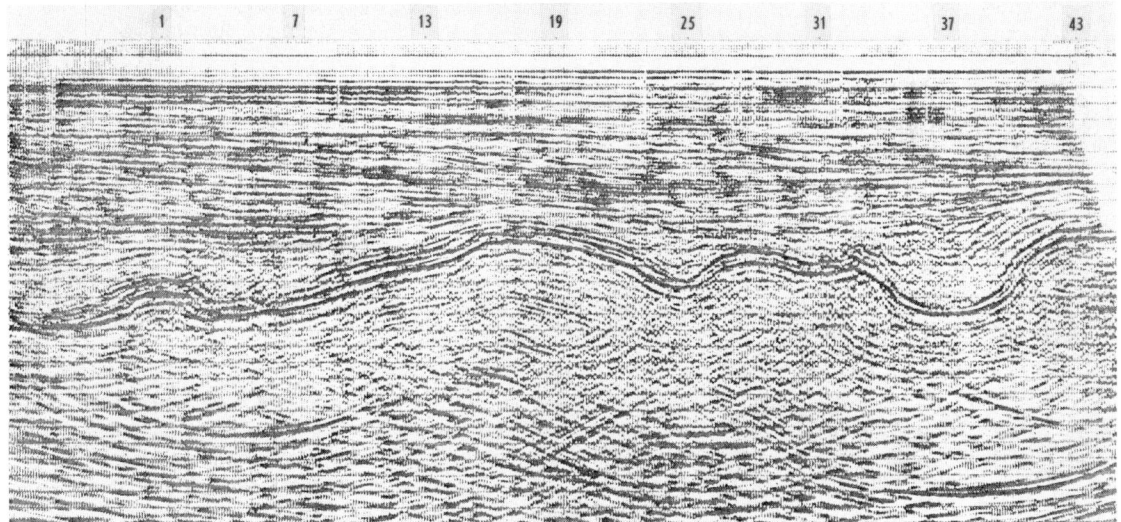

Figure 6. Migrated seismic section.

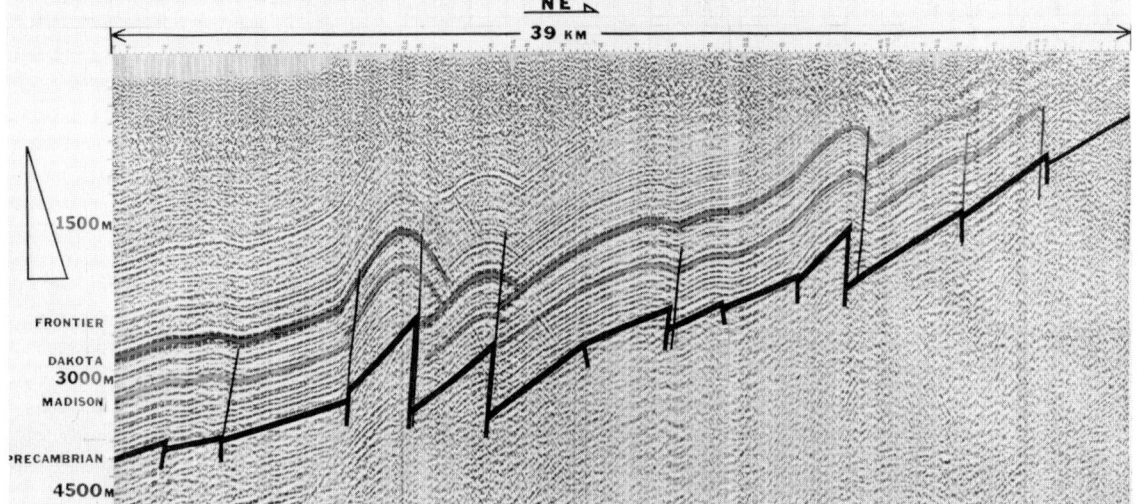

Figure 7. Seismic section, northern Bighorn Basin, Wyoming.

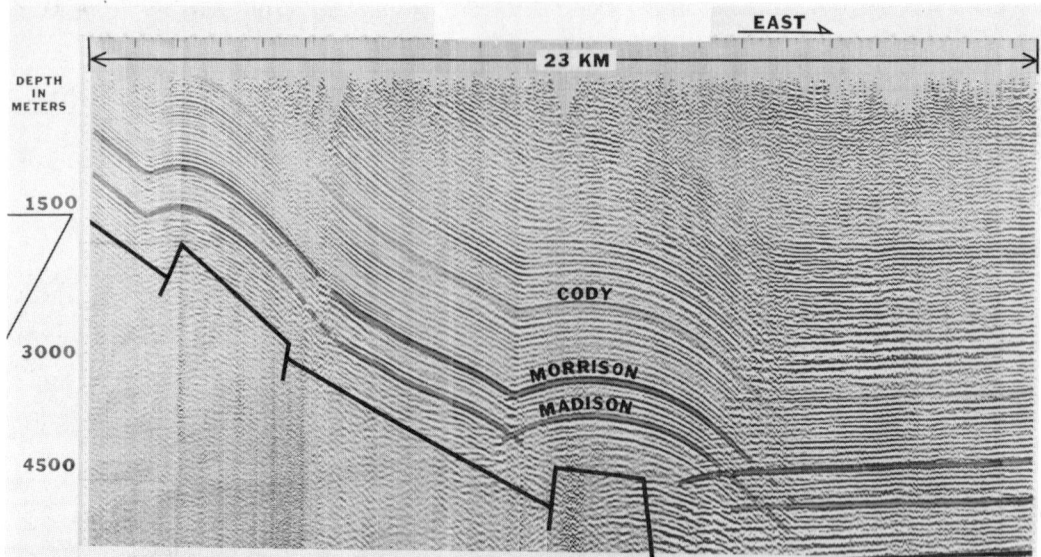

Figure 8. Nonmigrated (time) seismic section, western Powder River Basin, Wyoming.

blocks are where one would normally interpret faults in the sedimentary section. The presence of these faults seems reasonable on the nonmigrated section; however, in reality, the steep sides of the asymmetric drape folds are probably not faulted or, if faults do exist, they are of minor displacement.

Figure 8 is a nonmigrated time section recorded in the western Powder River Basin. The drape folds are quite similar to those in the Bighorn Basin (Fig. 7). The migrated version of this line (Fig. 9) illustrates how the structural highs and lows change configuration after migration. Note particularly the continuity of events at the Cody and Morrison horizons after migration as compared to the nonmigrated section. The breaks in continuity at the Madison level are probably caused by interfering seismic events and incorrect migration velocity rather than faults or discontinuities in the Madison Formation. The migrated depth section (Fig. 10) is the truest representation of the subsurface geology if correct velocities are used and all processing manipulations are properly achieved. In Figure 10 incorrect velocity interpolations have caused false dip-rate changes.

Figure 11 is a nonmigrated seismic line across the Brady structure located on the east flank of the Rock Springs uplift in southern Wyoming. The reflections at the Madison horizon and between the Morrison and Madison horizons are discontinuous across the syncline on the west side of the structure. It would seem reasonable to interpret a fault where the discontinuities occur. However, the migrated section of the same line (Fig. 12) shows nearly continuous reflections at all horizons down to and including the Madison. It is concluded, therefore, that little or no faulting is present where this seismic line crosses the Brady structure. Figures 11 and 12 show another phenomenon common to many deeply buried drape-fold structures. The shallow reflections have little or no critical west dip, and the amount of west dip increases with increasing depth until a maximum is reached at the deepest observable reflections.

Figure 13 shows a drape fold in the southwestern Wind River Basin. Displacement of the basement is approximately 2,500 m, and the overlying drape fold has probably separated by faulting.

Figure 14 is another section from the southwestern Wind River Basin. The basement

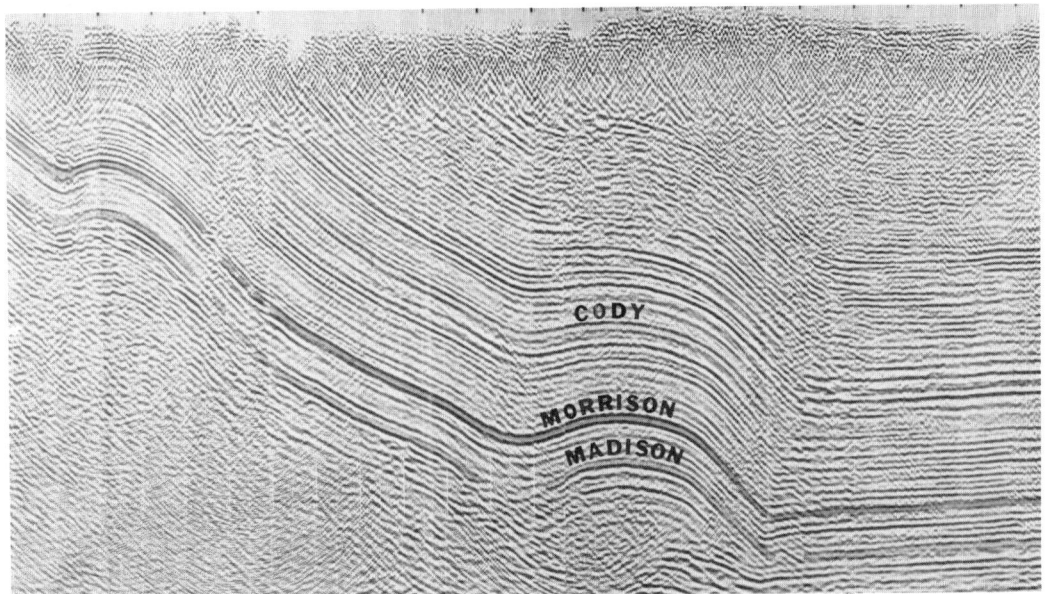

Figure 9. Migrated (time) seismic section, western Powder River Basin, Wyoming.

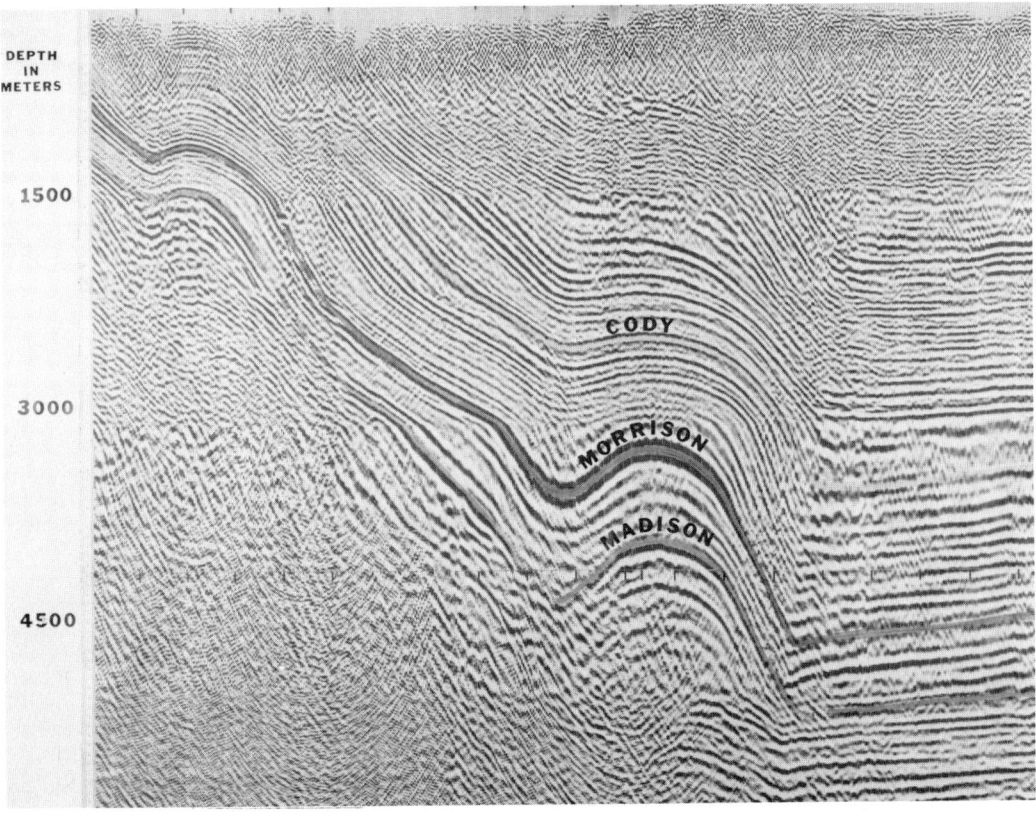

Figure 10. Migrated (depth) seismic section, western Powder River Basin, Wyoming.

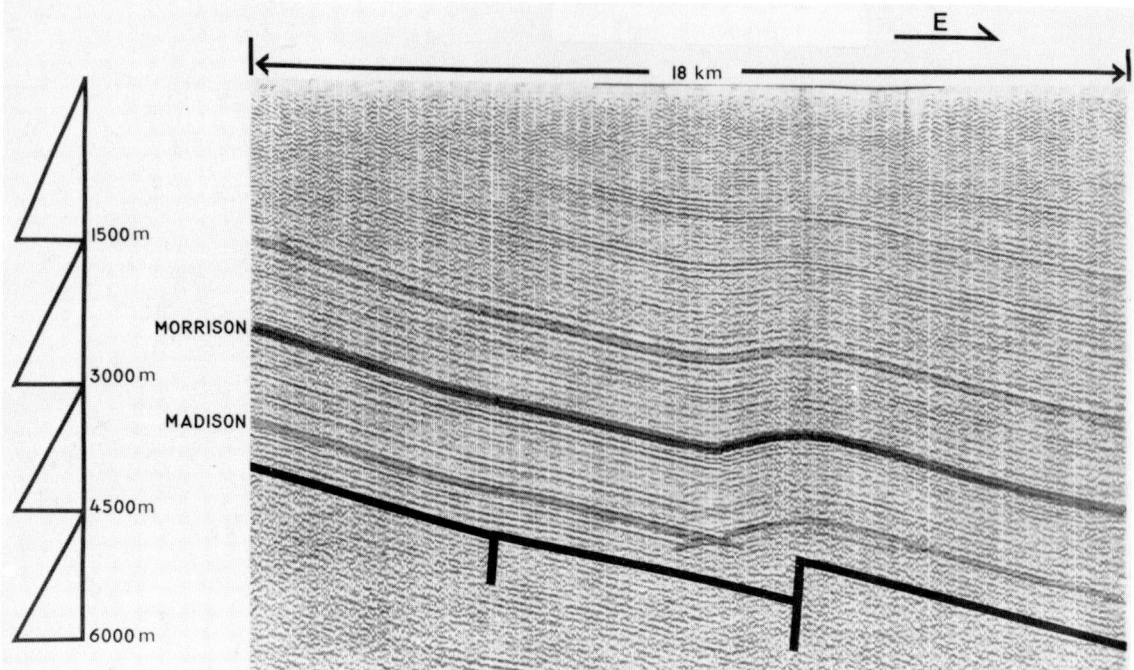

Figure 11. Nonmigrated seismic section, Brady structure, southern Wyoming.

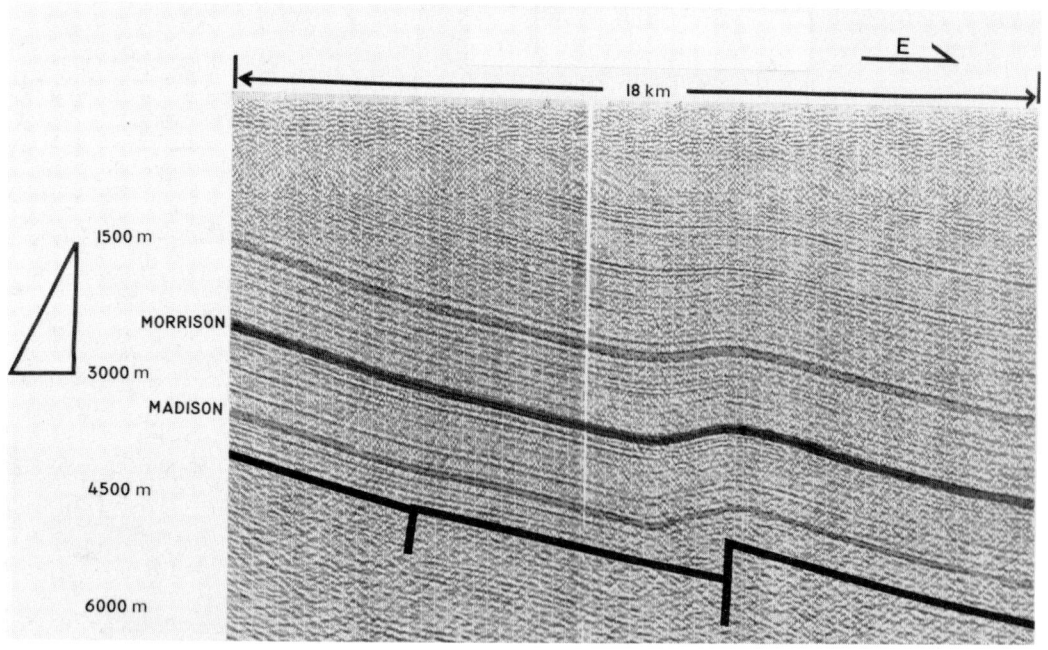

Figure 12. Migrated seismic section, Brady structure, southern Wyoming.

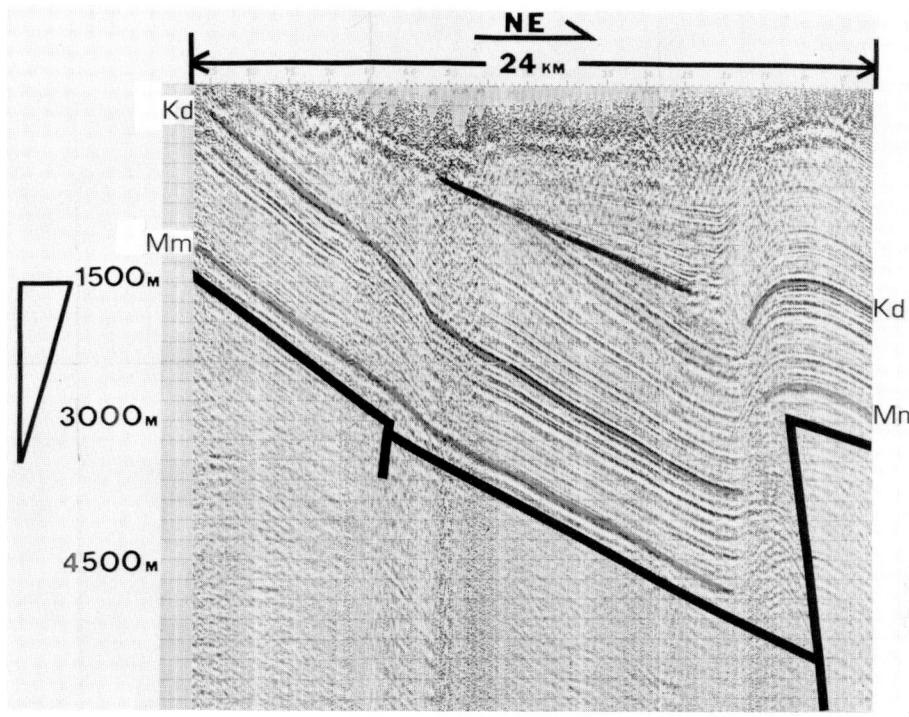

Figure 13. Seismic section, southwestern Wind River Basin, Wyoming.

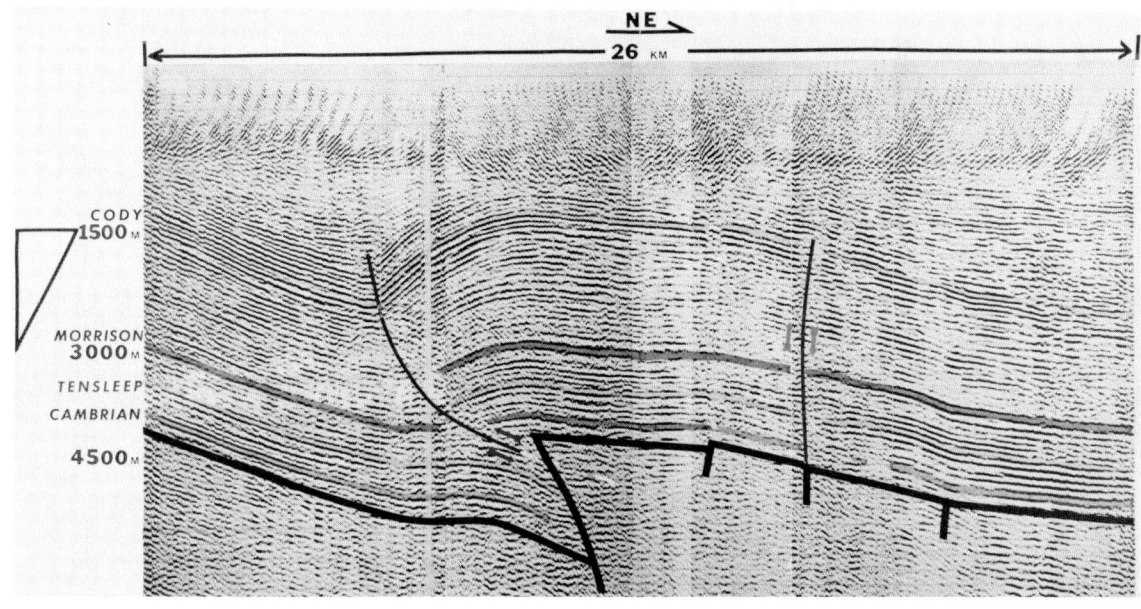

Figure 14. Seismic section, southwestern Wind River Basin, Wyoming.

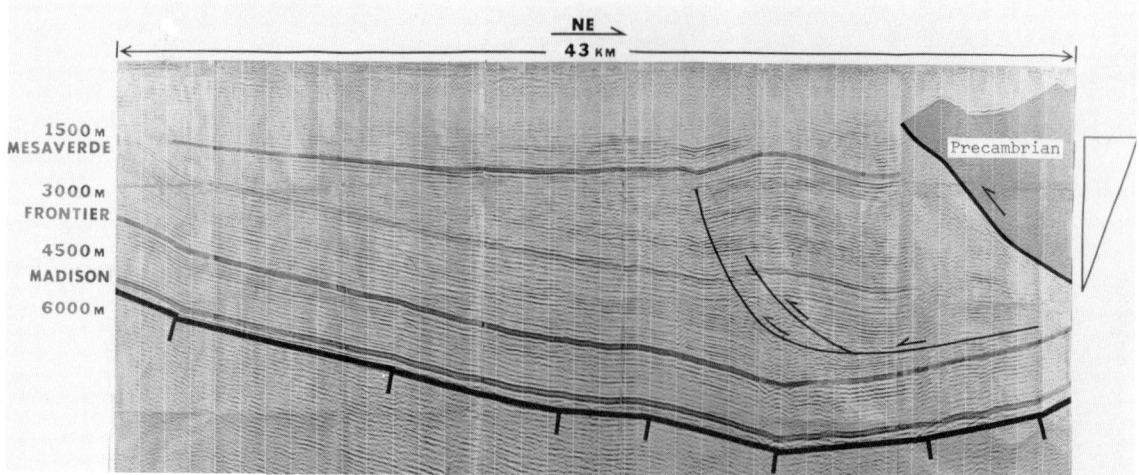

Figure 15. Seismic section, northern Green River Basin, Wyoming.

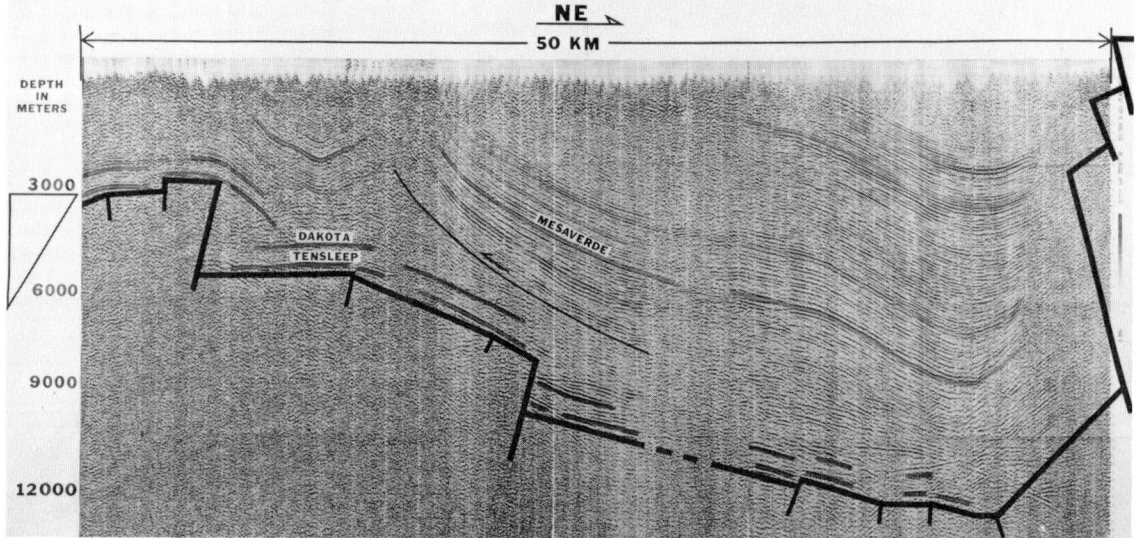

Figure 16. Seismic section, Hanna Basin, Wyoming.

fault is at a lower angle than on previous sections, and compressional forces have developed in the sediments. An extension or tension zone has formed on the gentle-dip side of the drape fold in which a normal fault has developed. Also note the probable velocity pull-up at the Tensleep and Cambrian horizons beneath the left side of the drape fold.

Figure 15 is a seismic section from the northern Green River Basin and illustrates a basement block that has emerged at a low angle (approximately 20°). The shaded area in the upper right is a part of the Wind River Mountain block. The sedimentary layers to the left of the basement block have been end-loaded, and a compressional fold and thrust faults have formed. Note that there is no basement uplift directly below the compressional fold, and no drape folds are present.

Figure 16 is a seismic line from the Hanna Basin of southeastern Wyoming.

When compared with other Wyoming basins, the Hanna Basin is unique in several respects. It is small in size (approximately 2,600 km^2). It has the thickest combined Upper Cretaceous–Tertiary sedimentary section of any basin in the Rocky Mountains foreland. Differential vertical displacement of the basement surface from the deepest part of the Hanna Basin to the Shirley and Seminoe Mountains is approximately 16,000 m in a distance of 20 km. Steeply dipping compressional folds such as that near the left side of Figure 16 occur in the Upper Cretaceous rocks on the west, south, and east flanks of the basin. These folds have no basement uplift beneath them so are unrelated or at least indirectly related to basement block fault movements. The placement of the basement surface on the right side of Figure 16 is interpretive because of deterioration of the seismic data along the steep north flank of the Hanna Basin.

CONCLUSIONS

Structural interpretation should incorporate all information sources available. The use of a model that can be observed at the surface as a base from which to analyze buried structures that can be observed only indirectly is a reasonable approach to interpretation. Seismic lines in the Rocky Mountains foreland support the basic concepts established by surface study of drape-fold structures. However, as illustrated by the seismic sections, structures other than drape folds certainly do exist in the Rocky Mountains foreland.

ACKNOWLEDGMENTS

I thank Amoco Production Company for granting permission to publish the model study and seismic sections shown in this paper. Ray Gilley of Amoco did the Bighorn Basin model study, and GeoQuest International, Ltd., provided the migration examples shown in Figures 5 and 6.

Manuscript Received by the Society June 27, 1977
Manuscript Accepted August 25, 1977

Printed in U.S.A.

Geological Society of America
Memoir 151

Some two-dimensional kinematic analyses of the drape-fold concept

DAVID M. WEINBERG*
Department of Geology and Center for Tectonophysics
Texas A&M University
College Station, Texas 77843

ABSTRACT

The vertical basement-block movements and overlying folds in the sedimentary veneer that characterize the deformation style of the Wyoming province are amenable to simple two-dimensional kinematic analyses. These analyses examine the lateral transport needed when no thickness change of the folded sedimentary rocks occurs, and they are based on simple geometric constructions that approximate the folded shape of the uppermost bedding surfaces of the Mesozoic and Paleozoic sections (about 3,900-m total thickness). Lateral-displacement requirements of these rocks are significantly different, and this difference implies that the rocks on the downthrown basement block are put into a state of shear. Additionally, these differences suggest that significant volume problems are created within the fold during the folding process. The first models show that, within the Paleozoic carbonate section (450 m thick), lateral displacements of the top and bottom bedding surfaces differ by nearly 100% at one point during folding but are nearly identical at other times. Refinements of the simple models indicate that some stratigraphic horizons in folds with 1,200 m or less displacement over vertical faults in the basement may not require large lateral movements. If several bedding-plane detachments were present in the Paleozoic carbonate blocks, volume problems within the hinge zones could be significantly reduced.

Geometrically scaled laboratory specimens were experimentally deformed, and their final geometry is comparable to that of natural counterparts. Comparison of calculated and observed lateral displacements agree within 7%, based on the simplest analytical model, but only agree within 50% for another model.

It is concluded that the kinematic approach to drape folding is useful, because it provides information about the required lateral displacements of large rock masses during folding. It also illuminates other problems that must be investigated before a better understanding of drape-fold structures is possible.

---

*Present address: Continental Oil Company, Research and Development, P.O. Box 1267, Ponca City, Oklahoma 74601.

INTRODUCTION

Stearns (1971, and this volume) has proposed a mechanical system by which folds in the Wyoming province (as defined by Prucha and others, 1965) were formed. His principal conclusion is that there are at least two styles of deformation involved: (1) faulting of the brittle basement, and (2) folding of the overlying, more ductile sedimentary layers into a rather consistent configuration that he terms "drape fold." He has used Rattlesnake Mountain, west of Cody, Wyoming, as his "type" structure (Fig. 1), and he has shown that the Paleozoic carbonate rocks fold as rigidly rotated blocks. He also has listed many other examples throughout the Wyoming province that have similar geometries. He has further pointed out that the specific configuration of and deformation mechanisms within the sedimentary layers depend on the rock types involved and their position within the total sedimentary section.

Stearns (1971, p. 136) showed that his drape-folding concept requires the Paleozoic carbonates to be transported laterally about 1,500 m, and, as they are not measurably thinned, this results in a volume problem.

The purpose of this study is to investigate kinematically the two-dimensional displacement field of rock masses involved in drape folding. The functional relations between the distance that rocks must be transported laterally to accommodate folding and the vertical displacement and dip of the fault in the basement are also examined.

Analytic expressions for these displacement relationships were derived from simple geometric constructions. Although Stearns (1971) was primarily concerned with deformation of Paleozoic rocks, analytic models were also developed for the Mesozoic rocks in drape folds of the Wyoming province.

Three assumptions that pertain to these models of both Paleozoic and Mesozoic rocks are that (1) there are no movements of rocks parallel to the fold axis; (2) faulting of the sedimentary rocks does not occur; and (3) all rocks above the upthrown basement block do not move laterally perpendicular to the fold axis. Other assumptions used in the development of the Paleozoic models are that (4) the Paleozoic rocks fold mostly by rigid-body rotation and translation, and the hinges between these blocks are fixed with respect to internal coordinates; (5) the Paleozoic carbonate rocks do not thin across the fold; (6) the Paleozoic carbonate rocks can move laterally along at least one bedding-plane detachment surface above the downthrown basement block; (7) the displacement paths of the hinges are in two fixed directions—one parallel to the detachment surface, the other parallel to the basement fault; and (8) there is a particular sequence in which the carbonate blocks are formed and rotated. In addition to assumptions 1, 2, and 3 above, to develop the Mesozoic models it is further assumed (from seismic evidence) (9) that the shape of the Mesozoic rocks across a drape fold is approximated by either one or two circular arcs.

From existing field data and the analytic expressions, values for lateral transport (y) are plotted against increasing values of vertical fault-displacement (d). This is a means of investigating the lateral displacements of the sedimentary rocks during drape folding. The mathematical formulations of the models are presented for completeness, but they can be omitted by the reader without greatly impairing the understanding of this study.

Although movements parallel to drape-fold axes do occur (Cook and Stearns, 1975; Vaughn, 1976; Weinberg and others, 1976), the analyses in this paper deal *only* with the two-dimensional movement paths of the sedimentary beds because they are simple to deal with and because many natural folds have lengths very

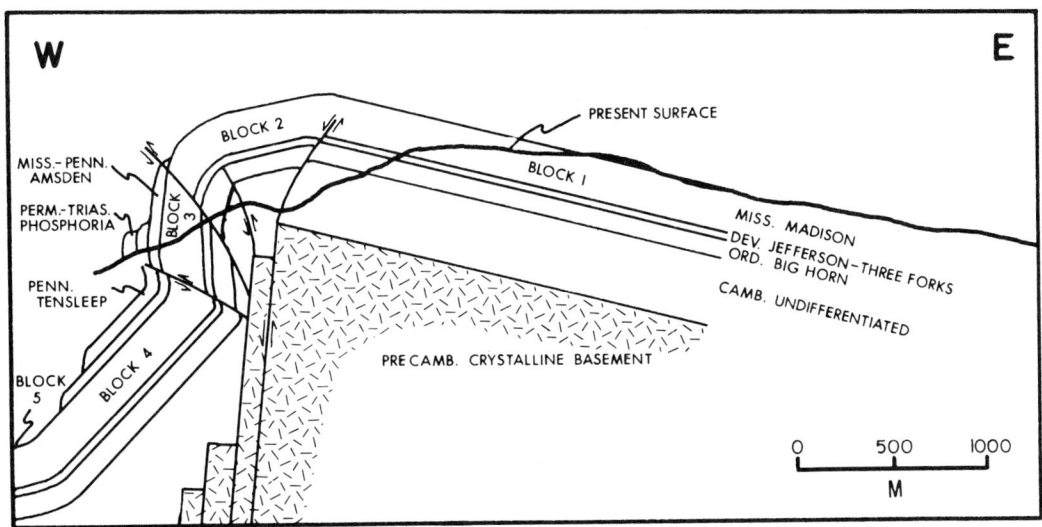

Figure 1. Geologic cross-section of Rattlesnake Mountain (modified from Stearns, 1971, Fig. 5, p. 131).

much greater than their widths. Therefore, the models are only an approximation, but they represent one important aspect of folding.

MODELS OF THE PALEOZOIC ROCKS

Stearns (1971) has made several particularly germane observations of the Paleozoic carbonate rocks in drape folds. First, these rocks fold by rigid rotations of blocks separated by hinges fixed with respect to internal coordinates. He has delineated the different blocks by numbering them from 1 through 5 (Fig. 1), and this terminology is retained for consistency. His block 2, however, has been divided into two smaller blocks (2a and 2b), because its shape is better approximated by two straight lines. Stearns's (1971) second important observation was that the Paleozoic carbonates do not thin across the fold. Third, these rocks have moved laterally toward the basement fault along a bedding-plane detachment surface that occurs at the top of the basal carbonate bed in the Bighorn Dolomite. Stearns's (1971) evidence for movement along the detachment (layer-parallel gouge zones and truncated faults) indicates that significant lateral movements occurred only in the beds on the downthrown side of the fault. He further stated that small-scale structures are continuous across the detachment horizon on block 1 so that these rocks must have been pinned. The rotation of block 1 is ignored in this section so that all models apply to plateau-type uplifts in which the upthrown block is not tilted.

This information, combined with assumptions 7 and 8 (see Introduction), can be used to construct a very simple model for the uppermost bedding surface of the Paleozoic carbonate layer (Fig. 2). The lines and circles represent the rigid blocks and hinges, respectively (t_0, Fig. 2). If the right-hand line is displaced vertically upward while remaining horizontal, two important phenomena are observed. First, the movement of the right-hand circle is vertical and parallels the displacement imparted to the right line (t_1, Fig. 2). At the same time, the second circle moves to the right along a horizontal path. Second, only the center line (λ) must rotate so long as the vertical displacement is less than the length of the line itself (t_2,

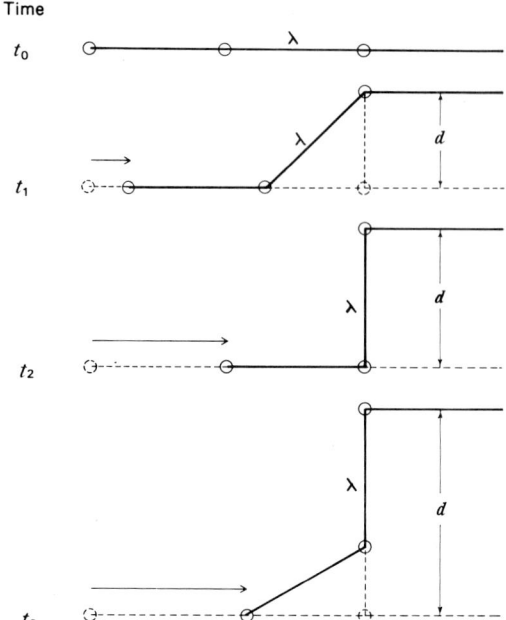

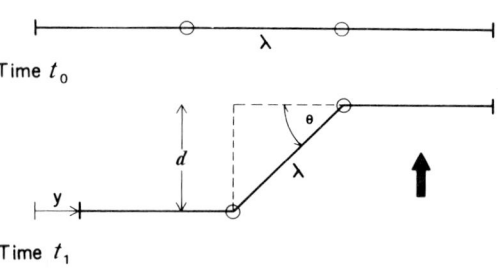

Figure 3. Construction to develop the initial equations for lateral displacement.

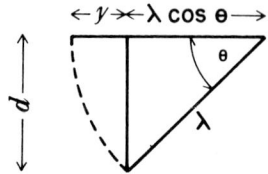

Figure 4. Construction showing the relation of dip (θ) to length (λ) of block and amount of lateral movement (y).

Figure 2. Schematic diagram of the line-circle model. Lines represent the uppermost bedding surface of the (rigid) Paleozoic carbonates; circles represent the fixed hinges. The sequence from t_0 to t_3 shows the displacement-path restrictions and order of block rotations.

Fig. 2). Moving the right line to a point higher than the length of the central line (λ) causes the left-hand line to begin rotating after the central line becomes vertical (t_3, Fig. 2).

This simple exercise emphasizes the assumptions that the hinges are restricted to two, and only two, possible movement paths, and that a block will achieve its maximum rotation before the next block begins to rotate. Geologically, this means that the blocks in a drape fold rotate sequentially, and for most models, only one block at a time is rotated. In natural drape folds the maximum rotation of every observed Paleozoic carbonate block seldom reaches 90° (Fig. 1); therefore, I assume that the present observable dip of the blocks are the maximum dips ever achieved.

To investigate the functional relation between the vertical fault displacement (d) and the distance (y) that the downthrown end of the block must move laterally, a construction is made (Fig. 3) where λ is the length of the rotating block and θ is the dip of the block at time t. Figure 4 illustrates how d and y can be written in terms of λ and θ:

$$d = \lambda \sin \theta, \tag{1}$$

$$y = \lambda (1 - \cos \theta). \tag{2}$$

Generalizing these expressions to describe more than one rotated block,

$$d_t = \sum_{i=1}^{n-1} (\lambda_i \sin \theta_i) + \lambda_n \sin \theta_n, \qquad (3)$$

and (for y in terms of d)

$$y_t = \sum_{i=1}^{n-1} [\lambda_i - (\lambda_i^2 - d_i^2)^{1/2}] + \lambda_n - (\lambda_n^2 - d_n^2)^{1/2}, \qquad (4)$$

where d_t and y_t are the total vertical and horizontal displacements, respectively, and n is the number of the rotating block ($n = 1, 2, 3, 4, 5$ usually represent blocks 2a, 2b, 3, 4, and 5, respectively).

In principle, by choosing appropriate lengths and dips, the final value of y for any drape fold may be found. Most models investigated here will incorporate the length and dip values from Stearns (1971, Fig. 5, p. 131), as shown in Figure 5.

MODELS OF THE MESOZOIC ROCKS

Because very few descriptions of the configurations of the Mesozoic rocks in drape folds of the Wyoming province are available, two hypothetical models that are partially substantiated by seismic data are made to characterize the uppermost bedding surface of this sequence.

Both constructions are parts of circular arcs. The first is a quadrant of a circle, the center of which is the uppermost hinge in the Paleozoic rocks and the radius of which is the total thickness of the Mesozoic rocks (Fig. 6A). The second model is much the same, except that it consists of two arcs: one each for the anticlinal and synclinal segments (Fig. 6B). These models are referred to as the single-arc and double-arc, respectively. Here the derivation of the single-arc model only is shown. In Figure 7, T is the total thickness of the Mesozoic rocks, d is the vertical displacement of the fault, L is the original length of the folded section (with respect to external coordinates), L' is the folded arc length, and ϕ (in radians) is the rotation of T (the radius) for a given d. Of interest is y, the difference between the original straight-line length and the resulting arc-length after a displacement d, that is

$$y = L' - L. \qquad (5)$$

From Figure 8 it is seen that

$$\phi = \cos^{-1}(1 - d/T), \qquad (6)$$

and

$$L = T \sin \phi. \qquad (7)$$

From Figure 7,

$$L' = T\phi, \qquad (8)$$

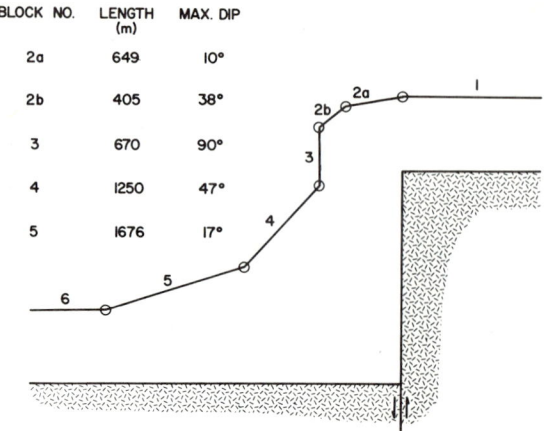

Figure 5. Schematic diagram of Rattlesnake Mountain showing the line-circle model (modified from Stearns, 1971, Fig. 11, p. 139).

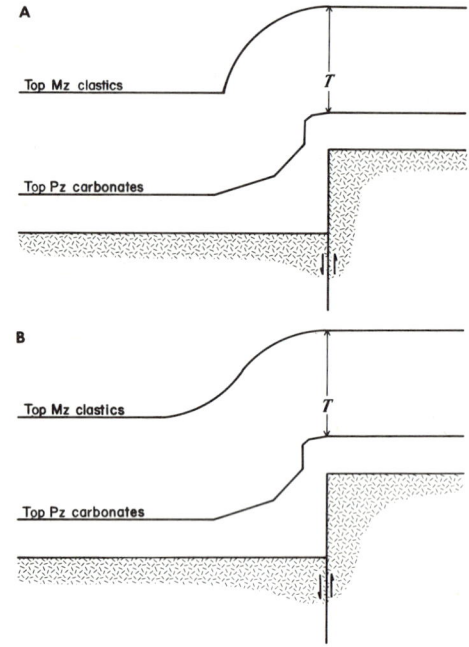

Figure 6. Schematic diagram of the models of the Mesozoic rocks. (A) single-arc model; (B) double-arc model.

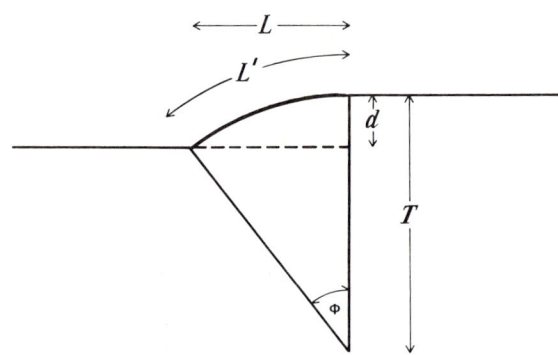

Figure 7. Schematic diagram of the parts of the model in Figure 6 needed for computation.

and substituting from equation 6

$$L' = T\left[\cos^{-1}(1 - d/T)\right]. \tag{9}$$

To find L in terms of d, a triangle with the angle ϕ and the adjacent side $1 - d/T$ is constructed (Fig. 9). The opposite side (k) is

$$k = \sin\phi = (1 - \cos^2\phi)^{1/2}, \tag{10}$$

but from equation 6,

$$\cos^2\phi = (1 - d/T)^2; \tag{11}$$

therefore,

$$\sin\phi = [1 - (1 - d/T)^2]^{1/2}. \tag{12}$$

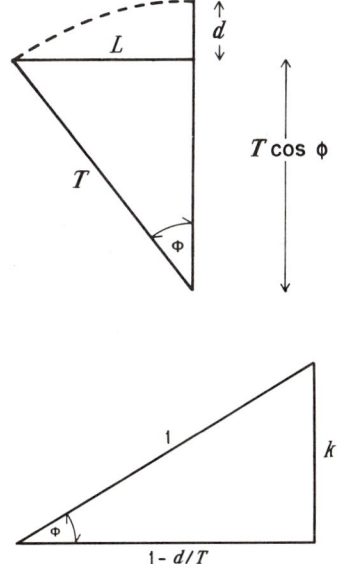

Figure 9. Relations of φ to d and T.

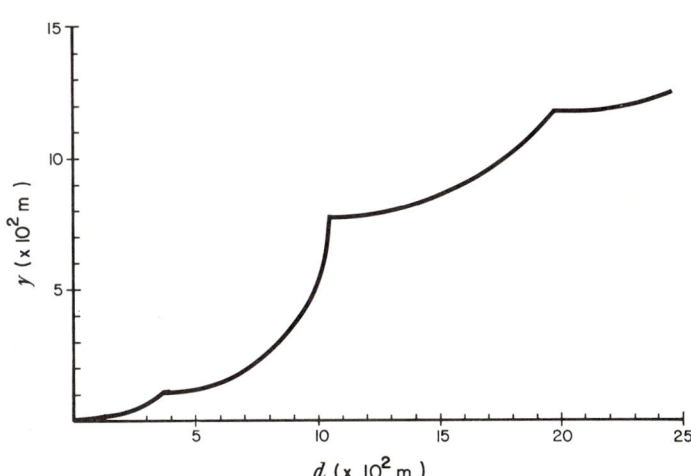

Figure 8. Definitions of parameters in equations describing d and y for the single-arc model.

Figure 10. Plot of equation 4 from data shown in Figure 5.

Squaring and expanding inside the radical gives

$$\sin \phi = \{1 - [1 - 2d/T + (d/T)^2]\}^{1/2}, \tag{13}$$

and simplifying yields

$$\sin \phi = [2d/T - (d/T)^2]^{1/2}. \tag{14}$$

Substituting equations 7 and 8 into equation 5 gives

$$y = T(\phi - \sin \phi), \tag{15}$$

and using equations 6 and 14 in equation 15 yields

$$y = T\{\cos^{-1}[1 - (d/T)] - [2d/T - (d/T)^2]^{1/2}\}. \tag{16}$$

But from equation 14,

$$\phi = \sin^{-1}[2d/T - (d/T)^2]^{1/2}, \tag{17}$$

and the final equation is

$$y = T\{\sin^{-1}[2d/T - (d/T)^2]^{1/2} - [2d/T - (d/T)^2]^{1/2}\}. \tag{18}$$

The general expression for the double-arc model is derived by replacing T in equation 19 by $2T$, which results in

$$y = 2T\{\sin^{-1}[d/T - (d/T)^2]^{1/2} - [d/T - (d/T)^2]^{1/2}\}. \tag{19}$$

Figure 6 shows that for the single-arc model, the maximum displacement allowable is equal to the total thickness of the Mesozoic sequence, because only one quadrant of the circle is available. Similarly, the maximum displacement for the double-arc model is twice the thickness because the maximum arc-length available is two quadrants (one each for the anticlinal and synclinal portions). These restrictions do not apply to most real drape folds because the vertical displacements on the basement faults are commonly less than about 2,400 m. The thickness of the Mesozoic section at Rattlesnake Mountain at the time of uplift was probably about 2,500$^+$ m, and, assuming that faulting of the Mesozoic rocks begins at approximately the same vertical-displacement interval as it does in the Paleozoic rocks below, the models can be safely used for displacements of about 2,400 m and less. Should faulting occur at lesser vertical displacements, a different model would be more appropriate.

MODEL CURVES FOR THE PALEOZOIC ROCKS

Equations 3 and 4 can be used to compute vertical (d) and lateral (y) displacements for any drape fold if block lengths (λ) and dips (θ) are known. Given Stearns's (1971) vertical displacement (about 2,400 m) as a suggested maximum before the Paleozoic rocks are faulted and values for λ and θ, the values for d and y can be plotted. The vertical displacement (d) is always plotted on the abcissa as a positive number, increasing to the right. Note in Figure 2 that the horizontal displacements for this simplest model are always toward the uplifted block, and they are assigned positive values. Under certain circumstances, however, y can be negative when movement is away from the uplifted block. By starting the d values at zero and computing y as d increases, a curve is generated that represents the horizontal displacement for a chosen drape-fold model.

Vertical Fault: Plateau Uplift

Given λ and θ values from Rattlesnake Mountain (Fig. 5), equation 4 yields a curve (Fig. 10) that is made of arcs: *one for each rotated block*. In fact, for vertical faults, these are arcs of circles of radius λ, equal to the length of each block. The length of each arc simply depends on the block length and the angle of rotation of the block.

Figure 11 shows some important results: (1) The number of arcs of the curve equals the number of *rotated* blocks (Fig. 11A). (2) Very small y values for a given large d can only be achieved by rotating a long block through a small angle (Fig. 11B). (3) For a given d, relatively large y values can be achieved by rotating blocks equal to, or shorter than, d in length through 90° [Fig. 11C (b)]. (4) For a given pair of d and y coordinates, many models can be used, but a different set of lengths and angles and, therefore, a different number of arcs are required (Fig. 11C). (5) Although two multiple-block models may be similar in general shape, the resulting curves will be different unless all block lengths and dips are exactly the same (Fig. 11D).

Nonvertical Faults: Plateau Uplift

Because many basement faults in the Wyoming province dip less than 90°, the contribution to the total horizontal displacement of the sedimentary layers owing to horizontal displacements of the uplifted basement blocks must be understood.

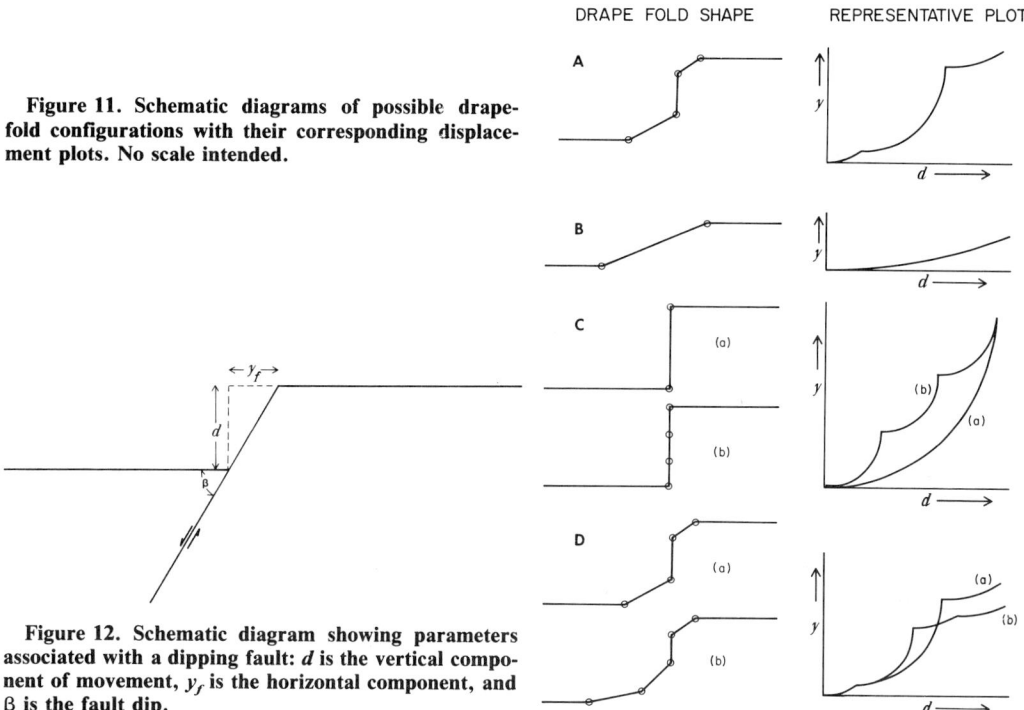

Figure 11. Schematic diagrams of possible drape-fold configurations with their corresponding displacement plots. No scale intended.

Figure 12. Schematic diagram showing parameters associated with a dipping fault: d is the vertical component of movement, y_f is the horizontal component, and β is the fault dip.

By the stated sign convention, the horizontal component of basement uplift is positive for normal faults and negative for reverse faults. The exact value, added or subtracted, is a function of the fault dip and the amount of vertical displacement. This relation can be derived from Figure 12 as

$$y_f = d/\tan \beta. \tag{20}$$

This calculation is validly applicable to both the sedimentary layers and the basement fault, so long as the sedimentary rocks on the upthrown side of the fault are pinned, and no thickness changes occur in these rocks (Fig. 13).

Within the models outlined so far, only reverse faulting can generate negative values for y. This is because the Paleozoic blocks are assumed to behave perfectly rigidly (Fig. 14), and obviously reverse faulting itself gives rise to movements away from the uplifted block. Note in Figure 14, however, that when blocks are allowed to rotate to 90°, although the initial movements are negative, the rotation of the block beyond its being perpendicular to the fault results in positive movements, which can be much greater than the negative ones. The net y value for this case is positive. Displacement plots for faults dipping 60° normal, vertical, and 60° reverse are shown in Figure 15. Note that the arcs in curves A and C (Fig. 15) are no longer circular.

From the geologic viewpoint, curve B (Fig. 15) shows that the Paleozoic rocks on the downthrown side of the basement fault always move in a direction toward the fold axis. Furthermore, the layered rocks beyond the last hinge (block 6 by extrapolation of Stearns's 1971 nomenclature) have been transported over 1,200 m. This amount of lateral transport may be difficult to imagine, but it is the

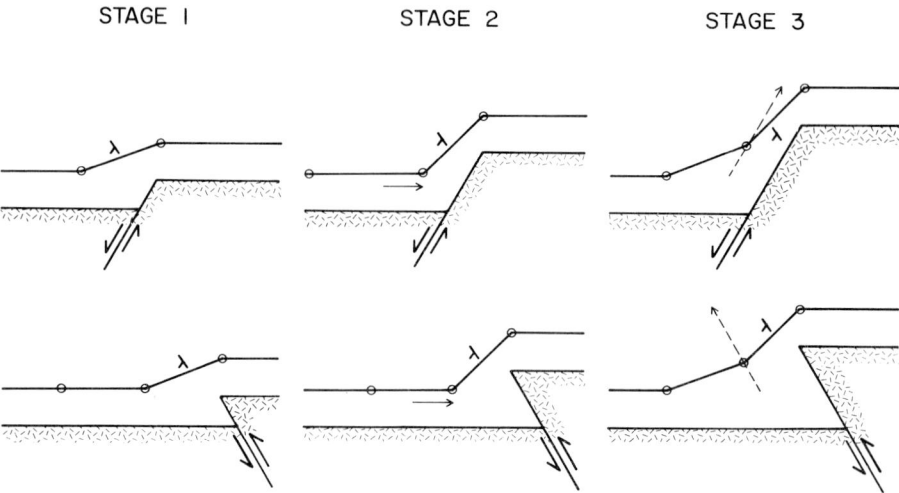

Figure 13. Schematic representation of path restrictions for hinges when faults are not vertical.

amount required to form a drape fold the size and configuration of Rattlesnake Mountain. In fact, this value is conservative, because the effects of horizontal movements of the basement have been ignored. The Rattlesnake Mountain frontal fault is normal and dips 85° to 89° as determined from fault-parallel fractures (Stearns, 1974, personal commun.). From Figure 15 (curve A), it is obvious that any component of dip in the "normal" sense will only increase the y value for any particular displacement on the fault.

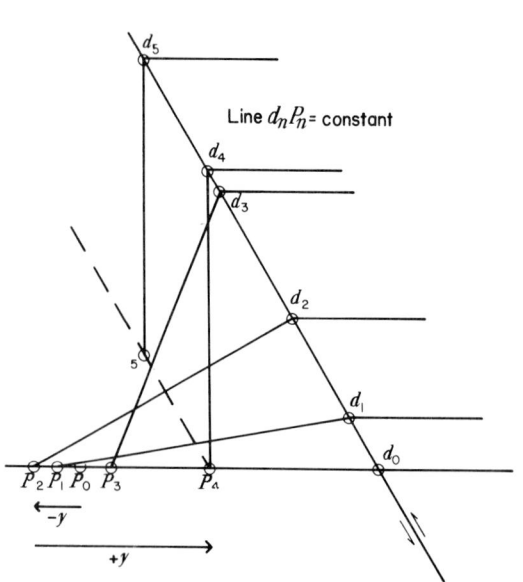

Figure 14. Schematic diagram showing movement directions associated with a 60°-dipping reverse fault (plateau uplift).

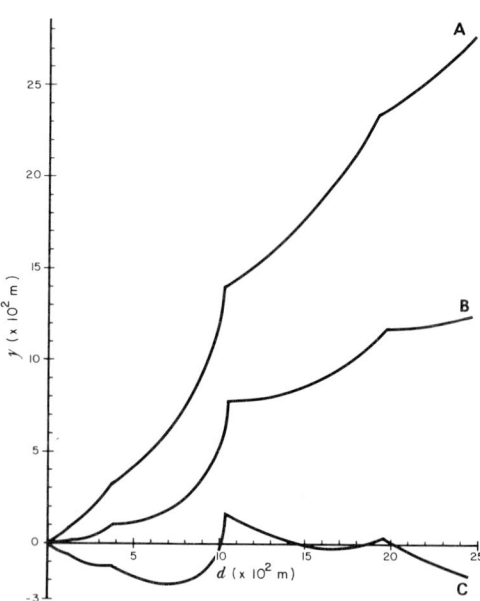

Figure 15. Displacement plots for faults dipping 60° normal (A), vertical (B), and 60° reverse (C). Data used for lengths and dips are shown in Figure 5.

Another unexpected feature of the curves is the extremely high slope in the 900- to 1,100-m fault displacement range; that is, very small fault displacements can lead to very great horizontal movements.

Curve C (Fig. 15) reveals equally interesting problems. Although here the final y value (about -180 m) is much smaller than for curve B, the rocks have moved in two opposite directions. During part of the fault-displacement history, they move away from the fold axis, but at other times they move toward it.

If lower y values are assumed to be more reasonable geologically, curve C (Fig. 15) appears to be more acceptable than either of the others. However, the differences of y values for these curves is solely due to differences in fault dip. Therefore, curve C (Fig. 15) is not applicable to many drape folds in the Wyoming province, regardless of how reasonable the curve might at first appear. I am not aware of any drape folds over any basement block whose fault dips less than 75° and the Paleozoic rocks are not faulted through. In fact, Berg (1962, 1976), Stearns (1970), and Stearns and Weinberg (1975), among others, show that when a low-angle reverse fault occurs in the basement, such as in the Wind River or Owl Creek Mountains of Wyoming, little draping has occurred, but rather, the fault is propagated through the sedimentary section as a thrust fault. I do not know of any published descriptions of drape folds over lower-angle (less than about 75°) normal faults. I would expect, however, that the rocks overlying such a fault would have been subjected to extension and would have been faulted through. The work of Handin and others (1967) on the brittle-ductile transition of rock might be used to support this idea, as rocks in extension ($\sigma_3 < \sigma_1 = \sigma_2$) fail by fracture (remain brittle) at significantly higher mean stresses than do rocks in compression ($\sigma_1 > \sigma_2 = \sigma_3$). For these reasons, most remarks about the effects of basement-fault dip on lateral displacements will be confined to faults that dip 75° or more.

MODEL CURVES FOR THE MESOZOIC ROCKS

In a manner similar to the preceding one, equations 18 and 19 are used to construct displacement curves for the uppermost bedding surface of the Mesozoic rocks (Fig. 16). Note first that these curves are smooth with none of the cusps displayed by those for Paleozoic rocks. This dramatically illustrates an important

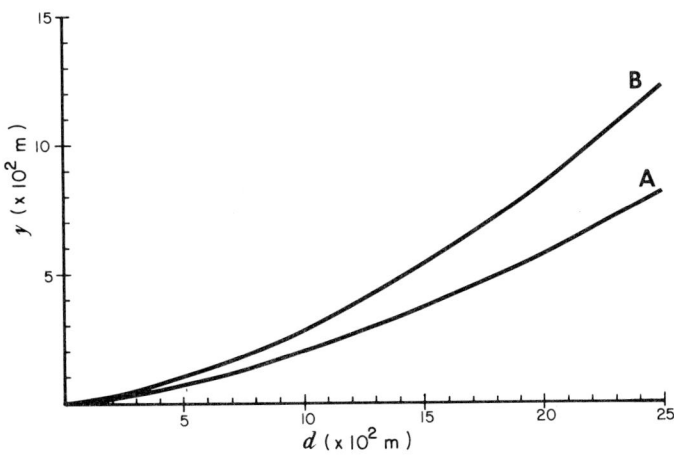

Figure 16. Displacement plots of the Mesozoic double-arc (curve A) and single-arc (curve B) models. Fault is vertical, and Mesozoic thickness is 3,048 m.

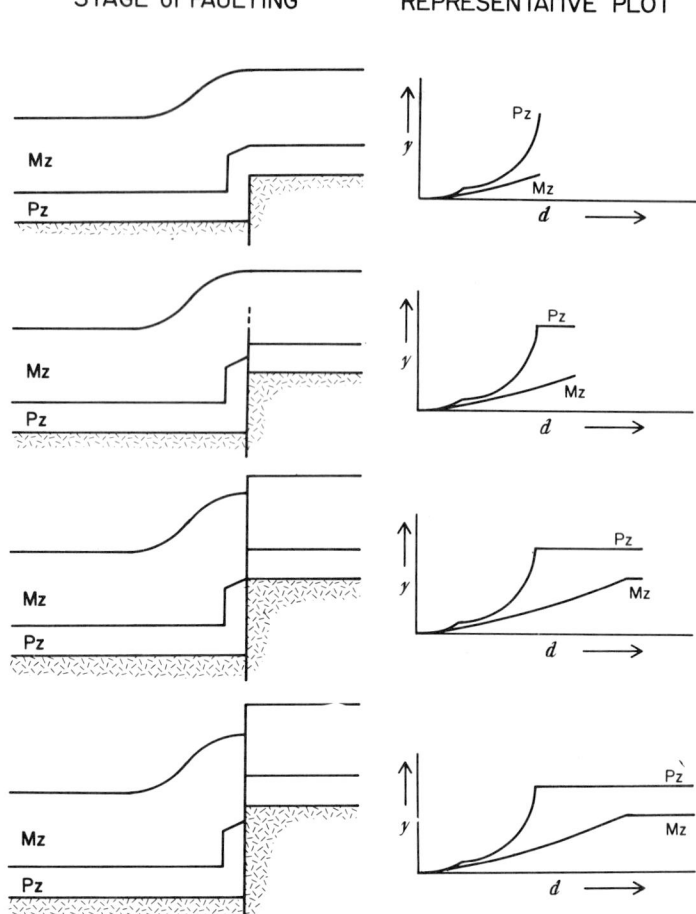

Figure 17. Schematic diagrams of the effects of faulting of the sedimentary layers on displacement plots.

point made earlier: the displacement curve is very sensitive to the shape of the modeled surface. The plots for the Mesozoic models are smooth curves, because the shapes of the surfaces modeled are also smooth curves. Even though the shapes of these curves differ little, the final y value of curve A (Fig. 16) is nearly 40% less than that of curve B.

The effect of fault dip on these curves is similar to that for the Paleozoic models; that is, for a normal fault, y values are greater than those for a vertical fault, and for a reverse fault, they are lesser. This generalization holds for all models, because the normal fault extends the section horizontally between two points on opposite sides of the fault, whereas the reverse fault shortens it.

Although these models for the shape of the Mesozoic upper boundary are idealized, they may well represent possible "end members." The actual shapes of subsidiary bedding surfaces within the Mesozoic section may vary significantly throughout the sequence, owing to vertical variations in ductilities of these clastic rocks; therefore, the y value associated with a particular fault displacement probably resides somewhere between curves A and B (Fig. 16). For simplicity, however, only the double-arc model will be treated in subsequent diagrams. The lack of vertically continuous outcrops of Mesozoic rocks in drape folds of the Wyoming province requires methods other than direct field observation such as well-log and seismic data to better delineate the variations in shape.

EFFECTS OF FAULTING IN THE SEDIMENTARY ROCKS ON DISPLACEMENTS

One assumption is that faulting within the sedimentary rocks during the drape-folding process can be ignored. This is naïve, because faulting of the Paleozoic rocks at Rattlesnake Mountain (Stearns, 1971) and elsewhere does occur. Many structural cross sections of drape folds in the Wyoming province show several faults of varying displacements in the sedimentary rocks.

A sequential development of a faulted drape fold and the effect of the faulting on the y values is shown in Figure 17. Whatever the last pre-faulting y value may be, it then remains constant throughout the remaining vertical-displacement history, but an important assumption is implied. After a unit is faulted, all subsequent vertical and horizontal components of movement are taken up by the fault(s), and consequently no further folding of the beds occurs on the downthrown side of the fault. This might be taken as an easy way to explain away the large y (1,200 m) predicted for Rattlesnake Mountain. However, the field evidence reported by Stearns (1971) does not indicate that faults cutting the Paleozoic rocks have significantly affected the lateral transport of material into the fold. The total vertical fault displacement in the Paleozoic carbonate blocks at Rattlesnake Mountain amounts to less than 100 m (Stearns, 1971, Fig. 5, p. 131). As the total vertical displacement of the basement fault is about 2,300 to 2,500 m, most of it had to be accomplished by folding, not faulting. Thus a large y value is still required.

These analyses of Rattlesnake Mountain further assume that any faulting in the veneer occurs late in the development of the fold. Had early initiation of faulting and contemporaneous folding and faulting been assumed, the total throw on all the faults would still be small (< 100 m) compared to the vertical displacement accounted for by the folding (2,400 m).

ENVELOPE CURVES

For the analyses, I have assumed the shape of the Mesozoic upper boundary and the rotations of the carbonate blocks as being one at a time and in a specific sequence (block 2a before 2b, 2b before 3, and so on). Actually, the shape of the Mesozoic rocks is mostly unknown, as is the real sequence of development of the Paleozoic carbonate blocks. Thus none of the curves gives the "true" displacement path of any particular drape fold. However, field evidence does show that certain end members are probable within which the real horizontal displacements may fall.

This approach is used in constructing Figure 18, which shows two possible curves for different sequential developments of the carbonate blocks. Curve A is the same as in Figure 10; curve B differs only in that block 3 is rotated first, followed by 2a, 2b, 4, and 5 in that order. Since in both models blocks 4 and 5 are rotated in the same order and at the same d value, their curves coincide. Any other order is plotted by simply transposing the arc representing any particular block and linking it onto the end of that of the preceding block. There are important geologic implications of these envelopes between end members. The y values are identical for both models up to about 350 m and beyond about 1,000 m of vertical displacement, but between these two fault displacements, the y values may differ by as much as 100% (Fig. 18).

As mentioned by Stearns (1971), there is little field evidence to use in determining which of the blocks is the first to form. However, many drape folds do have a nearly vertical block 3 and very little development of block 4. A good example is reported by Stearns and Weinberg (1975) at the Dry Fork anticline in the northeastern Big Horn Mountains.

The models do provide some insight into how drape folds grow. Although the amount of lateral transport predicted by a particular model may be unreal, reasonable assumptions can be made and upper and lower limits calculated.

DIFFERENCE CURVES

So far only models of the top surfaces of the Paleozoic carbonate and Mesozoic clastic sequences have been treated, but nothing has been said about the rocks between them. Some appreciation of the lateral displacements of the entire section can, however, be gained by simply subtracting the curve of a given Mesozoic model from that of the corresponding Paleozoic model while the fault dip remains constant. By plotting the differences of y against d, the difference curve (Δy) shown in Figure 19 is obtained.

The cusps in curve C (Fig. 19) are a reflection of the cuspate nature of the Paleozoic plot. Another feature of the Δy plot is worth noting. Exclusive of reverse faulting in the basement, this is the first appearance of negative slopes. If the lowermost Mesozoic clastic rocks are not welded to the uppermost Paleozoic carbonate rocks and, therefore, are not carried "piggyback," the negative slopes, then, indicate a startling phenomenon: there are periods during basement uplift when, for a given increase in vertical displacement, the uppermost Paleozoic rocks are moving farther into the drape fold than are the uppermost Mesozoic rocks, but at other displacement intervals the reverse is true. No *time* connotation is implied, because the "rate of change" is of lateral displacement (y) with respect to vertical displacement (d). Figure 20 shows some implications.

The sign, positive or negative, of Δy does not reveal the direction of movement of the rocks between the limiting horizons; it merely indicates which curve has the greater y for any given fault displacement and for any given model. Difference (Δy) plots are made for three different fault dips and the same Paleozoic and Mesozoic models (Fig. 21). The resulting Δy curves are identical, because the values are found by subtracting y for a Mesozoic model (y_{Mz}) from y for the Paleozoic model (y_{Pz}) at a particular fault displacement. When the fault is not vertical, a component of horizontal movement due to basement-block motions is within the equations of both y_{Mz} and y_{Pz}. Because the dip of the fault is assumed constant, and the vertical displacement is the same for both equations of y,

$$\Delta y = (y_{Pz} \pm d/\tan \beta) - (y_{Mz} \pm d/\tan \beta). \tag{21}$$

The fault-dip term ($d/\tan \beta$) cancels out, and the result is the same as that calculated for the vertical fault.

If there are any "second-order" structures (disharmonic folds, small thrust faults, and so on) formed on a drape fold solely because of the "shear couple" created by Δy, such features may not be significantly different in orientation, geometry, or scale between folds as a function of basement-fault dip. The idea that the sedimentary section on drape folds is in a state of shear is not new (Z. Reches,

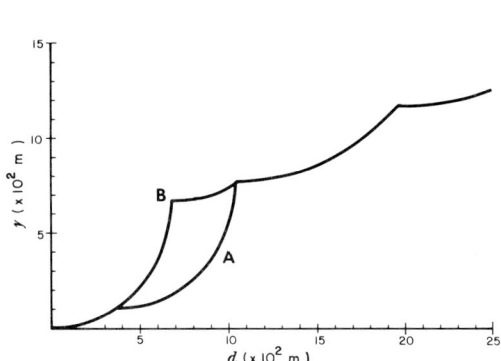

Figure 18. Envelope curves for differing sequential development of Paleozoic carbonate blocks. Curve A is standard sequence of block rotation (2a, 2b, 3, 4, and 5); curve B uses the sequence 3, 2a, 2b, 4, and 5.

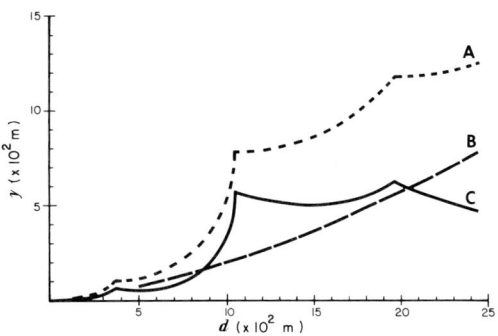

Figure 19. Displacement plot for y and Δy. Curve A is the Paleozoic model, curve B is the Mesozoic (double-arc) model, and curve C is Δy (Paleozoic minus Mesozoic). Fault is vertical.

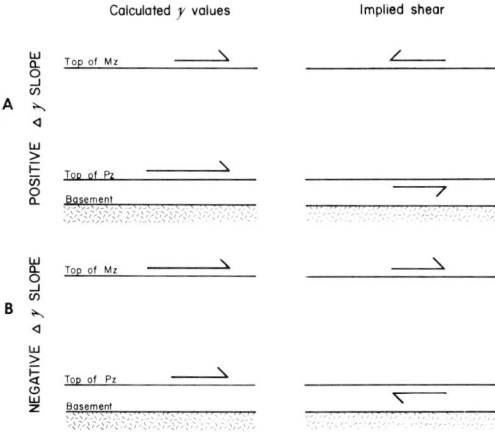

Figure 20. Implied changes in sense of shear as seen from changes in slope of Δy curve: A shows implied shear for positive slope, B shows implied shear for negative slope. Schematic diagram is only of the downthrown block, with the upthrown block off to the right.

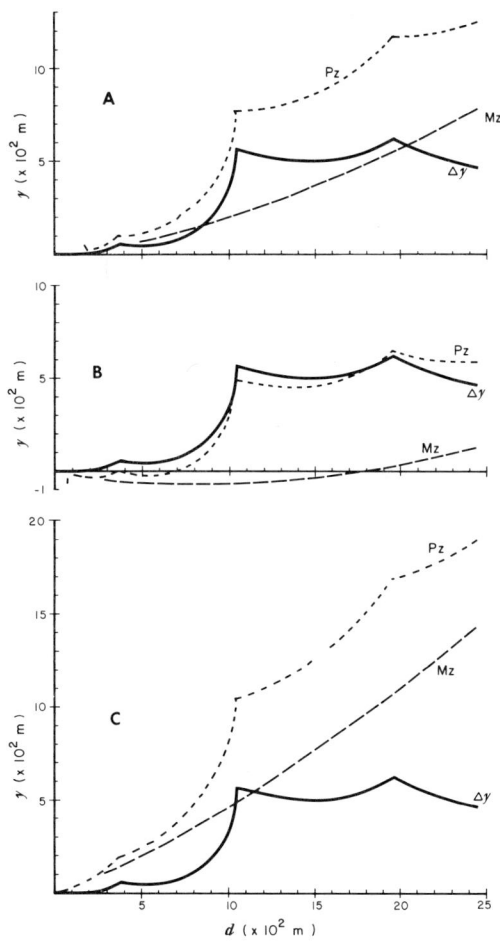

Figure 21. Displacement plots showing that Δy values are independent of fault dip: (A) vertical fault, (B) 75°-dipping reverse fault, (C) 75°-dipping normal fault.

1976, personal commun., and this volume), but previous analytic models have not been based on the kinematics of the system, but rather on stress-field boundary-value solutions (Min, 1974; Gangi and others, 1977).

PALEOZOIC TOP VERSUS BOTTOM

The concept of using Δy to illustrate the lateral-displacement requirements of two horizons and the possible kinematic implications of these differences is, perhaps, more easily acceptable when considering the Mesozoic clastic rocks (assumed ductile) rather than the Paleozoic carbonate rocks (assumed rigid). However, there is a difference between the shape of the top of the carbonate section and its bottom (Fig. 22A). Figure 22B shows the displacement curves for both the top and bottom of the Paleozoic carbonate rocks. Clearly, if only the final y values for each had been calculated, the difference would have been very small (about 40 m), but when all values are shown, between about 450 and 980 m of vertical fault displacement (the rotation of block 3), there is a drastic difference in the lateral transport required of the two horizons. This difference implies a sense of shear within the Paleozoic section much like that shown in Figure 20A, because, for a given d value within this interval, the bottom of the carbonate section has a larger value than does the top.

It could be argued that the Paleozoic rocks are strong enough to "absorb" the first 150 m of lateral movement by elastic distortion. However, when the lateral displacements at the top and bottom of the section begin to strongly diverge, the detachment surface observed by Stearns (1971) must be created if the rock is not internally deformed (thickened or thinned). Additionally, several bedding-plane detachments, other than the one in the lower Bighorn Dolomite (as were alluded to by Stearns, 1970, p. 65), might be formed because of these displacement differences. In my opinion, a single detachment surface is insufficient to account for all the lateral transport required of the Paleozoic rocks. I suspect that other slip surfaces do operate, but none has yet been documented as far as I know.

SUMMARY OF SIMPLE MODELS

To reiterate, the shapes of naturally formed drape folds can be approximated with simple geometric constructions. By using straight-line segments and hinges for the Paleozoic rocks and circular arcs for the Mesozoic rocks, analytical expressions can be derived for both vertical and lateral displacements during drape folding. Of particular interest are the expressions for lateral displacements.

By using field data in these equations and plotting them, displacement curves can be generated that represent the displacement path of the chosen model(s). To develop these models, many assumptions have been made, but the models do illustrate that the lateral-displacement requirements of drape folding can illuminate certain phenomena that are not observable in conventional stress-field solutions. Although the deformation history of natural drape folds is complex, certain assumptions can reduce these complexities (variation in basement-fault dip, faulting of the sedimentary rocks, and so on) so that their effects can be isolated and investigated.

Of interest also are the implications that are recognized when the differences of the lateral displacements required by two horizons on the same fold are examined. Whether these are Δy or top-bottom plots, the kinematic analyses indicate that

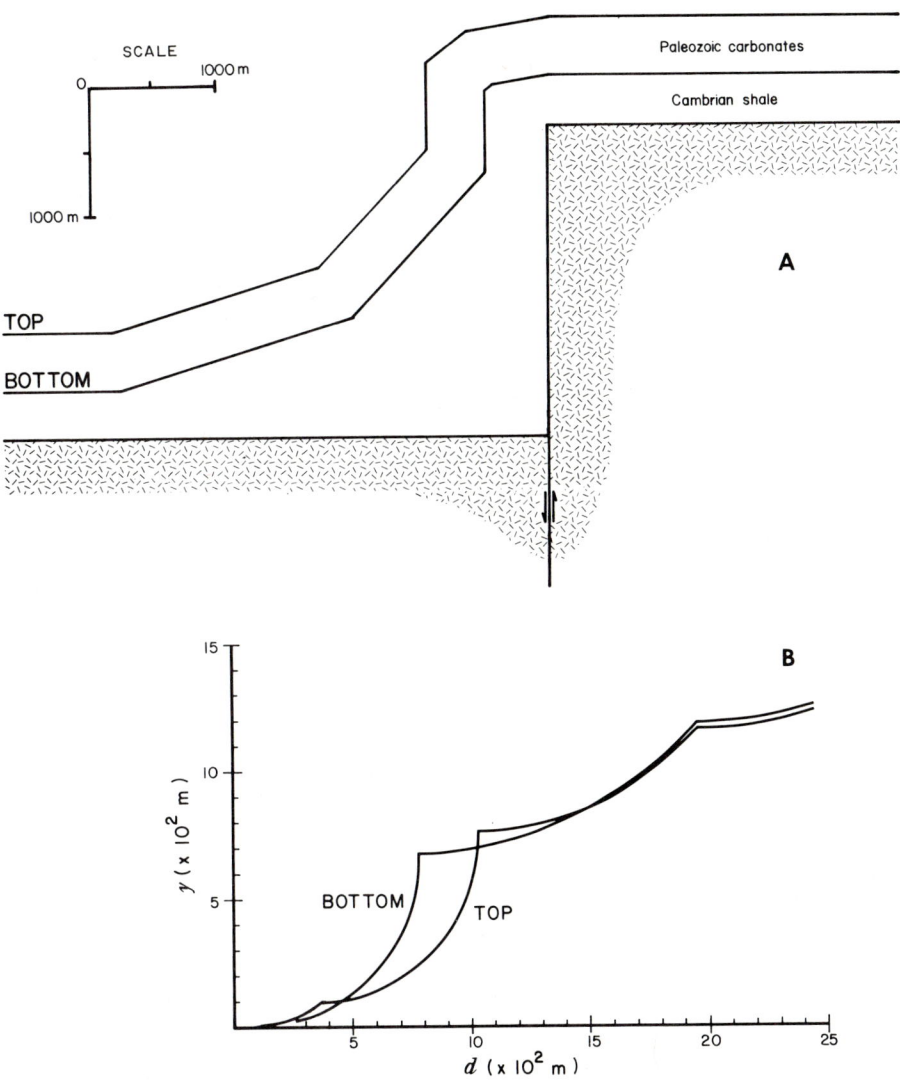

Figure 22. Schematic cross section of top and bottom of Paleozoic carbonate layer (A), and their displacement plots, (B).

the rocks between these horizons are placed in a state of shear. Geologic implication and extrapolation can then be used to examine what types of deformation in natural drape folds may be related to either the lateral-displacement requirements or the general displacement history.

REFINEMENTS OF THE MODEL

Although certain geologic insight into the drape-fold concept is gleaned from the simplified kinematic models, more insight becomes possible by modifying some assumptions. Only the Paleozoic models will be treated further, as the Mesozoic models are speculative at best, but the work of Stearns (1970, 1971, and this volume) has clearly delineated the geometry of the Paleozoic carbonate "strut."

Relaxing the condition that only one block at a time rotates, two possible cases are that (1) all blocks rotate together so that their final dips are reached at the same time, and (2) some blocks may rotate synchronously while others do not. These two significant cases are designated case 2 and case 3, respectively (case 1 is the original model).

Equation 4 is used to calculate the displacement curves for both cases. For case 2, the dip of all blocks is synchronously increased by some small increment of the final dip (0.01), and the resulting y values are summed. This calculation is repeated until the maximum d is reached. However, to make the calculations for case 3, a dip, or percent of the final dip, of each carbonate block is needed, so that when the next block in the sequence begins to rotate ("is triggered"), its contribution can be added to the next value of y. In Figure 23 the "trigger angles" are arbitrarily chosen, but recall that in many drape folds, block 3 is nearly vertical while block 4 is still inconspicuous. As of now, only trigger angles that seem to best fit the particular drape fold of interest can be chosen.

The next modification is suggested by the fact that most drape folds in the Wyoming province form over *rotated* basement blocks, not plateau uplifts. The rotational component of block 1 is now introduced into the analysis.

Block 1 rotates from the horizontal as displacement on the fault increases. As Stearns (1970, 1975) has pointed out, the rotation of rigid basement blocks requires curved faults. To construct adequate kinematic models, one must know the length of block 1. This quantity is not always easy to measure in the field, but it can be calculated if the dip of block 1 and the vertical displacement on the basement fault are known:

$$\Lambda = d/\sin \psi, \tag{22}$$

where ψ is the dip of block 1, d is the vertical displacement on the basement fault, and Λ is the length of block 1. Another question arises, however, because all carbonate blocks do not rotate 90° (Fig. 1). The reason for this is not understood, so, in the following analyses, two possible cases are treated.

The first, designated case 4, rests on the idea that a block rotates until its dip is that observed today, before the next block is activated. At this point, the next block begins rotating. For example, block 2a rotates to its maximum (10°) before block 2b begins to rotate. Since block 1 is rotating in the opposite sense, there would be a tendency for block 2a to be rotated in a direction that would decrease its dip if the block-1-block-2a hinge were to lock. Under the assumption of case 4, the carbonate block (2a, for example) would stop rotating at its present dip, the hinge would *not* lock, and the continuing rotation of block 1 would have no effect on the dip of the frontal carbonate block.

The second, case 5, is the opposite of case 4. In this model a block must have been rotated to a dip greater than that now observed. The hinge then becomes locked, so that subsequent rotation of block 1 reduces the dip. The equations for both cases do not differ significantly from equation 4. Figure 24 shows the displacement plots for both cases.

That all three curves converge at the maximum fault displacement is an artifact of the models, since all must agree with the real geometry of Rattlesnake Mountain. The final y value is nearly 300 m greater than that calculated earlier for a vertical fault (Fig. 15, curve B). The dip of the normal fault, plateau-uplift plot (Fig. 24, curve C) was found by striking a chord of the arcuate fault (used in cases

4 and 5) such that the chord passed through the upthrown and downthrown blocks where they intersect the fault. This is supposed to represent the average dip of the curved fault. Clearly, the plateau-uplift model, with the appropriate fault dip, closely approximates the models invoking rotation of block 1. For this reason, the remaining refinements of the Paleozoic model will be limited to plateau-uplift models.

Another refinement of the models relates to the hinge areas between the rigidly rotated Paleozoic carbonate blocks. If the Paleozoic carbonate sequence were treated as a layer, the thickness of which nearly equals the length of the blocks, it may more closely approximate the real structure at Rattlesnake Mountain where the Paleozoic section is almost 450 m thick (see Fig. 5 for block lengths).

Only when the hinge points (as seen in Fig. 25) go from the bottom of the block (anticlinal) to the top (synclinal) is the thickness of the block important. Referring to Figure 26, it is seen that this "offsetting" of the hinges occurs only through block 3. When both hinges at either end of a block are on top (or bottom), the thickness term (t) does not appear (as in equation 4). The offset-hinge effect is only considered during the rotation of one block. From Figure 27

$$d = d_1 + d_2, \tag{23}$$

where

$$d_1 = \lambda \sin \theta \tag{24}$$

and

$$d_2 = t(1 - \cos \theta); \tag{25}$$

therefore,

$$d = \lambda \sin \theta + t(1 - \cos \theta). \tag{26}$$

It is also seen that

$$y = \lambda - (\lambda \cos \theta + t \sin \theta), \tag{27}$$

or

$$y = \lambda(1 - \cos \theta) - t \sin \theta. \tag{28}$$

Curve B (Fig. 28) shows the significance of offsetting the hinges. When the final y values of the two models are compared, it can be seen that there is a reduction of the final y value by almost 36%, using the offset-hinge effect. This value is still relatively large (nearly 800 m), but it seems more reasonable than does the 1,200 m required by case 1 (curve A, Fig. 28). The portions of the curves representing the rotation of blocks 2a, 2b, 4, and 5 are identical in both cases; the difference in y is due solely to the offset-hinge effect. Cusps do not occur at the same vertical displacements, because the line connecting the offset hinges in block 3 is longer than the block itself; this results in a greater arc length (curve B, Fig. 28) for the block 3 segment.

Several difficulties arise when the offset-hinge effect is used, the most obvious of which are the "holes" created in the hinge zones. The rocks could hardly

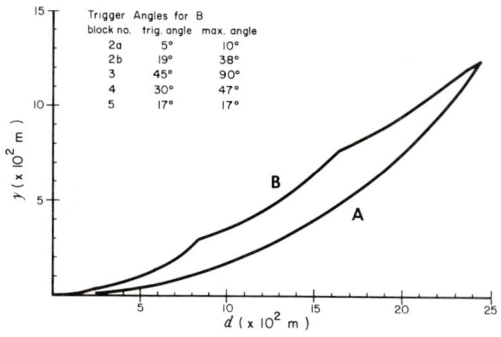

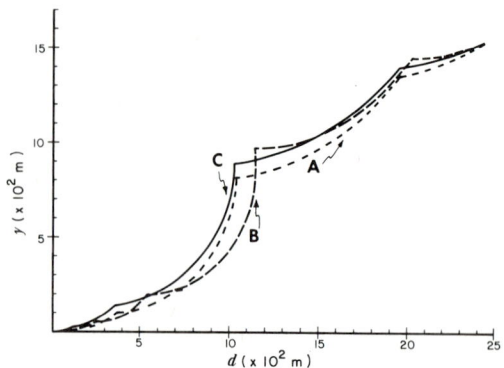

Figure 23. Displacement plots for Paleozoic models with synchronously rotating blocks: curve A, case 2; curve B, case 3. The cusps between blocks 2a and 2b, and between 4 and 5 are present in curve B but are very subdued because of synchronous rotations.

Figure 24. Displacement plots for case 4 (curve A), case 5 (curve B), and 83°-dipping normal fault (curve C). Dip of normal fault is that of chord of the arcuate fault.

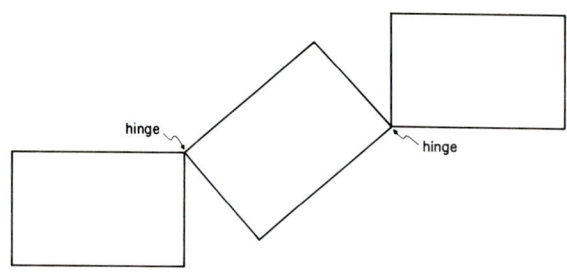

Figure 25. Offset of hinges from bottom to top across a block having different fold axes on either side.

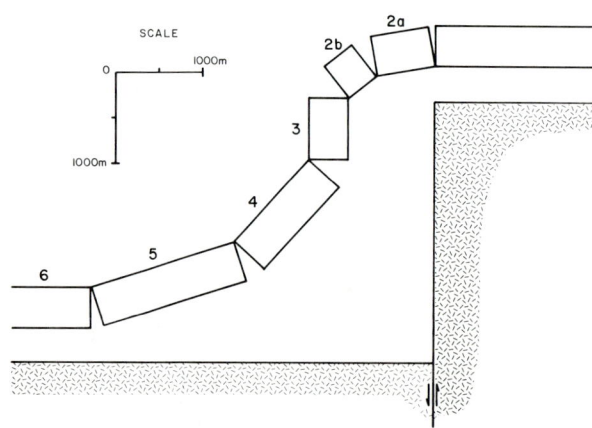

Figure 26. Schematic diagram of Rattlesnake Mountain. Carbonate blocks are assumed to be perfectly rigid. Note that offsetting of hinges occurs only in block 3.

withstand the stress concentrations that would occur at the corners of the blocks. There is, however, a possible solution.

Figure 29 shows how one detachment surface might occur within the carbonate sequence. The void volume is thereby significantly reduced (about 50% for this particular case). Imagine that a slip plane is created every 30 m vertically throughout the carbonate section. The holes would then become so small that they might, in fact, be filled with crushed rock from the area immediately adjacent to the sliding blocks (see App. J, Weinberg, 1977). This mechanism could result in less

KINEMATIC ANALYSES OF THE DRAPE-FOLD CONCEPT   71

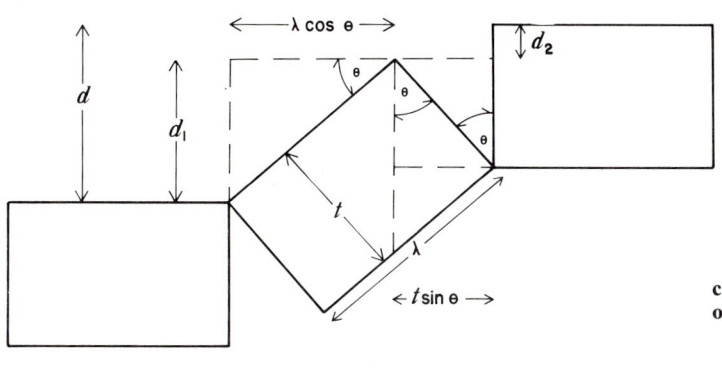

Figure 27. Parameters used to calculate the effect on y for the offset-hinge case.

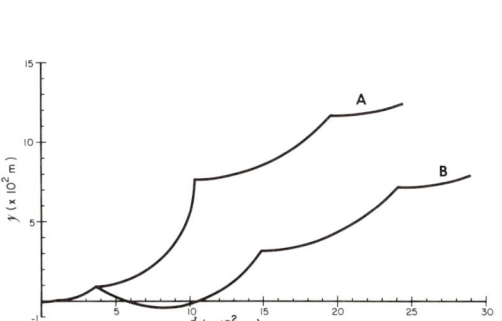

Figure 28. Displacement plots of original (case 1) Paleozoic model (curve A) and case 1 with the offset-hinge effect (curve B). Block lengths and dips, and fault dip (vertical) are the same for both curves.

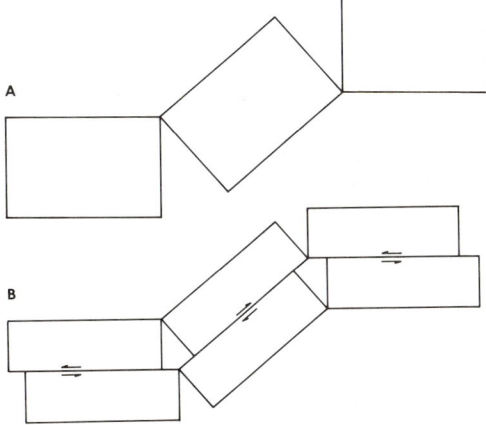

Figure 29. Schematic diagrams of the offset-hinge case. (A) Without internal slip plane; (B) with internal slip plane.

internal deformation within the carbonate block, thereby preserving some of the original bedding features. The observation of Stearns (1971, p. 134) that in the hinges "relict bedding is sometimes still distinguishable, and in these beds there is no appreciable change in thickness" might be explained by this mechanism.

Our understanding of how drape folds can form over faults with nearly 1,200 m of throw without horizontally transporting large rock masses for great distances (Fig. 28) may be improved by considering the offset-hinge effect with internal slip.

SUMMARY OF REFINEMENTS

By relaxing some of the original assumptions used in the original model of the Paleozoic rocks (case 1), three significant effects are observed in the displacement models. First, by allowing two or more carbonate blocks to rotate synchronously (cases 2 and 3), a noticeable dampening, or even complete disappearance, of the cusps occurs (Fig. 23). Second, the lateral-transport requirements are not significantly affected by including the rotation of block 1 in the analysis (cases 4 and 5), regardless of the hinging mechanism (case 4 *or* 5) chosen (Fig. 24). Third, by considering the Paleozoic section as a layer of finite thickness and using the offset-hinge effect, the final lateral transport required is reduced appreciably (Fig. 28). Not only have these three refinements allowed a very complex natural structure to be examined by simple methods, but they have also given further insight into some problems that still persist (such as locking of hinges and holes in the hinge zones).

EXPERIMENTAL WORK

Testing the applicability of the kinematic models to the natural drape-folding process is difficult. Choice of the model that most closely approximates real folding still depends on knowledge of the lateral transport (y value) that has occurred. Observations of a natural bedding-plane detachment show that the measuring y in the field is virtually impossible, so an alternative method has been attempted.

Geometrically scaled laboratory specimens were assembled so as to approximate the stratigraphic section of the Rattlesnake Mountain area. These rock beams were placed on a steel forcing-block as shown in Figure 30.

The experimental materials (lead, Blair dolomite, and Indiana limestone) were chosen so that their respective ductilities provided approximately the contrast that exists between the Cambrian shales, Paleozoic carbonate rocks, and the Mesozoic clastic rocks of the northern Bighorn Basin. The thickness ratios of the individual beams also approximate those of the three distinct stratigraphic units. The rock-deformation apparatus is described by Handin and others (1972).

The experiments were run in compression, on dry samples at room temperature, 0.5 kb confining pressure, and at a displacement rate of 6.6×10^{-3} cm/s. The experiments were terminated when the vertical displacement on arc 1 (Fig. 30) reached approximately the scaled equivalent displacement of the frontal fault at Rattlesnake Mountain. The confining pressure was chosen so that the dolomite remained brittle, but the lead and limestone were ductile (Fig. 31).

Several interesting features are obvious. First, there are rigidly rotated blocks of dolomite. Comparison of Figures 31 and 26 shows some striking similarities. The number and relative lengths of the dolomite blocks, as well as the angles of rotation are somewhat different, but the overall geometric similarities are obvious.

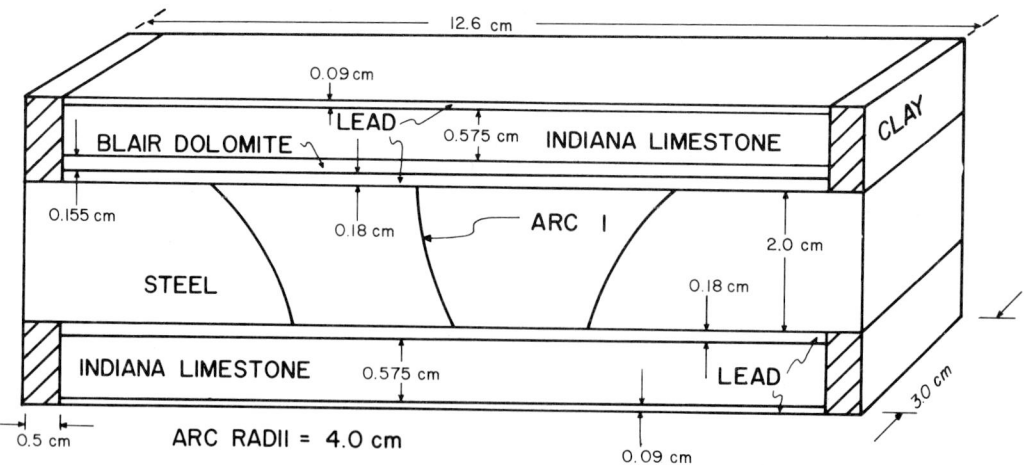

Figure 30. **Specimen configuration in drape-folding experiments. Vertical scale is 1 mm ≈ 300 m.**

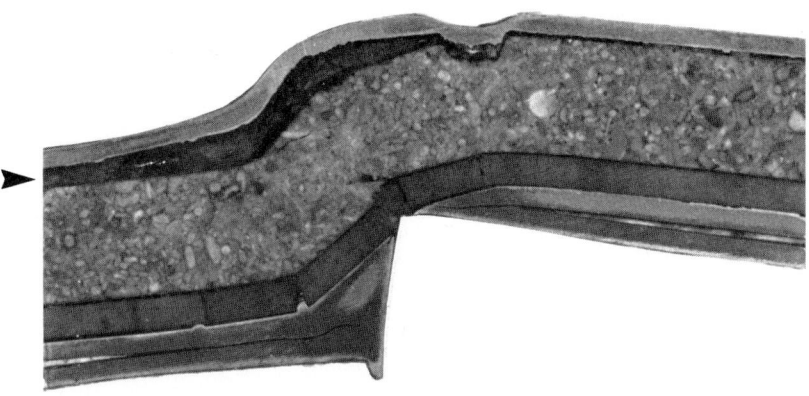

Figure 31. Photograph of drape-fold experiment (region above arc 1 in Fig. 30). Note rigidly rotated dolomite blocks, fixed hinges (in dolomite), and circular-arc shape of limestone upper surface. Dolomite layer is 1.55 mm thick. (Upper lead layer separated from specimen during cutting. Arrow shows position of epoxy fill.)

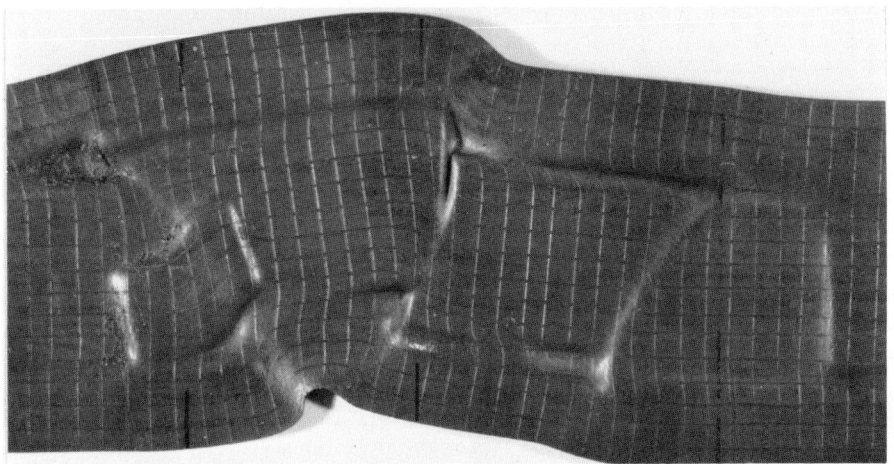

Figure 32. Photograph of distorted grid pattern on lead side jacket. Note that lateral movements on both upthrown and downthrown blocks are toward fold axis.

Second, the upper surface of the limestone layer can be approximated by circular arcs.

On one outside lead side jacket, an approximately 2.5-mm-square grid pattern was scribed before deformation. When the specimen was deformed, this grid was distorted (Fig. 32). The vertical lines have been displaced laterally. In particular, the displacements on the downthrown side of arc 1 are toward the upthrown block, and the lines on the upthrown block are offset in the opposite direction. From equation 4 a y value for any particular model can be calculated from the length and dip of the rotated blocks. For the scaled Paleozoic model, y is about 3.61 mm, and the fault displacement is about 6.14 mm. These compare favorably with the observed values (from the grid offsets) of 3.9 and 6.11 mm, respectively.

Since the simplest model (case 1) was used first (excluding the offset-hinge effect), the calculated y from the experiment includes the space of the holes. Using the offset-hinge effect gives about 1.13 mm, and the observed y is 2.35 mm, a nearly 50% difference, as opposed to a 7% difference using case 1. The reasons for these discrepancies are not known. It should be noted, however, that the contribution of the farthest left dolomite block (Fig. 31) was not included in the calculations, because its total length could not be measured. Also, when the specimen is deformed, the steel block to the left of arc 1 (Fig. 30) rotates in a counterclockwise manner (about the leftmost arc, Fig. 30), so the "true" horizontal datum cannot be established. If the contribution of that dolomite block could have been included, the total y value would have been larger and hence closer to the calculated y for the offset-hinge effect. This contribution would also increase the y value for case 1 and exceed the observed y value as is indicated in Figure 28. To date, four experiments have been run for reproducibility, and all show the rigidly rotated blocks and lateral transport toward the fold axis.

DISCUSSION

The main qualification of all implications that must be made for any Paleozoic model is that only at a few localities is the upper part of block 4 observed in the field (Rattlesnake Mountain being the best exposed), and, furthermore, block 5 (both its length and dip) is completely speculative. Therefore, only those portions of the displacement curves representing the upper three blocks (2a, 2b, and 3 that are commonly completely exposed) are regarded as forming a sound basis upon which geologic inferences can be made. This fact along with other uncertainties and complications (sequence of block rotation and possible faulting of the sedimentary veneer) requires that the curves be viewed only as approximations of the displacement paths, and that no individual curve be considered as uniquely displaying the actual displacement path of any real drape fold. Within this context, however, several aspects of the models are important.

The geologic implications based on these displacement analyses are of two kinds. The first concerns the lateral transport (both amount and direction) that a point on any one bedding plane may undergo during drape folding; the second type relates to the behavior of the rocks between any two particular horizons.

The very high slopes of the displacement curves generated by the rotations of blocks to steep dips (usually block 3) show that for small incremental increases in vertical displacement (d), very large lateral motions (y) are required. It can be speculated then that the rocks being transported laterally during the high-slope periods may be experiencing a significantly greater displacement rate (and possibly a greater strain rate) than when much lower slope conditions exist. Cases 2 and 3 indicate that the high-slope periods can be eliminated by allowing synchronously rotating blocks (Fig. 23), but field evidence indicates that block 3 does achieve dips approaching 90° before block 4 is developed. This observation must be considered when deformation features within individual drape folds are correlated to other drape folds of differing vertical displacement. It is postulated that the deformation of the downthrown rocks on a drape fold of 800 m or less vertical displacement (low-slope region) differs significantly from those drape folds vertically displaced 100 m or more (high-slope region). From Figure 18, the exact boundaries of low-slope to high-slope regions will vary according to the model chosen. However, although the exact paths are not known, these large slope changes were probably a part of the formation of most drape folds in the Wyoming province.

Because the displacement curves are very sensitive to the shape of the modeled surface, and the precise shape of natural drape folds is not commonly known, it is necessary to develop an "envelope" within which the true displacement curve probably resides. Additionally, if the dip of the underlying basement fault is unknown, as well as the rotation sequence of the blocks, the need for envelopes becomes very important.

One useful feature of any displacement curve is that it may represent the displacement path of a drape fold. Many palinspastic reconstructions are routinely done in the Wyoming province by comparing final folded lengths to initial unfolded lengths so that only the final lateral displacement (y) value is determined. No definition of displacement path is possible using this technique. This more common method can, therefore, mask many important aspects of drape folding such as changes in direction of lateral movements (Figs. 15 and 21B) or the differences in transport between the top and bottom of a single unit (Fig. 22).

Previous questions about the lateral-transport requirements of closely spaced drape folds (Pryor Mountains and portions of the Big Horn Mountains) and the nature of the hinge zones between carbonate blocks may be irrelevant if the offset-hinge effect approaches reality. Vertical fault displacements of less than 1,200 m may not require much lateral movement (Fig. 28), and if there are multiple detachment surfaces within the carbonate sequence, much of the drape-folding process may be volumetrically "self-compensating" in the hinge areas.

Because few, if any, drape folds have exactly the same shape from the bottom to the top, there must exist a difference of y values (Δy) within the unit(s) bounded by the different shapes so that layer-parallel shear is imparted to the unit(s). It is from this concept that several important geologic inferences are drawn.

Questions concerning the nature of bedding-plane slip have not been examined directly here, although bedding-parallel movements are discussed at length. I am not aware of any published systematic treatment of the causes and/or field evidences of bedding-plane slip. Therefore, I visualize bedding planes as being either perfectly free to slip or completely welded at a given instant in the folding process. If all bedding planes are free to slip, no significant volume problems arise within a drape fold. However, because block 1 (Paleozoic) rocks are pinned (Stearns, 1971) and my field observations within many stratigraphic units (mostly Mesozoic) show no evidence to support bedding-plane slip over very thick intervals, I feel that welding of the bedding planes is common, at least in many portions of the Mesozoic clastic sequence. This situation can produce very significant volume problems within drape folds.

If each bedding plane in the Mesozoic units were welded to the beds above and below it and in turn welded to the top of the Paleozoic carbonate section so that the Mesozoic rocks were laterally displaced the same amount as the underlying Paleozoic rocks, a large excess of Mesozoic rocks would be forced into the drape fold during the folding process (Fig. 19). This extreme case is not intended to approximate reality, but does emphasize one implication of the Δy curve. If welding of the bedding planes were to occur throughout some, but presently unknown, thickness of rocks, the shape difference between the upper and lower boundaries (bedding planes that have slipped) would create a Δy and thereby a volume problem. Such volume problems have already been recognized.

Cook (1975) and Cook and Stearns (1975) described a disharmonic anticlinal flexure (Teepee structure) that occurs within the Weber Sandstone in Dinosaur National Monument, Utah. Vaughn (1976) reported a similar disharmonic structure (Casper-Alcova fold) on the southeast flank of Casper Mountain, Wyoming.

V. Matthews III (1976, personal commun.; Matthews and Work, this volume) has discussed similar disharmonic features occurring in drape folds of the Colorado Front Range. I have seen many such structures in the Big Horn, Wind River, and Powder River Basins of Wyoming. Collectively, these disharmonic structures are here named "drape-subsidiary folds" (DSFs; modified from Dennis, 1967, p.

75); they are further defined as disharmonic folds whose axes parallel or subparallel the major fold axis and that generally occur in the hinge areas of drape folds (for further discussion of DSFs, see Weinberg, 1977). Although the causitive mechanism(s) of DSFs remains unknown, I speculate that these features are due to localized volume accommodation that is related to lateral displacements of large rock masses (see also Laubscher, 1976). Obviously, the effects of shear (Fig. 20), bending stresses, and so on play a part in the formation of DSFs, but at this time there are too many unknowns to speculate on their relative importance.

Kinematic analyses of drape folding provide a functional relation of lateral transport of the sedimentary veneer and basement-fault displacement. Use of kinematic analyses, in conjunction with more conventional techniques, may provide a better understanding of when and how: (1) fracture porosity and permeability are developed, (2) small internal-deformation features (faults and folds) are formed, (3) significant faulting of the sedimentary rocks may begin, (4) bedding-plane detachments might be initiated, and (5) "self-compensating" volume problems within drape folds (DSFs and offset-hinge effect) may be solved.

These discussions, by no means, cover all the possible geologic implications that arise from kinematically analyzing the drape-fold concept. The kinematic approach to drape folding is useful, but more field and analytic studies must be done, for without additional data, further speculation is fruitless.

CONCLUSIONS

1. Simple kinematic analyses can improve our knowledge of the drape-folding process.

2. The lateral transport of rock masses depends on the shape of the fold as well as the displacement and dip of the basement fault.

3. The direction of lateral transport is not always intuitively obvious and may not always be in the same direction for a particular fold throughout its vertical-displacement history.

4. The amount of lateral displacement needed in drape folds does not vary linearly; there are vertical-displacement intervals that require much greater lateral displacements than others.

5. Differences in the shapes of layers above or below stratigraphic units affect the lateral transport needed. The difference (Δy) can put the rocks into a state of shear.

6. Although the mechanism of hinging in the Paleozoic carbonate rocks is not well understood, in drape folds with less than 1,200-m vertical displacement, there may be no "room" problem within the hinge zones, as the system may be "self-compensating" even though no thickness change occurs.

7. Laboratory experiments on geometrically scaled rock specimens confirm that lateral displacement does occur, but its magnitude cannot yet be predicted from analytic models.

8. Much work remains to establish the shape of the Mesozoic sequence in drape folds.

ACKNOWLEDGMENTS

I thank D. W. Stearns, J. M. Logan, and A. F. Gangi for many helpful suggestions during the development of this study and, in particular, J. N. Shapiro without whose help this study would have been most difficult. I also thank many fellow graduate students, particularly M. Fahy, J. Morse, and W. Jamison for their many helpful discussions. The Department of Geophysics of Texas A&M University provided the computing facilities, and the Center for Tectonophysics made its rock-deformation equipment available. L. Billingsly, P. Bowman, and G. Merrifield capably assisted me in the field.

Financial assistance was provided by the Department of Geology of Texas A&M University, the Jensen/Mark Corporation, and the National Science Foundation through grants GA-36127X and DES 74-22954 to the Center for Tectonophysics.

The manuscript was read and criticized by R. R. Berg, J. W. Handin, and T. J. Parker, and their assistance is most gratefully acknowledged.

This study represents research done in partial fulfillment for the degree of Doctor of Philosophy at Texas A&M University.

REFERENCES CITED

Berg, R. R., 1962, Mountain flank thrusting in Rocky Mountain foreland, Wyoming and Colorado: Am. Assoc. Petroleum Geologists Bull., v. 46, p. 2019–2032.

——1976, Deformation of Mesozoic shales at Hamilton Dome, Bighorn Basin, Wyoming: Am. Assoc. Petroleum Geologists Bull., v. 60, p. 1425–1433.

Cook, R. A., 1975, Mechanisms of sandstone deformation: A study of the drape folded Weber Sandstone in Dinosaur National Monument, Colorado and Utah [Ph.D. dissert.]: College Station, Texas A&M Univ., 123 p.

Cook, R. A., and Stearns, D. W., 1975, Mechanisms of sandstone deformation: A study of the drape folded Weber Sandstone in Dinosaur National Monument, Colorado and Utah, *in* Bolyard, D. W., ed., Deep drilling frontiers of the central Rocky Mountains (symp.): Denver, Rocky Mtn. Assoc. Geologists, p. 21–32.

Dennis, J. G., ed., 1967, International tectonic dictionary: Am. Assoc. Petroleum Geologists Mem. 7, 196 p.

Gangi, A. F., Min, K. D., and Logan, J. M., 1977, Experimental folding of rocks under confining pressure: Pt. IV. Theoretical analysis of faulted drape-folds: Tectonophysics (in press).

Handin, J., Heard, D. C., and Magouirk, J., 1967, Effects of the intermediate principal stress on the failure of limestone, dolomite, and glass at different temperatures and strain rates: Jour. Geophys. Research, v. 72, p. 611–740.

Handin, J., Friedman, M., Logan, J. M., Pattison, L. J., and Swolfs, H. S., 1972, Experimental folding of rocks under confining pressure: Buckling of single-layer rock beams: Am. Geophys. Union Geophys. Mon. 16, p. 1–28.

Laubscher, H. P., 1976, Geometrical adjustments during rotation of a Jura fold limb: Tectonophysics, v. 36, p. 347–365.

Matthews, V., III, and Work, D., 1978, Laramide folding associated with basement block faulting along the northeastern flank of the Front Range, Colorado, *in* Matthews, V., III, ed., Laramide folding associated with basement block faulting in the Western United States: Geol. Soc. America Mem. 151 (this volume).

Min, K. D., 1974, Analytical and petrofabric studies of experimental faulted drape-folds in layered rock specimens [Ph.D. dissert.]: College Station, Texas A&M Univ., 90 p.

Prucha, J. J., Graham, J. A., and Nickelson, R. P., 1965, Basement-controlled deformation in Wyoming province of Rocky Mountains foreland: Am. Assoc. Petroleum Geologists Bull., v. 49, p. 966–992.

Reches, Ze'ev, 1978, Development of monoclines: Pt. I. Structure of the Palisades Creek branch of the East Kaibab monocline, Grand Canyon, Arizona, *in* Matthews, V., III, ed., Laramide folding associated with basement block faulting in the Western United States: Geol. Soc. America Mem. 151 (this volume).

Stearns, D. W., 1970, Drape folds over uplifted basement blocks with emphasis on the Wyoming province [Ph.D. dissert.]: College Station, Texas A&M Univ., 118 p.
——1971, Mechanisms of drape folding in the Wyoming province: Wyoming Geol. Assoc., 23rd Ann. Field Conf., Guidebook, p. 125–144.
——1975, Laramide basement deformation in the Bighorn Basin—The controlling factor for structures in the layered rocks: Wyoming Geol. Assoc., 27th Ann. Field Conf., Guidebook, p. 149–158.
——1978, Faulting and forced folding in the Rock Mountains foreland province, *in* Matthews, V., III, ed., Laramide folding associated with basement block faulting in the Western United States: Geol. Soc. America Mem. 151 (this volume).
Stearns, D. W., and Weinberg, D. M., 1975, A comparison of experimentally created and naturally formed drape folds: Wyoming Geol. Assoc., 27th Ann. Field Conf., Guidebook, p. 159–166.
Vaughn, P. H., 1976, Mesozoic sedimentary rock features resulting from volume movements required in drape folds at corners of basement blocks—Casper Mountain area, Wyoming [M.S. thesis]: College Station, Texas A&M Univ., 93 p.
Weinberg, D. M., 1977, Two-dimensional kinematic analyses of selected aspects of folding in the Rocky Mountain foreland, and their geologic implications [Ph.D. dissert.]: College Station, Texas A&M University, 111 p.
Weinberg, D. M., Holyfield, P. E., and Couples, G., 1976, Fractures associated with folding: An addendum: Geol. Soc. America Abs. with Programs, v. 8, p. 71.

MANUSCRIPT RECEIVED BY THE SOCIETY JUNE 27, 1977
MANUSCRIPT ACCEPTED AUGUST 25, 1977

Experimental folding of rocks under confining pressure: Part VI. Further studies of faulted drape folds

JOHN M. LOGAN
M. FRIEDMAN
M. T. STEARNS
Center for Tectonophysics and Department of Geology
Texas A&M University
College Station, Texas 77843

ABSTRACT

Experimental studies of faulted drape folds differ significantly from traditional tests on short, right-circular cylinders because of the presence of deliberately introduced heterogeneous stress and strain fields. These experiments differ from other experiments on the folding problem (for example, buckling) in that (1) a well-developed analytical basis was not formulated when the study was initiated and (2) the experimental program was designed to gain insight into problems posed from field relations. Experimental data are only one tool available to the geologist, and they are best utilized in conjunction with theoretical and/or scale-model investigations. Coupling of these different approaches aids extrapolation of the results to nature. Specifically, the problem associated with geometric scaling becomes more tractable when analytic studies are used to interpret the experimental results. Differences of deformation mechanisms with scale and rock type require careful consideration, but even here experiments can provide insights into critical field problems. Some of these problems are exemplified by studies of mass transport, corner drape folding, and differences between high-angle and vertical faulting in the forcing member. In all these examples, the most valuable use of the experiments is perhaps to provide insights into which processes are truly critical. Experiments can guide and narrow the field investigation to critical questions, which may in turn be answered even better by further relevant experiments, suggested by the field work.

INTRODUCTION

Since the earliest recognition of folds, geoscientists have sought an understanding of the processes that operate when rocks are highly deformed. The inability to observe the formation of natural folds has logically led to laboratory experiments designed to fold rocks under conditions analogous, in part, to those in the Earth's crust (Handin and others, 1972, 1976; Friedman and others, 1976a, 1976b; Means, 1973; Means and Williams, 1972; Patterson and Weiss, 1962, 1966, 1968). This new approach to experimental rock deformation has been encouraged by field geologists (for example, Stearns and Weinberg, 1975).

Our program of experimental folding of rocks under confining pressure dealt first with the buckling of monolithologic, single-layer rock beams (Handin and others, 1972) and then of multilithologic layered rock sequences (Handin and others, 1976). In Friedman and others (1976a, 1976b) we studied a third set of experiments in which faulted drape folds are formed by forcing an essentially rigid block of sandstone along a lubricated saw-cut surface against an intact sedimentary veneer consisting of one to five layers of limestone, sandstone, and rock salt. The experimental and observational work then was followed by analytical study of the stresses in these specimens (Gangi and others, 1977). In the present paper we extend the study of experimental drape folding to include draping about "corner uplifts," folding over vertical faults of progressively increasing displacement, and the consideration of mass-balance problems in cross sections. This new work is followed by discussion of the philosophy of this type of experimental rock deformation.

PREVIOUS STUDIES

Briefly summarized here are the major results from our previous work reported in Parts III, IV, and V (Friedman and others, 1976a; Gangi and others, 1977; and Friedman and others, 1976b, respectively). The drape folds and reverse faults are produced experimentally at confining pressures to 2.0 kb and shortening rates of 10^{-3} to 10^{-6} s^{-1} by displacing a block of brittle Coconino sandstone (2 by 3 by 12.6 cm) along a lubricated saw cut into one to five initially intact layers (0.2 to 1.0 cm thick and as much as 12.6 cm long) of Indiana limestone, Coconino sandstone, and rock salt (see Fig. 1a). The veneer is varied to investigate the effects of ductility contrasts, variations in rock type, and loading conditions. Once the specimen is assembled, a 2-mm grid pattern is stamped with ink on all four sides. Lead strips are attached, and MoS_2 is applied to the ends that will be in contact with the piston. Tests with and without the lead strips show that the lead does not influence the deformation; these strips merely facilitate handling the specimen after deformation (Fig. 1b). The assembled specimen is jacketed with two layers of heat-shrink polyolefin tubing.

The basic ideas which are postulated for the experiments are that (1) a layered sequence of sedimentary rocks overlies more-rigid forcing blocks; (2) the rigid forcing blocks are faulted at high angles to the layer boundaries; (3) the dominant displacement takes place at high angles to the layer boundaries; (4) body forces may be neglected; (5) under conditions of confining pressure, simulating some suitable depth of burial, the rock would deform in a manner analogous to that in nature; and (6) the stress and strain fields would be independent of scale so that the results of the experimental studies could be compared to theoretical analyses and eventually to field relations with confidence.

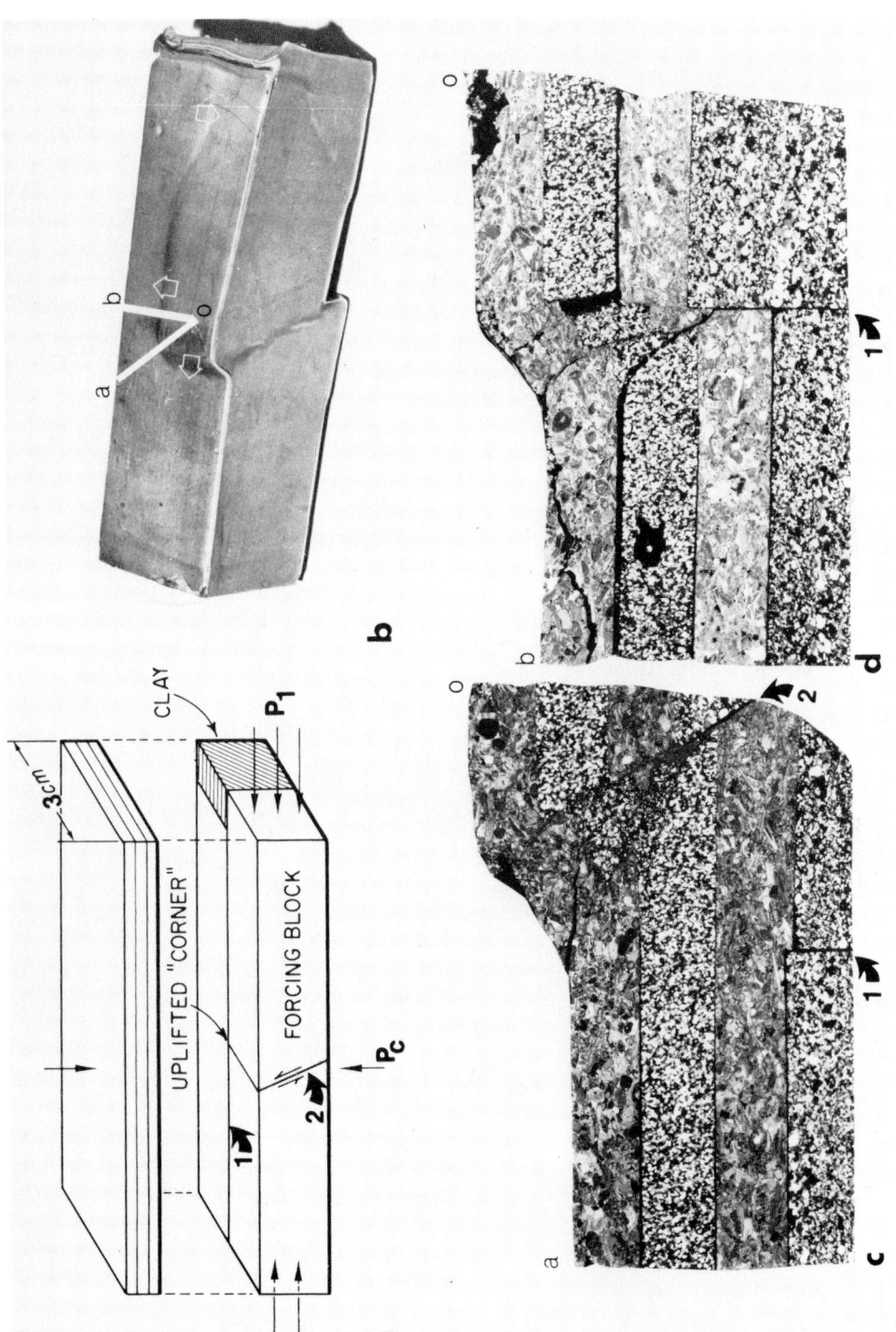

Figure 1. Experiment on corner drape folding. (a) Loading conditions; precut surfaces 1 and 2 are shown. (b) Specimen 351 with lead strips in place; white outlined arrows show three directions of draping; white lines show orientations of thin sections (c and d) below. (c) Thin section cut along cross section a–o shows vertical precut 1, the 65°-dipping precut 2 (see diagram a), and the deformed limestone-sandstone-limestone layers. The lower limestone layer is thickened between precuts 1 and 2. (d) Thin section cut along cross section b–o shows draping along vertical precut 1. Layer thickness is 0.3 cm.

Major results include the following:

1. Observational studies indicate that reverse faults, curving concave downward, propagate upward from the saw cut in the forcing block. With increasing displacement along the precut faults, the faults and associated gouge zones in the layers steepen and become progressively younger toward the upthrown block. The faults are preceded by swarms of extension microfractures that form throughout the deformation and that are the best clues to the stress trajectories. The downthrown layers are thickened by uniform flow and by repetition caused by the faulting. They are displaced away from the faults by bedding-plane slip. The maximum deformation of the downthrown block occurs when the saw cut is inclined at about 65° to the layering. The upthrown layers are all extended parallel to the layering and perpendicular to the fold axes, as indicated by extension fractures, thinned layers, and development of calcite twin lamellae.

2. Trajectories of the greatest principal compressive stress, σ_1, are inferred from the faults (assumed to be inclined at 30° to σ_1), extension microfractures, and calcite twin lamellae. The σ_1 trajectories in specimens with progressively increasing displacement along the lubricated 65°-dipping reverse faults are inclined at low angles to the layer boundaries near the faults and become perpendicular to these boundaries away from the faults.

3. Principal strains calculated from calcite twin lamellae (after Groshong, 1972, 1974) are within an average of 0.01 of those calculated from layer-thickness changes. The strain technique permits clear resolution of individual events in domains of superimposed deformation.

4. The fabric data and corresponding inferred stress trajectories are in good agreement with the results of analytical solutions for spatially varying boundary stresses obtained from use of the convolution theorem (Gangi and others, 1977). In combination they demonstrate that the major features of the deformation are dictated early in the history of displacement of the forcing block.

5. Observational studies provide structural details of the extension of the upthrown layers, the reverse faults, and the complexities in the downthrown layers, which should aid in the recognition of this structural style in the field and in the subsurface.

6. Some of the experiments demonstrate how large rigid-body rotation of a layer boundary can result from cataclastic flow and faulting in an adjacent domain.

Although these specimens are not scale models of natural prototype structures, our results do shed light on the mechanics of deformation in this structural style, especially with regard to sequences and relative locations of certain structural features. They suggest mechanisms—for example, the rigid-body rotation of units by cataclasic flow in adjacent domains—that need to be investigated in the field.

EXPERIMENTAL RESULTS

To illustrate one of the most valuable uses of the experimental studies, that is, to suggest problems and features for the field geologist to investigate, three new examples have been selected for discussion herein. These are corner drape folds, deformation associated with vertical faults in the forcing block, and mass-transport problems.

Corner Drape Folds

Block uplifts like those in the Wyoming province are characterized by rather sharp corners, manifested by abrupt changes in strike of the bounding basement faults. Some good examples occur in the Prior and the Big Horn Mountains, north and east of Lovell, Wyoming; the southeast end of Rattlesnake Mountain near Cody, Wyoming; and the Clarks Fork corner of the Beartooth Mountains. The Paleozoic sedimentary veneer is draped over these corners, and it seems to conform exactly to the shape of the forcing block. The mechanisms by which the carbonate rocks of the Madison Formation drape at a corner in the field currently are being studied (Stearns and Stearns, this volume). Their preliminary data suggest that bed-thickness changes, intragranular deformation, and fracturing are of only minor importance as mechanisms of draping. In an attempt to gain insight into how rocks deform at these corners and what the accompanying stress and strain fields are, we have investigated corner draping experimentally.

In the experiment, two precut surfaces are introduced into a sandstone forcing block that is overlain by a veneer of three initially intact layers of limestone, sandstone, and limestone (Fig. 1a). The thickness of the veneer is 0.9 cm; that of the forcing block is 2.0 cm; and the specimen is 10.0 cm long and 3.0 cm wide. Precut surfaces 1 and 2 (Fig. 1a) are inclined at angles of 90° and 65°, respectively, to the veneer; both are lubricated with molykote (MoS_2) before assembly of the specimen, which is provided with thin lead strips on four of its sides and jacketed with heat-shrink tubing. The test is run dry, at room temperature, at a displacement rate of about 10^{-3} cm/s, and 1.0-kb confining pressure. Once the axial load $P_1 > P_c$ is applied and the frictional resistance along the precuts is exceeded, the forcing block is displaced into the veneer and rotated somewhat so that draping occurs around the uplifted corner and across the vertical precut (Fig. 1b). Three corner-drape experiments were made with similar macroscopic results; only specimen 351 was studied in detail.

Seven thin sections were cut from specimen 351: one perpendicular to the high-angle reverse fault (precut surface 2, Fig. 1a), four perpendicular to the vertical fault (precut surface 1, Fig. 1a), and two parallel to the line a–o near the corner (Fig. 1b). Petrofabric analyses included preparation of cross sections showing the locations and orientations of faults and microfractures and the percentage change in bedding thickness within the three-layer veneer (thinning, that is, shortening, denoted as positive), determination of principal-stress orientations from calcite twin lamellae (Turner, 1953), and the determination of principal strains from the calcite twin lamellae (Groshong, 1972, 1974). Discussions of these techniques, espcially their accuracies and limitations, are given elsewhere (Friedman and others, 1976a, 1976b). Here it is sufficient to discuss the cross sections and strains observed along the line of section a–o (Figs. 1c, 2b) and the cross section and associated inferred stress trajectories across the vertical fault (Fig. 1d) for three different magnitudes of vertical displacement (Figs. 3, 4).

Thickness changes along the section a–o are most spectacular in the lower limestone bed across the trace of the vertical fault (fault 1, Fig. 2b). The bed is thinned 3% on the upthrown side and is thickened to 15% in the downthrown block. The top of this layer is essentially flat across the vertical fault, and the thickening increases progressively toward the 65°-dipping reverse fault. Elsewhere, the limestone beds are thinned on the upthrown block of the 65°-dipping fault and thickened on the downthrown block as in previous specimens at these experimental conditions. The amount of thinning on the upthrown block is about the same here at the corner as it is in the section parallel to the XZ plane (Fig. 2a) away from the corner; that is, no excessive thinning occurs because of the presence of the corner alone.

The downthrown block is characterized by four small thrust faults, developed in front of the 65°-dipping reverse fault, which is continuous with precut surface

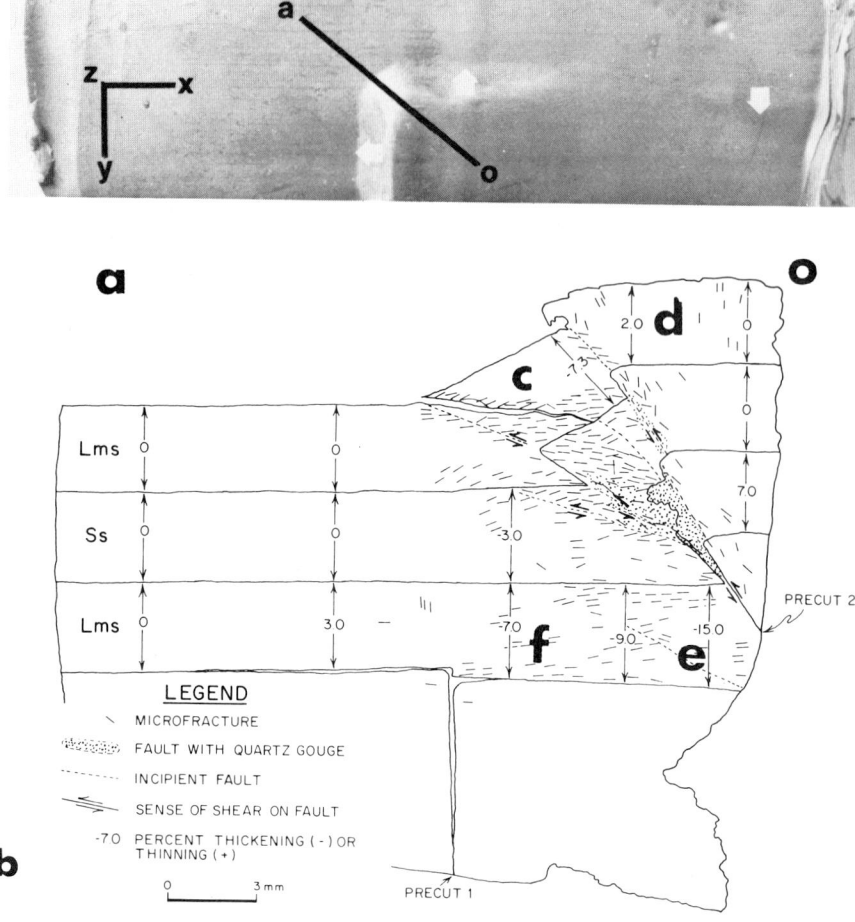

Figure 2. Diagrams illustrate type and location of deformation features and results of strain analyses in corner drape-fold experiment on specimen 351. (a) Photograph of XY surface of specimen shows orientation of thin section along a-o. (b) Sketch of thin section shows faults, microfractures, and layer-thickness changes (in percent).

2 (Fig. 2b). Microfractures in both the limestone and sandstone are much more abundant in the downthrown block, and they are oriented either subparallel to the layering or at 20°–30° to the small thrusts in their immediate vicinity. As argued previously (Friedman and others, 1976a), these are extension fractures, that form before faulting and microscopic feather fractures (Conrad and Friedman, 1976) that develop during faulting. In the upthrown block the microfractures away from the 65°-dipping reverse fault are nearly normal to bedding, and they reflect the layer-parallel extension of the block. The microfractures in the upper limestone layer of the upthrown block are oriented nearly perpendicular to the a-o section, which is inclined some 40° to the XZ plane (Fig. 2a). Yet in thin sections cut away from the corner and perpendicular to precut faults 1 and 2, the corresponding microfractures are also nearly normal to the thin section. This means, therefore, that the strikes of the microfractures and the upper beds in the draped region probably are parallel and mutually bend around the corner. A similar change in strike of microfractures in the leading edge of the sandstone forcing block is not

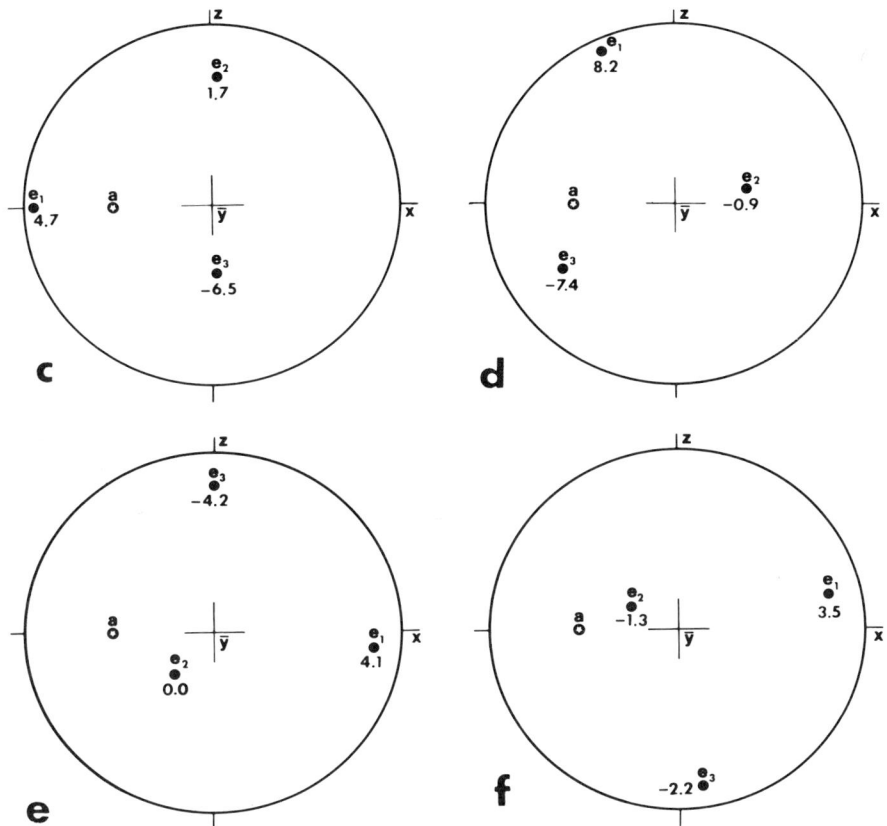

Figure 2 (continued). Equal-area, lower-hemisphere projections of principal strains determined from calcite twin lamellae in the four areas marked c, d, e, and f in diagram b. Shortening strain is denoted as positive. Point a and locations of X, Y, and Z are for orientation.

apparent in the existing thin sections because they parallel the strike of precut surface 2 in the a–o and XZ sections, and they also parallel the strike of vertical fault (1) in YZ sections. Thus, microfractures that might similarly change strike right at the corner in the forcing block were not sampled. It would appear, therefore, that the influence of the corner on the deformation does increase upward.

Strains deduced from the calcite twin lamellae by Groshong's method (1972, 1974) are shown for four domains (Fig. 2c through 2f). In these diagrams, e_1, e_2, and e_3 are the greatest, intermediate, and least principal shortenings, respectively, with elongations denoted as negative. As in most previous work (Friedman and others, 1976b), there is a good correlation between e_1 and σ_1 (the greatest principal compressive stress) deduced from the calcite twin lamellae, and this aspect of the work will not be discussed further. First the orientations of the principal strains will be considered. Shortening e_1 is essentially parallel to the X-axis at c, e, and f (Fig. 2), and it is subparallel to Z at d. In each domain, therefore, there is a good correlation between e_1 and the extension microfractures (that is, the trajectories of the local σ_1) and the changes in bedding thickness. Shortening e_3 is subparallel to Z in the lowest limestone layer (locations e and f) where thickening of the layers is the greatest. In the upper limestone layer at location c, there is elongation subparallel to Y, while in the upthrown block at location d, the

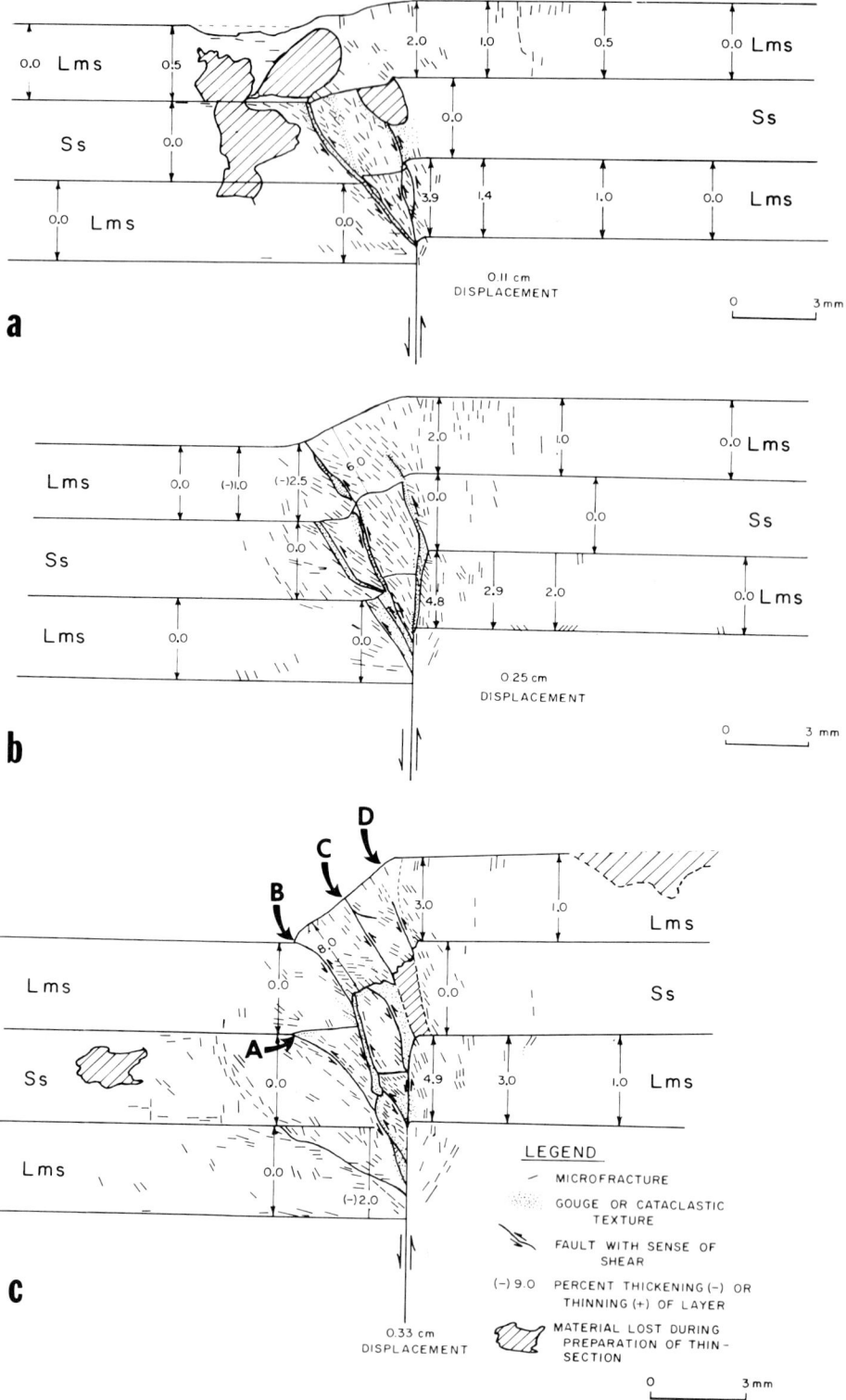

Figure 3. Nature of deformation shown in thin sections cut perpendicular to the vertical fault in specimen 351 with displacements of 0.11 cm (a), 0.25 cm (b), and 0.33 cm (c).

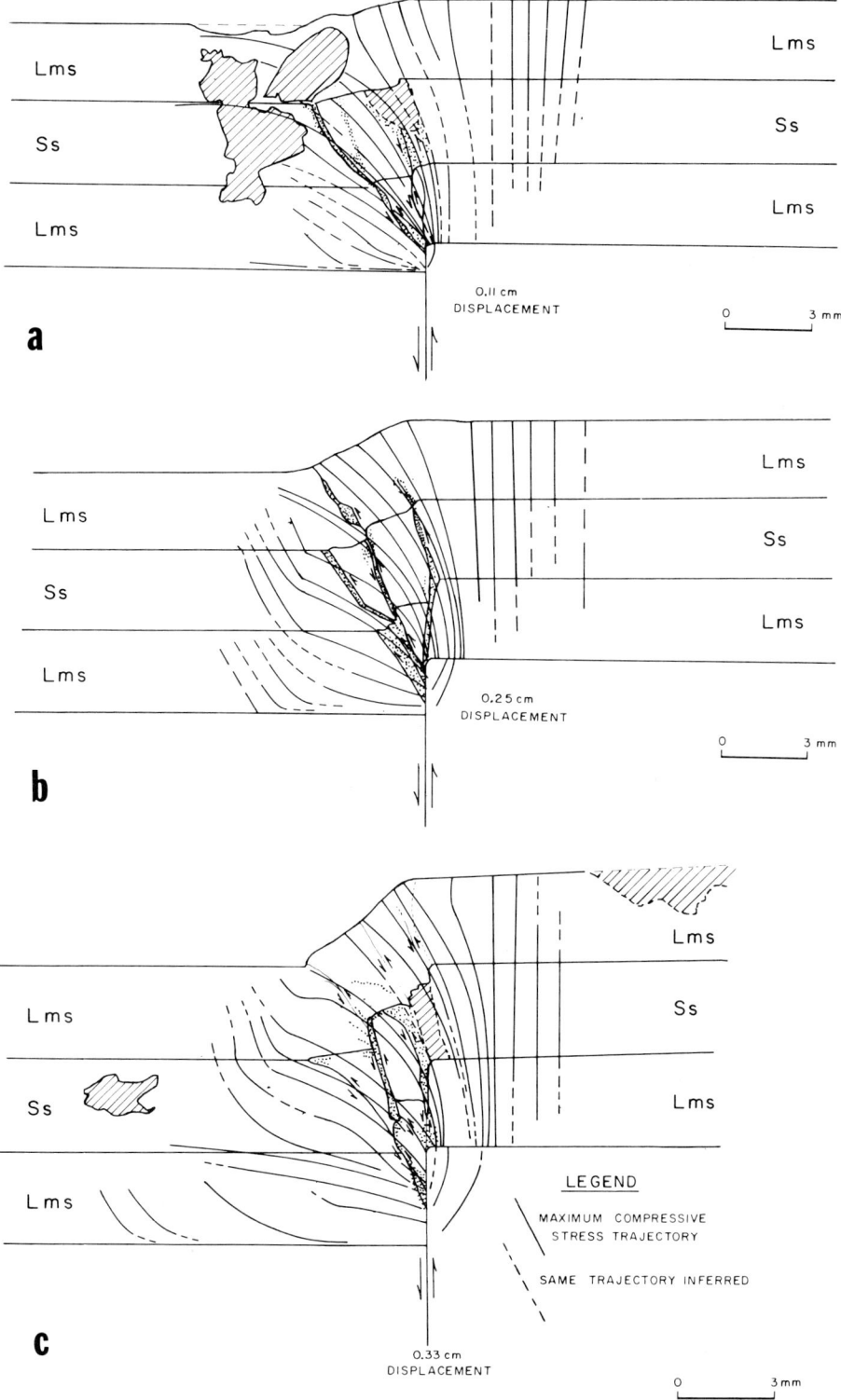

Figure 4. Trajectories of the maximum principal compressive stress derived from the faults and microfractures illustrated in Figure 3.

elongation is inclined at <25° to the dip of the beds at the corner. These orientations and the corresponding relatively large strains (−6.5% and −7.4%) suggest internal deformation of the layers as they stretch around the corner. Note that it would not be possible to observe microfractures that would augment these strains in the a–o thin section as they would be oriented nearly parallel to the plane of the section. In the lower limestone layer, e_3 increases from −2.2% to −4.2% at locations f and e, respectively. This increase parallels that reflected in the bed thickening (−7.0% to −15.0%), but the calcite twinning accounts for only about 30% of the thickening. Microfractures are oriented here so as to enhance the thickening; this arrangement perhaps accounts for all or part of the remaining strain.

In summary, (1) beds in the upthrown block of the 65°-dipping reverse fault near the corner are thinned as a function of distance from the corner; (2) bedding thickness changes are most pronounced in the downthrown block between the vertical and 65°-dipping reverse fault; (3) extension microfractures in the upper limestone change strike around the corner and thus parallel the changing strike of the layer; (4) a similar change in strike of microfractures in the leading edge of the sandstone forcing block is not observed [this point together with (3) suggests that the influence of the corner on the deformation of the layers increases upward]; (5) orientations of principal strains calculated from calcite twin lamellae are in excellent agreement with layer-thickness changes and the orientation and type of microfractures and faults; (6) greatest elongations in the upper limestone layer are relatively large (−6.5% and −7.4%) and along with their orientation suggest a stretching of the layer around the corner so that the greatest elongation is subparallel to the strike and dip of the layer in the downthrown and upthrown blocks, respectively; (7) strains calculated from calcite twinning in the lower limestone layer account for only 30% of the observed layer thickening; and (8) separations along suitably oriented microfractures in this layer may account for all or part of the remaining thickening.

Vertical Faults

Deformation associated with differential displacement along the vertical fault (Fig. 1a) is observed in four thin sections. The smallest displacement (0.04 cm) occurs in the a–o section (Figs. 1c, 2b), and displacements of 0.11, 0.25, and 0.33 cm occur in three sections cut parallel to o–b (Fig. 1b, 1d) and located within 3.0 cm of the corner along the vertical fault. Data consist of cross sections (Fig. 3) and accompanying inferred stress-trajectory diagrams for these three sections (Fig. 4). The cross sections are prepared by transferring the microscopic details of the deformed layers and forcing block, the faults and gouge zones, and the microfractures to overlays on 20-by-25 cm photomicrographs of each thin section. On the cross sections the microfractures are in part idealized in that although their relative numbers, orientations, and locations are accurate, their actual lengths and absolute numbers are schematic. The cross sections also show the percentage changes in layer thickness at specific localities as calculated from eyepiece-micrometer measurements. The corresponding σ_1 stress trajectories (Fig. 4) are prepared by assuming that the microfractures are precursive extension fractures or microscopic feather fractures (Friedman, 1975; Conrad and Friedman, 1976). It is assumed that the trajectories intersect the faults at angles of 30° and are appropriate to the sense of shear established from obvious offsets.

The cross sections and stress trajectories indicate the following:

1. With increasing displacement along the main vertical fault in the forcing block,

(a) the abundance of microfractures, the number of discrete faults, and the amount of associated gouge increase in each layer; (b) deformation increases upward and toward the upthrown block; (c) the magnitude of the layer-thickness changes increases in each domain; and (d) microfracturing also increases in the leading edge of the forcing block.

2. Faults in the veneer are for the most part high-angle reverse faults with dips >62°. Only at 0.33-cm displacement does one (fault A, Fig. 3c) dip at a low angle (25°). The faults exhibit discrete segments within individual layers, and they curve both upward and downward. Each fault seems to be triggered by the leading edge of the forcing block or by faulted corners of higher layers. They differ from faults in the veneer of specimens with only a 65°-dipping precut fault in the forcing block (Friedman and others, 1976a) by being high-angle reverse, segmented relative to layering, and curving both upward and downward.

3. The sequence of fault development is clearly evident in the three specimens of 0.11-, 0.25-, and 0.33-cm displacement (Fig. 3a, 3b, 3c). The upper limestone layer is not faulted at the smallest displacement, and it becomes progressively faulted, thinned, and rotated as the displacement along the vertical fault increases. This fact is detailed by observing the progressive development of faults A, B, C, and D (Fig. 3c) with increasing displacement. Thus at 0.11-cm displacement, fault A is completely developed, but fault C exists only in the lower two layers. At 0.25-cm displacement, fault B, an offshoot of A, has propagated to the upper surface, and fault C has advanced into the upper limestone layer. At 0.33-cm displacement, more movement has taken place along B and C, and fault D has developed. Fault A is essentially "dead" after 0.11 cm of displacement. These facts prove clearly that the sequence of faulting is from the lower limestone in the downthrown block to the upper limestone in the upthrown block. Further support comes from the fact that the very earliest deformations associated with the vertical displacement as observed along fault 1 (Figs. 1c, 2b) are microfractures skewed toward the downthrown lower limestone layer and oriented from 0° to 30° to the layer boundary.

4. The stress trajectories (Fig. 4a, 4b, 4c) inferred from the microfractures and faults have the same configuration at each magnitude of vertical displacement. All three specimens show layer-parallel σ_1 trajectories curving concave upward in the downthrown block and nearly vertical trajectories in the upthrown block that curve concave downward toward the fault zone. In general, these trajectories are very similar to those inferred from the deformation in specimens with only 65°-dipping faults (Friedman and others, 1976a, Figs. 5, 6). The sole difference is the zone of faulting where the trajectories (and subsidiary faults) arising from the vertical faulting are inclined at a larger angle to the layer boundaries.

Mass-Balance Problems

In drawing structural cross sections the geologist may be confronted with thickness changes that can be supported only by invoking dramatic compaction, dilatancy, or the transport of material along directions inclined to the plane of the cross section. This problem is also encountered in the faulted drape-fold specimens, where, in principle, it can be solved precisely, because the specimens are deformed under determinable boundary conditions, and dimensions of the specimen before and after deformation are known exactly. Let us discuss the significance of these problems in the laboratory and suggest possible solutions.

For example, in the cross sections represented by thin sections cut exactly perpendicular to the strike of the 65°-dipping precut in the forcing block (dip sections; Fig. 5a, 5b), we find that in the vicinity of the faults there is an increase in area as compared to that occupied by the layers prior to deformation. This effect was first discovered when attempts at palinspastic reconstruction of the original configuration failed. All parts of the deformed, central sandstone layer in each of eight specimens could not be fitted back into a single horizontal layer of known initial dimensions. In every specimen there is an excess of area. Either a significant dilation (porosity increase, with constant unit thickness) has occurred, or there has been a net addition of material by transport to the deformed domain as observed in the dip sections.

To appreciate the magnitude of this effect, consider first the undeformed area, which can be determined after deformation on the enlarged photomicrograph (Fig. 5a, 5b). Three measurements are taken directly: the thickness Z of the three-layer sequence, the layer-parallel distance X extending from the undeformed boundaries on each side of the faulted region, and the distance Y measuring the layer-parallel throw along the fault (Fig. 5c, 5d). The total undeformed area is then $(X + Y)Z$. Next, the total deformed area within the X and Z boundaries is measured directly from the photomicrographs with a planimeter. Then, if the areas of slightly deformed domains 1 and 2 (specimen 338; Fig. 5c) and 3 and 4 (specimen 336; Fig. 5d) are measured separately and subtracted from the total, the increases in area of the highly deformed domains become 41% and 29%, respectively, for the two specimens. These enlarged areas include those occupied by the limestone and the sandstone layers. To determine changes in the sandstone layer alone, one calculates its undeformed area by using the quantity $(X + Y)$ as before, and setting Z equal to the undeformed thickness. The total deformed sandstone area is measured with a planimeter and the sandstone areas in domains 1 and 2 and domains 3 and 4 are also remeasured. The increases of area in the sandstone alone are then 13% and 8% for specimens 338 and 336, respectively.

Accordingly, either dramatic increases in porosity or significant additions of material to the deformed regions must occur. Since the initial porosities of both rocks are about 10%, the final porosities in the highly deformed regions in toto would have to average 51% and 39% (or 23% and 18% in the sandstone member alone) if dilation were the sole mechanism by which the areas could grow. Such huge porosities would be conspicuous in thin section, and although some dilatancy surely occurs during microfracturing, faulting, and gouge formation, nothing like 20% to 50% is observed. Cursory estimates indicate that dilatancy could account for at most 25% to 50% of the increase of area actually measured. Thus, addition of material into the deformed regions, that is, mass transport, must be invoked.

Where did this material come from? Consider contributions derived laterally (within the plane of the dip section) from regions 1 through 4. One possibility is transport into the highly deformed region by rigid-body translation of the layers through "bedding-plane" slip. However, these are either zero or opposite in sense to that required for specimens 338 and 336 (Friedman and others, 1976a, Table 2, Fig. 4). Alternatively, transport from the source regions is possible through mechanisms of intragranular flow in the limestone or cataclastic flow in the sandstone. These should be reflected in measurable bedding-thickness changes. The sandstone layer, for example, would have to thin uniformly 8% to 9% in regions 1 through 4. The fact is, however, that it does not thin except immediately adjacent to the faults, being elsewhere constant along the complete length of the specimen. The limestone layers do show thickness changes within regions 1 through 4 (for example, Fig. 5c in Friedman and others, 1976a), and strain determinations from calcite twin lamellae do show compressive principal strains that in some domains are subparallel and in others are inclined to the dip sections (Friedman and others,

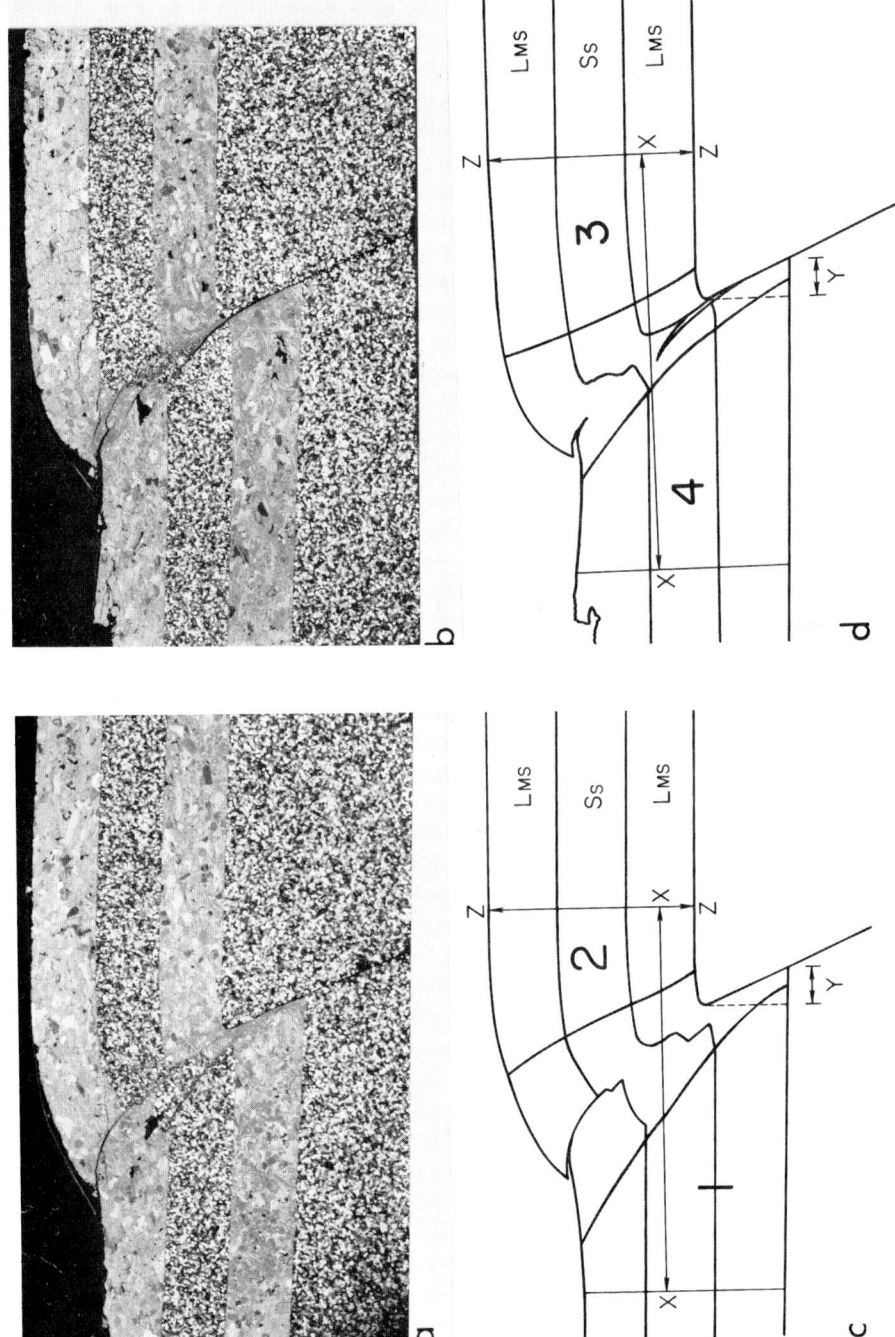

Figure 5. Photomicrographs (a and b) illustrate the dip-section configuration of specimens 338 and 336, respectively. Tracings of these specimens (c and d, respectively) define dimensions X, Y, and Z and regions 1 and 2, and 3 and 4. Layer thickness is 0.3 cm.

1976a, Figs. 11, 12). Accordingly, the growth of the area in the limestones could be explained by some combination of dilatancy and intragranular flow. However, the effect in the sandstone can be ascribed only in part to dilatancy; the remainder (at least 4% to 7%) must be due to mass transport along directions inclined to the dip sections.

In order to investigate potential mechanisms for this transport, thin sections are cut parallel to the strike of the 60°-dipping precut in a number of specimens (Fig. 6), see also Friedman and others, 1976a, Fig. 10). In these strike sections, the thrust and high-angle reverse faults (Fig. 6a) are clearly curved, that is scoop-shaped along strike (Fig. 6b, 6c). We do not yet understand this fault configuration (also observed in nature), but it could provide a mechanism for mass transport parallel to strike that would increase the area of the deformed region as observed in dip sections. This can be appreciated best by noting that strike-slip along a planar fault would simply displace material points uniformly without increasing area. On the other hand, strike-slip on a curved fault surface results in rotations of the hanging block that can increase the area in dip sections. Consider rotations about a vertical axis and a horizontal axis lying in the dip section. Two effects are obtained:

1. The Section-Orientation Effect. Upon either of these rotations, a dip section after deformation always exposes a larger area of the hanging block than does an undeformed dip section.

2. The Taper Effect. Since the fault surface is scoop-shaped, the thickness of the hanging wall decreases in both directions along strike from some maximum value at the center. Thus, upon strike-slip displacement (and rotation), a larger area is exposed in some dip sections and smaller ones in others relative to the sense of shear and position along strike.

These effects occur on all the several scoop-shaped faults in all the specimens, and the relative influence of each factor depends on the exact fault configuration and the amount of displacement and rotation relative to a given dip section. Thus, effect 1 always yields an increase of area, but effect 2 may average to a net change of zero for the specimen as a whole. We did find increases in all eight specimens. These sections were cut within 1 cm of a free boundary, that is, they do not represent a random sampling of dip sections for the complete 3.0-cm length of the specimen along strike. Thus, we can not discriminate between the section-orientation and the taper effects.

Evidence was sought for strike slip along the curved faults in the sandstone in the form of grain offsets or microscopic feather fractures (Conrad and Friedman, 1976), but none was found. The grain offsets and microfractures seem to be dominated by the major dip-slip character of the faults. Moreover, the orientation of microscopic feather fractures coincides with that of precursive microfractures developed prior to faulting (Friedman and others, 1976a, p. 1063) so that the microfractures cannot be uniquely identified and used for dynamic interpretations. On the other hand, there is macroscopic evidence for small strike slip at the upper surface of the top limestone layer. A grid inked on this surface before deformation is offset as much as 1.5 mm in several of the specimens.

In the experimentally deformed specimens, therefore, areas do increase in deformed regions during faulting. Although some of this growth no doubt arises from dilatancy, mass transport of material from directions inclined to the plane of dip sections also must occur. In the limestones this transport could arise from intragranular flow. Strike slip along faults that are curved along strike may be a mechanism in the sandstones. Although there is evidence for small strike slip at the top of the upper limestone layer, it cannot be demonstrated that such

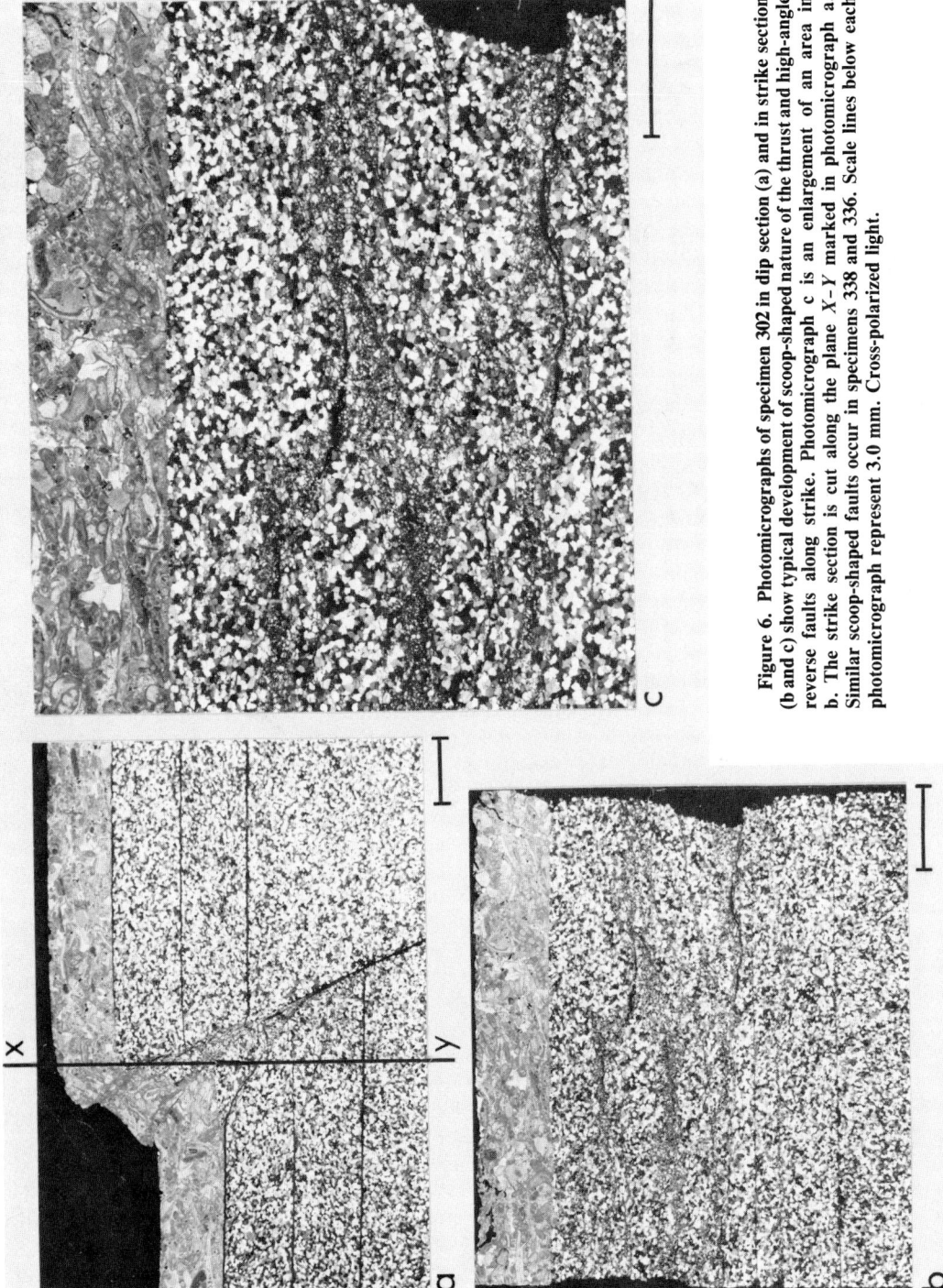

Figure 6. Photomicrographs of specimen 302 in dip section (a) and in strike section (b and c) show typical development of scoop-shaped nature of the thrust and high-angle reverse faults along strike. Photomicrograph c is an enlargement of an area in b. The strike section is cut along the plane $X-Y$ marked in photomicrograph a. Similar scoop-shaped faults occur in specimens 338 and 336. Scale lines below each photomicrograph represent 3.0 mm. Cross-polarized light.

displacement actually occurs within the sandstone. Thus, even in controlled laboratory specimens it is difficult to account exactly for every aspect of the deformation. It is hardly surprising, therefore, that mass-balance problems in the field are also difficult to solve (Stearns and Stearns, this volume).

PHILOSOPHY OF EXPERIMENTAL STUDIES

In conversations with field geologists, we find some misconceptions about and misuse of the experimental data. Therefore, it is appropriate here to consider the philosophy of the design of such experiments. At the outset it should be emphasized that the experimental folding of real rocks is perhaps the newest approach to a long history of inquiry into the folding process, primarily through theoretical and model studies (Handin and others, 1976, p. 1035). All three approaches—experimental, theoretical, and physical modeling—have particular advantages and disadvantages. Theoretical studies involve well-defined boundary conditions and rheological behavior and do not suffer from the scaling problem. But the idealizations necessary to provide mathematical tractability frequently limit their usefulness. Also, the results are often expressed as stress- and strain-trajectory diagrams, and so they are difficult for geologists to translate into deformation mechanisms that can be recognized in the field. Scaled and, particularly, photomechanical models offer an opportunity to look at more complex and realistic situations (Gallagher and others, 1970; Sowers and others, 1972; Currie and others, 1962). Unfortunately, materials scaling limits the type of problem that can be approached, and no information concerning the deformation mechanisms in real rock is provided. Experimental studies of real rocks do, however, provide direct insight into the mechanical behavior of these rocks under appropriate simulated conditions, and they are done under reasonably well-known boundary conditions. Extrapolating the results to the field must be done cautiously. The best use of any one of these methods is in conjunction with the other two, as they are complementary. Theoretical studies, for example, can extend greatly the applicability of the experimental results.

Ideally, the problem to be investigated experimentally should already have been identified in the field. That is, the geometrical configuration of the laboratory specimen and boundary conditions on it should be postulated from the field relations. A theoretical analysis should then follow, and the experiments should be designed and interpreted with this information in hand. The theoretical model and the experimental data then are evaluated by comparison to the original field evidence. This reasoning may appear obvious, but it is frequently overlooked. Experimental work is virtually useless if it cannot be applied to real problems.

Prior to our folding experiments, most experimental rock deformation had been on right-circular cylinders to determine the mechanical properties or deformation mechanisms such as fracture and flow (for example, Griggs, 1936; Handin and Hager, 1957, 1958; Handin and others, 1963; Heard, 1960, 1963; among many others). In all these previous experiments, attempts were made to achieve homogeneous states of stress and strain. Thus, the idea of folding rocks in the laboratory is a distinctly different approach for it deliberately extends experimental work into specimens in which states of stress and strain change in space and time. This departure from the traditional work is useful only because the petrofabric techniques for mapping stress and strain fields have developed to a relatively sophisticated level (Friedman, 1963, 1969; Friedman and others, 1976b; Gallagher and others, 1974; Groshong, 1972, 1974; Teufel, 1975; Turner, 1953). These techniques are necessary because of the technical difficulties of directly measuring forces and displacements in small domains under conditions of high confining pressure and temperature. In addition, they help to bridge the gaps between experiment, theory, and field, for they define actual deformation mechanisms and the states of stress and strain in the specimen and thus permit correlation with the theory on the one hand and the naturally deformed counterparts on the other.

Before we consider these new experiments on drape folding, let us recall our earlier studies of beams loaded parallel to their long axes (Handin and others, 1972, 1976). Our idea to "buckle" beams of rocks developed from well-established field and theoretical knowledge. Rocks folded by layer-parallel loading have long been recognized by field geologists, and theoretical treatments of this problem are extensive (see Handin and others, 1976, p. 1035). The experimental effort was then logically extended to supplement this body of knowledge and especially to determine the mechanisms of deformation and the stress and strain history of the folds in the laboratory where we could observe the entire process. Comparison of the experimental data with the theoretical analysis provides a firmer basis for extrapolation to the field. The laboratory beam is only 20 cm long, but the agreement between experiment and theory, which does not depend on scale, encourages applications of the results to much larger natural folds. Even the question of whether buckles (instabilities) ever develop in nature (Fahy, 1976) is amenable to experimental investigation.

In contrast, experiments on drape folding have been designed even though interpretations of the field relations are still controversial [compare Stearns (1971) with Osterwald (1961) and Foose and others (1961)], and the boundary conditions are only partially known from field data. Furthermore, theoretical analyses had not been completed. Hafner (1951) and Sanford (1959) treated the general problem of vertical tectonics, but not the specific problem of drape-fold development. Thus, we had to depart from the ideal sequence of the scientific method. The basic hypothesis of formation of drape folds was developed from the field interpretation; the experiments were done; and then a theoretical analysis of the experiments was carried out. Finally, the results were compared with the field data.

We made no attempt to test a scaled model. The theory of properly scaled models is well developed (Bridgman, 1922; Hubbert, 1937), and clearly the choice of real rocks in the experiments precludes such scaling. The theories of partially scaled models (Spalding, 1963) are just developing and, at best, our experiments only approximate these conditions. The term "model" has many meanings, most of which do not apply to our experiments. In particular, geometric scaling seems to be inherent in the usage of structural geology. For this reason the term "model" is eschewed as misleading.

Let us consider some aspects of the specific scaling conditions. It is obvious that a 10-by-20 laboratory specimen is very different in size from a natural fold which may be many kilometres long. The advantage of the theoretical analysis is to permit an evaluation of this geometric scale difference. Very similar stress-trajectory fields are deduced from petrofabric examinations of the laboratory specimens and elastic solution of the relevant boundary-value problem (Friedman and others, 1976a, p. 1064; Min, 1974; Gangi and others, 1977). Min (1974) also showed that this field does not change with geometric scale. With this theoretical support, the application of petrofabric studies of our laboratory specimens (for example, the problem of apparent volume changes along cross sections of the deformed specimens) to natural features can be attempted more confidently. The vast difference in size between the laboratory experiments and nature results in a great difference in body forces. Obviously those problems in which body forces are significant cannot be treated in our experiments. The stratigraphic sequence overlying the forcing block of a natural drape fold cannot be scaled in detail in the laboratory. At best, one can only grossly idealize large sequences of rock (see Weinberg, this volume). Cataclasis may be the significant mechanism in the laboratory geometry, whereas bedding-plane slip may be the easiest deformation mode in the field. It is necessary to extrapolate mechanisms very cautiously. At the microscopic scale in a sandstone, a single grain of quartz may behave similarly in the laboratory and in nature, for its immediate surroundings are the same. But whether cataclasis occurs on the scale of grains in the laboratory or of blocks metres in size in the field may well be a function of the scaling of the rock sequence in the laboratory experiments. The sum total of the deformation will be the same in the laboratory and the field, but the relative contributions of the several possible deformation mechanisms may be very different. For instance, it would be expected that bedding-plane and interlayer slip would play the more important role in natural deformation, because the multiplicity of real bedding cannot be duplicated in the laboraotry.

Our experiments are not intended to represent depth to any specific geologic structure, nor the confining pressure, temperature, pore pressure, or time needed for deformation. These parameters all influence the active mechanisms of deformation. The test is designed to show the spectrum of possible behaviors, including ductility contrast, which the field geologist may try to fit to a natural counterpart.

Obviously, not all features of the experimental drape folds are directly correlatable to specific field problems. For instance, the sandstone and limestone layers in the specimens are chosen for their petrofabric tractability and contrast in ductilities. This contrast in properties is reflected in the operative deformation mechanisms and, therefore, in some aspects of the final configuration of a specimen. Sandstone layers favor fracture and cataclasis, whereas the limestone and, to an even greater degree, polycrystalline halite allow homogeneous flow. The experiments provide insights into the stress and strain fields accompanying the drape-fold process, but the behavior of rock in real structures can only be simulated approximately in the laboratory (Weinberg, this volume), for it is impossible to reproduce a particular, natural rock sequence in the laboratory. Even so, despite their limitations, the experiments do indeed provide insights into the critical problems of the processes of folding and provide the geologists with powerful new tools for field investigations of drape folding.

CONCLUSIONS

From our additional experiments on faulted drape folds, the following conclusions may be made:

1. The corner drape folds show elongation of the upper part of the limestone layer around the corner. This elongation is confirmed by both microfracture orientation and strain determinations from calcite twin lamellae, both of which apparently contribute to the strain. Thickening of the lower limestone in the downthrown block occurs simultaneously with this thinning.

2. Cross sections and inferred σ_1 trajectories for progressive displacement along the same vertical fault in the forcing member clearly show that (a) deformation within the sedimentary veneer increases with increasing displacement along the vertical fault; (b) in the main, the faults within the veneer are high-angle reverse and curve both upward and downward in contrast to those in specimens with a 65°-dipping precut fault in the forcing member, which are thrusts and all concave

downward; (c) the sequence of faulting in the veneer progresses from the lower limestone in the downthrown block to the upper limestone in the upthrown one with faults seemingly initiated by the leading edge of the upthrown forcing block; and (d) the stress trajectories inferred from microfractures and faults have the same configuration at each magnitude of vertical displacement. They differ from those with 65°-dipping precut faults (Friedman and others, 1976a, Figs. 5, 6) only in being inclined at larger angles to the layering.

3. Mass-balance problems do exist in experimentally deformed specimens. Areas within the deformed regions do increase during faulting. Some of the increase in area is probably due to dilatancy, but mass transport of material must also occur. For limestones this transport might be explained by intragranular flow, whereas within sandstones, strike slip along faults which are curved along strike may be a mechanism. Even in controlled laboratory experiments, more work must be done to resolve the problem.

4. Experimental studies on rocks when combined with petrofabric and analytical analyses have given considerable insight into the processes involved in the formation of drape folds. Stress and strain fields have been successfully mapped. Although deformation mechanisms will vary with differing rock types, the critical areas on which to concentrate field studies have been defined, and some insights into solution of existing problems have been gained.

ACKNOWLEDGMENTS

We thank John Handin and Dave Stearns for their many stimulating discussions and critical review of the text, and J. D. Morse and K. A. Yagishita for their help with the petrofabric analysis of the corner drape. Our research on folding has been generously supported by National Science Foundation Grants GA-2332, GA-36127X, and DES74-22954.

REFERENCES CITED

Bridgman, P. W., 1922, Dimensional analysis: New Haven, Conn., Yale Univ. Press, 112 p.

Conrad, R. E., II, and Friedman, M., 1976. Microscopic feather fractures in the faulting process: Tectonophysics, v. 33, p. 187–198.

Currie, J. B., Patmode, A. W., and Trump, R. P., 1962, Development of folds in sedimentary strata: Geol. Soc. America Bull., v. 73, p. 461–472.

Fahy, M. F., 1976, Are any folds buckles? Geol. Soc. America Abs. with Programs, v. 8, p. 19–20.

Foose, R. M., Wise, D. W., and Garbarini, G. S., 1961, Structural geology of the Beartooth Mountains, Montana and Wyoming: Geol. Soc. America Bull., v. 72, p. 1143–1172.

Friedman, M., 1963, Petrofabric analysis of experimental deformed calcite-cemented sandstones: Jour. Geology, v. 71, p. 12–37.

——1969, Structural analysis of fractures in cores from the Saticoy Field, Ventura County, California: Am. Assoc. Petroleum Geologists Bull., v. 53, p. 367–389.

——1975, Fracture in rock: Rev. Geophysics, v. 13, p. 352–358.

Friedman, M., Handin, J., Logan, J. M., Min, K. D., and Stearns, D. W., 1976a, Experimental folding of rocks under confining pressure: Pt. III. Faulted drape folds in multilithologic layered specimens: Geol. Soc. America Bull., v. 87, p. 1049–1066.

Friedman, M., Teufel, L. W., and Morse, J. D., 1976b, Strain and stress analyses from calcite lamellae in experimental buckles and faulted drape folds: Royal Soc. London Philos. Trans., ser. A., v. 283, p. 87–107.

Gallagher, J. J., Sowers, G. M., and Friedman, M., 1970, Photomechanical model studies relating to fracture in granular rock aggregates: Geol. Soc. America Abs. with Programs, v. 2, p. 285.

Gallagher, J. J., Jr., Friedman, M., Handin, J., and Sowers, G. M., 1974, Experimental studies relating to microfracture in sandstone: Tectonophysics, v. 21, p. 203–247.

Gangi, A. F., Min, K. D., and Logan, J. M., 1977, Experimental folding of rocks under confining pressure: P. IV. Theoretical analysis of faulted drape-folds: Tectonophysics (in press).

Griggs, D. T., 1936, Deformation of rocks under high confining pressures, I. Experiments at room temperatures: Jour. Geology, v. 44, p. 541–577.

Groshong, R. H., Jr., 1972, Strain calculated from twinning in calcite: Geol. Soc. America Bull., v. 83, p. 2025–2038.

——1974, Experimental test of least-square strain gage calculation using twinned calcite: Geol. Soc. America Bull., v. 85, p. 1855–1864.

Hafner, W., 1951, Stress distribution and faulting: Geol. Soc. America Bull., v. 62, p. 373–398.

Handin, J., and Hager, R. V., Jr., 1957, Experimental deformation of rocks under confining pressure; Tests at room temperature on dry samples: Am. Assoc. Petroleum Geologists Bull., v. 41, p. 1–50.

——1958, Experimental deformation of sedimentary rocks under confining pressure: Tests at high temperature: Am. Assoc. Petroleum Geologists Bull., v. 42, p. 2892–2934.

Handin, J. W., Hager, R. V., Jr., Friedman, M., and Feather, J., 1963, Experimental deformation of rocks under confining pressure: Pore pressure tests: Am. Assoc. Petroleum Geologists Bull., v. 42, p. 717–755.

Handin, J., Friedman, M., Min, K. D., and Pattison, L. J., 1976, Experimental folding folding of rocks under confining pressure: Buckling of single-layer rock beams: Am. Geophys. Union Geophys. Mon. 16, p. 1–28.

Handin, J., Friedman, M., Min., K. D., and Pattison, L. J., 1976, Experimental folding of rocks under confining pressure: Pt. II. Buckling of multilayered rock beams: Geol. Soc. America Bull., v. 87, p. 1035–1048.

Heard, H. C., 1960, Transition from brittle to ductile flow in Solenhofen limestone as a function of temperature, confining pressure and interstitial fluid pressure, in Griggs, D., and Handin, J. W., eds., Rock deformation: Geol. Soc. America Mem. 79, p. 193–226.

——1963, Effect of large changes in strain rate in the experimental deformation of Yule marble: Jour. Geology, v. 71, p. 162–195.

Hubbert, M. K., 1937, Theory of scale models as applied to the study of geologic structures: Geol. Soc. America Bull., v. 48, p. 1459–1520.

Means, W. D., 1973, Folding of wet salt-mica specimens at low strain rate [abs.]: EOS (Am. Geophys. Union Trans.), v. 54, p. 457.

Means, W. D., and Williams, P. F., 1972, Crenulation cleavage and faulting in an artificial salt-mica schist: Jour. Geology, v. 80, p. 569–591.

Min, K. D., 1974, Analytical and petrofabric studies of experimental faulted drape folds in layered rock specimens [Ph.D. dissert.]: College Station, Texas A&M Univ., 90 p.

Osterwald, F. W., 1961, Critical review of some tectonic problems in the Cordilleran foreland: Am. Assoc. Petroleum Geologists Bull., v. 45, p. 219–237.

Patterson, M. S., and Weiss, L. E., 1962, Experimental folding of rocks: Nature, v. 195, p. 1046–1048.

——1966, Experimental deformation and folding in phyllite: Geol. Soc. America Bull., v. 77, p. 343–373.

——1968, Folding and boudinage of quartz-rich layers in experimentally deformed phyllite: Geol. Soc. America Bull., v. 79, p. 795–812.

Sanford, A. R., 1959, Analytical and experimental study of simple geologic structures: Geol. Soc. America Bull., v. 70, p. 19–52.

Sowers, G. M., Savage, W. Z., and Heinze, W. D., 1972, Criteria for separating stable and unstable folding, I. [abs.]: EOS (Am. Geophys. Union Trans.), v. 53, p. 523.

Spalding, D. B., 1963, The art of partial modeling, in Ninth symposium on combustion: New York, Academic Press, p. 833–846.

Stearns, D. W., 1971, Mechanisms of drape folding in the Wyoming province: Wyoming Geol. Assoc., 23rd Ann. Field Conf., Guidebook, p. 125–144.

Stearns, M. T., and Stearns, D. W., 1978, Geometric analysis of multiple drape folds along the northwest Big Horn Mountains, Wyoming, *in* Matthews, V., III, ed., Laramide folding associated with basement block faulting in the Western United States: Geol. Soc. America Mem. 151 (this volume).

Stearns, D. W., and Weinberg, D. M., 1975, A comparison of experimentally created and naturally formed drape folds: Wyoming Geol. Assoc., 27th Ann. Field Conf., Guidebook, p. 159–166.

Teufel, L. W., 1975, Strain analysis of twinned calcite for two experimentally superposed deformations of Indiana Limestone: Geol. Soc. America Abs. with Programs, v. 7, p. 240.

Turner, F. J., 1953, Nature and dynamic interpretation of deformation lamellae in calcite of three marbles: Am. Jour. Sci., v. 215, p. 276–298.

Weinberg, D. M., 1978, Some two-dimensional kinematic analyses of the drape-fold concept, *in* Matthews, V., III ed., Laramide folding associated with basement block faulting in the Western United States: Geol. Soc. America Mem. 151 (this volume).

MANUSCRIPT RECEIVED BY THE SOCIETY JUNE 27, 1977
MANUSCRIPT ACCEPTED AUGUST 25, 1977

Printed in U.S.A.

Geological Society of America
Memoir 151

Laramide folding associated with basement block faulting along the northeastern flank of the Front Range, Colorado

Vincent Matthews III
David F. Work
*Amoco Production Company
Security Life Building
Denver, Colorado 80202*

ABSTRACT

Folds and faults along the northeastern flank of the Front Range were examined using NASA (U.S. National Aeronautics and Space Administration) high-altitude photographs, published geologic maps, and new field studies. Detailed analysis of each structure reveals that block faulting of brittle basement is the dominant tectonic feature within the area. Twenty-three of the twenty-six folds were created by passive draping of sedimentary rocks over the edges of basement blocks. The other three folds were caused by local compression generated by anomalously tilted blocks.

INTRODUCTION

The Colorado Front Range contains one of the largest expanses of exposed Precambrian basement rocks in the Western United States. Along its northeastern flank, the Paleozoic and Mesozoic sedimentary strata are folded into a variety of Laramide structures including monoclines, symmetrical and asymmetrical anticlines and synclines, as well as domes and basins. The contact between the basement rocks and the folded sedimentary rocks is well exposed in many places. Thus, the area provides an unusual opportunity for studying the relationship of the Precambrian basement to the overlying old structures in the sedimentary rocks.

Previous workers have shown the relationship between the Precambrian rocks and the sedimentary strata in various ways. Most workers have depicted the Precambrian basement folded with the overlying sedimentary rocks. Some suggest that folds formed by drag along faults that cut both the basement and sedimentary

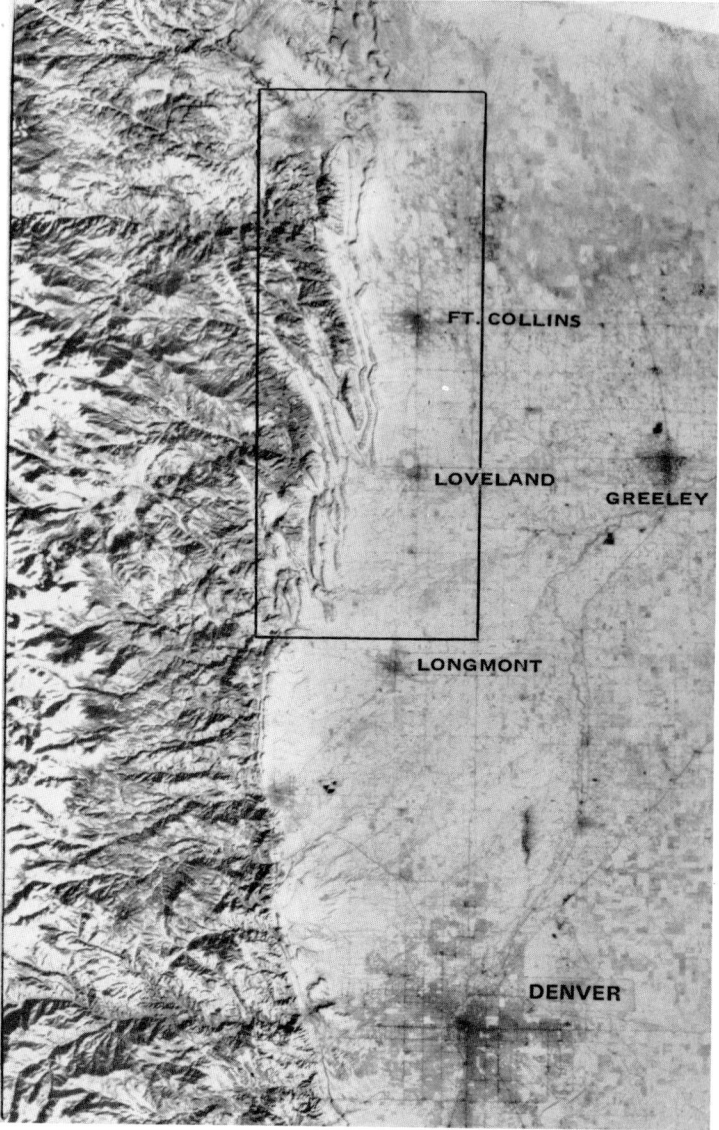

Figure 1. ERTS image of the northeastern flank of the Front Range, Colorado. Rectangular outline delineates the area discussed in this paper.

rocks. Warner (1956) suggested that folds were caused by draping of sedimentary strata over basement faults. Boos and Boos (1957) suggested that all three kinds of folds are present. In many places the evidence is equivocal. Indeed, workers studying individual, selected fold structures along the northeastern flank of the Front Range reached differing conclusions on the stresses responsible for the folding. Prucha and others (1965) concluded that a synclinal structure in sedimentary rocks formed from "differential, principally vertical, movement of three discrete basement fault blocks," whereas LeMasurier (1970) concluded that a nearby anticline formed from horizontal compressive stress that caused the Precambrian basement to arch into the core of the fold.

In our study we examined each of the individual fold structures along the northeastern flank of the Front Range (Fig. 1) in an effort to determine whether

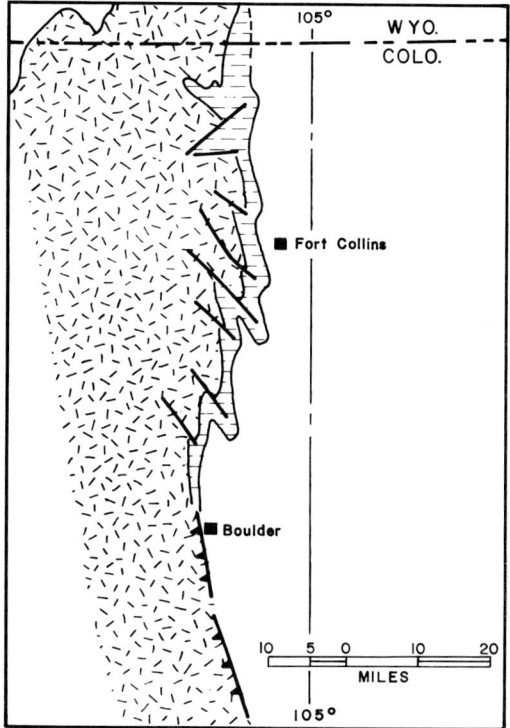

Figure 2. Generalized geologic map of the northeastern flank of the Front Range illustrating a common way of depicting the fault pattern, that is, a predominance of northwest-striking faults that die out along strike into the Precambrian core of the range (from Warner, 1956).

there is a common mechanism of formation. The study combined interpretations of ERTS (Earth Resources Technology Satellite) imagery, NASA high-altitude (60,000 ft; 18,000 m) aerial photography, and conventional low-altitude aerial photography with examination of published and unpublished geologic maps as well as new or additional field studies where necessary.

We conclude that the dominant tectonic process was differential, vertical uplift and rotation of basement fault blocks. Most fold structures in the study area were caused by draping of the sedimentary strata over differentially uplifted basement blocks. Three of the anticlines were formed by local compressive stress generated by anomalously tilted blocks.

NATURE OF DEFORMATION IN THE PRECAMBRIAN BASEMENT

The northeastern flank of the Front Range is usually described as a series of en echelon, north-to-northwest–trending folds and faults. The close association between the folds and faults has often been noted. Commonly, the northwest-trending faults have been depicted as extending into the Precambrian core of the Front Range and eventually dying out to the northwest along strike (Fig. 2). However, Boos and Boos (1957) showed that there are also a multitude of northeast-trending faults and that the faults in the area are interconnected.

The area included in our study (Fig. 3) can be divided into three different terranes: (1) a western terrane composed almost solely of Precambrian igneous and metamorphic rocks, (2) a central terrane where Paleozoic sedimentary rocks are in contact with the Precambrian basement, and (3) an eastern area of folded Paleozoic and Mesozoic sedimentary rocks with very little exposure of Precambrian basement.

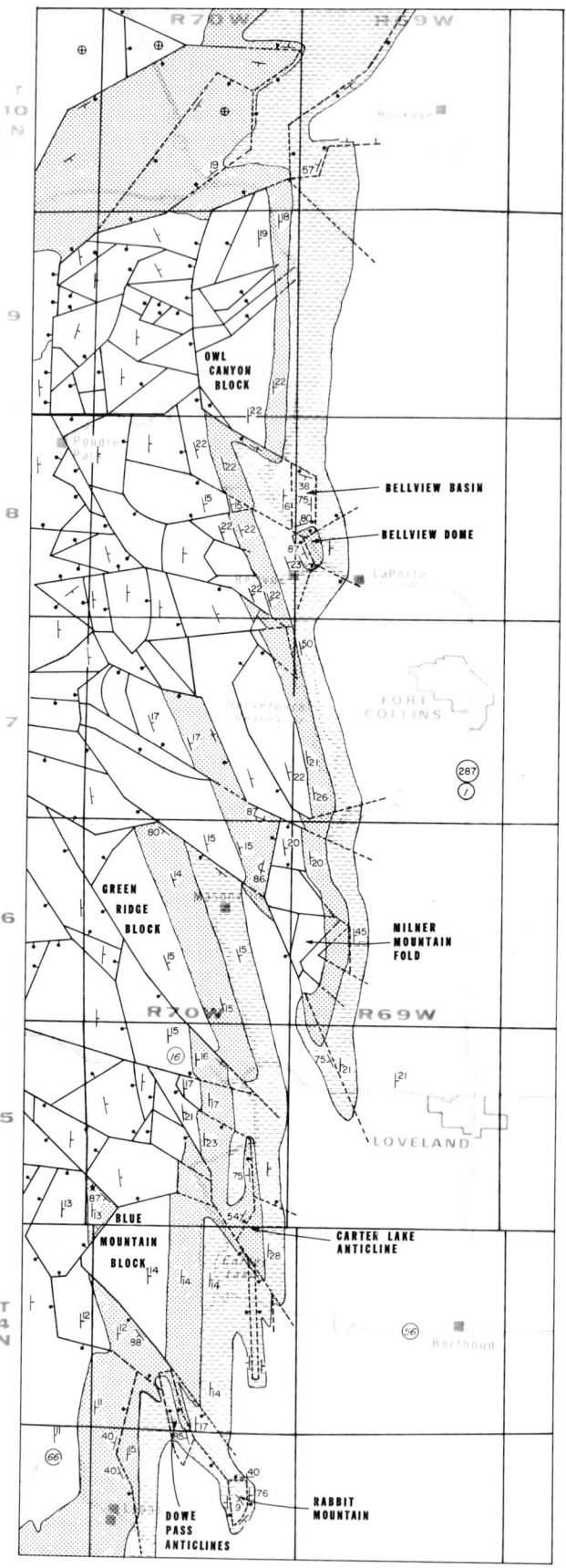

Figure 3. Regional geologic map of the northeastern flank of the Front Range. White areas on the west represent the Precambrian basement complex. Strike and dip symbols in the Precambrian area represent the attitude of the pre-Fountain erosion surface. Heavy lines are faults (ball on downthrown side). Heavy dashed lines represent inferred faults under the sedimentary strata. Shaded areas are Paleozoic strata; ruled areas are Mesozoic strata. Geology compiled by V. Matthews III.

We studied the western terrane of Precambrian basement using ERTS imagery and stereopairs of NASA U-2 photography. These studies of high-altitude photography disclosed that the topography of the western terrane reflects a series of discrete tilted fault blocks (Fig. 3 and 4). Using the stereopairs we delineated each major block, qualitatively determined its displacement relative to neighboring blocks, and qualitatively determined the attitude of the disrupted and tilted pre-Fountain erosion surface that is developed on the Precambrian rocks. Subsequently, quantitative displacements as well as quantitative attitudes of tilted blocks were determined from 7½-minute topographic quadrangle sheets. Comparisons of our photo-interpreted faults with faults mapped by conventional methods show a striking correlation. Almost all of the previously mapped faults coincided with our photo-interpreted faults. However, many segments of our photo-interpreted faults were not mapped as faults by earlier workers. In some instances where two or more workers had mapped the same area and not agreed on the location of faults, we found that their collective faults coincided with our photo-interpreted faults. And yet, there would be additional photo-interpreted faults in the area which field checking confirmed were indeed faults. In other words, the interpretation of high-altitude photography appears to give a more accurate picture of the extent of Laramide faulting in the western terrane than conventional geologic mapping. There are probably several factors which explain this.

First, the topographic expression of the differentially uplifted and rotated blocks, which is so obvious on the high-altitude stereopairs, is difficult to detect on 7½-minute topographic maps. Second, many of the shear zones at the margins of the blocks are in valleys and thus are commonly concealed by alluvium. Third, although the displacement on the faults is large, the separation of beds across the fault is small because of the geometric relations of the faults and offset strata, that is, the faults are steeply dipping and the dikes and metamorphic layering are steeply dipping.

During the Laramide orogeny the basement–sedimentary rock interface typically was buried to depths of 3,500 to 4,000 m. Laboratory experiments predict that basement rocks should deform brittlely (Borg and Handin, 1966; Heard, 1962) at equivalent confining pressures. Indeed, our study of the western terrane shows that during the Laramide orogeny, deformation of the northeastern flank of the Front Range can best be characterized as brittle deformation. The basement was broken into discrete blocks that were differentially uplifted and tilted toward the east. Paleomagnetic data support this conclusion (Kellogg, 1973).

We recommend that in regard to the Front Range the term "en echelon belt" be dropped from usage because it incorrectly describes the tectonic pattern along the northeastern flank of the Front Range.

RELATIONSHIP OF PRECAMBRIAN BASEMENT BLOCKS TO OVERLYING SEDIMENTARY STRATA

Having determined that the pattern of deformation in the Precambrian basement of the western terrane is one of differential uplift and rotation of rigid blocks, we then examined the central terrane where the Precambrian basement is in contact with the Paleozoic sedimentary rocks.

The sedimentary strata are planar over most of the central terrane, and deformation in the strata is restricted to narrow, linear zones. Where the tilted erosion surface on the Precambrian rocks projects under the sedimentary strata to the east, the strata are planar and parallel the attitude of the erosion surface within 1° or 2° of dip (Fig. 3). Where the edges of the fault blocks project under the sedimentary cover, the sedimentary strata are deformed (Fig. 3).

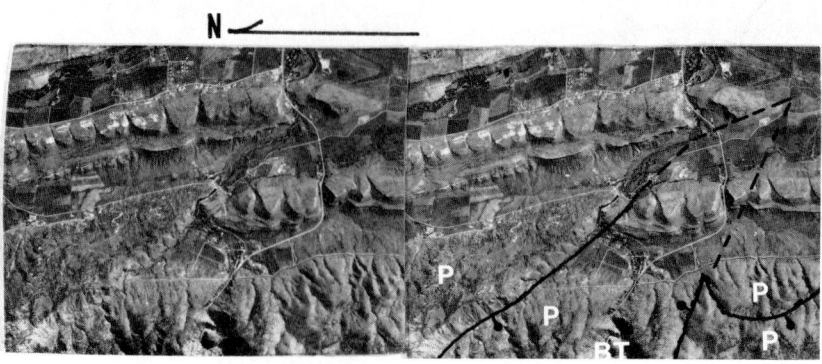

Figure 4. Examples of fault-block topography in Precambrian rocks along the northeastern flank of the Front Range. (a) Stereopair of NASA U-2 photographs near mouth of Big Thompson canyon (BT) illustrating that topography in western terrane resembles a series of tilted fault blocks. Note east-dipping, planar surface on Precambrian rocks (P) and that deformation in sedimentary strata to the east is confined to the block edges. The surface on the Precambrian rocks is best preserved near the Precambrian basement-sedimentary rock contact where the surface has recently been exhumed. To the west, the surface has been exposed to erosion for a longer period of time and is more deeply dissected. Even so, the geometry of the larger blocks can be readily determined for distances of 15 to 30 km westward from the contact (see also Fig. 4h). (b) View northward of 14°E-dipping surface (S) on Blue Mountain block.

Figure 4 (continued on facing page). (c) Oblique aerial view northward showing tilted (15°E) erosion surface (S) on Green Ridge block (photograph courtesy of Lew Dakan Photos). (d) View looking north at southern end of Owl Canyon block showing tilted (22°E) egosion surface (S). (e) View looking north at fault block 5 km north of Carter Lake showing 21°E-dipping erosion surface (S) and southwest-facing, fault-line scarp (F). (f) View northward of tilted erosion surface (S) on Milner Mountain. (g) Oblique aerial view northward of three differentially uplifted and rotated fault blocks (A, B, and C) near southwest end of Horsetooth Reservoir. (h) View southward from Colorado-Wyoming border showing tilted fault block just east of Owl Canyon block. Although this block is deeply dissected, the block geometry is still apparent (S = tilted erosion surface; F = fault-line scarp).

The deformation in the layered sedimentary rocks is manifested in two ways. The lower part of the Fountain Formation is faulted with the basement rocks that immediately underlie it. However, the faults die out upward through the Fountain Formation. From the Ingleside Formation upward through the Mesozoic strata, the deformation is manifested as an anticline. The basement faults consistently project vertically and laterally under the hinges of the anticlines. An examination of the entire length of the contact between the basement and layered rocks reveals that all deformation is restricted to block edges and all block edges project along strike into deformed sedimentary rocks (Fig. 3). Two of the larger blocks nicely illustrate these relationships.

Owl Canyon Block

The Owl Canyon block (Figs. 3 and 5) is a large (13 × 5 km) east-dipping block that is broken into several secondary blocks. From north to south, each secondary block is rotated slightly more to the east than the one to the north so that the northern part of the block dips 19°E whereas the southern part of the block dips 22°E (Fig. 4d). The displacement along the western edge of the block is approximately 300 m.

Except for the slight, but detectable, changes over the edges of the secondary blocks, the Paleozoic strata are planar and east-dipping over a north-south distance of 13 km. At the southern end of the block the east-dipping strata abruptly change attitude over a few tens of metres to strike N75°W and dip nearly vertical. The southern edge of the block projects along strike right under the hinge of the anticline, and we interpret the anticline to have resulted from draping over the edge of the block. However, in this area one could debate that the fold is a result of drag because the area to the south of the anticline is an alluvium-covered valley and the true structural relations are not revealed. Although the hinge of the fold is right along strike with the southern edge of the block, other workers have preferred to divert the fault farther south into the alluvium-filled valley just south of the anticlinal hinge (Boos and Boos, 1957; Hunter, 1955; Braddock and others, 1973a). However, at the northern end of the block the relationships are much better exposed. The planar east-dipping strata abruptly change attitude from due north, 18°E to N60°E, 50°NW; the northern edge of the block projects along strike under the hinge of the anticline. Moreover, at this end of the block it is not possible to argue for a drag-fold origin for the anticline because the Ingleside Formation is continuously exposed and clearly is not faulted (Fig. 6). On this northern end of the block the anticline is caused by drape over the edge of the tilted fault block.

Blue Mountain Block

The Blue Mountain block (Fig. 3) is one of the larger blocks in the area (50 km^2). The erosion surface on the tilted block of Precambrian basement dips 14°E

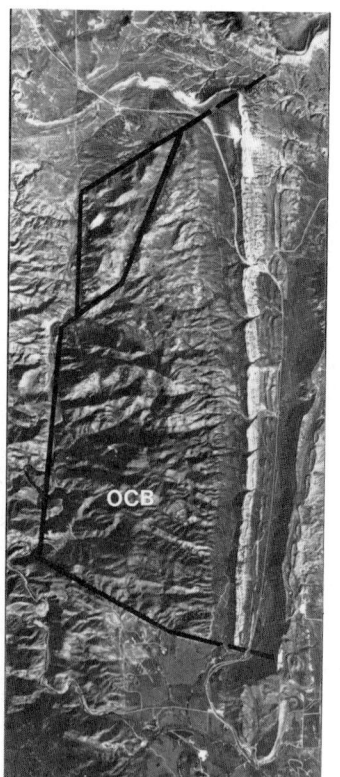

Figure 5. NASA photograph of Owl Canyon block (OCB). North is toward the top.

Figure 6. NASA U-2 photograph of north end of Owl Canyon block. Strata of Ingleside Formation drape over north end of block and are unfaulted. (P = Precambrian Sherman Granite. I = Ingleside Formation).

(Fig. 4b). Where the tilted surface of the block projects under the sedimentary rocks, the strata are planar over an area of 28 km^2 (Figs. 3, 7, and 8).

To the north the beds become horizontal on Flatiron Mountain where they pass onto another block. At the southern edge of the Blue Mountain block along the Little Thompson River the strata abruptly become vertical and strike N40°W (Fig. 8). Here, the relationship between the basement fault and the folds in the upper

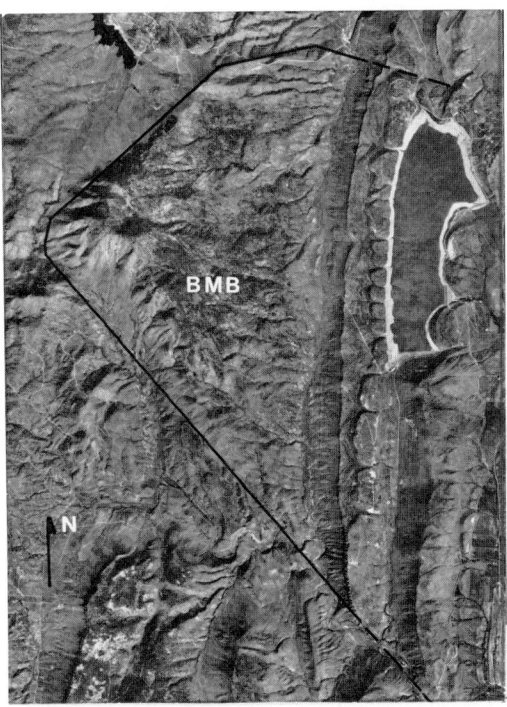

Figure 7. NASA U-2 photograph of planar, east-dipping strata resting on Blue Mountain block (BMB).

Figure 8. View northward of southwestern edge of Blue Mountain block. Planar, gently dipping Lyons (Lp) strata are resting passively on tilted block. Steeply dipping Lyons strata (Ld) are draped over the southwestern edge of the Blue Mountain block.

stratigraphic units is quite clear. Figure 9 shows the Fountain Formation in fault contact with vertical beds of the Lyons Sandstone. This fault between the Fountain and the vertical Lyons strata can be traced southeast for 0.75 km where it dies out in an anticline in the Lyons (Fig. 10). The anticline extends up through the Mesozoic strata.

Origins of the Folds in the Central Terrane

In areas of limited or poor exposure within the central terrane, the origin of the folds might be debatable. At some outcrops an origin by drag folding or drape folding might be argued equally well. However, where the exposures are good and the relationships are clear, the drag-fold model is not a viable explanation for the origin of the folds.

The concept that the basement is folded with the overlying sedimentary rocks does not appear to have any validity in the central terrane. Everywhere along the basement-sedimentary rock contact the basement is brittlely deformed, and the geometry of the folds reflects the block configuration in the basement. The tilted blocks create asymmetrical folds in the overlying sedimentary strata. North of the area discussed in this paper, the basement blocks are differentially uplifted but are not tilted. Where these untilted blocks pass under the sedimentary strata to the east, a series of drape-fold monoclines are formed (Matthews and Sherman, 1976).

Our analysis of the structures in the central terrane shows that one model for the origin of the folds can best explain their characteristics—a model of drape folding or forced folding. The important tectonic process is the differential uplift and tilting of the basement fault blocks. The geometry of the fold in the sedimentary cover simply reflects the shape of the basement forcing member.

ANALYSIS OF FOLDED SEDIMENTARY ROCKS OF THE EASTERN TERRANE

Using the model of block faulting and drape folding developed in the western and central terranes, we attempted to apply it to the eastern terrane where the

Figure 9. View along southwest edge of Blue Mountain block showing vertical Lyons strata (L) in fault contact with planar beds of Fountain Formation (F). Strata abruptly become planar in right side of outcrop.

Figure 10. View of southwest edge of Blue Mountain block showing how fault between Fountain (F) and Lyons strata projects vertically and laterally under unfaulted anticline (A) in Lyons strata.

basement is mostly concealed. Our interpretations of the folding are based on the following tenets or assumptions: (1) The pre-Fountain erosion surface that developed on the Precambrian basement is essentially planar and was horizontal at the time of deposition of the sedimentary rocks. (2) The Paleozoic and Mesozoic sedimentary formations were deposited parallel to the erosion surface. (3) At the time of Laramide deformation the basement–sedimentary rock interface was under approximately the same confining pressure (depth of burial) in the western, central, and eastern terranes of the study area. (4) The pattern of block faulting in the buried Precambrian of the eastern terrane is similar in style and scale to that observed farther west. (5) Where dips in the sedimentary rocks are less than 30° and are planar over large distances, the strata are resting passively on a tilted fault block. (6) Where dips exceed 30° in a narrow, linear zone, the rocks are draped over the faulted edge of a basement block.

In the following sections the model is applied to five areas in the eastern terrane: Carter Lake anticline, Belleview dome and basin, Milner Mountain anticline, Rabbit Mountain "box fold," and the Dowe Pass anticlines (Fig. 3).

Carter Lake Anticline

East of Carter Lake Reservoir is a doubly plunging, asymmetrical anticline that has no Precambrian basement exposed in its core (Fig. 11). The structure is exposed over an area 6.5 km long and 3 km wide at its widest point. The east limb of the anticline is gently dipping and planar. The west limb is steeply dipping to overturned. The attitude of the east limb is uniform (due north, 27°E) over a distance of 4 km, whereas the attitude of the west limb ranges widely over the same distance. Strike of the west limb ranges from N60°W to N30°E, and the dip angle varies from 53° through vertical to 74° overturned.

We divided the fold into four segments (A, B, C, and D), each characterized by a different fold-axis geometry. The highest point on the fold axis is at the bend in the southern part of the fold (Fig. 11b). From the bend the axis plunges both northeast and southeast at the uniform rate of 100 m/km. One kilometre northeast of the bend the axis abruptly flattens out into segment B. In segment B the axis trends due north and is horizontal for 1 km; it then plunges abruptly to segment C at a rate of 200 m/km. The axis trends due north and is horizontal for 1.5 km in segment C but then plunges abruptly to segment D at a rate of 450 m/km over a distance of 0.5 km. The axis is horizontal for 1 km in segment D, but plunges again at the northernmost exposure of the fold.

Figure 11f illustrates the block configuration that could cause the observed features of the fold. If sedimentary rocks were draped over these blocks, they would take the shape of an asymmetric, doubly plunging anticline with a steep western limb and gentle eastern limb. The strike and dip of the eastern limb would be planar and uniform except for a minor deviation where the strata pass from block A to B. Because of the configuration of the front edge of the uplifted block, the western limb would be steep and have diverse and abrupt changes in strike.

Belleview Dome and Basin

Northeast of Belleview is a four-sided structural dome (Fig. 12). Just north of it is a four-sided structural basin. These two structures each have one gently dipping, planar limb and three steeply dipping limbs.

The planar, east limb of the Belleview dome dips 17°E. The other three limbs dip steeply to the north, south, and west, respectively (Fig. 12). Most previous

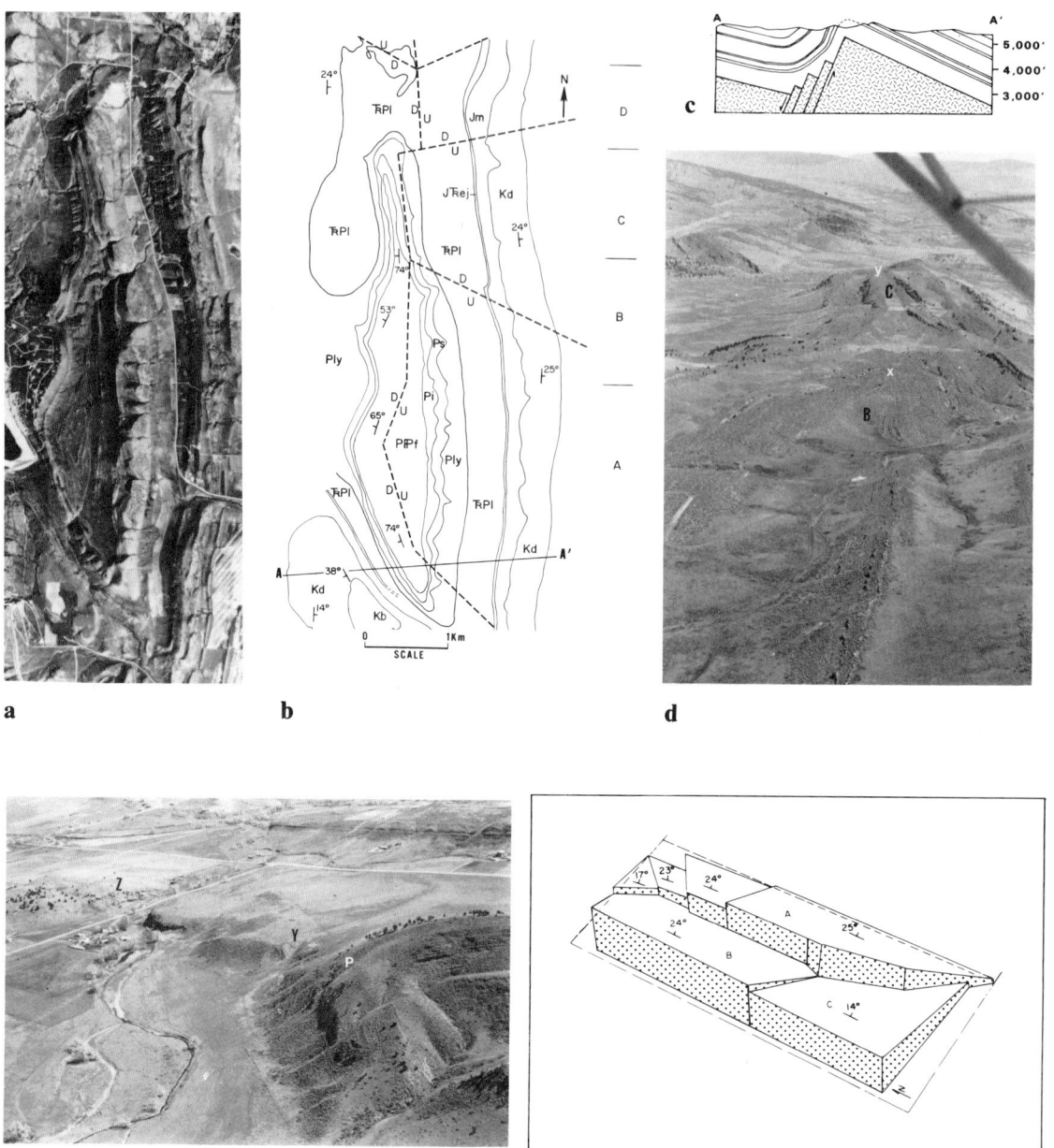

Figure 11. (a) NASA U-2 photograph of Carter Lake anticline. (b) Geologic map of Carter Lake anticline showing subdivision of fold into four segments (A, B, C, D; see text for explanation). Modified from Rowlinson (1957). Base map, USGS 7½-minute Carter Lake quadrangle. (c) Structural cross section A-A'. (d) Oblique aerial photograph looking north at segments B and C of the Carter Lake anticline. At point X, the axis of the fold plunges abruptly at 200 m/km, then is flat in segment C for 1.5 km, and plunges abruptly at 450 m/km at point Y. (e) Oblique aerial photograph showing plunge of axis at 450 m/km (P); plunge then flattens out from point Y to Z. (f) Block diagram showing interpretation of differentially uplifted and rotated blocks responsible for the Carter Lake anticline.

workers showed a fault along the west flank of the dome although exposures along the west flank are poor except for one outcrop of steeply dipping (N20°W, 88°SW) Paleozoic rocks. Braddock and others (1973a) showed a fault west of this outcrop, whereas Boos and Boos (1957) and Hunter (1955) placed it east of this outcrop. Our interpretation is that the basement and lower Fountain Formation are faulted but that the upper Fountain and younger strata are folded (Fig. 12c). The geometry of this dome appears to reflect an uplifted basement block that is shaped like block D in Figure 12g. The gently dipping, planar east limb is resting passively on the tilted block and the steeply dipping, north, west, and east limbs are draping over the edges of the uplifted block.

The planar west limb of the basin dips 15°E. The north, east, and south limbs are steeply dipping (Fig. 12). The axial part of the basin dips 8°S. Within the basin are several small-scale anticlines and synclines that are present only in the upper Dakota strata. These are superficial features that are probably related to landsliding on the weak Dakota shales. Similar folds and block-glide landslides are abundant at this stratigraphic horizon in nearby localities (Braddock and Eicher, 1962; Braddock and others, 1973a). The geometry of the basin suggests an underlying block configuration like that shown in Figure 12g. Block B would underlie the axis of the basin, and the gentle west limb would rest on block C. The steeply dipping north and east limbs would result from draping over the edges of the Owl Canyon block (Fig. 12f). The south limb would result from draping over the north edge of the block underlying Belleview dome.

It is significant that these two structures each have one planar, gently dipping limb and three steeply dipping limbs. This is exactly the geometry to be expected from the drape-fold model. Indeed, two other structural basins in the area have similar geometries. One of the basins is located 2 km northeast of Masonville (Fig. 3; Prucha and others, 1965), and the other is just northwest of the Blue Mountain block (Fig. 3; Kelly, 1967). Both of these basins are three-sided. Both have two steeply dipping limbs and one planar, gently dipping limb. In both places erosion has exposed the tilted block of Precambrian basement upon which the planar gently dipping limb is resting. All of the drape-fold structures in this area, whether they are anticlines, synclines, domes, or basins, have one planar gently dipping limb and one or more steeply dipping limbs.

Rabbit Mountain "Box Fold"

The Rabbit Mountain structure is a broad asymmetrical anticline with rocks no older than Mesozoic exposed in its core. Masters (1957) described it as a box fold. The steeply dipping west limb is separated from the crest of the anticline by a 1,200-m-wide area where the strata are planar and dip 9°SW (Fig. 13a). At the crest of the anticline the broad planar part of the structure abruptly gives way to a steeply dipping, east limb (Fig. 13c). At the north end of the structure the strata dip steeply to the north. In the southern part of Rabbit Mountain the broad, gently dipping, planar part of the structure narrows, and the steep limbs on either side merge.

The geometry of this fold leads us to conclude that it is underlain by a basement fault block that is uplifted and tilted toward the southwest (Fig. 13e). The large area of planar, southwest-dipping strata represents the attitude of the tilted erosion surface on the block of Precambrian basement. The steeply dipping strata on all sides of the planar area are draping over the edges of the block.

Within our study area, Rabbit Mountain block is the only block that is tilted toward the west. The block to the west of Rabbit Mountain is tilted toward the

Figure 12. (a) NASA U-2 photograph of Belleview dome (D) and basin (B). (b) Geologic map of area around Belleview dome and basin (PreЄ = Precambrian basement complex; P|Pf = Fountain Formation; Pi = Ingleside Formation; Ps = Satanka Formation; Ply = Lyons Sandstone; ŦPl = Lykins Formation; J-Ŧrej = Entrada and Jelm Formations; Jm = Morrison Formation; Kd = Dakota Group). Modified from Braddock and others (1973a). Base map, USGS LaPorte 7½-minute quadrangle.

Figure 12 (continued on facing page). (c) East-west structural cross section A-A′. (d) North-south structural cross section B-B′. (e) Oblique aerial photograph of Belleview dome. (f) View northward from U.S. 287 of east flank of Belleview basin showing planar beds of Dakota (Dp) resting on block A (Fig. 12g) and steep beds of Dakota (Ds) draping onto block B. (g) Block diagram showing interpretation of differentially uplifted and rotated basement blocks responsible for the dome (D) and basin (B) structure in the sedimentary strata.

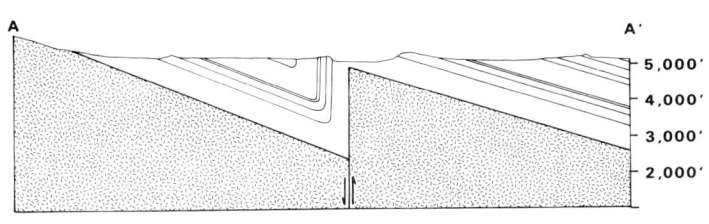

c

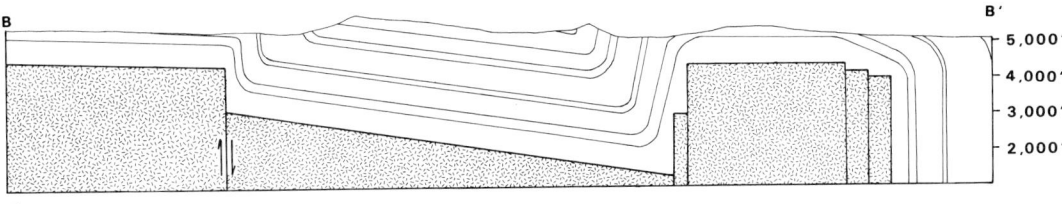

d

e

f

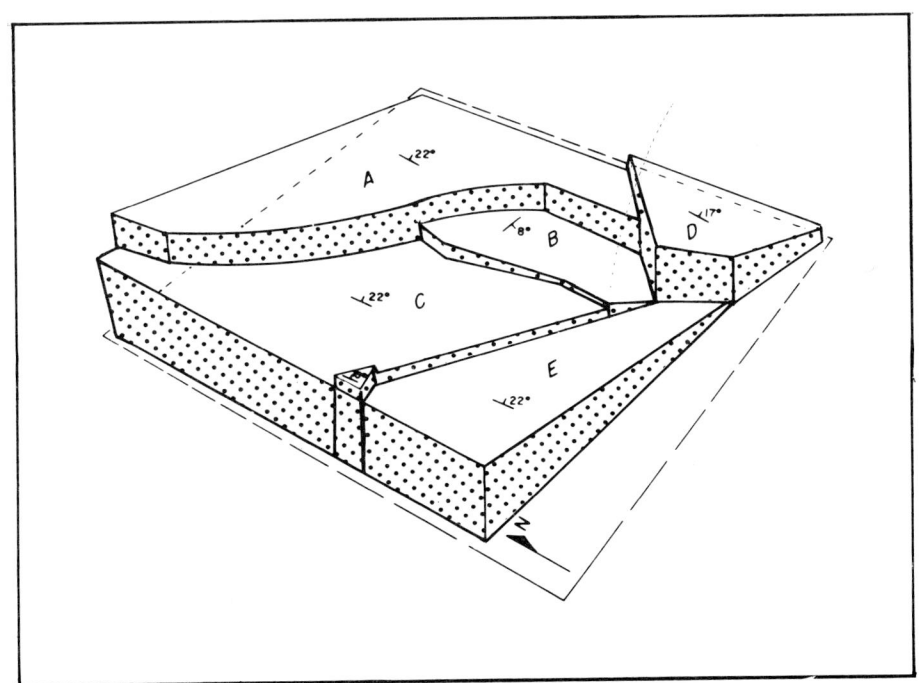

g

east. Therefore, these two blocks have opposing tilts that at the time of deformation should create a state of compression in the strata between them. One might expect to find a compressional structure between the two blocks, and indeed there is an anticline along the western limb of Rabbit Mountain that is atypical of most other folds in the area.

A tight, symmetrical anticline parallels the northern part of the steep western limb of the Rabbit Mountain structure for 1,500 m (Figs. 13a and 13d). This anticline has a sharp hinge, and its limbs dip at 60°. The anticline is most pronounced at the northern end and dies out to the south as the throw on the west edge of the Rabbit Mountain block decreases. Clearly, the geometry of this anticline does not appear to reflect a buried basement block. The fact that it is along the west flank of the only west-tilted block in the area must be more than happenstance. We interpret it as a compressional anticline existing only in the sedimentary veneer over the two oppositely dipping blocks (Fig. 13b).

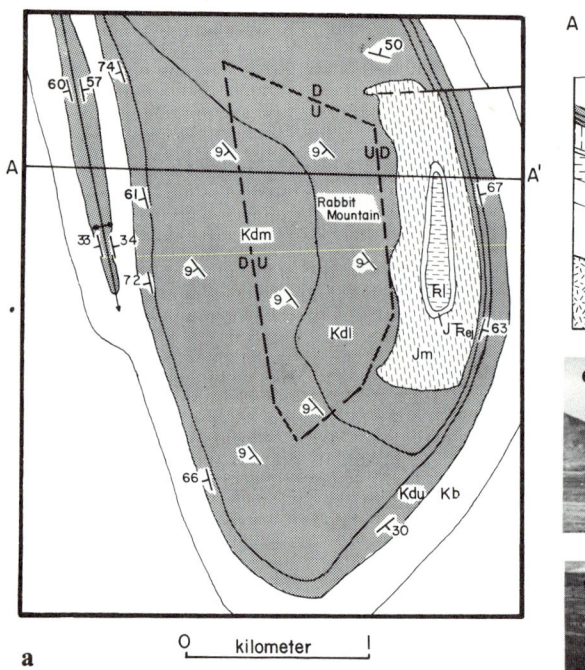

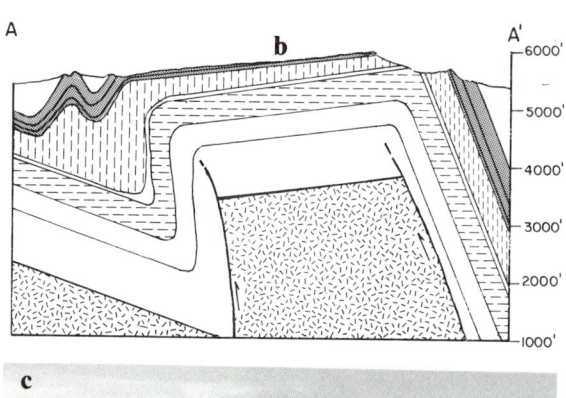

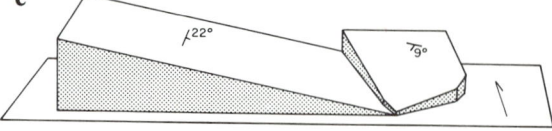

Figure 13. (a) Geologic map of Rabbit Mountain area (modified from Masters, 1957; base map, USGS 7½-minute Hygiene quadrangle). ŦRl = Lykins Formation; JŦRej = Entrada and Jelm Formations; Jm = Morrison Formation; Kdl = lower part of the Dakota Group; Kdm = middle part of the Dakota Group; Kdu = upper part of the Dakota Group; Kb = Benton Formation. (b) Structural cross section A-A'. (c) View northward of Rabbit Mountain showing steeply dipping (Ds) east limb and planar, southwest dip on Dakota sandstone (Dp). (d) Oblique aerial view southward of planar strata of Rabbit Mountain (B) and symmetrical anticline (A) along west side. (e) Block diagram showing interpretation of uplifted and rotated block that is responsible for Rabbit Mountain structures (drawn by T. Patton).

Milner Mountain Anticline (Loveland Fold, Big Thompson Anticline)

The Milner Mountain structure is an asymmetrical, plunging anticline located 5 km west of Loveland (Fig. 14). The anticline is 16 km long, plunges to the southeast, and has a core of Precambrian basement rocks. The Precambrian core of the structure is bounded on the west by the Milner Mountain fault that has a maximum displacement of 1,200 m. The northern boundary of the structure is an extension of the Buckhorn Creek fault and has <100 m of displacement.

The west limb of the structure dips steeply toward the southwest and in places is steeply overturned. In the southern part of the structure, the west limb is unfaulted. Near the southwestern exposure of the Precambrian core, the west limb is faulted. In the northern part of the structure the west limb is completely faulted out.

In most places the eastern limb dips gently toward the east. In the central part of the east limb the younger strata form a fairly smooth curve that is convex to the east, and the dips steepen to 45°.

The Precambrian rocks of the core are crisscrossed by many small Laramide faults. Maximum displacements on any one of these faults is 200 m, and most have much smaller displacements.

LeMasurier (1970) studied the area around the southernmost exposure of the Precambrian core and concluded that the features he observed were not compatible with the drape-folding model. To the contrary, our analysis of the Milner Mountain structure leads to the conclusion that the features of this structure are quite compatible with the drape-folding model.

Our interpretation is that the asymmetrical anticline in the sedimentary strata was caused by draping over a rigid block of Precambrian basement that was uplifted and rotated toward the east (Figs. 14c and 14d) about a N15°W-trending axis. Paleomagnetic data on Precambrian dikes in this structure support such a rigid-body rotation of the basement (Kellogg, 1973). Furthermore, as this large block was uplifted and rotated, it broke into many smaller secondary blocks that were themselves differentially uplifted (Fig. 14b). This secondary break-up of the large block is expressed in several ways at Milner Mountain.

A longitudinal profile of Milner Mountain (Fig. 14e) shows that the crest does not plunge uniformly to the southeast. Rather, there are several discrete steps that increase in elevation to the north. However, the northernmost step is lower than the next block to the south. The edges of these steps are shear zones, and the flat surfaces are expressions of the differentially uplifted secondary blocks.

The secondary blocks are also obvious along the eastern limb near the Precambrian rocks–Fountain Formation unconformity. From the nose of the fold northward to secondary block A (Fig. 14b), each successive block is uplifted relative to the one to its south. From secondary block A northward, each successive block is downdropped. Thus, secondary block A is uplifted more than the other secondary blocks along the unconformity in the east limb. The lower Fountain strata form straight-line segments that correspond to the various secondary blocks. Because displacement of the secondary blocks is not very large, the younger strata tend to smooth the straight-line segments into a curve. However, dips in any segment of the eastern limb are planar and uniform on any given block (except block A) for distances of 3 km eastward from the unconformity. Eastward of block A the dips increase to a maximum of 45°. These 45° dips in the east limb are present only between the projection of the edges of block A. We interpret these steep dips to be a result of draping over the buried eastern edge of secondary block A (Fig. 14b).

An interesting feature of this fold is the disharmonic folding in the Lykins

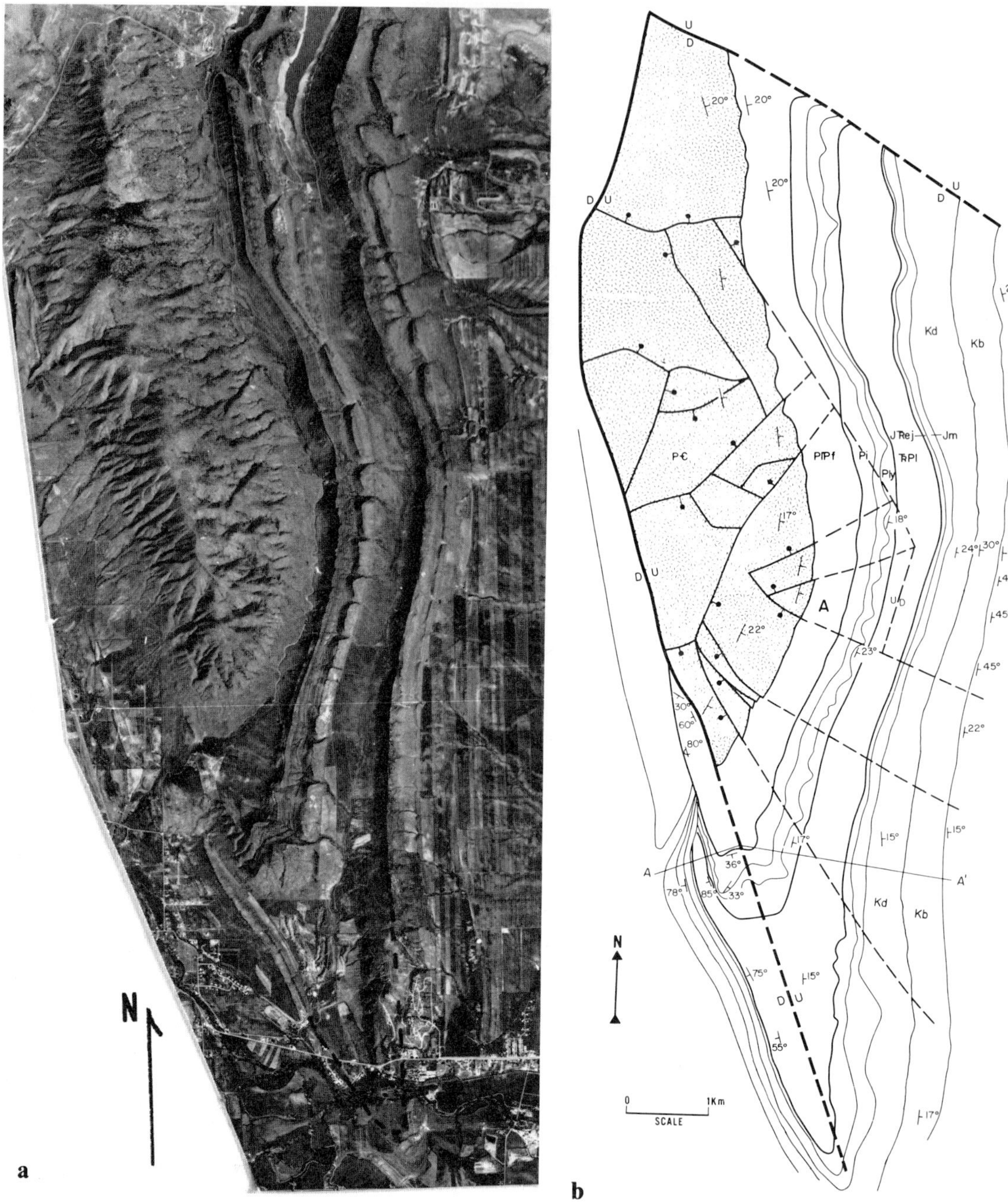

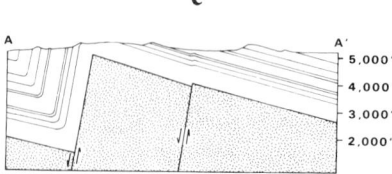

Figure 14. (a) NASA U-2 photograph of Milner Mountain anticline. (b) Geologic map of Milner Mountain anticline (modified from Braddock and others, 1970, 1973b). Base map, USGS 7½-minute Masonville and Horsetooth Reservoir quadrangles. PЄ = Precambrian basement complex; PIPf = Fountain Formation; Pi = Ingleside Formation; Ps = Satanka Formation; Ply = Lyons Sandstone; ᵼRPl = Lykins Formation; JᵼRej = Entrada and Jelm Formations; Jm = Morrison Formation; Kd = Dakota Group. (c) Structural cross section A-A′.

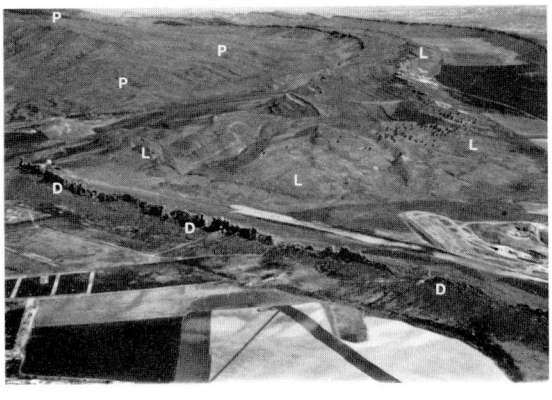 

d   e

Figure 14 (continued). (d) Oblique aerial view from southwest of Milner Mountain. P = Precambrian block tilted toward the east. L = Lyons Sandstone. D = Devil's Backbone hogback of Dakota sandstone that is the draped western limb of the Milner Mountain anticline. (e) View from east of longitudinal profile of Precambrian core of Milner Mountain anticline illustrating the steplike plunge of the crest.

Formation and sharpening of the hinge in the Mesozoic strata. The Paleozoic strata have two hinges and blocks 1, 2, and 3 (in the geometric terminology of Stearns, this volume) even though they are clastic rather than carbonate rocks. The lower Lykins strata conform to the geometry of the underlying Paleozoic strata, but the upper Lykins strata have only one sharp hinge. The single, sharp-hinge geometry persists up through the Mesozoic strata. Similar disharmonic folds occur in other folds along the Front Range but are not as well preserved. The origin of these disharmonic folds is unclear, but similar disharmonic folds are also present in many drape folds in Wyoming and are named "drape subsidiary folds" (Weinberg, this volume). Indeed, the disharmonic folding in the Wyoming drape folds commonly occurs in the Chugwater Formation, which is the stratigraphic equivalent of the Lykins Formation.

LeMasurier (1970) felt that the drape-folding theory was difficult to reconcile with the following relationships at Milner Mountain: (1) the arch of the Precambrian rocks–Fountain Formation contact lies mainly east of the zone affected by faulting; (2) the concentric structure of the fold, expressed by an increase in dip away from the fold axis, can be measured in the sedimentary section to at least 2.5 mi (4.17 km) east of the Big Thompson fault zone; and (3) there is no evidence either for a fracture system or for mass-flowage phenomena that relieved tension in the crest of the fold (LeMasurier, 1970, p. 432).

LeMasurier's (1970) concern that the arch of the Precambrian rocks–Fountain Formation contact is mainly east of the fault zone does not rule this structure out as a drape fold. Indeed, this situation is observed in many drape folds in Wyoming and is even to be expected. As pointed out by Stearns (1971), the sharp corners of the uplifting basement blocks are subjected to considerable stress and tend to break off and become rounded. Therefore, the crest of the fold would not be expected to be located exactly at the fault zone.

LeMasurier's (1970) contention that the concentric nature of the fold can be measured to at least 2.5 mi (4.17 km) east of the fault zone is somewhat misleading. The dips steepen *only* east of secondary block A (Fig. 14b). North and south of secondary block A the dips in the sedimentary section are planar and uniform for 3 km east of the Precambrian rocks–Fountain Formation contact. These relationships seem more compatible with drape folding than concentric folding.

LeMasurier's (1970) third objection is based on a misinterpretation of the results of Sanford's (1959) sand-box experiments. LeMasurier (1970) interpreted Sanford's (1959) work to indicate that strata should be thinned in the crest of a drape fold. However, Sanford (1959) specified that the material that he was deforming represented a 5-km-thick layer of uniform properties that was "perfectly elastic, linearly elastic, homogeneous, and isotropic." Clearly, this is not a description of the sedimentary strata at Milner Mountain or any other place in the Rocky Mountains. Sanford's (1959) work is important in understanding deformation in the basement but should not be applied to the anisotropic, ductile sedimentary package overlying the basement. Indeed, if thinning is present in a drape fold, it would be expected in the steep, draped limb of the fold and that is just where it is found in the Milner Mountain anticline.

Throw on the Milner Mountain fault increases from the southeast to the northwest. In the southeastern part of the west limb, thinning and faulting of the sedimentary section are not apparent. To the northwest the strata of the west limb become thinned, disrupted, and eventually are completely faulted out. As LeMasurier (1970) noted, the thickness of the west limb is systematically decreased. The attenuation is such that the incompetent units are completely removed; only the competent units are still present and in proper stratigraphic order (not shown in Fig. 14b because of scale). These relationships appear quite compatible with the interpretation that the Milner Mountain anticline is a drape fold.

Dowe Pass Anticlines

Dowe Pass (Fig. 15) is bounded on the east and west by symmetrical anticlines that are atypical of most of the folds in the study area. With the exception of these two anticlines, the configuration of most of the strata in the vicinity of Dowe Pass is due to one or more underlying tilted blocks. These two symmetrical anticlines are interpreted to have been caused by an underlying-block configuration that is unique to the rest of the region.

The relationships in the southern third of the map of the Dowe Pass area (Fig. 15a) indicate the usual drape-fold geometry. In the southwestern part of the area the Dakota Group is planar and dips 15°E. Toward the south-central part of the area, the dip steepens to 55° in a linear north-trending zone; this indicates drape off the back side of block A (Fig. 15c). In the southeastern part of the area the Dakota Group is planar and dips 15°E. Near the south-central part of the area the dips reverse and steepen to 85°SW in a linear northwest-trending zone; this indicates drape over the front edge of block C (Fig. 15c). The geometries in the southern part of the area indicate the presence of a V-shaped graben formed by the downdropping of block B (Fig. 15c).

The geometric relationships in the northern part of the mapped area are much the same as in the southern part except for the two closed, symmetrical anticlines. In the northwestern part of the area the Dakota Group is planar and dips 15°E. Toward the central part of the area the strata do not simply steepen as they did in the southern part of the area. Rather, the strata reverse dip and then form a closed to tight, nearly symmetrical anticline (Figs. 15a, 15b, and 15d). This anticline abruptly dies out on the northwest and southeast.

In the northeastern part of the area the strata are planar and dip 15°E. Toward the central part of the area the strata steepen to 55°E and form a closed symmetrical anticline (Figs. 15a and 15e). At the level of the Lyons Sandstone the axis of this anticline is 135 m higher than in the drape fold to the southeast (Fig. 15e).

The axes of these two symmetrical anticlines are parallel, and their north and south terminations are coincident. This arrangement suggests that the two anticlines are genetically related. To the south they both change into normal drape folds. In the area under the two folds the graben formed by downdropped block B (Fig. 15c) is much narrower than it was to the south. Our interpretation is that the narrow graben created a space problem in the overlying strata, and the excess material was accommodated in a compressional anticline over the edge of each bounding block.

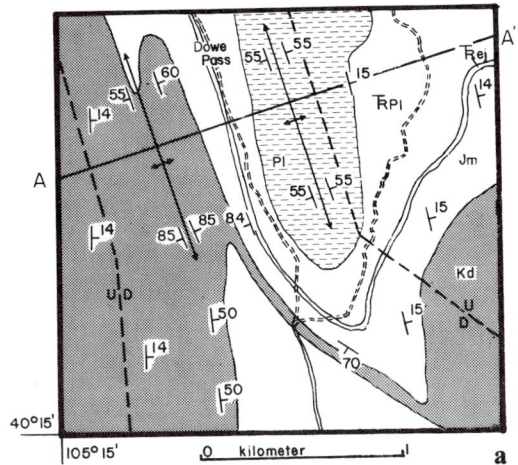

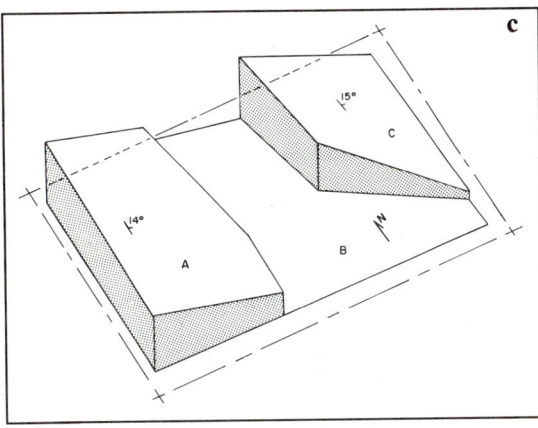

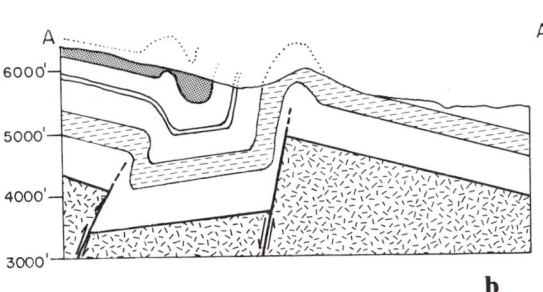

Figure 15. (a) Geologic map of Dowe Pass area (modified from Masters, 1957; base map, USGS 7½-minute Carter Lake quadrangle). Pl = Lyons Sandstone; ℞Pl = Lykins Fm; ℞Jej = Entrada and Jelm Formations; Jm = Morrison Formation; Kd = Dakota Group. (b) Structural cross section A-A'. (c) Block diagram showing interpreted configuration of basement blocks in the Dowe Pass area. (d) View northward along axis of symmetrical anticline west of Dowe Pass. The Dakota Group and Morrison Formation are exposed. (e) View northward along axis of symmetrical anticline east of Dowe Pass (DP). In foreground (A), strata of Lyons Sandstone are in normal asymmetrical drape-fold configuration. But geometry of fold changes abruptly to a closed, symmetrical anticline (B). Axis of anticline at the level of the Lyons Sandstone at point C is 135 m higher than at point A.

Figure 16. View southward of anticline just north of Lyons, Colorado. Strata of Lyons Sandstone are planar at A and B. At C strata of Lyons Sandstone are unfaulted and draped over a block with 200 m of displacement.

DISCUSSION

Thinning and Loss of Continuity in the Folded Strata

Unless there are detachment surfaces within the sedimentary column, the strata must be attenuated as they pass over the steep edges of the blocks. Most of the cross sections in this paper do not show thinning in the steep, draped limbs of the folds because the exact extent of thinning cannot be determined in most of the structures. In many structures it appears that thinning has occurred in the steep limb, but it isn't possible to measure the exact amount because of poor outcrop and/or obliteration of marker beds. It was pointed out in the section on the Milner Mountain anticline that increased displacement on the basement block causes thinning and eventual disruption of the strata in the steep limb.

The amount of displacement that is necessary to cause the strata of the steep limb to separate is also difficult to document. In many places where the strata of the steep limb are missing, it is not clear whether separation actually occurred or whether the strata are simply eroded through.

Complete exposure of a structure just north of Lyons (Fig. 16) shows that the Lyons Sandstone is continuously draped over a block with 200 m of displacement.

In other structures the strata are continuous at the present level of exposure but may be separated at depth. For instance, the strata in the steep limb of the Carter Lake anticline are thinned 20% relative to the gentle limb, and they are continuous. However, only the upper 100 m of the steep limb are exposed whereas the displacement on the basement blocks is almost 1 km. It is probable that the strata are broken through at depth.

Nature of Basement Deformation

It is noteworthy that all of the many workers who have studied Laramide deformation in the exposed Precambrian interior of the Front Range report only faulting (brittle deformation). None have proposed that the Precambrian rocks of the interior of the Front Range were folded during the Laramide orogeny. It is only when dealing with the buried Precambrian rocks along the eastern flank of the Front Range that workers are tempted to depict the basement as folded. And yet, during Laramide deformation the depth of burial for the Precambrian rocks of both areas was insufficient to cause ductile deformation of the basement.

Moreover, our analysis of the fold structures in the eastern terrane shows that their geometries are compatible with tilted fault blocks in the subsurface. Indeed, in the southern half of the study area (Fig. 3), a number of blocks containing exposed Precambrian rocks have a triangular shape with the apex pointing toward the west. Likewise, there are several folds in the eastern part of the same area that have similar triangular shapes, and their apexes also point west. In the northern half of the study area the exposed blocks are more quadrilateral in shape, and the folds to the east are also quadrilateral in shape.

The three anticlines that have geometries typical of compressional anticlines are located in areas where the block configuration is anomalous to the rest of the study area. Two of them are located over the edges of the only graben in the area, and the other one is located along the west side of the only westward-tilted block in the area.

Stearns (1975; this volume) reviewed the experimental and field evidence that supports the axiom "do not fold the basement" for conditions of Laramide deformation in the Bighorn Basin, Wyoming. Our analysis shows that this axiom also holds for the structures along the northeastern flank of the Front Range. The important and throughgoing tectonic element in this area is the differential uplift and rotation of basement blocks.

ACKNOWLEDGMENTS

Both of us were introduced to the model of forced folding by D. W. Stearns. We are grateful to him for visiting with us in the field and offering helpful suggestions.

Matthews gratefully acknowledges the donors of the Petroleum Research Fund, administered by the American Chemical Society, for financial support of the research. The University of Northern Colorado Faculty Research Committee and the UNC Administration provided release time to Matthews.

We thank J. J. Prucha for critically reviewing the manuscript.

REFERENCES CITED

Boos, C. M., and Boos, M. F., 1957, Tectonics of eastern flank and foothills of Front Range, Colorado: Am. Assoc. Petroleum Geologists Bull., v. 41, p. 2603–2676.

Borg, I., and Handin, J. W., 1966, Experimental deformation of crystalline rocks: Tectonophysics, v. 3, p. 249–368.

Braddock, W. A., and Eicher, D. L., 1962, Block-glide landslides in the Dakota Group of the Front Range foothills, Colorado: Geol. Soc. America Bull., v. 73, p. 317–324.

Braddock, W. A., Calvert, R. H., Gawarecki, S. J., and Nutalaya, P., 1970, Geologic map of the Masonville quadrangle: U.S. Geol. Survey Geol. Quad. Map GQ-832.

Braddock, W. A., Connor, S. J., Swann, G. A., and Wohlford, D. D., 1973a, Geologic map and cross sections of the Laporte quadrangle, Larimer County, Colorado: U.S. Geol. Survey Open-File Rept.

Braddock, W. A., Calvert, R. H., O'Conner, J. T., and Swann, G. A., 1973b, Geologic map and sections of the Horsetooth Reservoir quadrangle: U.S. Geol. Survey Open-File Rept.

Heard, H. C., 1962, The effect of large changes in strain rate in the experimental deformation of rocks [Ph.D. thesis]: Los Angeles, Univ. California.

Hunter, Z. M., 1955, Geology of the foothills of the Front Range in northern Colorado: Denver, Rocky Mtn. Assoc. Geologists (map).

Kellogg, C. J., 1973, A paleomagnetic study of various Precambrian rocks in the northeastern Colorado Front Range and its bearing on Front Range rotations [Ph.D. thesis]: Boulder, Univ. Colorado, 177 p.

Kelly, J. M., 1967, The geology of the Rattlesnake Park–Blue Mountain area, Larimer County, Colorado [M.S. thesis]: Boulder, Univ. Colorado.

LeMasurier, W. E., 1970, Structural study of a Laramide fold involving shallow-seated basement rock, Front Range, Colorado: Geol. Soc. America Bull., v. 81, p. 421–434.

Masters, C. D., 1957, Structural geology of the Rabbit Mountain–Dowe Pass area, Colorado [M.S. thesis]: Boulder, Univ. Colorado, 60 p.

Matthews, V., and Sherman, G. D., 1976, Origin of monoclinal folding near Livermore, Colorado: Mtn. Geologist, v. 13, p. 61–66.

Prucha, J. J., Graham, J. A., and Nickelsen, R. P., 1965, Basement-controlled deformation in Wyoming province of Rocky Mountains foreland: Am. Assoc. Petroleum Geologists Bull., v. 49, p. 966–992.

Rowlinson, N. R., 1957, Structural geology of the Carter Lake area, Larimer County, Colorado [M.S. thesis]: Boulder, Univ. Colorado, p. 45.

Sanford, A. R., 1959, Analytical and experimental study of simple geologic structures: Geol. Soc. America Bull., v. 70, p. 19–51.

Stearns, D. W., 1971, Mechanics of drape folding in the Wyoming province: Wyoming Geol. Assoc., 23rd Ann. Field Conf., Guidebook, p. 125–143.

——1975, Laramide basement deformation in the Bighorn Basin—The controlling factor for structures in the layered rocks: Wyoming Geol. Assoc., 27th Ann. Field Conf., Guidebook, p. 143–158.

——1978, Faulting and forced folding in the Rocky Mountains foreland, *in* Matthews, V., III, ed., Laramide folding associated with basement block faulting in the Western United States: Geol. Soc. America Mem. 151 (this volume).

Warner, L. A., 1956, Tectonics of the Colorado Front Range: Am. Assoc. Petroleum Geologists Rocky Mtn. Sec. Geol. Record, Feb. 1956, p. 129–144.

Weinberg, D. M., 1978, Some two-dimensional kinematic analyses of the drape fold concept, *in* Matthews, V., III, ed., Laramide folding associated with basement block faulting in the Western United States: Geol. Soc. America Mem. 151 (this volume).

MANUSCRIPT RECEIVED BY THE SOCIETY JUNE 27, 1977
MANUSCRIPT ACCEPTED AUGUST 25, 1977

Laramide structures and basement block faulting: Two examples from the Big Horn Mountains, Wyoming

JOHN C. PALMQUIST
Department of Geology
Lawrence University
Appleton, Wisconsin 54911

ABSTRACT

Two areas in the Big Horn Mountains are given as examples of the beauty and simplicity of the drape-folding-over-basement-block model of structural interpretation. The Piney Creek thrust is interpreted as an uplifted fault block with sedimentary strata draped over its edges. Structural cross sections of the Horn area show how sedimentary layers are continuously folded over a faulted basement.

In both areas, the role of a ductile stratigraphic unit, the Gros Ventre shale of Cambrian age, is important in partially decoupling the basement from the overlying sedimentary veneer and in providing a slide surface for gravity phenomena. Gravity glide masses probably are more widespread than is generally recognized in the foreland province and should be considered a characteristic element of the foreland tectonic style.

The construction of three-dimensional models of the basement surface, based on detailed geologic mapping, and the comparison of observed structures in the sedimentary cover rocks with experimental draping over the model are effective tools for gaining a mechanical understanding of the development of structures in the foreland province.

INTRODUCTION

Recent interest in drape folding over uplifted basement blocks has led to a re-examination of two illustrative areas—the Piney Creek area and the Horn area of the Big Horn Mountains (Fig. 1).

The Big Horn Mountains were described by Darton (1906, p. 91) as "an anticline

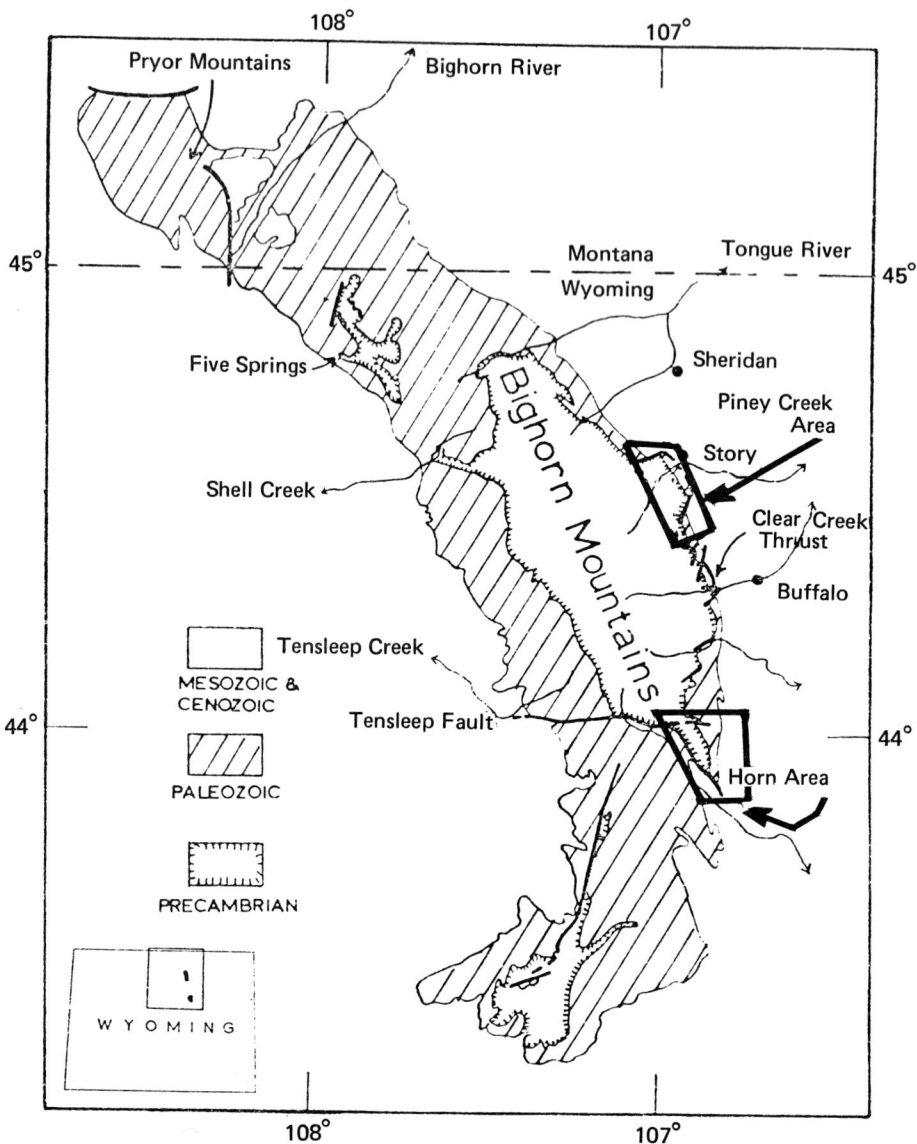

Figure 1. Index map of Big Horn Mountains showing location of Piney Creek area and Horn area.

rising steeply out of the nearly horizontal strata of the Great Plains." As a part of the foreland area of the central Rocky Mountain structural province (King, 1959; Osterwald, 1961; Blackstone, 1963; Eardley, 1963), the range exemplifies the variety of structural features that collectively define the structural style of the Wyoming province of the Rocky Mountains foreland (Prucha and others, 1965). Much structural description and interpretation was concentrated on the sedimentary cover rocks, and debate centered around the relative importance of tangential compression versus vertical uplift in the development of the folds and faults in Paleozoic and Mesozoic sedimentary rocks. The structural style has been variously interpreted, but all would concede that basement uplift has been an important part of the deformation. Basement boundary faults of the province, in general, and of the Big Horn Mountains, in particular, show orientations ranging from normal through vertical and upthrust to low-angle thrust. Lowell (1974) reviewed the hypotheses of foreland basement uplift and grouped them into three categories: (1) tangential compression with important vertical adjustments or with important lateral coupling, (2) vertical uplift, and (3) wrenching or strike-slip movements. It appears that elements of all three seem to be needed to satisfy the structural complexity of the Wyoming province, and additional work is needed to determine their relative importance on an area by area basis.

In a long-range program of research in the Big Horn Mountains, Hoppin has published a series of papers covering the role of the basement in Laramide deformation (1961; Hoppin and Palmquist, 1965; Hoppin and others, 1965), thrusting on the east flank (1961) and on the west flank (1970), and Cenozoic tectonic elements (Hoppin and Jennings, 1971). Prucha and others (1965) called attention to the role of the basement in controlling the style of Laramide deformation while outlining the main ideas and giving several field examples of relationships between basement blocks and sedimentary drape folds. More recently, Stearns (1975) and Stearns and Weinberg (1975) have further developed and elaborated the mechanical basis of drape folding by discussing the contrasting structural behaviors in the basement and cover rocks in the context of theory and experiment.

Recent work by Stearns (1971, 1975) has firmly established the mechanical basis for the drape-fold model, and the two examples described here are further illustrations of the applicability of the model. In the case of the Piney Creek area, uplift and draping has mimicked thrust and tear faults along the mountain front; the Horn, though less controversial, could be interpreted as an asymmetrical, thrusted anticline caused by crustal shortening, but the interpretation offered here is that of a basement block with a draped sedimentary cover.

PINEY CREEK AREA

In the Piney Creek area (Fig. 2), a block 13 km long juts eastward about 5 km from the main front of the Big Horn Mountains. This block is composed of Precambrian crystalline rocks and resistant Paleozoic sedimentary rocks, which together define the mountain front. The following discussion will evaluate two structural explanations for the excursion of the mountain front referred to as the Piney Creek lobe.

Structurally, the Piney Creek area has been widely accepted among previous workers as a thrust block. In fact, Billings (1972, p. 264) used the Piney Creek area as an example of tear faults bounding a thrust. His revision of Hudson's map (1969, Pl. 1) is reproduced here as Figure 2. In this paper an alternative interpretation will be proposed, namely, that the Piney thrust is actually a differentially uplifted basement block, bounded by high-angle faults in which actual relative movement was dominantly vertical with a minor amount of lateral (northeastward) displacement during the later stages of movement.

The Piney Creek lobe could be accounted for in one of three ways: (1) as a thrust block (Fig. 3A), (2) as an uplifted block (Fig. 3B), or (3) as a block moved both upward and laterally eastward. Because the third hypothesis is merely a combination of (1) and (2), it will be considered hereafter as a modification and not treated separately. The lack of evidence bearing on the geometry of the bounding faults and the difficulty of determining net slip insures that there will

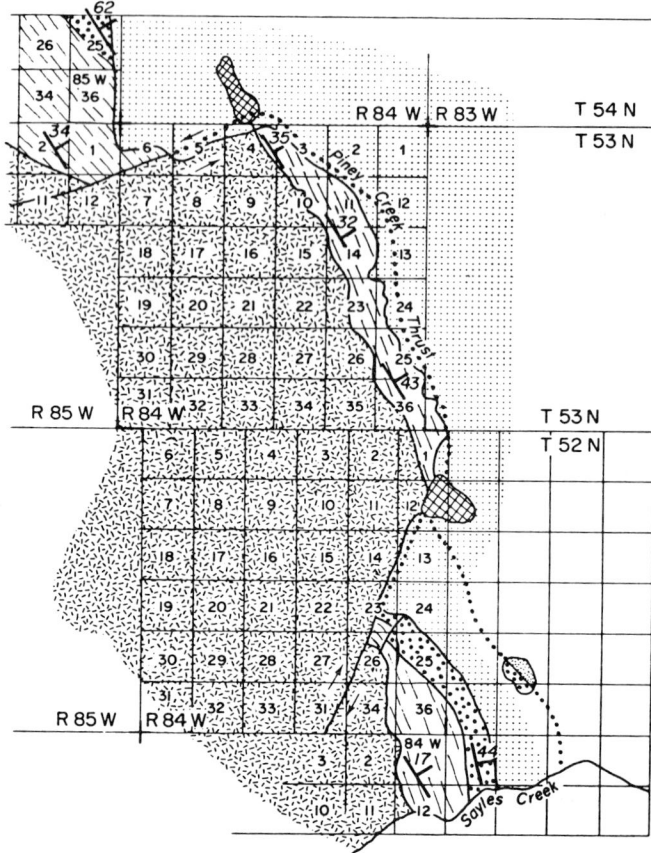

Figure 2. Geologic map of Piney Creek area. Precambrian basement shown by random dashes; diagonal dashes, Paleozoic; random dots, Mesozoic; linear dot pattern, Cenozoic. Cross-hatched areas at corners of block denote chaotic blocks of Paleozoic limestones. Map from Billings (1972) after Hudson (1969).

continue to be some controversy surrounding the interpretation of the Piney Creek lobe.

Thrust-Block Hypothesis

Billings (1972, p. 264) described a thrust block: "A block bounded by two strike-slip faults may move forward between relatively stationary terrains on either side. In many instances the moving block is also separated by a thrust fault from the stationary block beneath it. In a sense, therefore, the moving block is bounded by one scoop-shaped fracture, the bottom of the scoop corresponding to a thrust, the sides of the scoop to the strike-slip faults." Billings used the Piney Creek area as an illustration of tear faults bounding a thrust, based on the work of Hudson (1969). Hudson's main purpose was to evaluate the nature and extent of Precambrian control during Laramide deformation in the Piney Creek area (1969, p. 284), but his data on joints in Precambrian and Paleozoic rocks led him to conclude that a conjugate shear joint system and associated maximum principal stress oriented N52°E-S52°W represent early Laramide compressive deformation (p. 287). Other workers have established the existence of thrusting along the east flank of the central Big Horns structural segment (Bucher and others, 1934; Demorest, 1941; Hoppin, 1961), so the explanation of the Piney Creek lobe as a thrust block

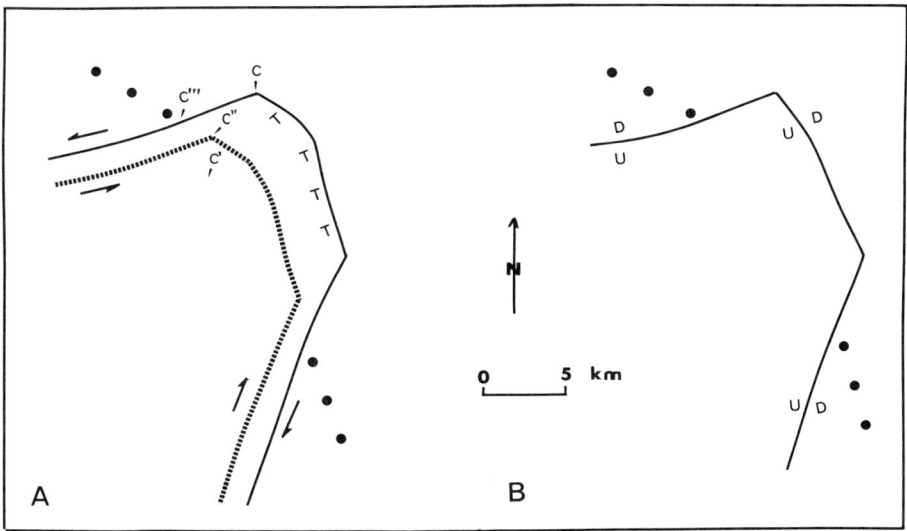

Figure 3. Two interpretations of the Piney Creek lobe. A. Thrust block hypothesis. Dotted line indicates position of mountain front; dashed line suggests prefaulting position of Piney lobe; solid line outlines the actual position of the Piney lobe (after faulting). The north outside corner of the lobe is designated C; possible prefaulting positions of the north outside corner are designated C', C", and C'''. A room problem results from moving the wedge-shaped rigid basement eastward along strike-slip faults. Overlap could be alleviated by low-angle thrusting at front of lobe (T marks upper part of thrust), but no solution to the room problem is possible at the sides if the block is truly rigid. B. Block uplift hypothesis. No room problem results if the basement is differentially uplifted along high-angle faults.

appears to be consistent with these studies. Strike-slip movement would be one way to explain the lateral separation of the mountain front.

Basement-Block Uplift Hypothesis

Lateral separation of the mountain front is equally well explained by an uplifted block of Precambrian basement rocks; uplift and concomitant erosion would cause the contact to migrate in a downdip direction and result in a lobate geometry of the mountain mass (see Figs. 3, 4). Uplift on the order of 3 km together with draping of the sedimentary veneer provides a basis for understanding the structure of the sedimentary rocks adjacent to the bounding faults. Details of the draped structures will be discussed later.

Comparison of Thrust-Block Hypothesis and Basement-Block Uplift Hypothesis

The block-uplift hypothesis avoids the room problem inherent in the thrusting of a wedge-shaped mass (see Fig. 3). In the thrusting hypothesis it is assumed that the mountain front was once continuous, and the lateral separation was accomplished by slip along a low-angle thrust (Fig. 4A). If thrusting involves lateral transport of a rigid basement to form a salient, there should be a corresponding recess on the west flank of the range. No recess exists. Instead, the Piney lobe is situated adjacent to the widest and highest part of the range. Also, if the thrusted

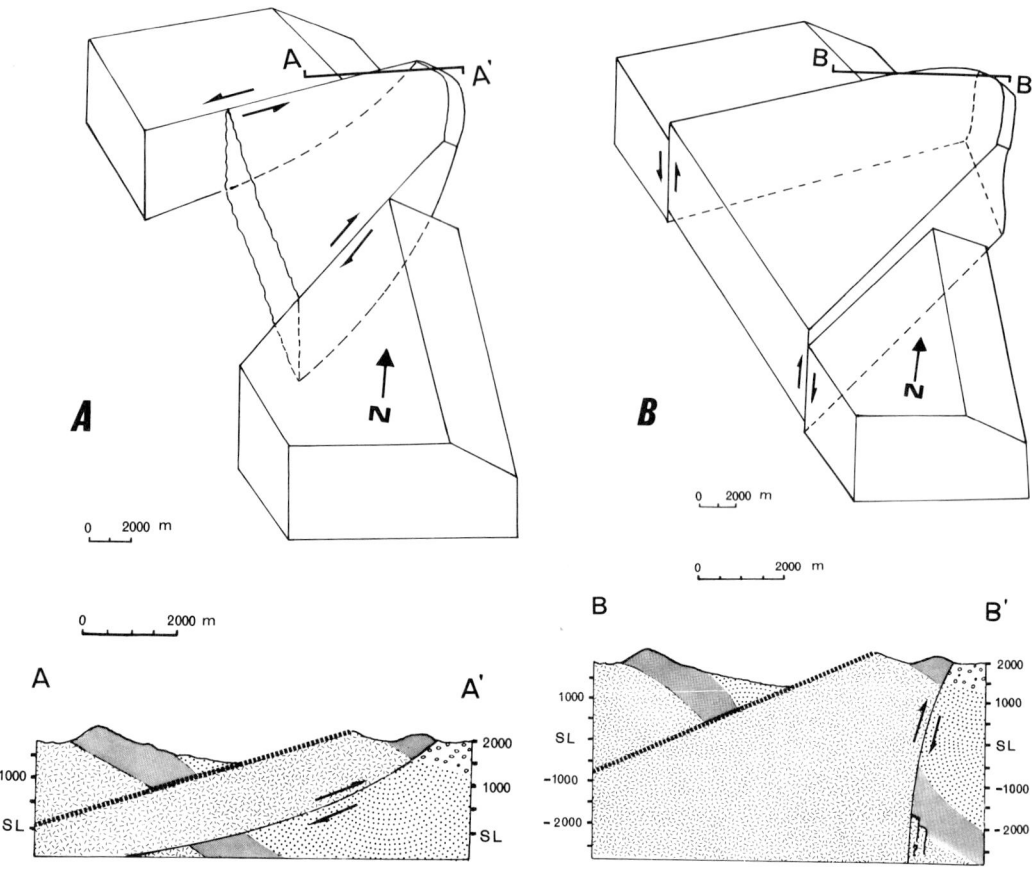

Figure 4. Two interpretations of the Piney Creek lobe in three dimensions. (A) Thrust-block hypothesis. Note how the wedge shape would require the north block to move toward the north and/or the south block to move toward the south. Also, a recess might be expected at some distance to the west. (B) Block-uplift hypothesis. The vertical separation of the sub-Cambrian surface is a function of its slope. Although the block diagrams are schematic, they are roughly to natural scale (in metres). The structure sections are at natural scale (in metres) and show the north block (upper left) and the Piney block (lower right) separated by a dashed line. Random dashes, Precambrian rocks; shading, Paleozoic rocks; circles, Cenozoic rocks; SL, sea level. Structure sections are modified after Hudson (1969).

basement block is rigid it would, owing to its wedge-shape, push laterally against the blocks adjacent on the north and south. The block-uplift hypothesis, on the other hand, assumes that the outline of the Piney lobe is defined by high-angle faults along which the slip was dominantly vertical (Fig. 4B). The mountain front, then, may never have been continuous in the Piney Creek area. Thus, the only room problem associated with block uplift is the vacated space below the uplifted block. At appropriate depths plastic flow would fill the space if, indeed, the intrusion of plastic material is not, in fact, the cause of the uplift in the first place.

The regional structual setting is one of uplift, not thrusting. The thrusts, "although impressive where they are exposed are only relatively locally developed along the flanks. They occur in restricted salients of more intense tectonic movements" (Hoppin and Jennings, 1971, p. 43). Flank thrusts in the Big Horns region steepen toward vertical with depth and "probably represent the product of a release in lateral pressure with block movement upward" (p. 43). This is in conflict with Hudson's earlier statement (1969, p. 286) that the fault is presumed to have a moderate to low dip in the subsurface; likewise, the scoop-shaped fault boundary of Billings (1972) is not consistent with the regional tectonics. Hudson's statement (1969, p. 286) that the frontal fault "dips at a high angle westward, as shown by its nearly straight trace in areas of moderate to high relief" is, however, consistent with the block-uplift hypothesis. Darton (1906) showed the frontal fault dipping at 90°.

Hudson's remaining evidence for thrusting, the joint pattern, is also subject to alternate interpretation. In basement block uplift, the basement is the active mover, whereas the sedimentary veneer is passively draped over the rising blocks. The Piney Creek basement is foliated and cut by fractures, shear zones, and dikes. Uplift of the Piney Creek block was probably guided along its southeastern bounding fault by a well-developed N17°E Precambrian direction of jointing, shearing, and fracture cleavage (Hudson, 1969, p. 294). Elsewhere, the block behaved as though it were isotropic. Uplift probably reopened old joints and generated new ones. It seems likely that stresses resulting from the drape process were transposed, owing to bedding anistropy, in such a manner that two of the three principal stresses were contained in the bedding plane. Joints, therefore, tend to be dominantly perpendicular to bedding. Sturgul and others (1976), using the finite-element method, have shown that the major principal stress parallels the mountain slope for the Hochkönig massif in Austria. This conclusion supports the hypothesis that stresses are self-gravitationally induced. If, in the Piney lobe, stresses originated because of the gravity of the overburden, then the joint data is consistent with the model of sediments draped over a faulted, tilted basement block. The practice of rotating beds around strike and presuming to obtain the original orientation of joints and slickenslides seems rather risky. Hudson (1969, p. 287) argued on this basis, using slickenslides that become horizontal when bedding attitude is rotated back to horizontal, that the principal stress of early Laramide deformation was (1) normal to bedding strike, (2) horizontal, and (3) initiated the uplift process. A scheme more consistent with the Big Horns structural development would begin with an uplift, producing the Eocene Kingsbury and Moncrief gravels. Minor thrusting follows the uplift, as a consequence of the release of pressure when the draped veneer is freed of lateral confinement. The slickenslides in this scheme are interpreted as they generally are, namely, recording only the latest movement—not the earliest. The evidence seems most consistent with the idea of dominant early vertical uplift and minor late lateral expansion or thrusting.

Details of the Folding Associated with the Uplifted Basement Block

A scale model of the basement configuration was constructed to compare simulated draping of material over the model with the structures of sedimentary veneer as observed in the field. A sketch of the basement-block model is shown in Figure 5; the draped model is depicted in Figure 6.

At the north inside corner, Cambrian and Ordovician rocks are strongly overturned (A in Fig. 5). At Dry Fork drape fold in the northern Big Horns and at Rattlesnake Mountain near Cody, Wyoming, the ductile Cambrian shales have been squeezed off the corner of the uplifted block (Stearns and Weinberg, 1975, p. 163). By analogy, similar behavior is inferred for the Piney Creek folding as shown in Figure 7. Bulging out of the shales is a speculation that helps explain the overturned beds observed at the north inside corner.

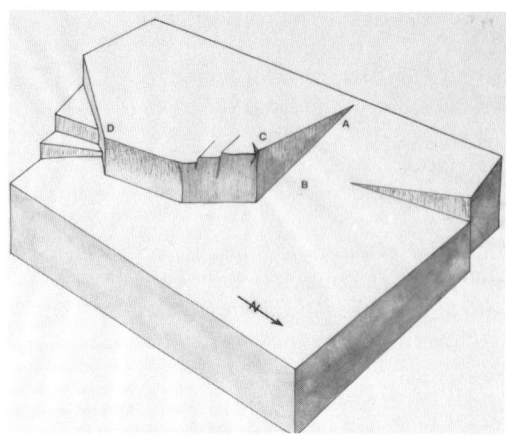 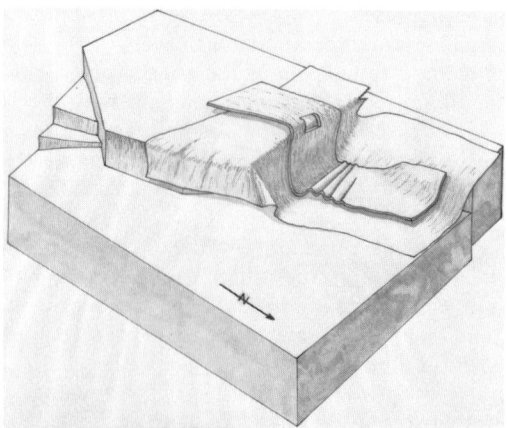

Figure 5. Sketch of model showing faulted Precambrian basement surface in Piney Creek area. Letters designate significant localities discussed in text: A, north inside corner; B, folds in Jurassic Sundance formation; C, north outside corner; D, south outside corner.

Figure 6. Partially draped model shows good agreement between model folds and observed structures in sedimentary veneer. "Floating" layer represents Jurassic Sundance Formation.

Farther east along the north boundary fault (B in Fig. 5), an exposure of Jurassic oolitic limestone (Sundance Formation) forms a succession of anticlinal and synclinal folds. Figure 8 shows the geometry of the folds and gives an axis plunging 36°, S86°E. This is to be expected as the drape process creates a synclinal bend along the north boundary (B in Fig. 5), and crowding in the higher part of the section produces a succession of folds as depicted in the draped model (Fig. 6).

At the outside corners (C and D in Fig. 5), the drape process produces conical folds just as one sees in a tablecloth hanging off the corner of a dining table. Erosion of resistant Paleozoic limestones in the hinge area of these folds would result in a chute floored with greasy Cambrian shales supporting unstable carbonate blocks ready to slide down the chute. Although these corner folds are completely eroded away, chaotic glide blocks are found at both outside corners (see Hudson, 1969, Pl. 1). Hudson also interpreted these chaotic Paleozoic blocks as gravity features but considered them as detached segments of the thrust plate. Previously, they had been mapped as an integral part of the thrust plate (Demorest, 1941; Mapel, 1959). The preferred interpretation offered here is that such masses are gravity-driven phenomena related to the uplift and the drape-folding process. More specifically, they probably represent the excess material caused by draping over an "outside" corner.

HORN AREA

The Horn, a Precambrian basement block, is the southernmost extension of the central Big Horn Mountains (Fig. 1). The tripartite division of the Big Horns established by Demorest (1941) utilizes the Tongue River fault and the Tensleep fault to divide the range into northern, central, and southern divisions. Structures in the sedimentary cover rocks generally verge west in the northern division, east in the central, and west in the southern division.

The Tensleep fault. Demorest's structural basis for separating the central and

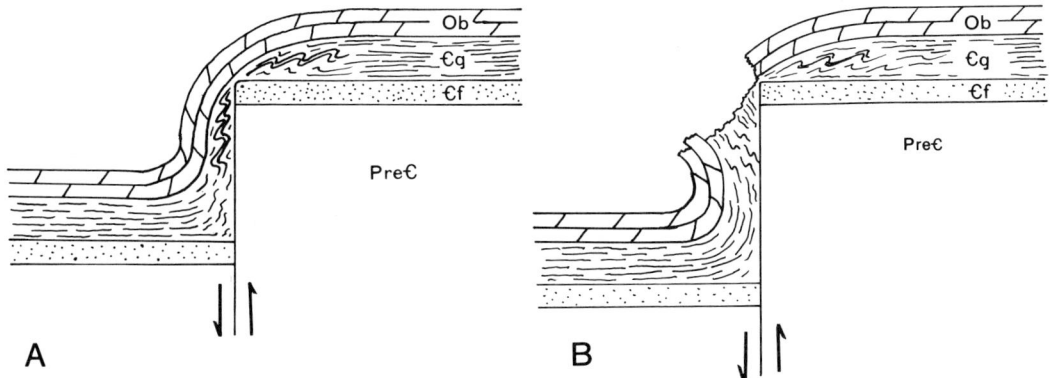

Figure 7. Block-faulted basement schematically showing the behavior of the Cambrian and Ordovician formations. PreЄ, Precambrian basement; Єf, Cambrian Flathead Formation; Єg, Cambrian Gros Ventre and Gallatin Formations; and Ob, Bighorn Dolomite. The role of the ductile Єg is important in absorbing the faulting and permitting the younger layers to deform smoothly and continuously over the broken basement blocks. The reversed drag folds provide a means of testing this idea in the field. (B) If the fold is "pinned" at the hinge and fault displacement occurs, the lower limb must lengthen and very likely break; this then would allow outflow of the ductile shales and overturning to occur.

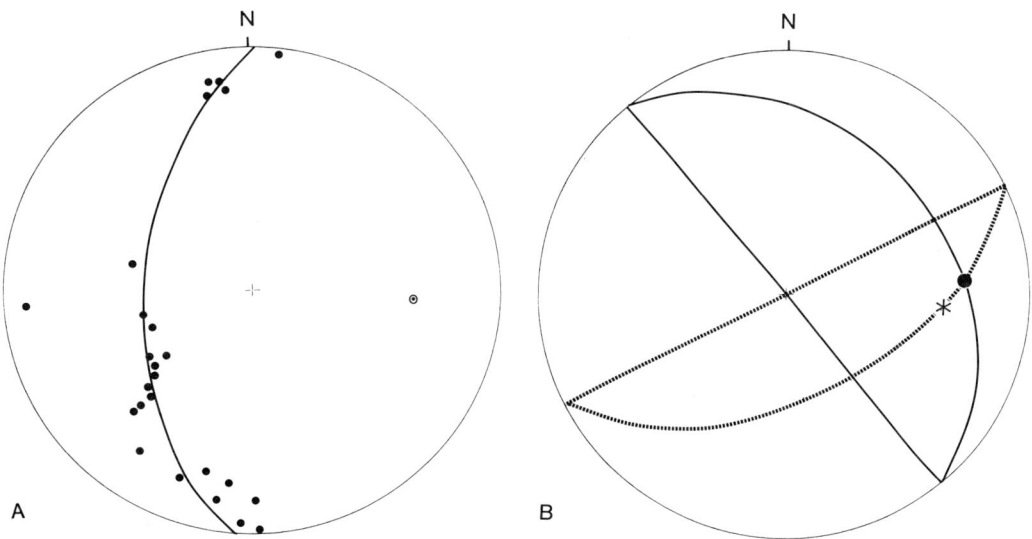

Figure 8. (A) Stereoplot of poles to bedding in oolitic limestone of the Jurassic Sundance Formation. Dots represent poles to bedding; solid line is great circle containing poles; circled dot is the pole to the great circle and indicates a statistical fold axis plunging 36° and trending at S86°E. (B) Comparison of fold axis predicted by drape model (large dot) with observed fold axis (asterisk). Solid great circle defines attitude of sedimentary veneer in normal mountain front north of the Piney Creek lobe (N40°W, 34°E); dashed chord shows strike of fault along north border of Piney Creek lobe (N63°E); dashed great circle shows beds overturned (55°SE) as expected in drape model. Observed axis, asterisk, is within 10° of predicted fold axis, large dot.

southern Big Horns, has obscured the essential unity of the Horn with the central Big Horns, which is striking when viewed on satellite imagery. Because it is south of the Tensleep fault and because its asymmetry (gentle on the east and steep on the west) is like that of the southern Big Horns, the Horn would seem to belong with the southern Big Horns. Evidence supporting the contention that the Horn is more correctly part of the central Big Horns is its continuity on the east flank with the central unit, the parallel trend with the central unit, and finally, the satellite imagery and regional maps that place the Horn as the southernmost extension along the axis of maximum uplift that defines the central unit. These and other considerations indicate that the time has come to revise Demorest's classification and develop a version that reflects knowledge gained since 1941 about the structural configuration of the Big Horn Mountains. Hoppin and Jennings (1971, p. 39) also stressed the need for revision and indicated their continuing work toward a revised structural classification of the mountains.

I have previously studied the Horn area (Palmquist, 1965, 1967). These earlier papers emphasized the petrology of the Precambrian core (1965) and analyzed the internal structure of the crystalline rocks and their role in influencing Laramide deformation (1967). This discussion will concentrate primarily on the Laramide deformation.

Structure of the Horn Block

The Horn area contains a mosaic of blocks as shown in Figure 9. The faults bounding the block(s) are well exposed. The Horn block itself may be described as bounded on the north by a scissor fault, with displacement ranging from 0 at its eastern pivot and increasing to its juncture with the Horn fault at the northwest corner of the Horn where a stratigraphic throw of 210 m is indicated. Along the Horn fault, stratigraphic throw is estimated at 1,050 to 1,500 m, with the maximum at the southwest corner of the Horn, where another scissors fault defines the southern boundary of the block. Thus, a crude first approximation of the Horn block would be to think of its upper surface as an unevenly raised trap door. The structural draping of the cover rocks was described and mapped previously

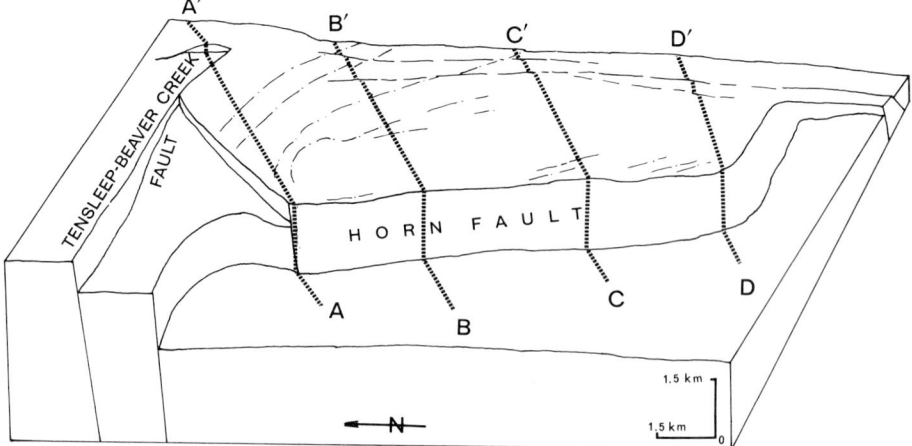

Figure 9. Sketch of a model showing the basement configuration in the Horn area. Dashed lines give the location of cross sections shown in Figure 10; dash-dot lines show the traces of foliation.

(Palmquist, 1967). Even though the basement was described as deforming en bloc, cross sections were drawn (1967, Pl. 2) with the basement surface bending smoothly and conformably with the overlying bedding. Hoppin (1970), p. 2410) suggested that the difficulty of finding definitive exposures is the reason for this widespread practice. In any case, revised cross sections that obey the axiom of "do not fold the basement" (Stearns, 1975, p. 150) are given as conjectural representations of the Horn structures—interpretations that are consistent with the field data and with current ideas about the mechanical properties of the basement and its sedimentary veneer.

The revised cross sections (Fig. 10) show the Horn fault as a series of steps rather than a single fault zone. Each step is capped by the Flathead Sandstone of Cambrian age, but the overlying shales are shown thinning over each outside corner and thickening in each inside corner; in this way the steps are smoothed so the overlying Bighorn Dolomite and successive layers can drape in a smooth curve. This speculation of flowing shales is supported, as stated earlier in this report, by examples of similar behavior elsewhere in the Big Horn Mountains. The speculated step faults were chosen as a means of explaining the apparent curvature of the basement surface. Stearns (1975) proposed curved fault surfaces to permit

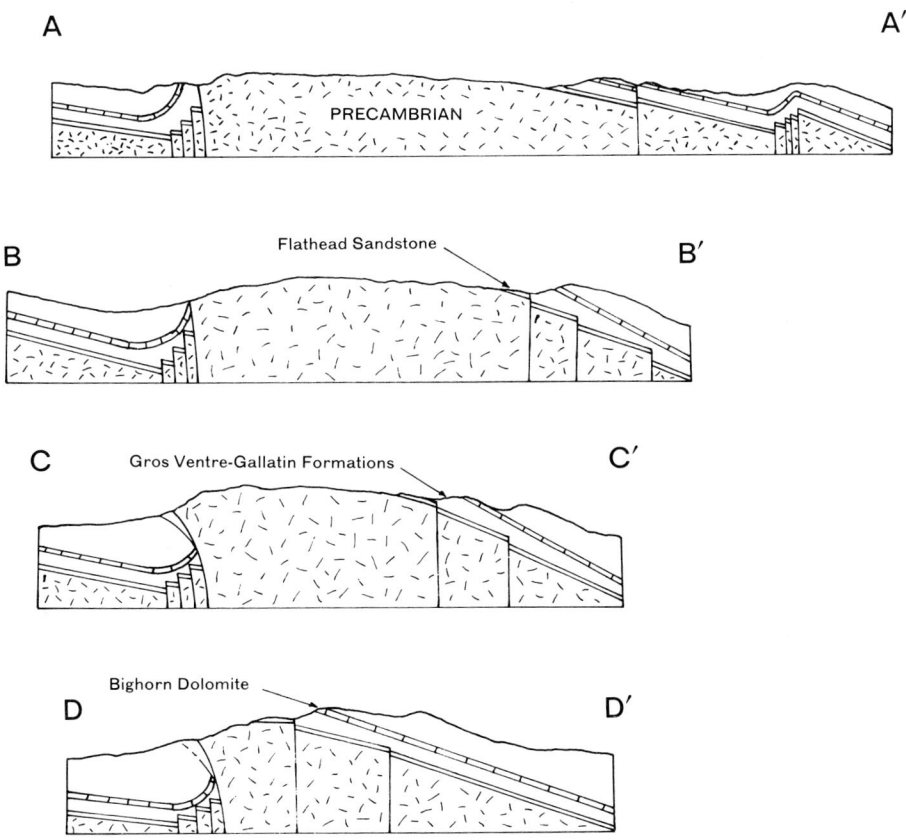

Figure 10. Structural cross sections of the Horn area. Only Cambrian and Ordovician sedimentary formations are shown above the Precambrian basement. Younger sedimentary units deform in accord with the Bighorn Dolomite. Faults that do not penetrate the surface are hypothetical. Natural scale.

rotation of basement blocks, particularly for Rattlesnake Mountain where distributive shear can be disproved. Hoppin (1970) demonstrated that pervasive shearing permitted rotation of adjacent basement blocks in the Five Springs Creek area of the Big Horn Mountains. In the Horn, the adjacent blocks are the east flank of the southern Big Horns and the east dipping Horn block. Therefore, the basement is not rotated, rather, the adjacent blocks have similar tilts. Where the basement can be examined along the Horn fault-line scarp, no evidence of intense shearing was seen. The zone between the blocks is, therefore, postulated to be a series of step faults that cut the basement surface but are lost in the ductile Cambrian shales. In this way, the overlying beds assume a continuous curve as they drape over the broken basement.

DISCUSSION

Block Uplift versus Crustal Shortening

Both the Piney Creek and Horn areas have been shown as block uplifts of the Precambrian basement: 3,048 m relative to the eastern block in the case of the Piney Creek block and 1,524 m at the south end of the Horn fault for the Horn block. The region has undergone differential vertical movement of at least 10,360 m (Prucha and others, 1965, p. 970). Thus, a vertical uplift of 3,048 m in the Piney Creek area, rather than low-angle thrusting, seems a more reasonable way to account for the 4,572 m of horizontal separation. The sequence of uplift followed by minor thrusting is suggested, rather than tangential compression preceding uplift. In the Horn area, classic corner draping is observable at both the northwest and southwest corners, structures difficult to reconcile with crustal shortening.

Because all beds tend to be stretched in the drape process, rather than shortened on the inner hinge and lengthened over the outer hinge as beds are in typical folds, the drag type parasitic folds might not be generated. Instead, upper layers may move farther downdip from a fixed hinge than lower layers and produce reversed drag folds as shown in Figure 7.

Gravity-Controlled Movements

Both areas also contain out of place rock masses interpreted as gravity phenomena. At both outside corners of the Piney lobe, and along the Horn fault, rock slides have occurred. Another area south of the Piney Creek area also contains significant rock sliding. Although these features are much smaller than the well-known Heart Mountain phenomenon, it is suggested here that such features are so widespread that they should be considered a characteristic feature of the foreland style of deformation.

Role of Gros Ventre Formation

The role of the ductile member near the base of the sedimentary veneer is twofold: (1) It has acted as a shock absorber and has helped transform basement faults into continuous, or modestly broken, drape folds at higher stratigraphic and structural levels. (2) It frequently has reactivated its ductile behavior in the form of landslides or rockslides whenever it has become exposed at the surface. This unstable unit has, most likely, been involved in aiding large-scale downslope movements since first exposed to erosion, and it continues to plague the Wyoming Highway Department and others today.

ACKNOWLEDGMENTS

Comments and suggestions by Richard A. Hoppin and Vincent Matthews III were helpful in improving the manuscript. The support of the geology departments of Lawrence University and the University of Illinois is also gratefully acknowledged.

REFERENCES CITED

Billings, M. P., 1972, Structural geology (3rd ed.): Englewood Cliffs, N.J., Prentice-Hall.
Blackstone, D. L., Jr., 1963, Development of geologic structure in the Central Rocky Mountains: Am. Assoc. Petroleum Geologists Mem. 2, p. 160–179.
Bucher, W. H., Thom, W. T., Jr., and Chamberlain, R. T., 1934, Results of structural research work in Beartooth–Big Horn region, Montana and Wyoming: Am. Assoc. Petroleum Geologists Bull., v. 17, p. 680–693.
Darton, N. H., 1906, Geology of the Bighorn Mountains: U.S. Geol. Survey Prof. Paper 51, 129 p.
Demorest, M. H., 1941, Critical structural features of the Bighorn Mountains, Wyoming: Geol. Soc. America Bull., v. 52, p. 161–176.
Eardley, A. J., 1963, Relation of uplifts to thrusts in the Rocky Mountains: Am. Assoc. Petroleum Geologists Mem. 2, p. 209–219.
Hoppin, R. A., 1961, Precambrian rocks and their relationship to Laramide structure along the east flank of the Bighorn Mountains near Buffalo, Wyoming: Geol. Soc. America Bull., v. 72, p. 351–368.
——1970, Structural development of the Five Springs Creek area, Bighorn Mountains, Wyoming: Geol. Soc. America Bull., v. 81, p. 2403–2416.
Hoppin, R. A., and Jennings, T. V., 1971, Cenozoic tectonic elements, Bighorn Mountain region, Wyoming-Montana: Wyoming Geol. Assoc., 23rd Ann. Field Conf., Guidebook: p. 39–47.
Hoppin, R. A., and Palmquist, J. C., 1965, Basement influence on later deformation: The problem, techniques of investigation, and examples from Bighorn Mountains, Wyoming: Am. Assoc. Petroleum Geologists Bull., v. 49, p. 993–1003.
Hoppin, R. A., Palmquist, J. C., and Williams, L. O., 1965, Control by Precambrian basement on the location of the Tensleep–Beaver Creek fault, Bighorn Mountains, Wyoming: Jour. Geology, v. 73, p. 189–195.
Hudson, Robert F., 1969, Structural geology of the Piney Creek thrust area, Bighorn Mountains, Wyoming: Geol. Soc. America Bull., v. 80, p. 283–296.
King, P. B., 1959, The evolution of North America: Princeton, N.J., Princeton Univ. Press, p. 189.
Lowell, James D., 1974, Plate tectonics and foreland basement deformation: Geology, v. 2, p. 275–278.
Mapel, W. J., 1959, Geology and coal resources of the Buffalo–Lake DeSmet area, Johnson and Sheridan Counties, Wyoming: U.S. Geol. Survey Prof. Paper 1078, p. 148.
Osterwald, F. W., 1961, Critical review of some tectonic problems in Cordilleran foreland: Am. Assoc. Petroleum Geologists Bull., v. 45, p. 219–237.
Palmquist, J. C., 1965, Petrology of the Horn area, Bighorn Mountains, Wyoming: Illinois Acad. Sci. Trans., v. 58, p. 241–254.
——1967, Structural analysis of the Horn area, Bighorn Mountains, Wyoming: Geol. Soc. America Bull., v. 78, p. 283–298.

Prucha, J. J., Graham, J. A., and Nickelsen, R. P., 1965, Basement-controlled deformation in Wyoming province of Rocky Mountains foreland: Am. Assoc. Petroleum Geologists Bull., v. 49, p. 966–992.

Stearns, D. W., 1971, Mechanisms of drape folding in the Wyoming province: Wyoming Geol. Assoc., 23rd Ann. Field Conf., Guidebook, p. 125–144.

——1975, Laramide basement deformation in the Bighorn Basin—The controlling factor for structures in the layered rocks: Wyoming Geol. Assoc., 27th Ann. Field Conf., Guidebook.

Stearns, David W., and Weinberg, D. M., 1975, A comparison of experimentally created and naturally formed drape folds: Wyoming Geol. Assoc., 27th Ann. Field Conf., Guidebook.

Sturgul, J. R., Scheidegger, A. E., and Grinshpan, Zvi, 1976, Finite-element model of a mountain massif: Geology, v. 4, p. 439–442.

MANUSCRIPT RECEIVED BY THE SOCIETY JUNE 27, 1977
MANUSCRIPT ACCEPTED AUGUST 25, 1977

Geological Society of America
Memoir 151

Geometric analysis of multiple drape folds along the northwest Big Horn Mountains front, Wyoming

MARTHA TIREY STEARNS
DAVID W. STEARNS
Department of Geology and Center for Tectonophysics
Texas A&M University
College Station, Texas 77843

ABSTRACT

Excellent exposures in an area of multiple drape folds in the northwest corner of the Big Horn Mountains provide data for the construction of a three-dimensional model of the upper surface of the Mississippian Madison Formation. In the model, the vertical and horizontal scales are the same. Comparison of deformed and undeformed Madison surfaces indicate that displacements required by the deformation vary from place to place on the surface in both magnitude and direction. Calculations made from field measurements show that there is not enough fracturing or thinning in the Madison layer to account for the indicated displacements. Also, reasonable shortening on the basement faults underlying the forced folds cannot account for the calculated displacements in the sheet. Any solution to the mechanics of the drape-folding process must, therefore, meet the described constraints and still account for the indicated nonuniform displacements.

Detachment of the sedimentary layers from the basement and intrastratal slip are two processes that meet the constraints and still account for the indicated nonuniform displacements.

INTRODUCTION

In the past decade, most structural analyses of the Rocky Mountains foreland have been directed toward the identification of the tectonic style. Only casual attention has been directed toward the actual movement of large volumes of rock required to accomplish the style. Several workers (Berg, 1962; Prucha and others, 1965; Stearns, 1971) have established that the layered sedimentary rocks fold in response to faulting of the basement and further that the resulting fold shape is forced by the shape of the basement block. To a limited extent, cross sections

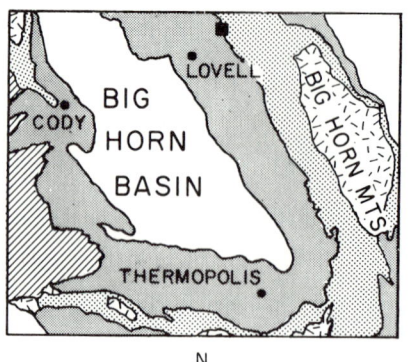

Figure 1. Study area in northwest corner of Big Horn Mountains, Wyoming.

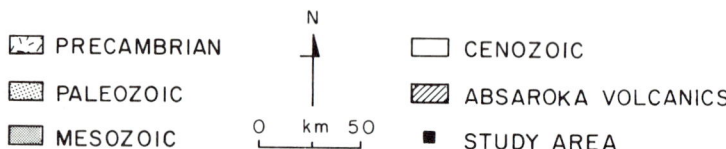

(that is, two-dimensional analyses) have been used to establish certain aspects of the required movements such as the necessity for fixed hinges and detachment near the base of the Paleozoic carbonate section (Stearns, 1975). These two-dimensional studies are useful, but are usually incomplete with respect to the total movements in the folding process. In many instances, cross sections so oversimplify the true movements that they are more misleading than helpful. Some workers lose sight of the fact that two-dimensional analysis is valid, at best, only near the centers of very large uniform folds; cross sections are very gross misrepresentations of reality near the ends of folds or where the fold geometry is complicated.

These comments are not meant as criticisms, but rather as statements about our stage of progress toward understanding this complicated style of folding in the foreland. It would seem, therefore, that to better understand drape folds, we should begin to pay attention to the three-dimensional movement patterns implied by the observational facts from the field. The defense of this proposition is the purpose of this paper.

To date there are only two studies that we know of that directly concern themselves with the three-dimensional aspects of drape folding. Cook and Stearns (1975) studied the Weber Sandstone (Pennsylvanian) in northern Colorado, and Vaughn (1976) studied movements in selected Mesozoic layers at Casper Mountain, Wyoming. Both of these studies demonstrated conclusively that complex movements within three-dimensional plates differ in magnitude and direction and cannot be understood by cross-sectional studies alone. Both studies showed that individual single layers in drape folds have some regions of excess material and some regions deficient in material and that differential lateral transport within the layer is required by the deformation. However, in both studies, limited exposures precluded answering certain critical questions.

Therefore the purpose of this effort is to study a single rock unit (Mississippian Madison Formation) in three dimensions in a region of good exposure to better define the types of movements that appear to be required in the drape-folding process. The area selected is in the northern Big Horn Mountains (Fig. 1), where the Madison Formation is well exposed across a region of multiple drape folds. Wherever possible, actual field observations will be the basis of all arguments. There is sometimes a tendency among geoscientists to only believe what they can understand. In this study, observations of what the rocks do will be used independent of our ability to understand how they do it. This means that certain aspects of the movement pattern will be established as absolute requirements and certain previously proposed aspects, no matter how appealing to our intuition, will be ruled out. However, the most important objective of this paper will be to define the problem sufficiently so that future work in analysis can concentrate on a solution. The goal then is one of problem definition rather than problem solution.

With this goal in mind we will first present the field measurements and observations with no attempt to interpret their meaning. Having established the data base we will then try to incorporate all the observations into the implied movement scheme. All of the observations must be considered collectively, because too often single observations can be explained only to find that the explanation contradicts other observational facts.

STUDY AREA

The area chosen for detailed field study is 8 km (5 mi) north of Wyoming Highway 14A and 22.5 km (14 mi) east of Lovell, Wyoming, in the northwest corner of the Big Horn Mountains (Fig. 1). The area will be referred to as the Cottonwood Canyon area and contains approximately 103 km^2 (40 mi^2). The boundaries of the study area are shown in Figure 2 as well as the area covered by the geologic map of Voldseth (1973), which is Figure 3. Within the study area, the Mississippian Madison Formation is well exposed across several drape folds of different size, strike, and vertical displacement. Among the drape folds present, two drape folds with different strikes join together to form one larger drape fold (point A, Fig. 2). At two points there are abrupt changes in strike of continuous drape folds as the folds form corners. This geometry may seem overly complicated for a definitive study, but such a complex geometry with continuous outcrop of carbonate rock is just what is needed to define the movement problems. Cross sections drawn through drape folds in areas which do not involve terminations or bifurcations of the folds do a satisfactory job of describing the shape of the fold. However, a cross section has only two dimensions and does not include information about motions at angles to the plane of the cross section. At terminations and bifurcations of drape folds, we must begin to think of the drape-folding process in three dimensions or, in other words, as plate deformations. In this three-dimensional study of the complicated geometry of Cottonwood Canyon, we hope to document the necessity of differential lateral transport of material within a layer as it is drape folded.

STRATIGRAPHY

In the deepest canyon in the area (Cottonwood Canyon), complete sections of Paleozoic rocks are exposed. Mesozoic rocks in the study area are eroded off the drape folds and are present only in the structurally low areas (Fig. 3). The Paleozoic rocks were deposited on the Wyoming shelf in shallow marine water. This shelf remained a stable tectonic element throughout Paleozoic time and, therefore, the sediments deposited are fairly uniform and widespread. The unit of particular interest here is the Mississippian Madison Formation. It is a carbonate deposit with alternating layers of crystalline limestone and dolomite (Richards, 1955). During Mississippian time the Wyoming shelf covered all of Wyoming and parts of Utah, Montana, and South Dakota (Haun and Kent, 1965). Because of

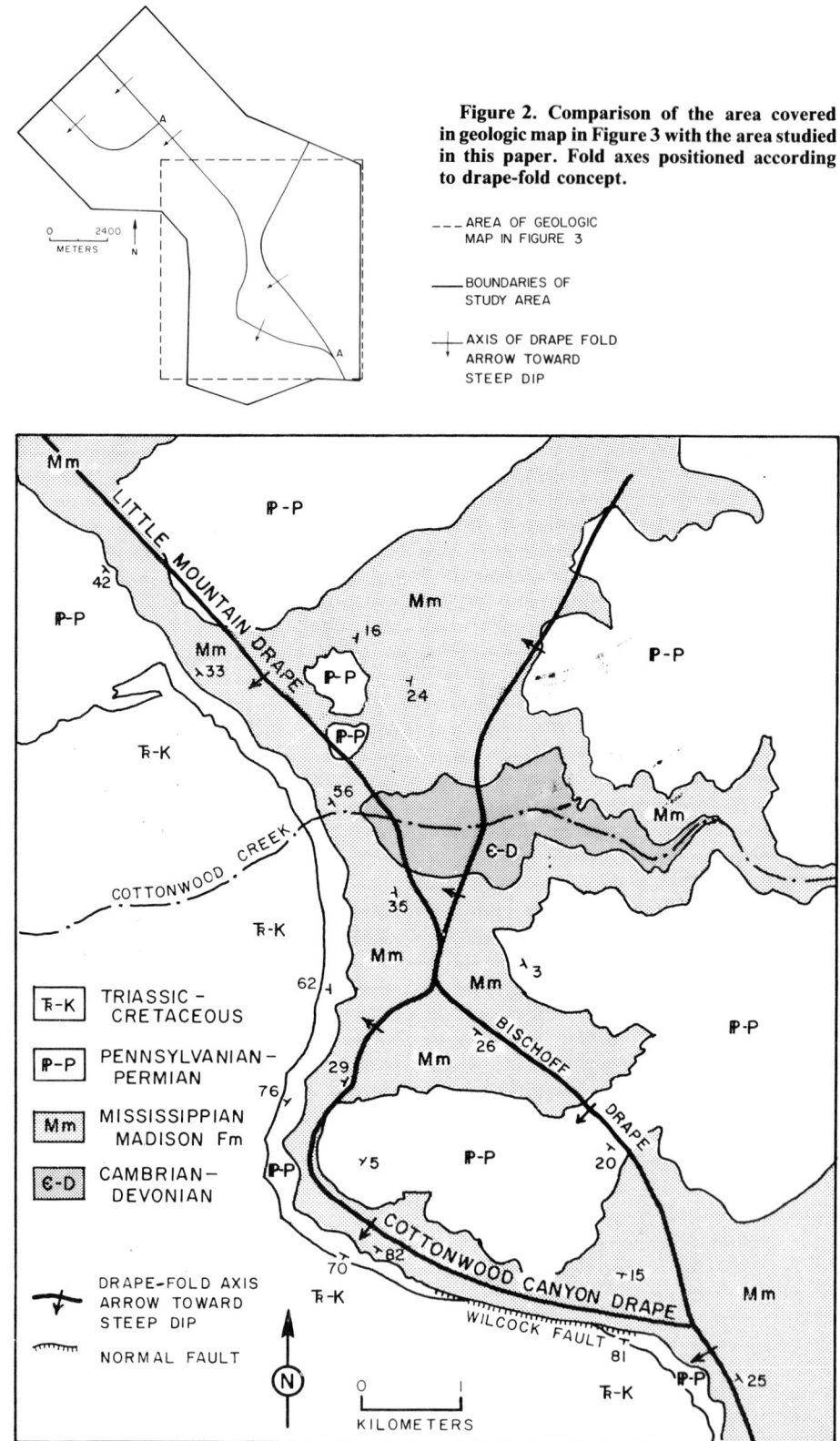

Figure 2. Comparison of the area covered in geologic map in Figure 3 with the area studied in this paper. Fold axes positioned according to drape-fold concept.

Figure 3. Geologic map of the Cottonwood Canyon area (after Voldseth, 1973).

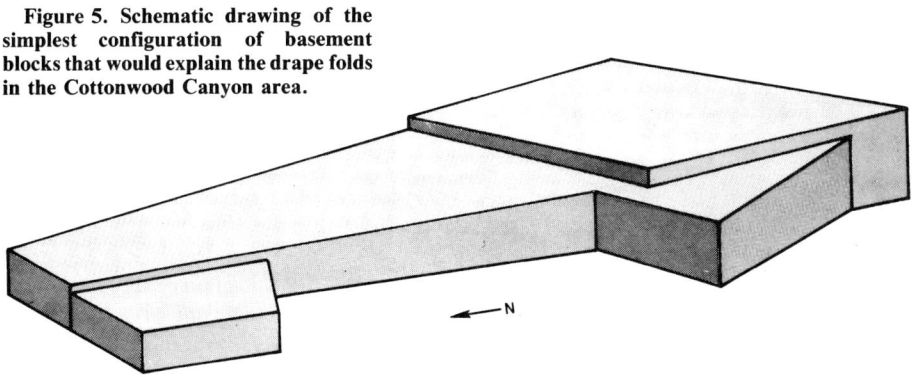

Figure 4. Partial stratigraphic column for the Cottonwood Canyon area.

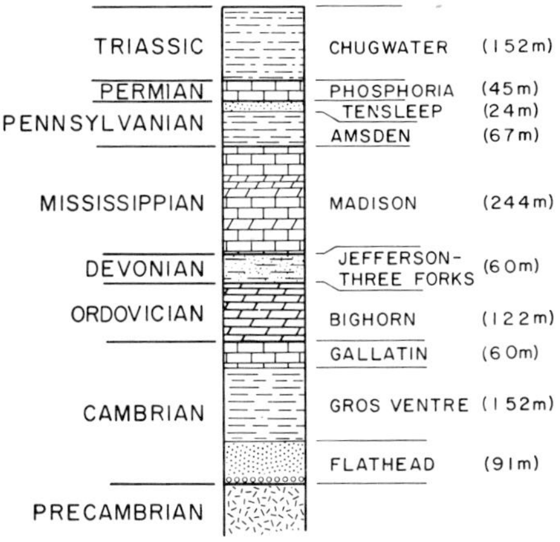

Figure 5. Schematic drawing of the simplest configuration of basement blocks that would explain the drape folds in the Cottonwood Canyon area.

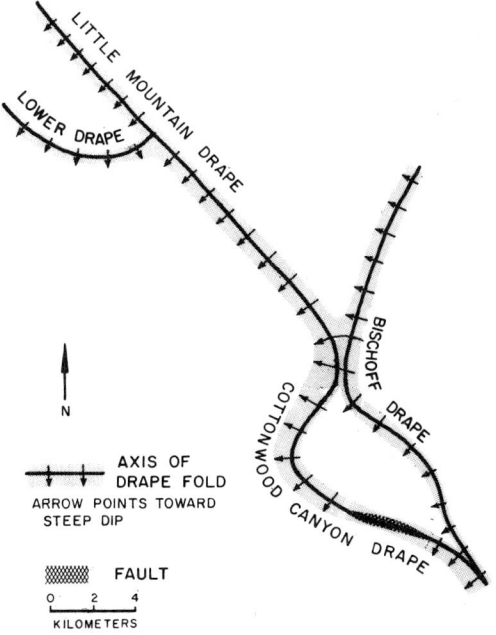

Figure 6. Axes and names of drape folds in the study area. The width of the shaded area is proportional to the amount of displacement on the Mississippian Madison upper surface. The cross-hatched area is scaled to the amount of displacement on the Wilcock fault (see Fig. 3).

the uniform deposition over such a broad area, there are no significant changes in the unit thickness over the 103-km^2 study area. Therefore, it can be assumed that the Mississippian Madison Formation was horizontal and of relatively uniform thickness before deformation. Figure 4 is a stratigraphic column showing rocks present in the study area and their undeformed thickness.

GEOMETRY

Laramide deformation forced the Mississippian Madison layer to deform into a series of multiple drape folds. Although there is no Precambrian basement exposed in the area, the regional geology necessitates that these folds be considered forced folds caused by uplift of the Precambrian basement beneath the sedimentary layers (Stearns, 1970). Figure 5 is a perspective drawing of the simplest configuration of basement blocks that could be responsible for the drape folds observed at the Madison level. Projection down to basement from the sedimentary layers indicates that the tops of all the basement blocks are nearly flat. Figure 6 shows the axis of each of the drape folds which overlie the faulted basement. If the block terminology of Stearns (1971; see Fig. 5 in Stearns, this volume) is used, the axis is in the hinge area between blocks 1 and 2. The arrows point toward the steep limb of the fold (in the direction of block 3). The lengths of the arrows and widths of the shaded areas are proportional to the amount that the upper surface of the Madison layer has been vertically displaced. Therefore, Figure 6 represents the map geometry of the folds and also the vertical displacement of each fold. In the Cottonwood Canyon area, most of the displacement is by folding, but some of the displacement is by discrete faulting along the small Wilcock fault (see Fig. 3). The cross-hatched area of the southeastern end of the Cottonwood Canyon drape (Fig. 6) represents the only place in the study area where the Madison Formation is faulted.

The Cottonwood Canyon drape is the largest continuous drape fold in the area. It changes strike abruptly from northwest to northeast to form a sharp outside corner; then just south of Cottonwood Creek it swings back to a northwest strike to make an inside corner. This folding occurs over a single irregularly shaped block (Fig. 5). To the northwest this feature is referred to as the Little Mountain drape (Fig. 6).

The second largest fold is the Bischoff drape fold (Figs. 3 and 6), which is structurally higher than the Cottonwood Canyon feature. This Bischoff fold forms over a separate basement block (Fig. 5) and like the Cottonwood Canyon fold forms a sharp outside corner just south of Cottonwood Creek. To the north this fold exhibits less throw because of the slight rotation of the upper surface of the Cottonwood block (Fig. 5).

The third and smallest feature in the area is called the Lower drape (Fig. 6) and represents a small bifurcation along the Little Mountain drape (Fig. 5). This feature is north of the area mapped by Voldseth (1973) and shown in Figure 3.

The region has drape folds that range in throw from 91 m (300 ft) (Lower drape) to 1,340 m (4,400 ft) (maximum throw on the Cottonwood Canyon drape). The relationship between the steepest dip on a particular fold and the total amount of throw is in accordance with the sequence of development for drape folds proposed by Stearns (1971). The folds such as the Lower drape have not yet developed a block 3 and are still gently (12°) dipping in the steepest parts. However, by the time the throw has increased to that of the Cottonwood Canyon drape, the steep beds are vertical and block 3 is fully developed. None of the folds has enough displacement to have developed a complete block 4, but it is interesting

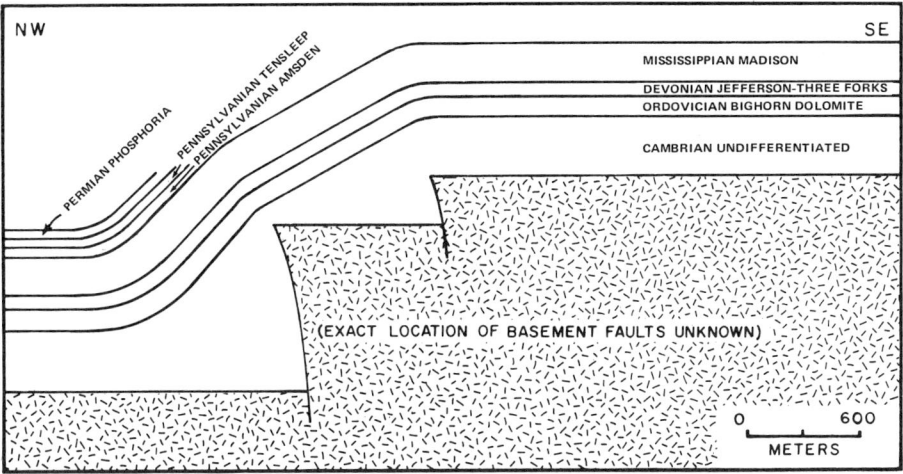

Figure 7. Cross section along Cottonwood Creek (see Fig. 3).

that the inception of a block 4 can be seen in front of Cottonwood Canyon where the total displacement is the greatest.

Another area that is geometrically interesting is in Cottonwood Canyon along Cottonwood Creek (Fig. 3). Here the Bischoff drape and the Cottonwood Canyon drape are separated by less than 1.6 km (1 mi). The distance between the potential lower hinge of the Bischoff fold and the upper hinge of the Cottonwood Canyon fold is too short for the Madison carbonates to form distinct hinges. Rather they simply rotate as a single rigid surface to accommodate the area between the folds (Fig. 7). There is not enough material available in the carbonate layer to develop upper and lower hinges as it does in either direction away from this area. It is presumed that this different, local behavior is accommodated by shale flowage in the Cambrian part of the section, but the outcrop does not permit observation deep enough in the section to be sure.

THREE-DIMENSIONAL MODEL

In order to more fully appreciate the present geometry, especially the interrelationship between separate drape folds, a solid three-dimensional model of a single surface in the folded region was constructed (Fig. 8). In many regions such a construction would involve too much extrapolation to place any reliance on the finished product. However, the region surrounding Cottonwood Canyon is sufficiently well exposed both horizontally and vertically to make a solid, scale model feasible. The upper surface of the Mississippian Madison Formation is the best surface to work with because of good outcrop. With such a surface it is possible to select any line across the model, compare its predeformation length (the horizontal projection of the line) to its postdeformation length measured on the model surface and calculate the displacements required by the deformation. Such calculated movements do not speak to how the displacement occurred or what cumulative deformation occurred above or below the layer. They do, however, establish what displacements must be accounted for in this single layer if we are to understand the folding process.

The area covered by the model is slightly larger than the area mapped by Voldseth (Fig. 3). Within the area of the model 20 cross sections were drawn. Many of

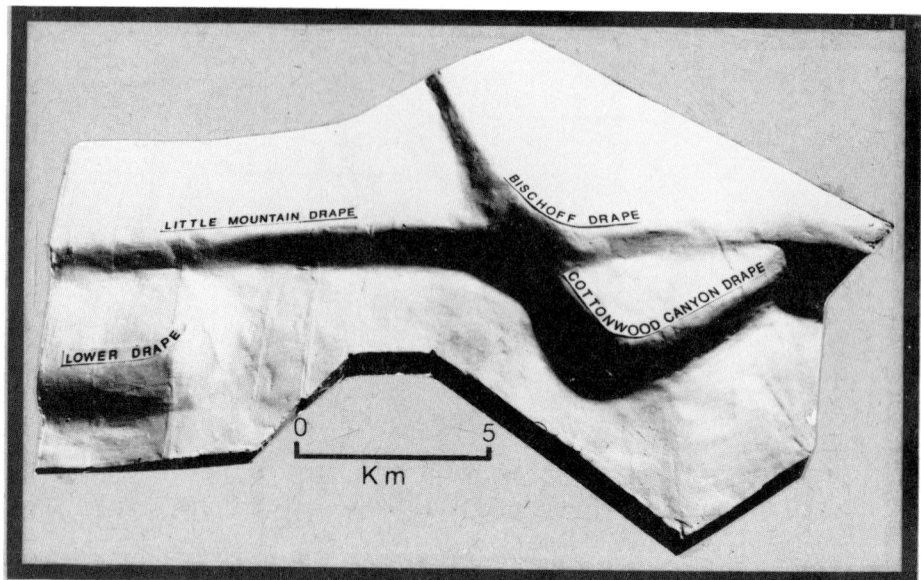

Figure 8. Photograph of the scale model of the upper surface of the Mississippian Madison Formation in the study area.

these cross sections were drawn parallel to canyons, and all are so controlled that confidence limits for the shape of the upper surface of the Mississippian Madison are high. The cross sections were then carefully cut out of Plexiglass sheets and mounted vertically on a base drawn to the same scale as the cross sections. This essentially produced a three-dimensional fence diagram, or ribbed solid surface. The areas between the Plexiglass cross sections were then filled with plaster of paris and smoothed to a solid surface, which is a three-dimensional representation of the upper surface of the Mississippian Madison (Fig. 8). To better appreciate the scale, Figure 9 shows four 1:1 cross sections for the Madison

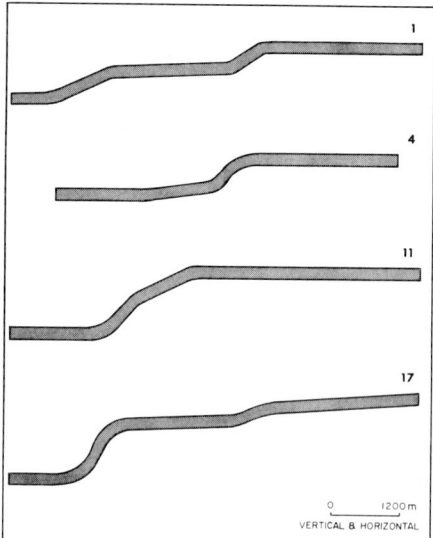

Figure 9. Cross sections of the Madison layer at four different locations in the study area; numbers refer to locations in Figure 12.

Figure 10. Photographs of different portions of the study area with their location relative to the model indicated by arrows.

layer 215 m (700 ft) thick. Field evidence demonstrates that this surface is a continuous, unfaulted surface with the exception of one small tear (Wilcock fault) in the southern part of the region (Fig. 3). This small discontinuity in the displacement field is relatively insignificant to the total displacements that occurred within the sheet. It can be seen from Figure 8 that, with the exception of this small fault, the displacement field from the predeformation surface to the deformed surface is a continuous displacement field and can, therefore, be treated without concern for discontinuities. From such a continuous surface, deformed line lengths in virtually any direction can be compared to their undeformed lengths, and the total displacement vector parallel to the line can be computed. The deformed length of a line is measured directly on the model with a pithometer. The undeformed length is the horizontal projection of the line onto a plane parallel to the model's base.

Figure 10 illustrates four different views of the drape folds in the study area with the locations of each photograph relative to the model indicated. It can be seen that there is a great deal of outcrop control for the modeled surface.

Comparison of Figures 3 and 8 shows that when a single surface through the drape folds is considered, the fold axes are not as sinuous as they appear on a geologic map. The model illustrates that the forcing member that caused the folding is composed of fairly regular blocks whereas the geologic map might imply irregularities along the edges of the forcing blocks. Figure 5 is the schematic representation of the simplest, but still realistic, block configuration for the basement that can be constructed from information from the model. This representation shows that there must be a minimum of three basement blocks. The true basement configuration must be something close to that shown in Figure 5. There is no attempt to draw the true dip for the frontal faults on the basement blocks in Figure 5. For purposes of illustration they are assumed to be vertical, and that they are, in fact, nearly vertical will be argued later.

The upper surface of the scale model can be considered an infinitely thin sheet at the top of the Madison Formation. The total amount of displacement that would be required in going from a predeformation, horizontal sheet the size of the study area to the deformed state represented by the model can be calculated along any line by simply taking the difference between the line length along the deformed surface and the length of the same line projected onto a horizontal plane. This is justified because field data clearly substantiate that there are no discontinuities in the deformed surface. Such implied displacements were computed along the lines shown in Figure 11. Figure 12 is the same line configuration projected onto a horizontal plane. Figure 13 shows arrows whose lengths indicate the amount of displacement calculated along each line in Figure 12. The calculated displacements in Figure 13 are all drawn on the downthrown side of the structures, but this

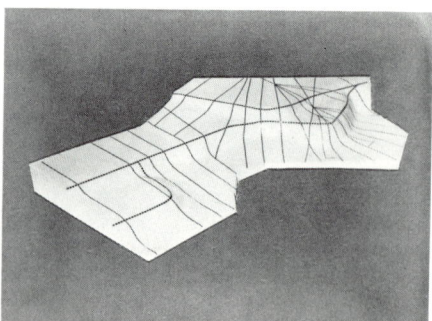

Figure 11. Scale model of the upper Madison surface showing the location of lines for which displacements were calculated. The dashed lines indicate the axes of the drape folds.

should not be interpreted as meaning that all of the displacement occurred here. It could have occurred at any place along the line of cross section. A given arrow in Figure 13 indicates nothing more than the cumulative total displacement needed along the particular line in order to have continuous displacement from a horizontal surface to the nonhorizontal configuration shown in Figures 8 and 11. These calculations do not demonstrate either where or how the displacement occurred. For example, the displacements could have been accomplished by thinning, fracturing, elastic strain, or any combination of these mechanisms that occurred within the Madison Formation during deformation. Certain of these mechanisms will be investigated later, but for the present time, Figure 13 shows what displacements are required by whatever cumulative mechanisms that were operating within the layered rocks during deformation.

It can be seen from Figure 13 that the total displacements needed are both large and varied in direction. This, then, becomes the definition of the problem involved, not the solution to the problem. If we are to thoroughly understand the deformation in this area, or similar areas, we must be able to explain displacements that occur both parallel to dip cross sections and oblique to dip cross sections. Figure 13 clearly indicates that the physical motions that occurred for various points within an infinitely thin layer at the top of the Madison Formation were not uniform in either direction or magnitude.

ARGUMENTS TO EXPLAIN CALCULATED DISPLACEMENTS

The simplest way to account for the excess material in the folded rock layers is to assume large-scale basement shortening. Either large-scale underthrusting or

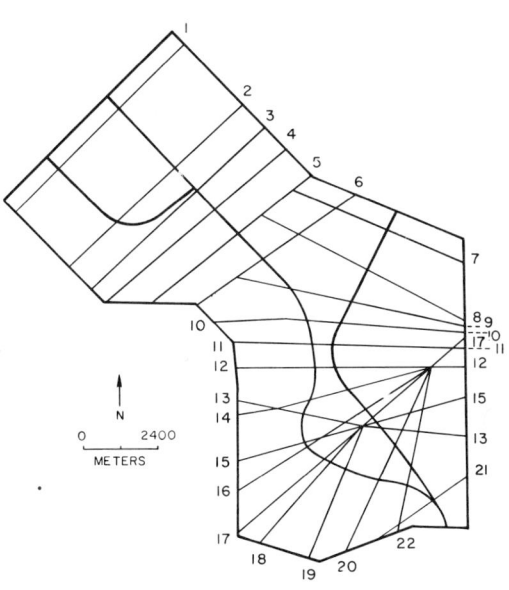

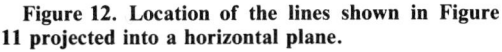

Figure 12. Location of the lines shown in Figure 11 projected into a horizontal plane.

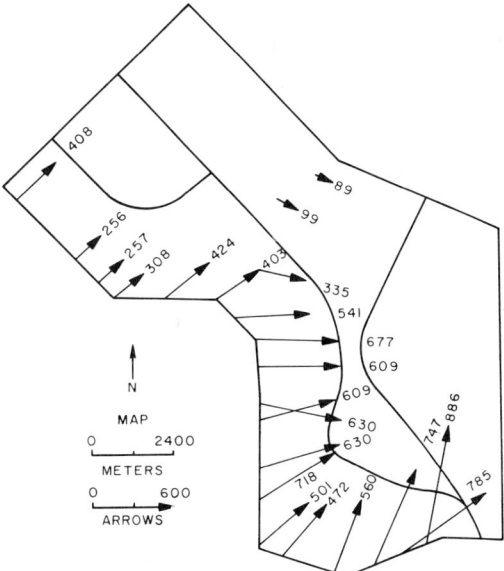

Figure 13. The calculated displacements along lines shown in Figure 11; calculations are based on vertical basement faults. Arrows are scaled to the amount of displacement for each line; displacements given are in metres.

overthrusting can easily be incorporated into any cross section as long as the basement is not exposed. There are three prime factors, however, which make the acceptance of large-scale horizontal motions in the upper surface of the basement untenable. The first is that the tops of the blocks in the Cottonwood area are nearly flat lying. Virtually any form of reasonable overthrusting or underthrusting would produce rotated blocks, and the blocks here are not rotated but rather are plateau-type uplifts. The second reason that large-scale horizontal movements are not acceptable in the upper basement of this region is that elsewhere in the Rocky Mountains foreland there is conclusive evidence (exposures or indisputable seismic data) in the sedimentary section that where the fault does leave the basement at a low angle, the fault continues to propagate up through the sedimentary rocks. Examples of low-angle basement faults that continue through the sedimentary rocks are along the southern front of the Owl Creek Mountains, the southwestern front of the Wind River Mountains, and either the north or south flank of the Uinta Mountains. However, in the study area the sedimentary rocks are unfaulted; this indicates that the basement faults are high angle. The third and most devastating argument against large-scale basement shortening is, as pointed out by Stearns (this volume), the three-dimensional aspects of the folds. If the long block fronts are low-angle faults in the basement, then the termination of these blocks, such as the end of the Lower drape or the corners in Cottonwood Canyon block (Fig. 5), would have to show evidences of lateral displacements along the faces that are perpendicular to the main fault front. There not only is no evidence for such displacement, but rather there is ample evidence that all displacements at block terminations are nearly vertical (Fig. 10). The fact that the draping in the layered rocks is continuous over the entire area as indicated in the model in Figures 8 and 11 speaks directly to the fact that the faults in the basements have to be high angle. Furthermore, the entire region surrounding the Cottonwood Canyon area (the northern Big Horn Mountains and Pryor Mountains) is characterized by high-angle faulting in the basement. Therefore, not only is it impossible to accomplish the displacement field indicated in Figure 13 by low-angle faults, it also calls on considerable coincidence to maintain that the unexposed basement faults here are completely different from any of those that are exposed in the region.

What dip limits, then, are acceptable for the basement bounding faults? Because of the many sharp, nearly right-angle corners in the area, it is not feasible to accept dips on the frontal faults of less than 75°. A normal fault in the basement places the overlying sediments in extension and thus causes faulting—not drape folding. Therefore, normal-fault movement can be excluded; the movement must be reverse. Even 75°-dipping reverse faults on both fronts of a square-cornered block are impossible unless the slip on the fault plane is oblique on each of the frontal faults. The lower the dip of the fault plane, the more oblique the movement has to be to accomplish both the flat-topped nature of the blocks and the corner configuration. Along the drape folds over these faults there is no evidence of oblique slip. Therefore, even accepting 75° as the lowest dip that is feasible is perhaps departing from reality. Nearly vertical faults are much more acceptable on the basis of evidence where the basement fault can actually be seen, such as at Rattlesnake Mountain or the Beartooth Mountains, which are both in the same region as the study area. In order to establish limits that are acceptable, all of the calculations that follow will be computed for two cases: (1) vertical faults in the basement and (2) a limiting condition of reverse faults dipping 75°.

Figure 14 illustrates the amount of apparent displacement along the same lines as shown in Figure 13 that would be required if all basement faults are reverse and dip at 75°. Displacement numbers are only low along lines where fault displacements are low. One number even becomes negative; this indicates that periods exist during the development of drape folds when there is excess material [as shown by Weinberg (this volume)]. However, there is still a considerable amount of displacement to be accounted for if we are to understand the generation of the surface shown in Figure 8, even if we assume the difficult case of 75°-dipping reverse faults.

The most obvious way to account for the apparent discrepancies in length between lines in predeformation and postdeformation surfaces is to thin the material across the fold, hold volumes constant, and thereby lengthen the surface. In fact this is precisely what layered rocks do over basement faults when the rock types involved have easily activated thinning mechanisms. Lightly cemented sandstones, for example, that overlie the uplifted basement blocks of the Uncompahgre show considerable thinning even over small-displacement basement faults (Stearns, 1970). The Pennsylvanian Weber Sandstone also thins in the drape folds of the eastern Uinta Mountains (Cook and Stearns, 1975). Vaughn (1976) likewise reported considerable thinning in the Mesozoic shale sections as they drape over the uplifted Casper Mountain block in Wyoming.

However, Stearns (1971) reported no thinning in the Paleozoic carbonates at Rattlesnake Mountain over a large drape. The Paleozoic section in Cottonwood Canyon is very similar to that studied by Stearns. Within the region of this study, complete sections of Paleozoic carbonates are not exposed any place except in the upper reaches of Cottonwood Canyon. However, the entire Madison Formation does crop out across the drape fold and the front of Cottonwood Canyon. In the upper reaches of Cottonwood Canyon where the beds are horizontal, an instrument-measured thickness for the Mississippian Madison is 216 m (709 ft). In the mouth of the canyon where the rocks are steeply dipping, the same interval measured 211 m (693 ft). The 5 m (16 ft) of indicated thinning could be stratigraphic, but even if it were all assumed to be structural it only amounts to 2.2% thinning within the entire Madison unit. No other place in the study area has exposures of complete sections of both undeformed and deformed Madison Formation.

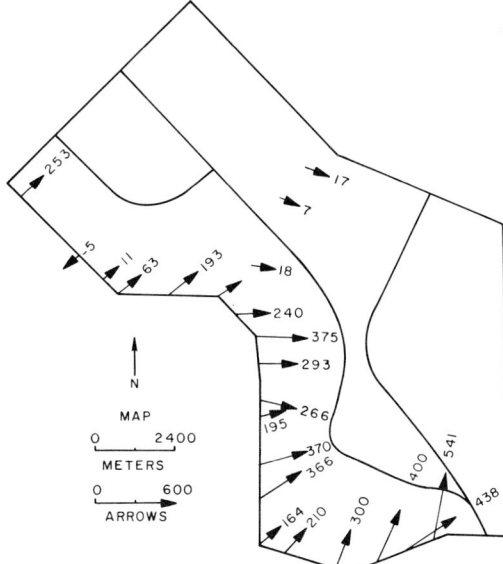

Figure 14. Amount of calculated displacement required if boundaries of basement blocks are 75°-dipping reverse faults. Arrows are scaled to the amount of displacement for each line; displacements given are in metres.

However, it should be pointed out that the greatest amount of total throw on the Madison surface occurs in Cottonwood Canyon, and if the rocks are not thinned there, it would be difficult to accept that they are drastically thinned in other places. The 2.2% thinning, even if it were structural and not stratigraphic, would not be nearly enough to satisfy the fold requirements by thinning alone. For example, the Cottonwood Canyon drape fold as exposed in Cottonwood Canyon would require a uniform thinning of 24% in the Madison. This thinning would have to start on the upper side of the top hinge and continue through the lower hinge. If the thinning was nonuniform, then certain areas would have to be thinned even more than 24%. Even if the basement fault is considered to be a 75°-dipping reverse fault, the uniform thinning would have to be 19% and the thinning would be greater if it were nonuniform. The amount of thinning, therefore, that would be needed to totally satisfy the displacements indicated in Figures 13 and 14 would be so large as to be obvious in the field. Such thinning is not present in the carbonate sections exposed in the canyons throughout the area.

To further substantiate the lack of thinning, a series of samples from the Madison Formation were collected across the face of the Cottonwood Canyon drape fold. Petrographic examination of many of the samples showed no calcite twins at all. In those samples in which calcite twins were present, the maximum strain that could possibly be accounted for through the twinning mechanism is 2%, and this only occurs in some of the samples. Therefore, it must be concluded that very little of the needed displacement indicated in Figures 13 and 14 can be accounted for by thinning of the thick carbonate sequences.

Another possible way to account for the needed displacements would be by volume increase (or density decrease) due to fracturing in the layered rocks. The carbonate rocks across the drape folds are fractured by ordered sets of fractures that Stearns and Friedman (1972) called type-1 and type-2 fracture assemblages associated with folding. The question then becomes, How many fractures would be needed to account for the indicated displacements in the folding process in the absence of thinning? Elkins (1953) calculated maximum and average fracture openings that could exist in Sprayberry Sandstone at depth. The maximum width that he reported was 0.33 mm (0.013 in.), and the average width was 0.051 mm (0.002 in.). Using the average fracture width calculated by Elkins and even considering that the fracturing is uniform for 0.5 km in either direction from the upper and lower hinge of the fold, it would still require 2,338 fractures per metre (713 fractures per foot) to accomplish the displacements if the basement fault is vertical. If the basement fault is dipping at 75°, 558 fractures per metre (170 fractures per foot) would be needed to accomplish the calculated displacements in the Madison. These figures are minimal because it is difficult to imagine that fracturing could occur in a uniform fashion as much as 0.5 km away from both the upper or lower hinge of the fold. Even under these minimal conditions, the number of fractures per metre that are required is too high to be acceptable. Therefore, although fractures may contribute to small-scale movements in the layers, they cannot be called upon to explain very much of the total displacements (Figs. 13 and 14).

Without being able to accomplish the needed displacements by thinning, fracturing, basement shortening, or by combination of these features, it would seem that the inescapable conclusion is that most of the displacement must occur by detachments and intrastratal slip within the layered carbonate rocks. This is the same conclusion that Stearns (1971) was forced to reach at Rattlesnake Mountain. The only difference is that Figures 13 and 14 clearly illustrate that major slip directions are not always perpendicular to structural strike. Further, the Cottonwood

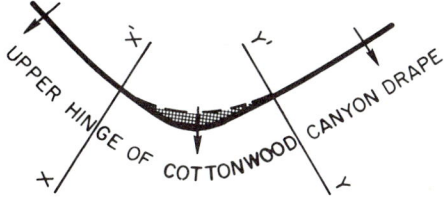

Figure 15. A comparison of the trace of the upper hinge of the Cottonwood Canyon drape fold to a circular arc. Lines XX' and YY' mark the ends of the linear parts of the drape fold.

Canyon area also demonstrates that displacements between layers cannot be uniform in either direction or magnitude. This means that to truly understand the deformation of these layers, it is necessary to use three-dimensional solutions. Two-dimensional cross-sectional solutions are insufficient, especially near any corners in the drape folds.

VOLUME CONSIDERATIONS AT THE CORNERS OF BLOCKS

If a piece of cloth is forced over a solid square corner, such as a table corner, a series of subsidiary folds that fan out from the corner form due to an excess of cloth at the corner. Such folds in rocks are frequently referred to as "tablecloth folds." It can be seen from the examination of the model in Figure 8 that no such folds occur at the sharp corners in either the Cottonwood Canyon, the Bischoff, or the Lower drape folds. Furthermore, such "tablecloth folds" are not seen at the corners of any of the major blocks in the Wyoming province. Their absence has long presented students of this fold style with yet another paradox in the fold mechanisms that occur in the carbonate sequences. Careful examination of the outside corner in the Cottonwood Canyon drape fold perhaps sheds some light on part of the solution to this enigma. On the outside corner of the Cottonwood Canyon fold, the upper and lower hinge traces are curved; but on either side of the corner, both the upper and lower hinges of the drape fold form fairly linear traces that are subparallel to each other (Figs. 8 and 10). Figure 15 shows the upper hinge trace for the Cottonwood Canyon fold with two lines, XX' and YY', drawn perpendicular to the end of the linear portions of the hinge trace. If extended, these lines would intersect on the flat upper surface of the fold. The dashed line in Figure 15 is an arc of a circle drawn from this intersection. It can be seen that the true trace departs slightly from the shorter circular arc. The departure of the true trace from the circular arc is a "bulge" in the hinge trace. Such bulging could be produced by flowage in the Cambrian shale sections. This is precisely the pattern of flowage that Vaughn (1976) observed at an outside corner at Casper Mountain. Here she found that Mesozoic shales away from the corner were thinned and the material was withdrawn into a thickened section at the corner that would produce a "bulge" in the stratigraphically higher rocks. It is also observable in the field area under study that shales units, such as the Amsden, do thin and thicken across the folds (the no-thickening rule applies only to the thick carbonate sections). However, the slight bulging that can be observed in Figure 15 is insufficient to satisfy the total absence of tablecloth folds at the corners. The amount of lateral shortening that would be required along XX' and YY' is 450 m (1,500 ft) in order to accommodate the steep folds at the Cottonwood Canyon corner. Transport into the corner along XX' and YY' for a distance of 450 m (1,500 ft) would produce a greater shortening than is reflected by the bulge in Figure 15. This observation once again attests to our limited knowledge of the mechanisms that operate in the folding of thick sections of carbonate rock.

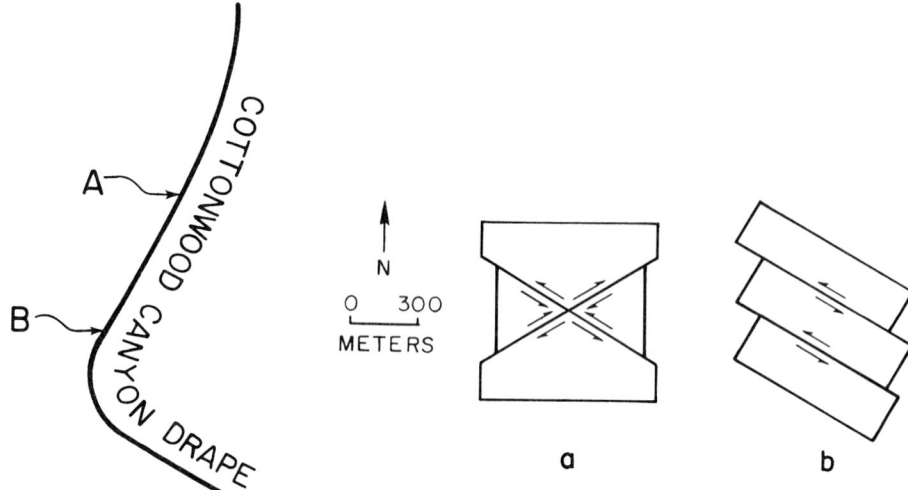

Figure 16. Location of fracture counts made from oblique aerial stereo photographs.

Figure 17. Schematic illustration showing the difference in mass transport directions along (a) equal development of two shear fractures of the same conjugate set and (b) overdevelopment of one shear-fracture direction.

There is one field observation that suggests, however, that the rocks are trying to move into the lower parts of corners. In the regions where the drape folds are linear, there is a good development of both the type-1 and type-2 shear fractures. Except for small local variations there is no obvious overdevelopment of either of the shear fractures that form the conjugate shear set for type 2. That is, each shear-fracture set seems to have about the same numerical development. However, near the corners there is a distinct preferential development of one of the two type-2 shear fractures. Macrofracture counts were made from stereo low-level oblique aerial photographs. Although there may be small fractures that cannot be seen on these photographs, the offset associated with them is too small to aid significantly in the lateral transport of large volumes of rock. Away from the corner in the Cottonwood Canyon drape fold (loc. A, Fig. 16), the two shear fractures of type 2 are developed such that 57% of the fractures belong to one shear set and 43% belong to the other. Closer to the corner (loc. B, Fig. 16) the relative shear-fracture development is unbalanced with 76% of the fractures belonging to the left-lateral shear of the type-2 set and only 23% belonging to the right-lateral shear. When both shear fractures are equally developed, the net motion is to elongate the material in a direction parallel to the bisector of the obtuse angle between the shear fractures (Fig. 17a). Overdevelopment of only one shear fracture has the net effect of transporting material in a lateral direction as shown in Figure 17b. These fracture patterns indicate that the rock has tended to move toward the corner. However, it should be pointed out that the total displacements on the shear fracture are not nearly sufficient to accommodate the indicated movements around the corners. They merely show the tendency of the rock to move in that direction, but how the movement is accomplished is still unsolved.

CONCLUSIONS

That the multiple drape-fold geometry in the Cottonwood Canyon area is accomplished without tearing the rock layer is an observational fact. Therefore, it is possible to compute the displacements along any line in the deformed layer that would be required in going from a horizontal layer the size of the study area to the nonhorizontal deformed layer described by the model. The calculated displacements indicate movement of material in a complex pattern which is nonuniform in magnitude and direction. These displacements cannot be accounted for by internal mechanisms such as thinning or fracturing. There is some indication of how volumes may be compensated in corners of drape folds, but there is no real solution as to how the entire sheet deforms. Therefore, it is concluded that physical detachments exist within the section and that lateral transfer in several directions is required. All of this points to the necessity, if we are going to understand this folding process, of doing three-dimensional mechanistic and displacement-field solutions. The displacements reported in this paper represent only the cumulative displacements; more work needs to be done on point-to-point displacements. Such a study is currently in progress for this area. The aim is to calculate the point-by-point displacement vectors needed in going from an undeformed sheet to the observed deformed sheet. The results will tell us precisely in which areas we have to call upon large-scale slip and which of the areas are internally compensated. Other studies are currently in progress to delineate the precise deformation and detachment requirements within the entire Mississippian carbonate section. We need to know, for example, how much of the displacement can be accounted for by elastic distortion. Can any of the motion be internally compensated by juxtaposition of anticlines and synclines across the entire fold? Other means available to accomplish the required displacements are presented by Weinberg (this volume), who considers the effects of shifting the hinge from bottom to top of a layer during early development of a drape fold. We have just begun to learn what the true kinematics of such folds are, and Weinberg (this volume) shows that detailed kinematic studies of a two-dimensional drape fold do not agree with our intuitive ideas of how a drape fold forms. He further demonstrates that there are critical periods in the fold growth that are more demanding on the displacement field than others. It will not be until we have integrated all such studies that we can even hope to gain an understanding of such a complicated deformation pattern. However, in the meantime we should not continue to deny the observational facts from the field simply because we cannot explain them with current methodology. Rather we should use field observations, such as the implied displacement patterns presented in this paper, as the basis for continuing studies of surface rocks that have been deformed into forced folds.

ACKNOWLEDGMENTS

The Department of Geology and the Center for Tectonophysics, Texas A&M University supported this work. Travis J. Parker and Robert J. Stanton, Jr., reviewed the manuscript and made suggestions for improvement that we have incorporated. Funding for field work was supplied by the National Science Foundation Grant DES74-22954. We thank Karen McBride for her assistance in the field during the summer of 1975.

REFERENCES CITED

Berg, Robert R., 1962, Mountain flank thrusting in Rocky Mountain foreland, Wyoming and Colorado: Am. Assoc. Petroleum Geologists Bull., v. 46, p. 2019–2032.

Cook, Robert A., and Stearns, David W., 1975, Mechanisms of sandstone deformation; a study of the drape folded Weber Sandstone in Dinosaur National Monument, Colorado and Utah, in Bolyard, D. W., ed., Symposium on deep drilling frontiers of the central Rocky Mountains: Denver, Rocky Mtn. Assoc. Geologists, p. 21–32.

Elkins, L. F., 1953, Reservoir performance and well spacing, Sprayberry trend area field of West Texas: Am. Inst. Mining and Metall. Engineers Trans., v. 198, p. 177–196.

Haun, John D., and Kent, Harry C., 1965, Geologic history of Rocky Mountain region: Am. Assoc. Petroleum Geologists Bull., v. 49, p. 1781–1800.

Prucha, John J., Graham, John A., and Nickelsen, Richard P., 1965, Basement-controlled deformation in Wyoming province of Rocky Mountain foreland: Am. Assoc. Petroleum Geologists Bull., v. 49, p. 6–32.

Richards, Paul W., 1955, Geology of the Bighorn Canyon–Hardin area, Montana and Wyoming: U.S. Geol. Survey Bull. 1026.

Stearns, D. W., 1970, Drape folds over uplifted basement blocks with emphasis on the Wyoming province [Ph.D. dissert.]: College Station, Texas A&M Univ., 118 p.

———1971, Mechanisms of drape folding in the Wyoming province: Wyoming Geol. Assoc., 23rd Ann. Field Conf., Wyoming Tectonics Symp., Guidebook, p. 125–143.

———1975, Laramide basement deformation in the Bighorn Basin—The controlling factor for structures in the layered rocks: Wyoming Geol. Assoc., 27th Ann. Field Conf., Geology and mineral resources of the Bighorn Basin, Guidebook, p. 82–106.

———1978, Faulting and forced folding in the Rocky Mountains foreland, in Matthews, V., III, ed., Laramide folding associated with basement block faulting in the Western United States: Geol. Soc. America Mem. 151 (this volume).

Stearns, D. W., and Friedman, M., 1972, Reservoirs in fractured rocks, in King, R. E., ed., Stratigraphic oil and gas fields: Am. Assoc. Petroleum Geologists Mem. 16, p. 82–106.

Vaughn, Patty H., 1976, Mesozoic sedimentary rock features resulting from volume measurements required in drape folds at corners of basement blocks—Casper Mountain area, Wyoming [M.S. thesis]: College Station, Texas A&M Univ., 93 p.

Voldseth, Mils, 1973, Geology of the Cottonwood Canyon area, Bighorn Mountains, Wyoming [M.S. thesis]: Ames, Univ. Iowa, 102 p.

Weinberg, D. M., 1978, Some two-dimensional kinematic analyses of the drape-fold concept, in Matthews, v., III, ed., Laramide folding associated with basement block faulting in the Western United States: Geol. Soc. America Mem 151 (this volume).

Manuscript Received by the Society June 27, 1977
Manuscript Accepted August 25, 1977

Geological Society of America
Memoir 151

Origin of Elk Mountain anticline, Wyoming

JAMES E. MCCLURG
Department of Geology
University of Wyoming
Laramie, Wyoming 82071

VINCENT MATTHEWS III
Amoco Production Company
Security Life Building
Denver, Colorado 80202

ABSTRACT

Analysis of the geometric relationships at Elk Mountain, Wyoming, indicates that the anticline is a forced fold that formed by the draping of ductile sedimentary strata over an uplifted and rotated basement block that is broken into three secondary blocks. Orientation of structures in the region seem more compatible with a stress field generated by a vertical force rather than an east-west, horizontal compressive force.

INTRODUCTION

The Elk Mountain structure is located at the northern end of the Medicine Bow Mountains, some 100 km northwest of Laramie, Wyoming (Fig. 1). It is a large (64 km^2) asymmetrical anticline with an exposed core of Precambrian granitic rocks. The structure was originally mapped by Beckwith (1941), who interpreted it to be underlain by a scoop-shaped thrust fault. He proposed that the structure formed from east-west compressive forces which bent the eastern part of the block into a curve convex to the east. He further depicted the Precambrian basement rocks as being folded along with the overlying sedimentary strata.

Our field work during 1976, experimental data on rock mechanics, and analysis of existing geologic information suggest a mode of origin involving vertical forces that caused brittle block faulting of the basement and drape folding of the ductile sedimentary cover.

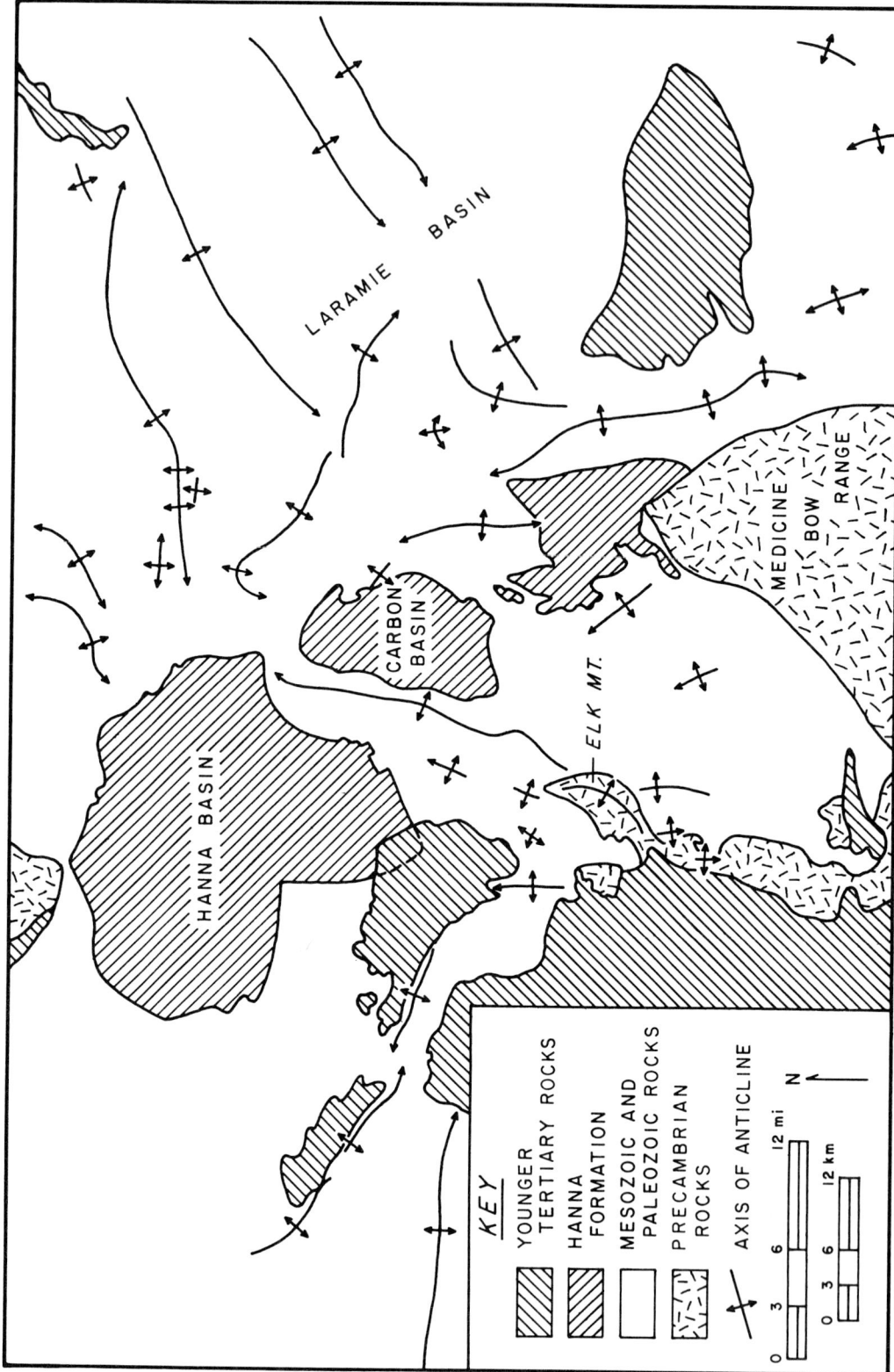

Figure 1. Regional map showing location of axes of anticlines (from Beckwith, 1941).

DESCRIPTION OF STRUCTURAL RELATIONSHIPS

Elk Mountain is a large, roughly triangular structure which is cored by Precambrian granitic rocks that are exposed over approximately 25 km^2. Although there are many small, complicated structures around its margins, the large anticline itself is rather simple. The southeast and northeast limbs are steeply dipping; the western and northwestern limbs are planar and gently dipping (Fig. 2). Foliation within the Precambrian granite trends uniformly east with dips ranging from 60°S through vertical to 60°N (Houston and others, 1968).

Sedimentary Strata

The northeastern limb is nearly linear and strikes N18°W. Strata of the Madison Limestone are vertical near the contact with the Precambrian granite. Away from the contact the strata are overturned toward the southwest. There is an abrupt change in strike of about 70° between the northeastern limb and the southeastern limb.

The southeastern limb is divided into three linear segments separated by abrupt changes in strike. Where the Madison Limestone is exposed near the contact with the Precambrian granite, it dips 55° to 65° toward the southeast. The strata progressively steepen away from the contact, and the Casper Formation is overturned in most places. Intense fracturing of the granite near the contact between the Precambrian rocks and Madison Limestone suggests that it is a fault contact rather than a depositional contact.

The strata in the west limb are planar over approximately 14 km^2, and the strata in the northwestern limb are planar over an area of 15 km^2. Along the northwestern edge of the northwestern limb, the strata abruptly steepen to 50°.

Precambrian Core

The planar, pre-Madison erosion surface that was developed on the Precambrian granite is well preserved in the exposed Precambrian core of the fold (Fig. 3). This pre-Madison erosion surface can be divided into three discrete planar sections that are separated by two northward-trending lineaments (Fig. 2). The attitude of the planar surface in the eastern section of the core is N63°E, 11°NW. In the central section the attitude is N34°E, 11°NW, and in the western section it is due north, 27°W.

The two linear boundaries between these three sections are marked by stream courses; this indicates zones of weakness. A search for outcrops along the two lineaments revealed no good exposures. However, many exposures in the near vicinity of the lineaments revealed intensely fractured, and/or brecciated granite, which suggests proximity to a fault zone. We conclude that these two lineaments are fault zones cutting across the core of the anticlinal structure. However, although the sedimentary strata on the northwest limb of the structure change strike where the faults project under them, they show no evidence of being faulted.

Where these lineaments terminate against the southeast limb of the structure, they correspond to the abrupt changes in strike of the three linear segments of the southeastern limb of the structure. Projection of these linear zones to the north under the sedimentary strata shows that they correspond to the boundaries of the west and northwest limbs of the fold. Also, the projection of the planar erosion surface under the strata of the west and northwest limbs shows a marked correlation with the planar attitudes of the strata.

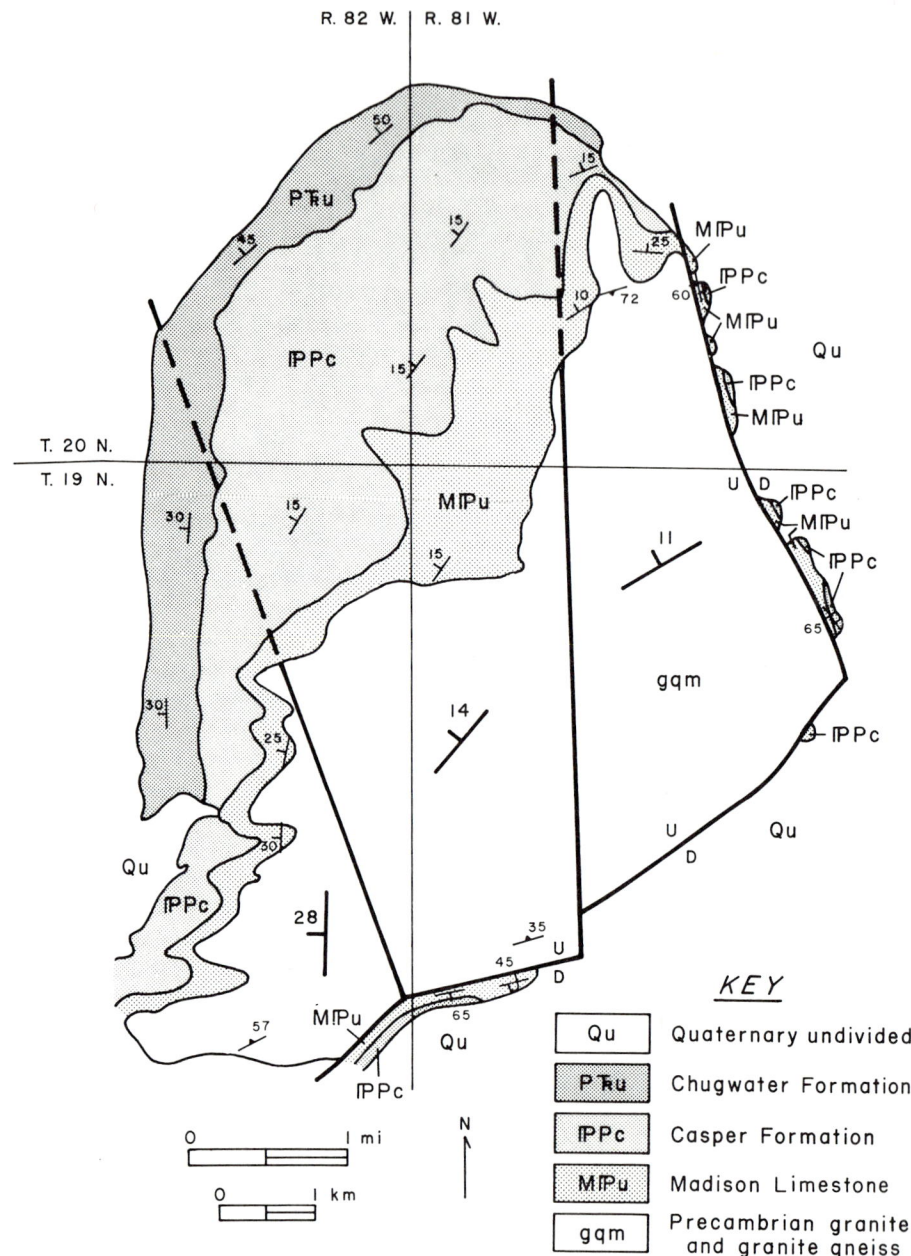

Figure 2. Geologic map of Elk Mountain [modified from Houston and others (1968) and Beckwith (1941)]. Strike and dip symbols in Precambrian rocks represent attitude of pre-Madison erosion surface determined from topographic maps.

Figure 3. Northeast view of Elk Mountain showing well-preserved, planar erosion surface (P) developed on the Precambrian granitic rocks. Field of view is approximately 7.5 km. C = Contact (nonconformable) between Madison Limestone and Precambrian rocks; M = planar strata of Madison Limestone; D = Paleozoic strata draping over the northeast edge of uplifted block of Precambrian rocks.

Location of Major Fault

Beckwith (1941), as well as Houston and others (1968), mapped a major thrust fault along the steeply dipping northeastern and southeastern limbs of the anticline. They placed the trace of the thrust fault in sedimentary strata approximately 500 m east of the Madison-Precambrian contact. However, it is nowhere exposed, and they show it as being concealed beneath Quaternary cover along the entire length of both limbs. They also mapped the contact between the Madison Limestone and the Precambrian granite as a depositional contact. We disagree with these interpretations.

The contact between the Madison Limestone and the Precambrian granite is not exposed along the northeastern and southeastern limbs, but we interpret it as a fault contact because the granite near the contact is intensely fractured. Moreover, we consider the dip of the Madison strata nearest the contact to represent the dip of the fault plane—that is, 55° to 65°SE along the southeastern limb and vertical along the northeastern limb. It is not clear to us why the strata in both limbs are overturned at some distance from the contact. At Rattlesnake Mountain,

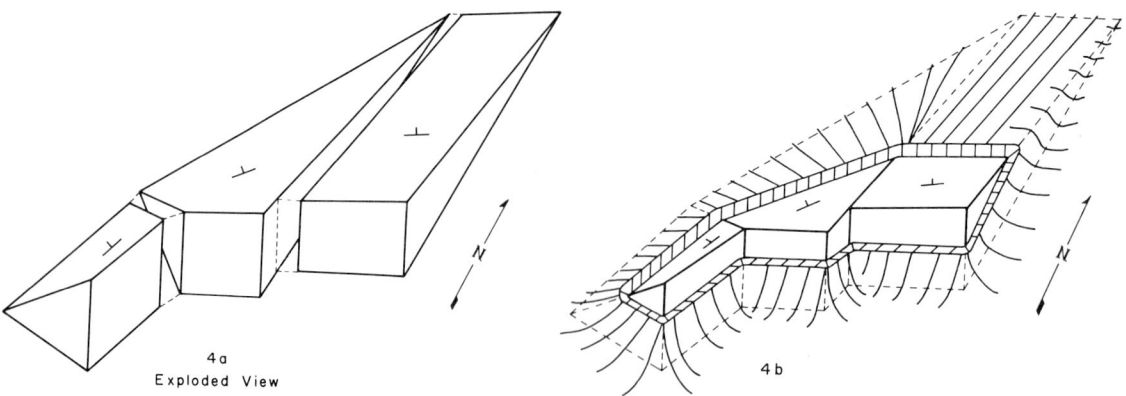

Figure 4. Block diagrams showing configuration of basement blocks (a) that are responsible for the fold geometry in the overlying sedimentary strata (b).

Stearns (1971) observed an analogous relationship that was caused by minor reverse faulting in the strata of the steep limb. Another possible cause might be gravitational instability of the steeply dipping strata. The exposures at Elk Mountain do not appear adequate to resolve the origin of this phenomenon.

ORIGIN

Our analysis of the Elk Mountain structure suggests a different origin than the one envisioned by Beckwith—one that entails vertical rather than compressive forces and fracturing rather than folding of the basement. In essence, it is a model utilizing the "drape folding" mechanism of Stearns (1971, and this volume).

Laboratory data suggest that crystalline rock that is deformed under confining pressures equivalent to the burial conditions during the Laramide orogeny will behave in a brittle fashion, whereas sedimentary rocks should behave in a ductile fashion and form a drape fold (Stearns, this volume). Our analysis of the Elk Mountain structure indicates that the sedimentary rocks did indeed behave ductilely, whereas the basement behaved brittlely. In our view, the structure fits a model wherein a large primary block of granitic basement was uplifted, rotated toward the northwest and broken into three secondary blocks (Fig. 4a).

The planar, gently dipping strata in the western and northwestern limbs are simply resting passively on the tops of the western and central secondary blocks. The abrupt steepening of dips in the northwest limb reflect draping over the northwestern edge of the central block (Fig. 4b).

The steep northeastern limb is caused by strata that are draped over the edge of the easternmost secondary block. The steep, southeastern limb of the fold is caused by strata draping over the southeastern edges of the secondary blocks. The strata of the southeastern limb are linear in front of each secondary block but change strike twice as they pass from one block to the next.

Regional Stress Field

If Elk Mountain is the result of vertical uplift along a normal fault rather than a scoop-shaped, low-angle thrust fault, then the need for an east-west, horizontal compressive force is gone. The uplift of Elk Mountain could easily be related to a stress field generated by a vertically directed force.

Indeed, an examination of the orientation of structures in the region raises serious questions about the validity of an east-west, horizontal force during Laramide deformation. In a stress field generated by east-west compression, one would expect a subparallel alignment of fold axes. However, Beckwith's regional tectonic map reveals no such strong orientation. Instead, one finds an almost random orientation (Fig. 1). We conclude that the diverse orientation of structures in this part of Wyoming is more compatible with vertically directed forces than with an east-west horizontal force.

CONCLUSION

In conclusion, we suggest that a model of drape-folded sedimentary rocks over an uplifted, rotated, and nonfolded basement block appears to explain more of the geometry of Elk Mountain than does one involving a folded basement and east-west compression. It is also more consistent with the diverse orientation of anticlinal structures in the region.

ACKNOWLEDGMENTS

We are grateful to L. T. Grose for reviewing the manuscript and making many helpful suggestions. Amoco Production Company granted permission to publish this paper.

REFERENCES CITED

Beckwith, R. H., 1941, Structure of the Elk Mountain district, Carbon County, Wyoming: Geol. Soc. America Bull., v. 52, p. 1445–1486.

Houston, R. S., McCallum, M. E., King, J. S., Ruehr, B. B., Myers, W. G., Orback, C. J., King, J. R., Childers, M. O., Matus, Irwin, Currey, D. R., Gries, J. C., Stensrud, H. L., Catanzaro, E. J., Swetnam, M. N., Michalek, D. D., and Blackstone, D. L., Jr., 1968, A regional study of rocks of Precambrian age in that part of the Medicine Bow Mountains lying in southeastern Wyoming: Wyoming Geol. Survey Mem. 1, 167 p.

Stearns, David W., 1971, Mechanics of drape folding in the Wyoming province: Wyoming Geol. Assoc., 23rd Ann. Field Conf., Guidebook, p. 125–143.

——1978, Faulting and forced folding in the Rocky Mountains foreland, in Matthews, V., III, ed., Laramide folding associated with basement block faulting in the Western United States: Geol. Soc. America Mem. 151 (this volume).

Manuscript Received by the Society June 27, 1977
Manuscript Accepted August 25, 1977

Printed in U.S.A.

Geological Society of America
Memoir 151

Laramide structure of the Black Hills uplift, South Dakota–Wyoming–Montana

ALVIS L. LISENBEE
Department of Geology and Geological Engineering
South Dakota School of Mines and Technology
Rapid City, South Dakota 57701

ABSTRACT

The 290-km-long Black Hills uplift extends from the South Dakota–Nebraska border to southeast Montana. A structurally higher east block of northward trend and a northwest-trending west block are separated by the west-facing Fanny Peak monocline. Paleozoic, Mesozoic, and Paleocene sedimentary rocks 2,220 to 3,000 m thick overlaid the anisotropic Precambrian igneous, metasedimentary, and metaigneous basement during Laramide uplift. Draping of the sedimentary section over basement fault blocks produced monoclines with, opposed to, and parallel to regional dip and ramps, terraces, and anticlines as subsidiary structures. Individual monoclines as much as 170 km long consist of linear segments commonly joined in gentle arcs. The dominant fold trend is northward, although one major segment of the Black Hills monocline trends northwest. The folds terminate at intersections with other monoclines, by decreasing stratigraphic offset along strike, or by splaying into several folds of lesser structural relief.

Gravity data suggest single basement faults beneath steep, narrow folds and multiple faults beneath wider structures. In a single locality the basement is exposed in fault contact with Paleozoic rocks that show a narrow zone of rotation in both fault blocks. Normal and reverse faults locally extend as high as the massive Mississippian and Pennsylvanian carbonate rocks, but elsewhere these units are folded into five linear segments. Mesozoic siltstone and shale are believed to be more rounded in section view due to flexural flow. In the inclined limb conjugate fault sets and horizontal tension cracks affect the brittle rocks, and ductile layers are thinned by flexural flow.

The uplift is separated from the Powder River Basin on the west by the Black Hills and the Fanny Peak monoclines which have structural relief as great as 1,670 m and dips as great as 90°. Gently inclined planar sedimentary units, inferred to parallel the surface of the basement, show changes of a few degrees dip across the folds. Such differences may result from rotation of rigid blocks of the anisotropic basement along curved faults in the style outlined by Stearns (1975).

The arcuate eastern boundary with the Interior Lowlands province forms a partial dome. Strain was distributed across at least 32 km, possibly by shear along the Precambrian schistosity. North- and east-trending folds, believed to reflect longitudinal and transverse faults in the arched basement, locally parallel the basement fabric. At this easternmost margin of the Wyoming province, the uplift and adjoining province were both elevated, although unequally, whereas the basin to the west of the uplift was an area of subsidence. This difference in absolute movement patterns is believed responsible for the asymmetry and composite structural style of the Black Hills uplift.

INTRODUCTION

The Black Hills of South Dakota and Wyoming are commonly referred to as a classic example of doming of the basement. Highly generalized geologic maps and birds-eye view sketches (Fig. 1) are used to illustrate the elliptical outcrop pattern and topographic expression of Cambrian through Tertiary strata surrounding a north-trending core of Precambrian granite and metamorphic rocks. The dome is regarded as a window to the basement resulting from erosion of an isolated fold in the otherwise flat-lying strata of the Great Plains region of the United States. The character of this outcrop pattern is misleading, however, in that the structural extent of the Black Hills uplift is greater than is the topographic expression, and the structural margin of the uplift is represented by monoclinal flexures for much of its length. The uplift is, in fact, the easternmost and least deformed of the Laramide uplifts of the Rocky Mountain region.

Based chiefly on a compilation of data from literature of the past 100 yr, this paper reviews the gross structural character of the Laramide uplift and attempts through the use of examples from four of the better known areas to show the detailed nature of folds and faults produced during this period of deformation. As the structural relief of this uplift is of a moderate nature compared to other uplifts of the Wyoming province (Fig. 2), it is hoped that such a study may be useful in understanding the early stages of their formation.

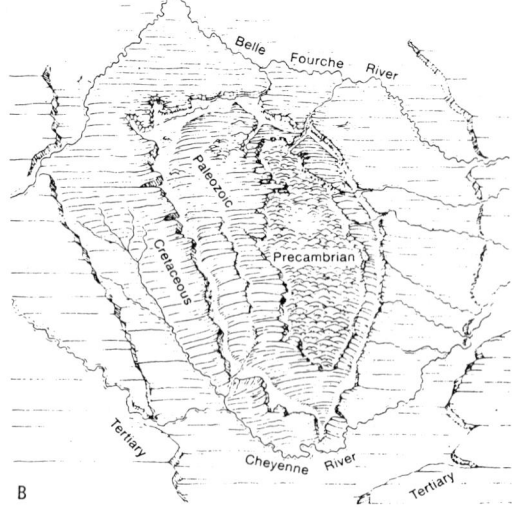

Figure 1. Diagrammatic illustrations of the Black Hills uplift. (A) "Birdseye view" from south (Newton and Jenney, 1880). (B) Oblique sketch from south (Hunt, 1972).

ROCK UNITS

Igneous and Metamorphic Basement

Precambrian basement is exposed only in the east-central part of the Black Hills uplift (Fig. 3). Redden (1975, p. 21) in the best synopsis of information to date states: "The Precambrian rocks . . . consist largely of metamorphosed sedimentary rocks and lesser amounts of metabasalt and metagabbro. Two small areas of older granite have also been metamorphosed, but the younger Harney Peak Granite has experienced little change. At least three and perhaps as many six separate episodes of deformation and probably two metamorphic events have affected the rocks." The metamorphic rocks are predominantly phyllites and schists derived from eugeosynclinal black shales and graywackes. The metamorphism reaches sillimanite grade at about the latitude of the Harney Peak Granite southeast of Hill City, South Dakota (Fig. 3). A low metamorphic grade trough extends northwest across the north-central part of the Precambrian terrane. The original sedimentary rocks, with a total thickness of at least 18,100 m, are believed to have been deposited between 2.5 b.y. ago, the age of the metamorphosed older granites, and 1.7 b.y. ago, the age of the Harney Peak Granite (Redden, 1975, p. 25). On the basis of an increase in metamorphic grade toward the northeast in the area of Deadwood, South Dakota, Noble and Harder (1948, p. 970) inferred the existence of another body of granite beneath the sedimentary cover in that direction. A test hole to the basement near Hulett, Wyoming, drilled by the U.S. Geological Survey penetrated 16.7 m (55 ft) of schist, granodioritic gneiss, and gabbroic gneiss (Blankennagel and others, 1977, p. 48). The granodiorite gneiss is similar to the older granite mentioned by Redden (P. H. Rahn, 1973, oral commun.). On the basis of analyses from five samples from drill holes, Steece (1975, p. 30) showed the basement beneath the area south of a line approximately through Edgemont and Hot Springs, South Dakota, to consist of gneiss and schist, with ages ranging from 1,410 to 1,460 m.y. B.P.

In his geologic map, Redden (1975) showed that the metamorphic units have

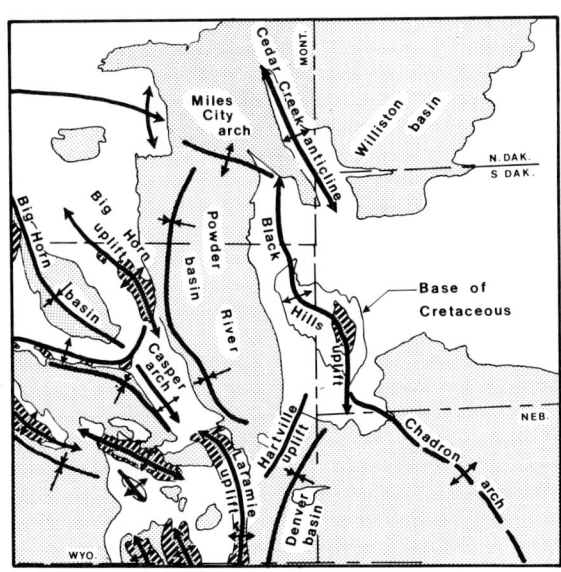

Figure 2. Regional tectonic setting of the Black Hills uplift in the Wyoming province. Diagonal pattern indicates Precambrian basement, dotted pattern indicates Tertiary deposits. Approximate location of axes of uplifts and basins shown with heavy line.

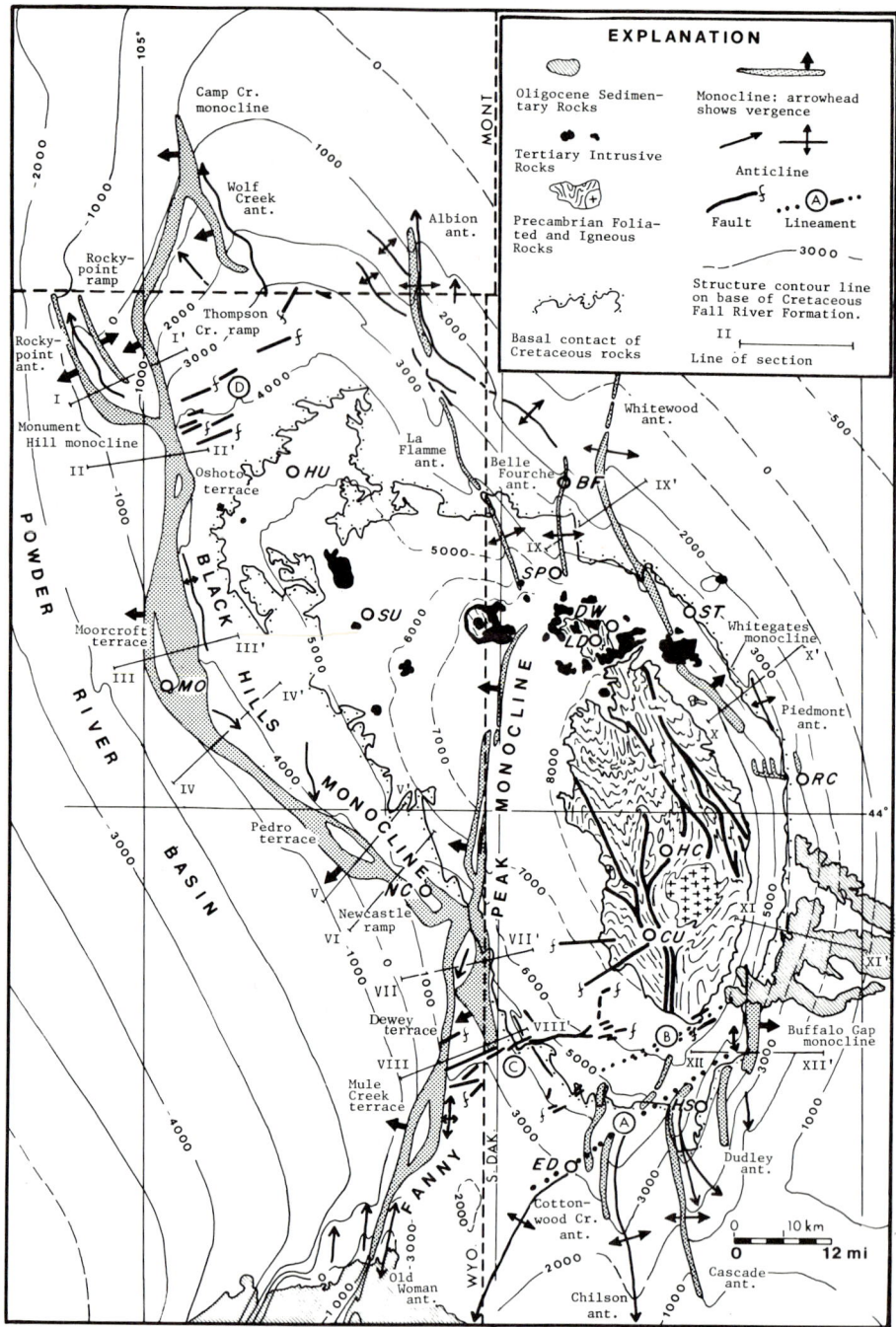

Figure 3. Tectonic map of the Black Hills uplift. Lineaments and fault zones are A, Edgemont; B, Long Mountain; C, Dewey; D, Little Missouri. Structure contour interval is 1,000 ft (302 m). Structure of Precambrian rocks by J. C. Redden.

a strong north-northwest trend although they are locally warped around the younger granite body (Fig. 3). He stated that for more than half of the area the dip of bedding and foliation is very steep or vertical. Thus, the basement of this region is strongly anisotropic both in rock type and in structural grain, but its exact nature at the margin of the uplift is unknown owing to lack of nearby exposures.

Sedimentary Cover

Phanerozoic sedimentary rocks of the Black Hills uplift are divisible into four sequences on the basis of origin and age (Lisenbee, 1975, p. 54). Sequence one is marine, sequence two is dominantly continental, sequence three is dominantly marine, and sequence four is continental. This sedimentation is tied to the broader tectonic picture of long-term epeirogenic warping interrupted by the basin and uplift tectonics of the Laramide. With the exception of the youngest rocks of sequence three, the first three sequences predate Laramide structural activity. Sequence four postdates Laramide deformation.

Sequence one consists of marine Paleozoic rocks of shelf facies that contain several disconformities. Silurian rocks are absent, and a karst surface is developed on Mississippian limestones. The basal units of this sequence are Cambrian and Ordovician sandstone, shale, and conglomerate that overlie the Precambrian basement complex along an essentially planar surface that has no more than 30 m of local relief (Mickelson and Kulick, 1963, p. 41). The upper part of the sequence contains massive limestone, dolomite, and sandy dolomite. Rocks beneath the Mississippian karst surface thin from approximately 670 m in southeast Montana to 0 m in the extreme southeast part of the area shown in Figure 3 (Mallory, 1972). Pennsylvanian rocks show variable thickness but, in general, thin from about 180 m on the west to 90 m in the east in the area of Figure 3 (Mallory, 1972, p. 115).

Sequence two is composed chiefly of continental clastic rocks of Permian through Lower Cretaceous age. Shale and siltstone dominate the sequence, which thickens from north to south across the Black Hills uplift. The Permian and Triassic units range from approximately 180 m on the north to 300 m on the southwest (Rasco and Baars, 1972, p. 146; MacLachlan, 1972, p. 169); Jurassic rocks are 150 to 210 m thick throughout most of the uplift, thickening slightly to the northwest (Peterson, 1972, p. 180), and Lower Cretaceous units reach a maximum thickness of 150 m in the southern part of the uplift (McGookey and others, 1972, p. 197). The Cretaceous rocks are fluviatile sandstone, siltstone, and mudstone (Gott and others, 1974), which represent a basal continental phase of the transgressive sequence of the Rocky Mountain geosyncline.

Clastic rocks of sequence three consist of a lower marine portion deposited in the Rocky Mountain geosyncline and an upper nonmarine portion. These rocks range in age from upper Lower Cretaceous to Paleocene. Marine shale grades upward through regressive sandstone and shale to sandstone, shale, and coal of continental origin. The Cretaceous rocks thicken across the uplift from approximately 1,500 m on the east to 1,950 m on the west (McGookey and others, 1972, p. 199, 207). Paleocene units reach a thickness of 1,570 m in northeastern Wyoming (Robinson and others, 1964, p. 99). Love (1960, p. 206) reported that Paleocene sediments derived from the Black Hills area occur in the Powder River Basin (Fig. 3). Part of the continental sedimentary rocks in the upper part of sequence three would, therefore, represent a transitional phase from a broad geosynclinal sedimentation to local basin and uplift patterns, at least in the immediate area of the Black Hills uplift.

Sequence four is composed of continental clastic rocks of Oligocene age, which lie with angular unconformity on all older rocks of the Black Hills uplift. In the immediate area of the uplift the rocks are conglomerate, sandstone, siltstone, and freshwater limestone. Their presence shows that uplift had ceased by Oligocene time.

Eocene Plutons

The third major group of rocks in the Black Hills uplift consists of alkalic intrusive bodies that postdate and have locally altered the original geometry of Laramide structures. Intrusion occurred in the Eocene along two zones, one in South Dakota and one in Wyoming, with individual intrusive bodies ranging from 39 to 59 m.y. in age (McDowell, 1966, p. 13–15). In the Precambrian basement the intrusive bodies generally have the form of dikes and stocks, although one ring dike complex is known (Welch, 1974). In the overlying sedimentary units dikes, sills, laccoliths, and stocks are present, and several domes are interpreted to be cored by intrusive bodies. Although individual masses may be as small as a few metres, the several centers of emplacement tend to consist of composite intrusive bodies and to range in diameter from 8 to 11 km.

The zone of intrusive centers in South Dakota extends from the area of Sturgis (Fig. 3) west 80 km roughly along a bearing of N75°W to the Wyoming–South Dakota state boundary. West of Sturgis these bodies lie approximately along a hinge line that separates more gently dipping strata on the south from more steeply inclined units on the north. East of Sturgis in the broad, easterly dipping limb of the Black Hills uplift, one igneous body is exposed at the surface and a second is interpreted on the basis of aeromagnetic data and physiography to underlie Cretaceous strata approximately 15 km southeast of the town (Redden, 1975, p. 45).

The zone of intrusive centers in Wyoming is approximately 72 km long and as much as 15 km wide. It extends S25°E from the Little Missouri fault zone (Fig. 3) to about 18 km southeast of Sundance. The largest of the complexes is in the Bear Lodge Mountains north of Sundance and at the apex of a triangular-shaped area underlain by gently northwest-dipping Mesozoic strata. The base of the triangle is the Little Missouri fault zone, and the legs are marked by abrupt changes of many degrees strike in the sedimentary units. Within the triangle the strata are deformed by small, circular domes too small to show at the scale of Figure 3 and several sills and plugs, the best known of which is Devils Tower. This body, along with the Missouri Buttes intrusive complex, lies along the west leg of the triangle. In a borehole in one of the domal structures, igneous rock was penetrated in the Minnelusa Formation at about 606 m depth (Robinson and others, 1964, p. 113), and the other domes are interpreted to have also formed over igneous plutons.

South of Sundance erosion has cut to strata of Triassic and Paleozoic age. Several plutons are exposed and several domes are present in the sedimentary layers.

STRUCTURAL UNITS

Historical Review

Although the diagram by Newton and Jenney (1880) presented here as Figure 1A suggests a domal pattern, cross sections (Fig. 4) and descriptions by these men clearly indicate their awareness of the flat-topped nature of the uplift and the presence of monoclinal flexures at the margins. They state: "It is observable almost everywhere on the slopes of the Hills . . . that in regions of elevation the inclining strata do not make their entire ascent in one continuous dip, but rise from their nearly horizontal position to their highest elevation by steps or waves, between which there may be considerable intervals with little or no inclination." This description relates to the area shown in Figure 1A, which is essentially restricted to South Dakota.

In the first quarter of this century Darton (1902, 1904, 1905), Darton and O'Harra (1907, 1909), Darton and Smith (1904), and Darton and Paige (1925), in a series of U.S. Geological Survey Folios that cover the Black Hills area at 1:125,000 scale, presented the only comprehensive study of the uplift yet available. Darton (1904, p. 7) considered the Black Hills uplift to represent an irregular dome rising on the northern end of an anticlinal axis that extends northward from the Laramie range. In a structure contour map of the top of the Permian Minnekahta Limestone, he shows essentially all the features present in Figure 3 of this paper; however, he gives only a very general discussion of them and does not name individual structures. The uplift as outlined in this map is 290 km long, extending from southeastern Montana to the South Dakota–Nebraska border. The maximum width is 110 km.

Noble (1952, p. 31) considered the uplift to consist of two nearly flat-topped blocks separated by a north-trending monocline, approximately along the Wyoming–South Dakota border. The eastern block is more elevated with exposures of Precambrian rocks in the core and shows a broad, but gentle, monoclinal flexure along its eastern margin. This is the area shown in the view of Newton and Jenney (Fig. 1A). The second, northwest-trending block is bounded on the west and southwest by monoclinal folds. The two blocks are separated by a monocline that is clearly defined in the structure contour map of Darton (1904), although he does not discuss its significance. Shapiro (1971a, 1971b) studied this northward-trending monocline and also reviewed the structural character of the margin of the uplift.

Regional studies of the northwestern, western, and southern parts of the uplift and the adjoining Powder River Basin have been published by Robinson and others (1964), Dobbin and others (1957), and Gott and others (1974). Additional maps and reports concerning local areas too numerous to mention individually are available for parts of the uplift, but, for much of the area, especially along the east side and north end of the eastern block, the work of Darton and his colleagues in the early part of this century remains the standard reference.

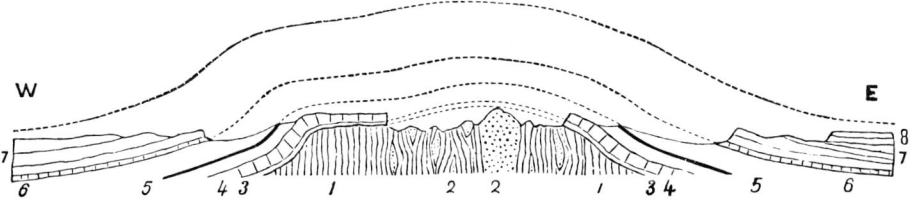

Figure 4. Diagrammatic cross section of the Black Hills uplift. Vertical scale is about six times the horizontal scale. Broken lines indicate the portion of the uplift removed by erosion. 1, Archean slates and schists; 2, granite; 3, Potsdam (resting unconformably on units 1 and 2); 4, Carboniferous strata; 5, red beds with included limestone; 6, Jurassic strata; 7, Cretaceous strata; 8, Tertiary White River strata (resting unconformably on unit 7). After Newton and Jenney (1880).

Definition of Terms and Tectonic Subdivisions

Most Laramide folds in the Black Hills uplift are divisible into two basic types—those that face toward the axis of the uplift and those that face away from the axis. Both have a narrow, steeper limb connecting long, more gently inclined planar limbs. Geometrically this results in either an asymmetric anticline or a monocline. Kelley (1955, p. 32) noted similar folds on the Colorado Plateau, both of which he regarded as monoclines. Those that face updip he called monoclines opposed to regional dip, and those that face downdip, the classic type of monoclinal flexure, he called monoclines with regional dip. This terminology, pertinent to the character of folds in the Black Hills uplift, will be used here. In addition, there are local examples of monoclines parallel to regional dip.

The monoclines of the margin of the Black Hills uplift are locally modified by subsidiary features called ramps and terraces in this paper. Inasmuch as it is the theme of this paper that all of these features are the result of displacement in the basement and draping of the overlying sedimentary layers, the block diagrams of Figure 5 are introduced to show the geometry of the subsidiary structures and the inferred shape of the underlying basement blocks. In ramp structures the surface of the gently inclined segment dips parallel to the strike of the major fold (Fig. 5a, 5b). Such ramps may be disrupted at their upper or lower ends by small monoclines that trend at an angle to the regionally developed fold (Fig. 5c, 5d). Terraces of two types, protruding (Fig. 5e) and recessed (Fig. 5f), are recognized on the basis of their location relative to the uplifted block.

As noted by Noble (1952, p. 31), the Black Hills uplift is broadly divisible into two north-trending, nearly flat-topped blocks. On the north and south ends of the eastern block, anticlines and synclines plunge away from the top of the uplift. The east flank of this block is a broad, curving, east-facing monocline convex to the east. The west block is bounded by abrupt monoclines on the south and west and plunges as a broad anticlinal nose on the north.

Laramide Faults and Lineaments

Faults of Laramide age are present on the Black Hills uplift, but individual examples are rather limited both in length and amount of offset. Four northeastward-trending zones of faulting or alignment of structural features are present. Local faults of east or northwest trend are also believed to have formed during Laramide deformation. The four northeast-trending zones as shown in Figure 3 are A, the Edgemont lineament; B, the Long Mountain fault zone; C, the Dewey fault zone; and D, the Little Missouri fault zone. These features are found on the western or central portion of the uplift.

The Edgemont lineament trends N60°E for 40 km from Edgemont, South Dakota, and is marked by northeastward bends and northward terminations of the Dudley, Cascade, Chilson, and Cottonwood Creek anticlines (Gott and others, 1974, p. 31). It lies along a concealed structure of Precambrian age, which is indicated by the sharp bend in a magnetic anomaly north of Hot Springs (Meuschke and others, 1963; Kleinkopf and Redden, 1975). Gott and others (1974, p. 31) inferred that the ancient structural zone was reactivated by Laramide stresses, although the exact nature of this deformation in the basement is unknown.

The Long Mountain fault zone (B in Fig. 3) is described by Gott and others (1974, p. 29) as consisting of small, northeast-trending normal faults exposed in rocks of the Cretaceous Inyan Kara Group and the Jurassic Sundance Formation. The zone begins 10 km north of Edgemont and trends N48°E for a distance of

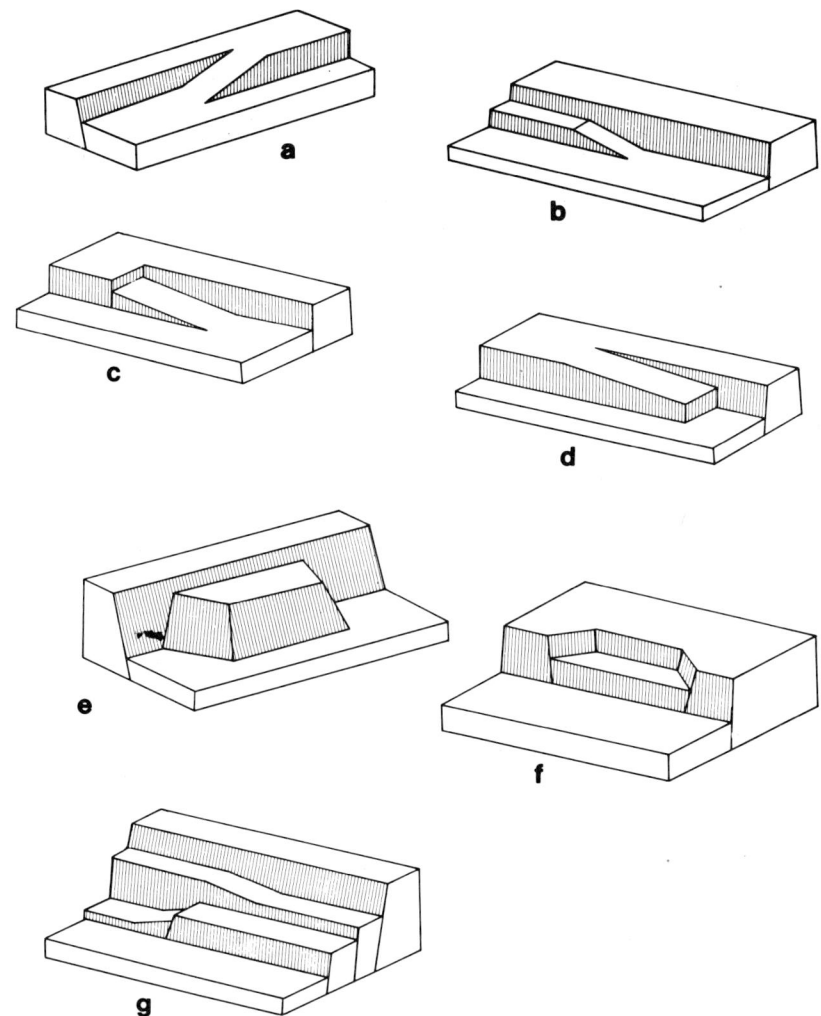

Figure 5. Diagrammatic block diagrams illustrating the character of ramp, terrace, and multiple fault surfaces in the Precambrian basement capable of forming the various structures of the monoclines bounding the Black Hills uplift. The various structures are *a*, ramp structure formed by en echelon overlap of two faults; *b*, ramp formed by termination of one fault; *c*, ramp disrupted at upper end; *d*, ramp disrupted at lower end; *e*, protruding terrace; *f*, recessed terrace; *g*, multiple fault surfaces capable of producing a broad monocline displaying changes in dip or width along strike.

at least 15 km. Individual faults as much as 2 km in length are parallel across an area 7 km wide. Displacement is a maximum of 12 m on individual faults, and there is as much as 18 m of structural relief due to folding of the sedimentary layers. Abrupt terminations or offset of magnetic and gravity features indicate a concealed Precambrian fault of northeast trend along a possible northeast continuation of this zone (Gott and others, 1974, p. 31 and Pl. 2). Gott and others (p. 30) stated that this is a zone of recurrent movement, as mild structural adjustments also influenced the Early Cretaceous sedimentation.

The Dewey fault zone (C in Fig. 3) begins on the west at the north end of Mule Creek terrace on the Fanny Peak monocline. It extends N70° to 75°E for approximately 25 km to a point at which it bifurcates into east- and northeast-trending segments. Although the geology is poorly known, the northeast segment appears to be traceable to the Precambrian-Paleozoic contact, and Redden (1975, oral commun.) stated that it can be followed in the Precambrian rocks at least to Custer. The east-trending segment is traceable for about 10 km (Gott and others, 1974, p. 29). In the western part of the zone Gott and others (Pl. 2) showed several faults in a left en echelon pattern with individual members as much as 13 km in length. Although this pattern suggests possible left slip, direct evidence of such movement is lacking (p. 29). Several smaller faults parallel the longer segments.

Facies studies indicate that the Dewey structural zone underwent minor movement in Middle to Late Jurassic time and again in Early Cretaceous time (Gott and others, 1974, p. 30). Jurassic strata of this age are thin or absent north of the zone, and deposition of channel sandstone deposits of the Early Cretaceous Inyan Kara Group also show effects of uplift along the north side of the zone. Laramide movement on the zone resulted in 150 m of vertical displacement by a combination of faulting and folding.

The fourth northeast-trending zone of faulting lies near the Little Missouri River in Wyoming and is here called the Little Missouri fault zone. From the Black Hills monocline on the west (Fig. 3) it extends N55°E for approximately 50 km. Robinson and others (1964, Pl. 1) showed the zone to be approximately 10 km wide along the rim of the monocline and to be represented by only single, discontinuous faults on the east. The faults are normal with maximum displacement of 30 m and most are downthrown on the north (Robinson and others, 1964, p. 115). The maximum length of any fault in the group is 5 km. The faults generally parallel the strike of the sedimentary units and separate more gently dipping beds on the uplifted block to the south from more steeply inclined beds in the Thompson Creek ramp on the north; in fact, they lie along a broad flexure separating the two areas. This flexure and the zone of faults connects on the west with a slight offset in the Black Hills monocline at the bifurcation of the Monument Hill monocline.

Folds of the West Flank of the Uplift

The boundary between the Black Hills uplift and the Powder River Basin consists of parts of two major and two minor monoclinal flexures whose combined lengths are 300 km in the area from Mule Creek terrace on the south (Fig. 3) to the end of the Camp Creek monocline in the northwest. The major folds, the Black Hills and Fanny Peak monoclines as defined by Brobst and Epstein (1963, p. 356), intersect 10 km east-southeast of Newcastle, Wyoming. The structural relief between the Powder River Basin and the adjacent uplift is 600 to 1,670 m for much of the length of the monoclines. Topographic expression is pronounced only where the monoclines expose the more resistant basal Cretaceous sandstone units, and locally the surface of the basin is topographically higher than the adjoining uplift. Figure 6 includes profiles across these structures at the level of the Cretaceous Fall River Formation, and Table 1 gives pertinent information regarding their geometries and subsidiary structures.

Fanny Peak Monocline South of the Black Hills Monocline Intersection. For 65 km south from the intersection with the Black Hills monocline, the Fanny Peak monocline separates the eastern block of the Black Hills uplift from the Powder River Basin. At the south end of the Mule Creek terrace, a saddle in the crest of the monocline marks the southern end of the uplift but not of the monocline. The latter structure continues southward into the Hartville uplift where it is expressed as a fault separating Precambrian rock on the southeast from Paleozoic rock on

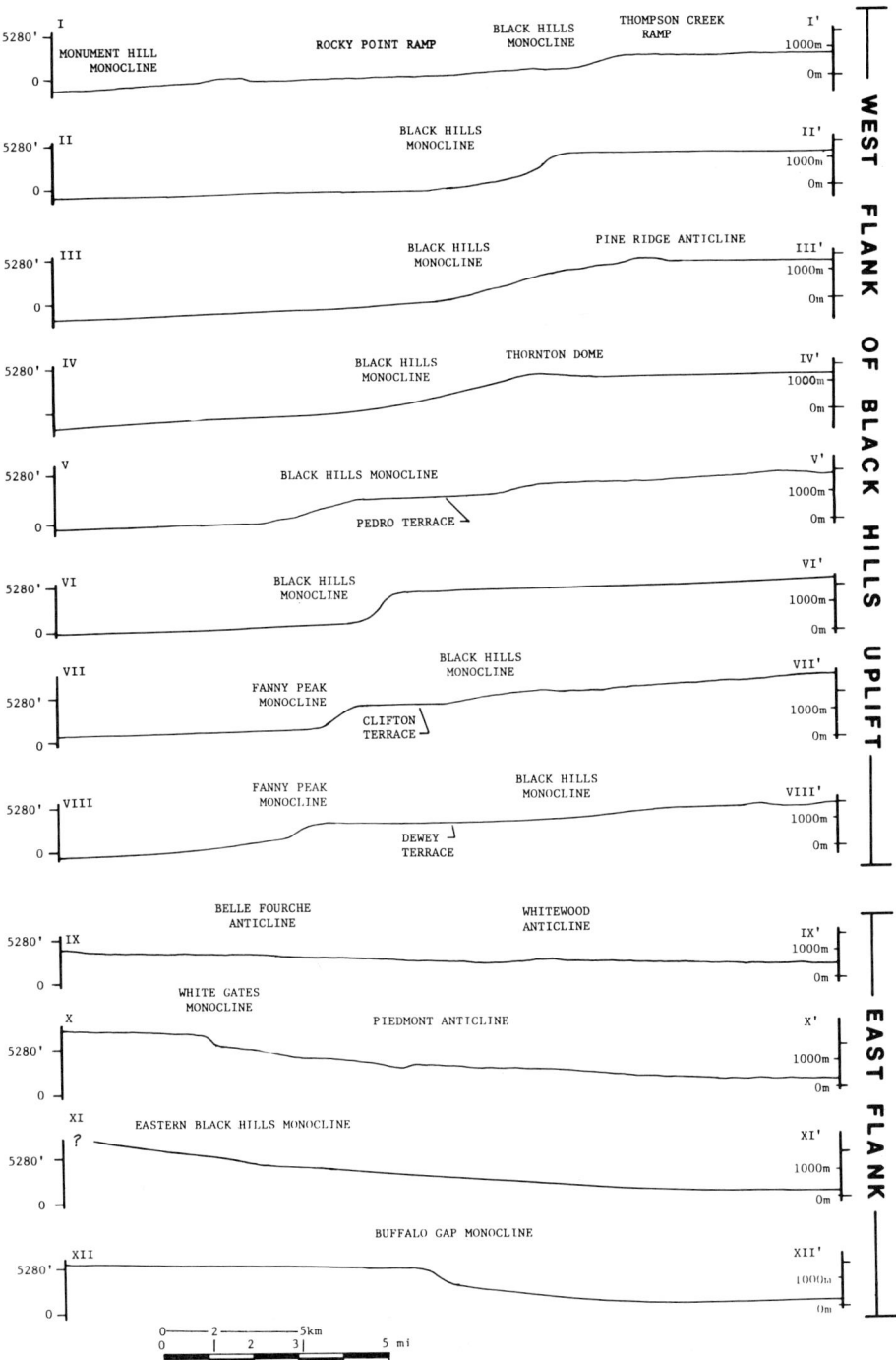

Figure 6. Profiles of margins of the Black Hills uplift. Profiles I through VIII compiled from structure contour data of Robinson and others (1964) and Dobbin and others (1957); profiles IX through XI adapted from Darton and Paige (1925); profile XII by Rawlins (at my request). All are drawn on the base of the Fall River Formation. Refer to Figure 3 for locations.

TABLE 1. GEOMETRIC CHARACTER AND SUBSIDIARY STRUCTURES OF THE MAJOR SEGMENTS OF MONOCLINES ALONG THE WEST FLANK OF THE BLACK HILLS UPLIFT

Monocline segment	Geometry					Subsidiary structures			Profile no. (Figs. 3, 6)
	Trend	Length (km)	Width (km)	Structural relief (m)	Max. dip	Ramp	Terrace	Parallel anticline	
Fanny Peak south of junction	N to N15°E	65	1.2 to 8	600 to 1,060	50°W		Mule Creek	Mule Creek	VII, VIII
Black Hills: Newcastle to Moorcroft	N45°W	67	2.5 to 10	970 to 1,200	75°W	Newcastle	Pedro	Moorcroft dome	IV, V, VI
Black Hills: Moorcroft to Little Missouri River	N5°E to N20°W	58	2 to 13	1,090 to 1,670	90°W		Moorcroft Oshoto	Oil Butte–Pine Ridge	II, III
North of Little Missouri River Monument Black Hills Hill monocline	NW	25	1 to 5	0 to 575	20°SW			Rocky Point	I
Black Hills monocline	N to NE	40	1 to 5	200 to 300	29°W	Rocky Point			I
Camp Creek monocline	NW	32	2	0 to 545	20°W	Thompson Creek		Wolf Creek	

continues northward for about 20 km, then turns to the northeast and joins the northwest-trending Camp Creek monocline. The general characteristics of these folds are given in Table 1.

Both the Monument Hill and Camp Creek monoclines are bordered on the northeast uplifted side by northwest-plunging anticlines whose axes roughly parallel the anticlinal bend of the monocline. The northeast flank of the Rocky Point anticline is locally monoclinal (Fig. 3).

Two disrupted ramps are present, separated by the Black Hills monocline (Fig. 6, profile I). On the west the Rocky Point ramp rises from the basin and abuts the monocline. On the east the Thompson Creek ramp descends northwest from the top of the uplift at the Little Missouri fault zone. It is bounded on the west and north by the Black Hills monocline and on the east by the Camp Creek monocline and is inclined 3° to 4° to the northwest.

Folds of the East Flank of the Uplift

The east flank of the uplift extends about 110 km from the Buffalo Gap monocline on the south to the crest of Whitewood anticline on the north. The strike of eastward-dipping sedimentary units changes along this distance from N20°E to N40°W. The topographic expression of the Black Hills begins at the basal sandstone units of the Cretaceous rocks and lies several kilometres inside the outer structural boundary.

Unlike the narrow monoclinal flexures that mark the western limit of the uplift, the eastern flank is as much as 32 km wide. Although the structural relief is as much as 2,100 m, dips are gentle, separating the essentially flat top of the uplift and the Interior Lowlands province to the east. Subsidiary structures consisting chiefly of monoclines opposed to regional dip, but including a few monoclines with, and parallel to, regional dip, and one example of basement faulting are exposed at the present erosional level. The monoclines opposed to regional dip are expressed as west-facing, asymmetric anticlines paralleled on the west by asymmetric synclines. The axes of these structures commonly trend at a slight angle to the regional strike. Detailed descriptions of two areas containing these structures are presented in later sections of this paper.

Folds of the North and South Ends of the Eastern Block of the Uplift

Folds are present at both the north and south ends of the eastern block of the Black Hills uplift. Each group of folds begins approximately at the level of the top of the uplift and plunges meridionally away for many kilometres. With the exception of the Cottonwood Creek anticline at the south, all of the folds are strongly asymmetric and face west. They are, in fact, monoclines opposed to regional dip.

On the south the folds begin on the south side of either the Long Mountain fault zone or the Edgemont lineament. The Dudley, Cascade, and Chilson anticlines (Fig. 3) are slightly curved so that their trace is convex to the west. Their greatest curvature is present either between the two previously mentioned structural zones or just to the south of the Edgemont lineament. The Cottonwood Creek fold is unlike the others in that it appears to be symmetrical, trends from N25°E to N60°E, plunges to the southwest, and represents the structural axis of the uplift in this area. In contrast, the gravity axis of the uplift follows the Chilson anticline (Gott and others, 1974, Pl. 2). A detailed description of part of the Cascade and Dudley structures is presented in a later section of this paper.

In the fold zone on the north end of the eastern block, folds of two trends are present. In one group anticlines begin at or near the zone of Eocene plutons (Fig. 3). Shapiro (1971a, p. 26) believed the folds to begin at an eastward-trending monocline whose appearance is strongly modified by the intrusive bodies. This zone marks the change from flatter dips on the top of the uplift to a northeastward dip of about 10°. The Whitewood and Belle Fourche anticlines have the same convex to the west pattern as found in the southern folds. La Flamme anticline, the westernmost of the three folds, also consists of curving segments in a very elongated S shape. It appears to connect on the north with the Colony-Albion anticline. These monoclines opposed to regional dip have steeper west limbs that vary in dip from 30° to 70° (Darton and O'Harra, 1909, p. 5–6).

North of Belle Fourche, South Dakota, folds of a northwest trend begin which cross the La Flamme–Albion structure on the west. They are also monoclines opposed to regional dip. For much of their length the folds of the northern Black Hills are found in Cretaceous shales and have no topographic expression.

Detailed Examples of Laramide Deformation

Introduction. The following section describes the character of Laramide deformation as evidenced in four of the better studied areas of the Black Hills uplift. Two occur on the east flank of the uplift and are selected to show the character of the Precambrian basement deformation and its geometric relation to monoclines in the adjoining Paleozoic and Mesozoic strata. The third area is in the fold belt in the southern part of the uplift and illustrates the effects of more pronounced structural offset across the steep limb of the folds. The fourth area lies at the intersection of two monoclines that comprise the boundary between the uplifted block of the Black Hills and the lowered block of the Powder River Basin.

The detailed review of the effects of deformation at different structural levels and to differing degrees will serve as the basis for summarizing the character of monoclines in the Black Hills uplift.

Rockerville Area. The Rockerville area, located about 10 km southwest of Rapid City, South Dakota (Fig. 3), is the only known locality in the Black Hills in which Precambrian basement is exposed in contact with Paleozoic rocks along a fault having the northward trend of most of the anticlines. The Precambrian rocks consist of slate, schist, metagraywacke, quartzite, and pegmatites [Rahn, 1978(?)] in which the schistosity strikes roughly N25°W and dips 60° to 90°W. As shown in Figure 8, the Paleozoic and Mesozoic sedimentary layers have northward strikes and dip 2° to 10°E except in the west limbs of folds. In the western half of the area a fault 6.5 kilometres long trends parallel to subparallel to this schistosity. Along the southern half the fault strikes approximately N12°W but in the north turns to parallel schistosity. The fault is poorly exposed through much of its length, but a straight outcrop pattern across the canyon of Spring Creek suggests a steep dip (P. H. Rahn, 1977, oral commun.). The maximum throw along the fault is 61 m [Rahn, 1978(?)] with the east side down. Sedimentary rocks within 120 m of the fault are locally rotated as much as 38° on the east side in beds of the Pahasapa Formation and 64° on the upthrown block in the basal Deadwood Formation. In part, such rotation occurs between the major fault and small, subsidiary faults which are not shown in Figure 6. Whether rotation is also aided by flow within shales of the Deadwood Formation or by warping or shear within the Precambrian rocks is unknown owing to poor outcrops. Nevertheless, the planar character of the sedimentary units at distances greater than 120 m from the fault show that the zone of deformation by shear or folding within the basement is

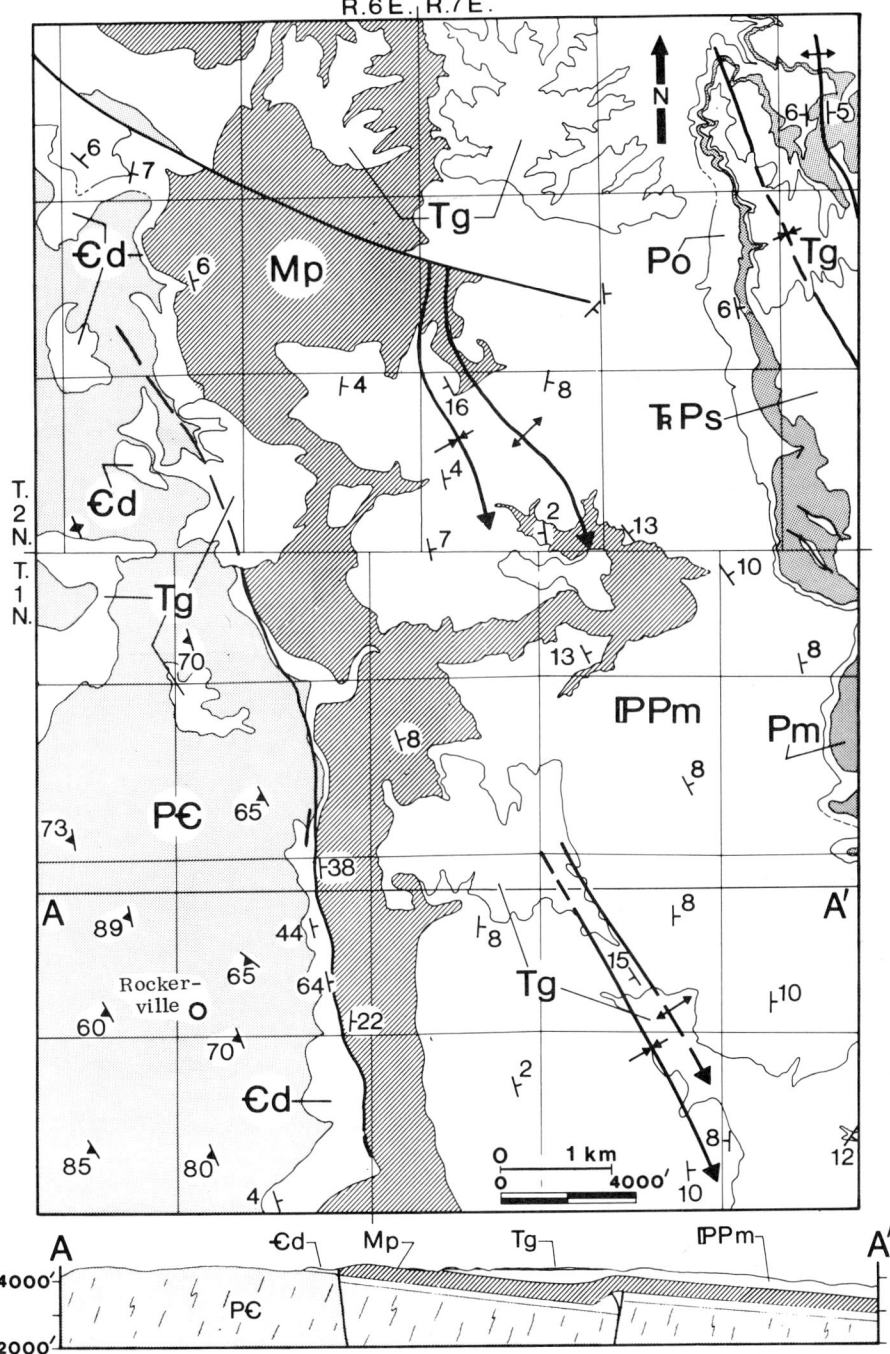

Figure 8. Geologic map and cross section of a part of the Rockerville quadrangle, South Dakota, showing fold patterns and fault orientations on the east flank of the Black Hills uplift (adapted from P. H. Rahn, 1978(?); Cattermole, 1969). Rock units are Precambrian schist and metagraywacke (pЄ); Cambrian Deadwood Formation (Єd), Devonian and Mississippian Englewood and Mississippian Pahasapa Formations (Mp), Pennsylvanian and Permian Minnelusa Formation (IPPm), Permian Opeche and Minnekahta Formations (Po and Pm, respectively), Permian and Triassic Spearfish Formation (ŦPs), and Tertiary gravels (Tg). Fold axes are indicated by arrows. Faults are shown in solid, heavy lines.

at most narrow and, locally, as in the area shown in the cross section of Figure 8, is essentially absent.

A second fault involving the Precambrian rocks enters the area at the extreme northwest corner (Fig. 8) and over a distance of 5 km changes strike from N45°W to N75°W. The fault dies out in the Minnelusa Formation apparently in an east-plunging anticline. Shapiro (1971a, p. 32) stated that R. Bailey informed him that the strike of the fault continues to change in the Precambrian exposures to the northwest and reaches N25°W, the same trend as the schistosity.

Three anticline-syncline fold sets in the area of Figure 8 strike northwest approximately parallel to the schistosity of the Precambrian basement. Folds of similar trend are common in adjoining areas [Cattermole, 1969; Rahn, 1978(?)]. The southeast-plunging anticlines are 2.5 to 5 km long with narrow west limbs dipping as much as 16°W. As shown in the cross section of Figure 8, they are actually monoclines opposed to regional dip. Structural relief is on the order of 60 m, and faulting is not recognized in the beds of Pahasapa, Minnelusa, Opeche, and Minnekahta Formations exposed in the fold limbs. The essentially planar character of the fold limbs is consistent with drape folding over basement faults that parallel the anisotropy of the metamorphic rocks. Such faults would die out upward in the Deadwood or lower Pahasapa Formations.

Regrettably, the basement fault on the west side of the area is not traceable along strike into clear-cut drape folding. On the south the fault appears to die out by a gradual decrease in throw, but the area is an alluvial valley. Exposures of Pahasapa Formation farther south along strike show only a very slight warping. In the north the fault dies out in the Precambrian rocks thus preventing any observation of the relation to formerly overlying Paleozoic strata.

Northwest Rapid City Area. In the northwest Rapid City area three northward-trending anticline-syncline fold couplets and an east-trending, south-facing monocline deform the Paleozoic and Mesozoic rocks (Fig. 9). The two eastern fold sets trend N5° to 20°W, and the anticlinal axes are approximately 1 km apart. The axes of the third set, 4 km to the west, change strike from N40°W in the south to north at the north edge of the area. Limestone of the Mississippian Pahasapa Formation crops out in the core of this anticline with dips to 43°W. In the eastern folds dips in the west limb reach 21°W in the Permian Minnekahta Formation. All three folds are asymmetric with gentle dips of 6° to 11°E in the east limb. Structural relief is about 60 m in the two eastern folds and 180 m in the western fold. As illustrated in the cross section of Figure 9, these folds are west-facing monoclines opposed to regional dip.

Each fold of the above group ends on the south at the south-facing monocline that parallels regional dip. This latter feature extends from the western fold at least 7 km east to the gap in the Cretaceous hogback formed by Rapid Creek. The monocline is composed of three en echelon parts in which individual segments end at the syncline bends of the eastern folds. Structural relief is at least 60 m in the areas of en echelon offset, but lack of marker beds and poor exposures of the Spearfish Formation prevent exact measurement. Only one small fault is known from the area, and it lies at the south end of the eastern fold in the monocline parallel to regional dip.

Figure 10 shows a diagrammatic block diagram illustrating a possible configuration of faults in the basement over which drape folding of the sedimentary units would yield the outcrop pattern observed in Figure 9. The curvature of the western fold axes to join the east-trending monoclinal flexure is similar to the change in fault trend observed in the Rockerville area and described in the preceding section.

Hot Springs Area. The Hot Springs area (Fig. 11) lies in the fold belt at the

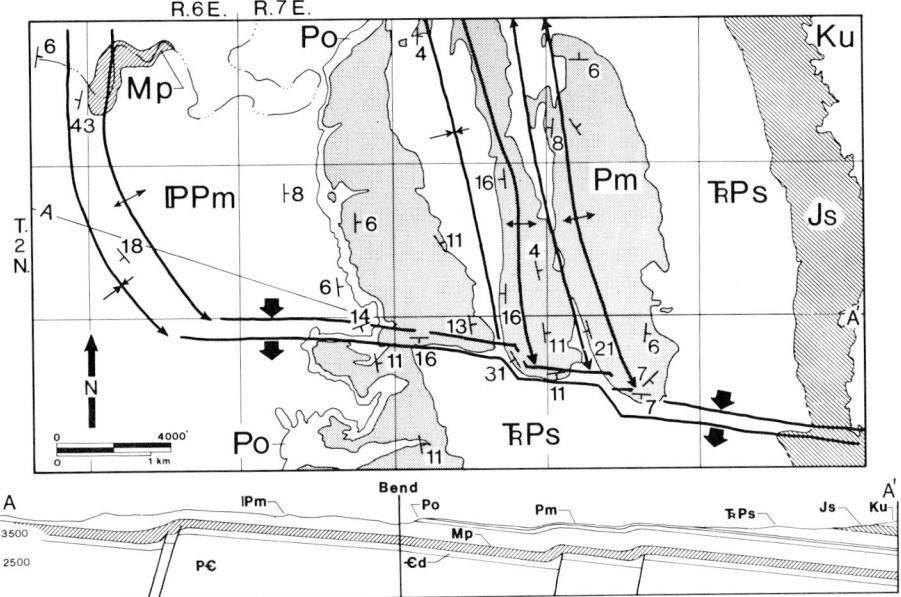

Figure 9. Geologic map and cross section of the northwest Rapid City area showing northward-trending asymmetric folds and east-trending monocline. Rock units are the same as given in Figure 8 plus the Jurassic Sundance and Unkpapa Formations (Js) and the Cretaceous Inyan Kara Group (Ku). Dark arrowheads show monoclinal flexures. Adapted from Cattermole (1969).

southern end of the Black Hills uplift. Regional dip is east-southeast so that the age of units increases from Pennsylvanian in the northwest to Cretaceous in the south and southeast. These strata are deformed into Z-shaped outcrop patterns by four south-plunging anticlinal folds. Only the easternmost of the group, which is 5 km long, is contained completely within the area shown. The others are traceable a short distance north of the map area and the largest fold, the Cascade anticline, continues south from the area at least another 19 km.

Each of the folds is asymmetric with a steep, narrow, west limb and a more gently dipping, broader east limb. In the Cascade anticline and the broader structure in the extreme northwest part of the area, the maximum width of the steep limb is 1.5 km or less. A conspicuous feature of the broader, east limbs is the continuity of bedding attitudes over large areas. This is very noticeable in the straight, evenly spaced structure contour line shown by Gott and others (1974, Pl. 1) and can be seen in the dip and strike values shown in Figure 11. A similar situation occurs on the east limb of the Chilson anticline whose axis lies west of the area shown in Figure 11. Such a pattern is not as well developed in outcrops of Minnelusa Formation in the northwest part of the area and may be due to the more deformed internal character of these beds or to a different style of deformation in this area.

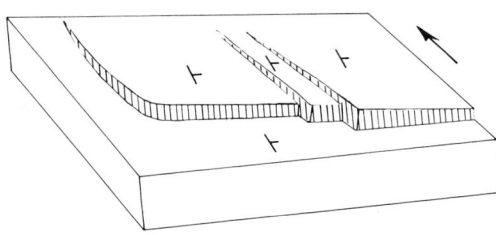

Figure 10. Block diagram illustrating possible Precambrian basement configuration beneath the draped folds and monocline shown in Figure 9. Fault surfaces are shown by diagonal pattern.

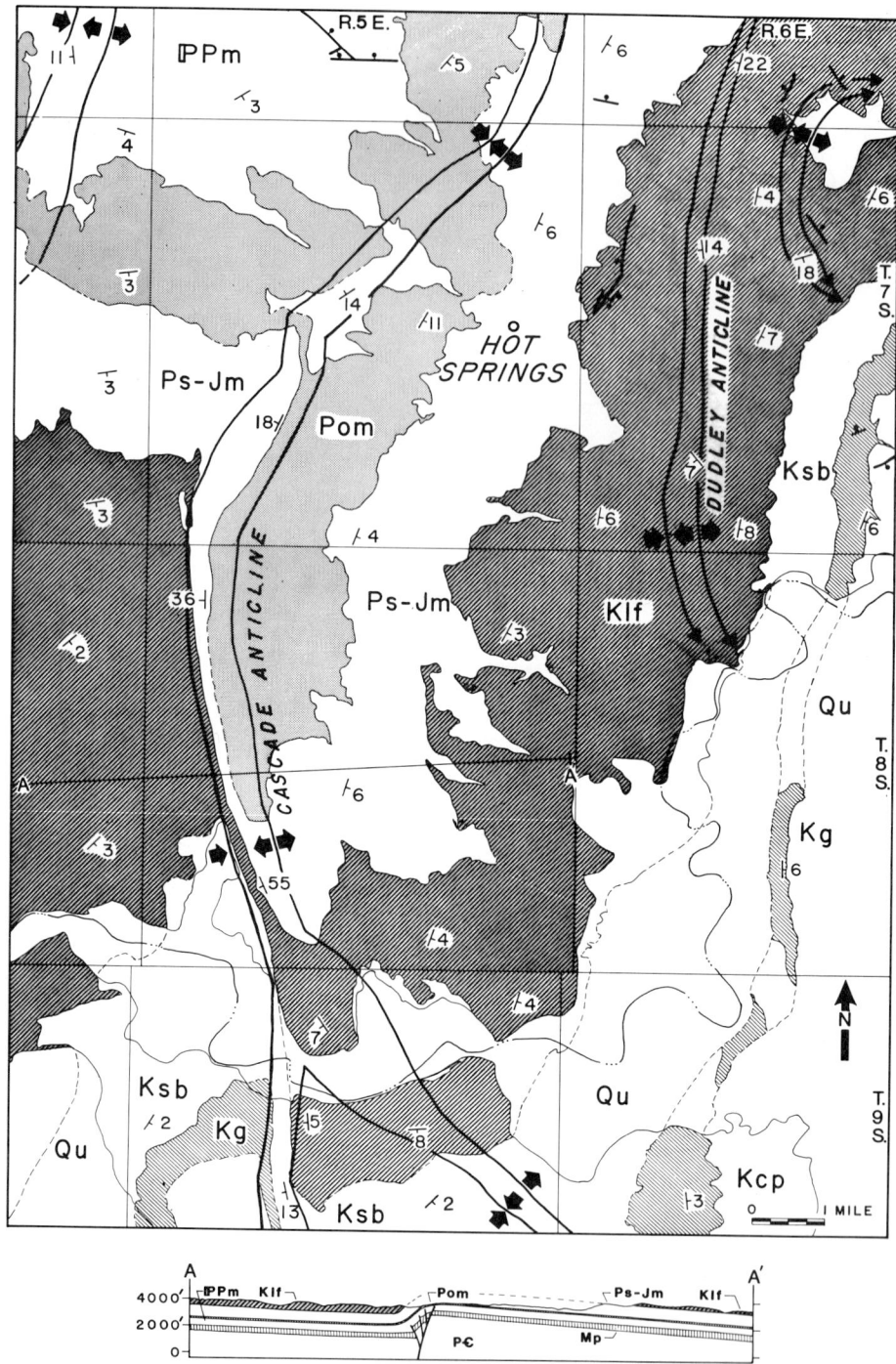

Figure 11. Geologic map and cross section of the Hot Springs area, South Dakota, showing the asymmetric Cascade Springs and Dudley anticlines (adapted from Gott and others, 1974). Dark arrowheads show flexures. Rock units are Minnelusa Formation (PPm), Opeche and Minnekahta Formations (Pom), Spearfish, Gypsum Spring, Sundance, Unkpapa, and Morrison units (Ps-Jm), Lakota, Fuson, and Fall River units (Klf), Skull Creek Shale, Newcastle Sandstone, Mowry Shale, and Belle Fourche Shale (Ksb), Greenhorn Formation (Kg), Carlile Shale, Niobrara Formation, Pierre Shale (Kcp), and Quaternary deposits (Qu).

The three eastern anticlines are curved to varying degrees in a C shape, which is particularly well developed in the smaller, eastern anticline-syncline couplet. The Cascade anticline also shows this pattern, but the curvature appears to result from a series of relatively straight segments rather than a smooth curve. At the join of these segments, west-plunging noses and chutes are developed. In the south the Cascade anticline bifurcates into south and southeast-trending segments. The maximum structural relief on the anticline, 495 m, and the steepest dips of 70°W are in the area of Cascade Springs just north of the split in the structure (Post, 1967, p. 493, 494). South of the split the western fold has about 390 m of structural relief and the eastern fold about 75 m. Closure on the Cascade anticline is over 120 m at the join of the north- and northeast-trending segments near the middle of the map area (Gott and others, 1974, Pl. 1) and extends at least 6 km to the north and 2.5 km to the south. Sandstone units of Jurassic and Cretaceous age in the inclined limb are intensely fractured. Conjugate reverse faults strike subparallel to the limb and dip 30° to 40° east and west, and subhorizontal joints spaced 1 to 20 cm are ubiquitous. The offset on individual faults is commonly less than 1 m. Their presence, as well as that of the horizontal tension cracks, illustrates the effect of stretching in the rotated limb.

The Dudley anticline has a maximum structural relief of 180 m (Wolcott, 1967, p. 440) and a closure of 30 m. The fold dies out 7 km south of the structurally highest point and decreases to 15-m amplitude at the north border of the area 4 km to the north. The small fold just east of the Dudley anticline has a closure of less than 30 m. The closed character of these folds is a function of the change in axial direction combined with a constant regional strike direction. Where the trend of the fold changes such that both segments are plunging, closure results at the area of the bend. In segments of continous strike, either northward or northeastward, closure is not geometrically possible in these folds.

The use of the term anticline by Post (1967) and Wolcott (1967) to describe the folds of the Hot Springs area is geometrically sound, but it belies the true character of these structures, that of monoclines opposed to regional dip. The cross section shown in Figure 11 illustrates clearly that at the present erosional level the narrow west limb of Cascade anticline is a monocline separating two homoclinally east-dipping blocks. This character is true of the other three folds of the area as well as the Chilson anticline to the west. Such folding by the mechanism of draping sedimentary layers over rotated basement rocks would be consistent with the observed outcrop patterns. Figure 12 is a block diagram illustrating a possible configuration for such basement faults. In this model changes in structural

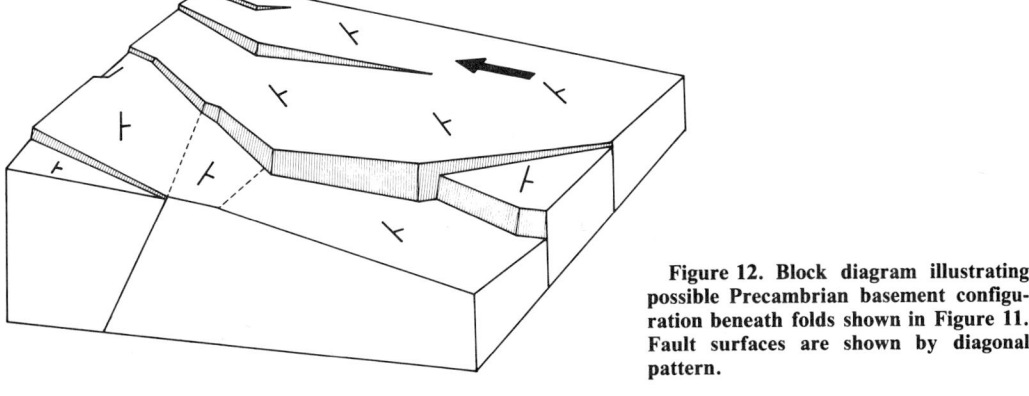

Figure 12. Block diagram illustrating possible Precambrian basement configuration beneath folds shown in Figure 11. Fault surfaces are shown by diagonal pattern.

relief along the strike of folds result from changes in throw on the basement faults, and the east-facing C pattern is a function of a trap-door fault mechanism in which throw decreases from a maximum in the middle of the C to zero at the ends. In the case of the Cascade anticline, the process is modified by the bifurcation at the south end of the area. Part of the throw is taken up by the splaying of a lesser fault that decreases in throw along its strike. South of this area the main south-trending segment also curves southeastward and decreases in throw over a distance of 16 km.

Newcastle Area. The junction of the Black Hills and Fanny Peak monoclines east of Newcastle, Wyoming (Fig. 13), hereafter referred to as the junction, is one of the better studied areas of the margin of the Black Hills uplift. Wulf (1955), Epstein (1958), Brobst and Epstein (1963), Shapiro (1971a, 1971b), and Gott and others (1974) have described the stratigraphy and the general character of the folds although they did not analyze the structures at the mesoscopic scale. Mapel and Pillmore (1963) mapped and described the Black Hills monocline in the immediate Newcastle area and Van Lieu (1969) mapped a segment of Fanny Peak monocline 21 to 46 km north of the junction. The following discussion is based on this work combined with my observations.

The character of both folds changes abruptly across the point of intersection. Both are narrower and have less structural offset on the uplifted block. Immediately north of the junction the Fanny Peak monocline is a west-facing structure boldly displayed in flatirons of the Pennsylvanian Minnelusa Formation. Structural relief is 600 m, dips are as much as 77°W, and the monocline is less than 1 km wide. One kilometre north of the intersection the Stockade Beaver Creek monocline separates from the Fanny Peak monocline (Fig. 13) but curves northward to parallel

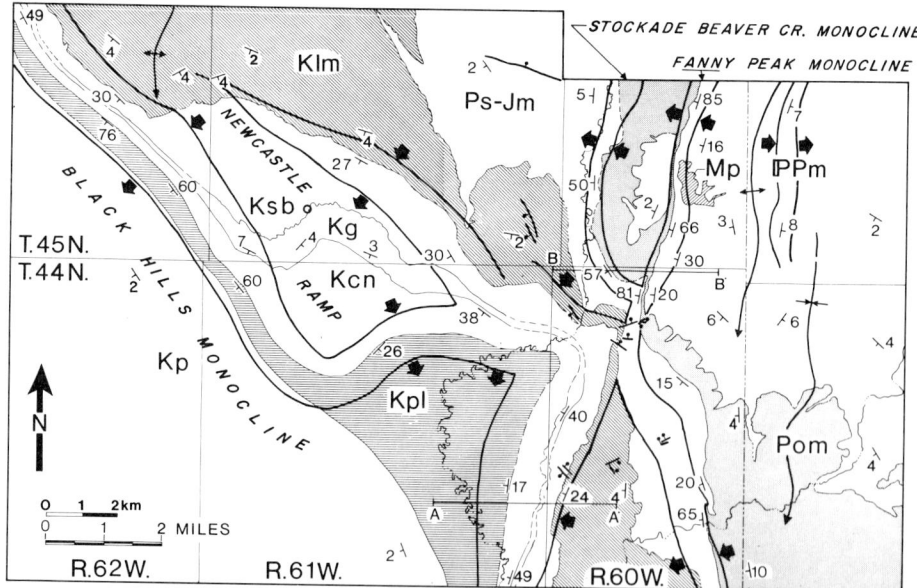

Figure 13. Geologic map of the Newcastle, Wyoming, area showing the intersection of the Black Hills and Fanny Peak monoclines (adapted from Mapel and Pillmore, 1963, and Gott and others, 1974). Monoclinal flexures shown with dark arrowheads. Rock units are the same as in Figure 11 for IPPm through Ksb. Mp is the Mississippian Pahasapa formation, Kcn is the Carlile shale and Niobrara Formation, and Kpl and Kp are subunits of the Pierre Shale. Monoclinal flexures are indicated by heavy arrows. Half dumb-bells indicate downthrown sides of faults.

it so that the two folds are separated by a disrupted ramp. The presence of these folds was recognized by Newton and Jenney (1880), and Figure 14 of this paper is their diagrammatic representation of the topographic and structural character of this segment of the folds. Stockade Beaver Creek monocline deforms Paleozoic and Mesozoic rocks at the present erosional level. It is less than 1 km wide, has westward dips to 60°, and as much as 225 m of structural relief.

South of the junction the monocline is expressed in a hogback of Cretaceous sandstone on the east and lowlands of Cretaceous shale on the west. The fold narrows from 3 km on the north to 2 km in the south along a distance of 10 km.

In the area of Figure 13 the Black Hills monocline is widest immediately west of the junction where the inclined limb is 3 km wide with dips of 20° to 40°S in exposures of Cretaceous strata. To the northwest 3.5 km, the Newcastle disrupted ramp results from the en echelon overlap of the monocline. The surface of the ramp rises 450 m from the lower end to the surface of the uplift in a distance of 0.5 km. Owing to the arcuate trend of the flanking folds, the width of the surface of the ramp decreases from 5 km at the lower end to 1.25 km on the northwest. Just northwest of the upper end of the ramp the Black Hills monocline has a stratigraphic separation of 1,330 m and dips to 75°SW across a horizontal distance of 1.5 km. On the basis of gravity data, Black and Roller (1961, p. 261) interpreted the monocline to be underlain by a single major fault offsetting the basement.

Southeast of the junction the Black Hills monocline is expressed as a west-facing dip slope on the Permian Minnekahta Limestone. In the area of Figure 13 the inclination increases from 15° in the north to 65° in the south.

The geometry of the Fanny Peak monocline in the area of Figure 12 and for an additional few kilometres to the north is illustrated in the cross sections of Figure 15. Approximately 19 km north of the junction the Fanny Peak monocline begins as a gentle anticline (Fig. 15, section D-D') just over 1 km east of the Stockade Beaver Creek fold (Shapiro, 1971a, Pl. 1). Along strike to the south the fold is succeeded by a normal fault 6 km long. At its southern end a second en echelon fault begins a few hundred metres to the west and extends an additional 10 km south. Although exposures are poor, Wulf (1955, p. 23) described this second fault as reverse with dips of 65° or more to the east beneath the upthrown block. The vertical dip shown at a deeper level in section C-C' is inferred and not seen in the field. Shapiro (1971a, p. 57) found a maximum stratigraphic separation of 273 m in the fault and associated folds across a width of approximately 1 km. In the general area of this section the Pahasapa Formation in the hanging wall is tilted gently west; the Minnelusa Formation of the footwall is strongly tilted

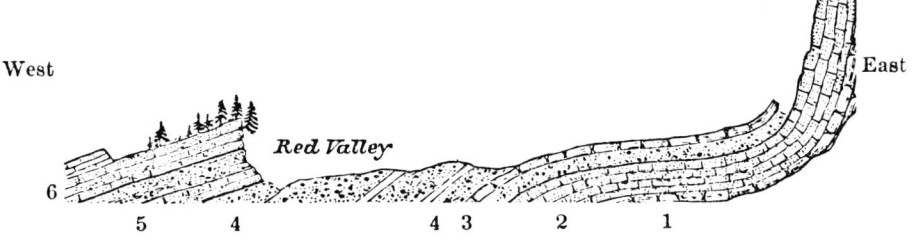

Figure 14. Diagrammatic cross section of the Fanny Peak and Stockade-Beaver Creek monoclines with the Fanny Peak structure on the east. 1, Carboniferous limestone; 2, red sandstone (Carboniferous); 3, purple limestone (red beds); 4, red clays with gypsum (red beds); 5, Jurassic strata; 6, Cretaceous strata. After Newton and Jenney (1880).

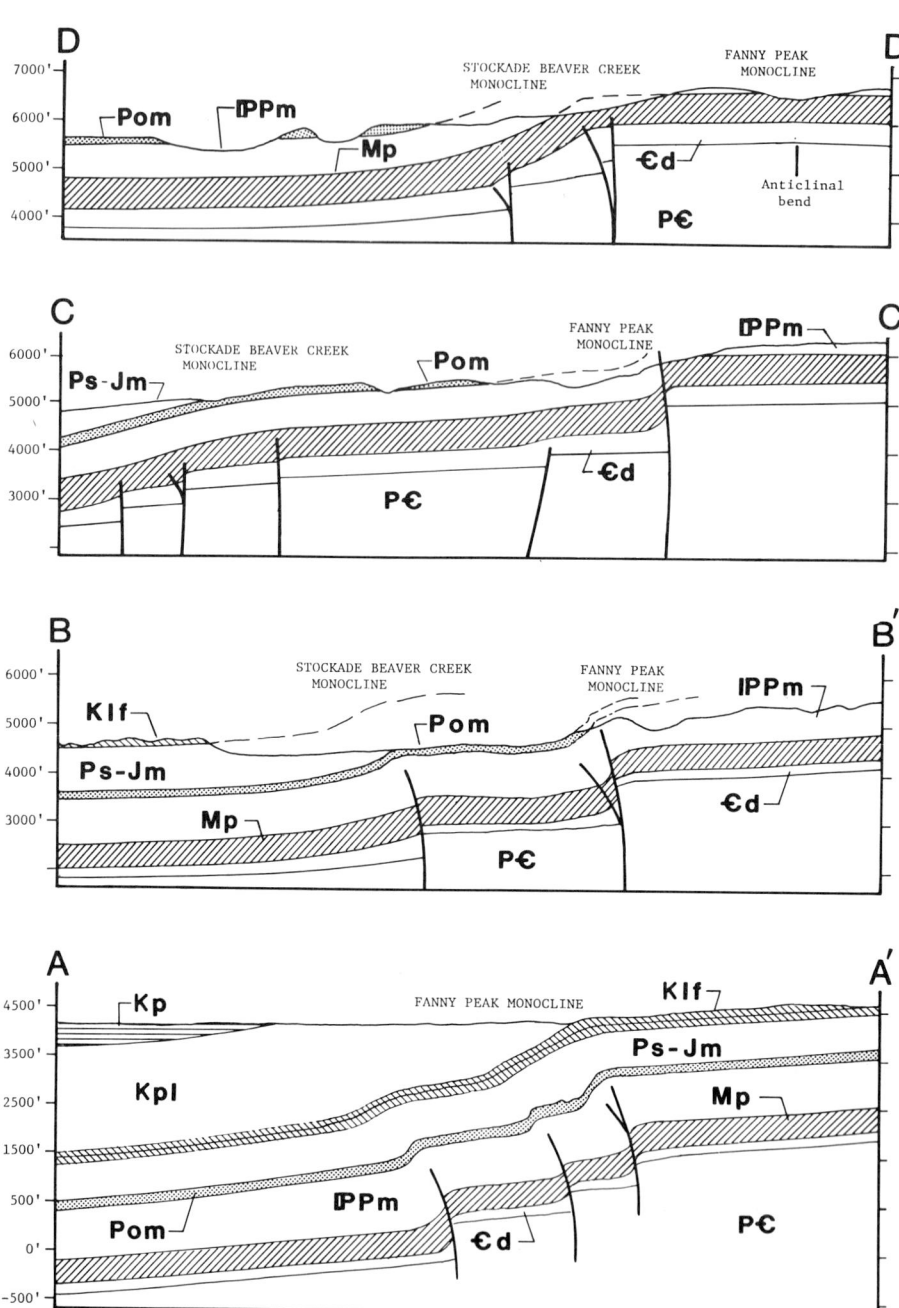

Figure 15. Cross sections of Fanny Peak monocline. The location of sections A–A′ and B–B′ are shown in Figure 13. Section C–C′ is 6.5 km north of the boundary of the area shown in Figure 13, and section D–D′ is 16 km north of the boundary.

and locally shows subvertical dips. At its southern end the reverse fault dies out in the steeply inclined limb of the monocline.

From the reverse fault south to the junction of the major folds, beds exposed in the monocline display a clear subdivision into segments of the style found by Stearns (1971) at Rattlesnake Mountain, Wyoming. Cross section B-B' illustrates this subdivision for the Opeche and Minnekahta Formations. Block 5 is the surface of the ramp separating the Fanny Peak and Stockade Beaver Creek monoclines. Block 4 dips 40°W, block 3 dips 75°W, block 2 flattens to approximately 30°W, and block 1 displays the 7°W dip of the surface of the eastern, uplifted block. At Rattlesnake Mountain, Wyoming, the monocline opposes regional dip and the inclination of block 1 is opposite that of the other blocks. The Fanny Peak monocline is with regional dip, and all segments as a consequence are inclined in the same direction. The reverse fault mapped to the north is assumed to be present at deeper stratigraphic levels beneath these segments.

The carbonate beds of the inclined limb are intensely fractured. Wulf (1955, p. 22-24) described numerous small faults that most commonly strike parallel to the monocline and dip 45° to 65°E. Locally conjugate shears are present that display this orientation in one set and a westward dip of 45° in the second. Subhorizontal joints are very widespread. In addition to the conjugates shears described above, abundant small, subvertical tear faults offset the Minnelusa Formation-Opeche Formation contact in the fold. These faults display right slip and trend at approximately 60° to the strike of the fold. Epstein (1958) mapped several faults having a similar sense of separation, but offsets to a few hundred metres, in the immediate area of the intersection of the Black Hills and Fanny Peak monoclines.

The thickness of the ductile Opeche Formation varies along strike in the inclined limb, and at the area of bifurcation of the Stockade Beaver Creek monocline the shale is completely squeezed out. Although the thick shale and siltstone beds of the Spearfish Formation are eroded from the Fanny Peak structure, they are present in the Stockade Beaver Creek monocline just to the west. Wulf (1955, Pl. VIII) showed a subsidiary fold of gypsum with greater than 5 m amplitude in the synclinal bend of the monocline, which suggests at least local flexural flow in this unit.

In the area of section A-A' (Fig. 15), south of the junction of the two major folds, Cretaceous shale and sandstone comprise the rocks exposed at the present erosional level. The sandstone is strongly jointed and locally sheared. Although flexural flow of the ductile shale is suspected to have accompanied folding, direct evidence is lacking in the poor exposures. A major fault is not seen in the inclined limb at this stratigraphic level, but Black and Roller (1961, p. 261) interpreted gravity data to indicate the presence of three high-angle faults cutting the basement beneath the monocline just to the south of the area of Figure 13.

The interpretation of gravity data by Black and Roller (1961) and the surface geology are used to prepare the block diagram of Figure 16. This figure illustrates an interpretation of faulting in the basement that could lead to draping of the overlying sedimentary strata in the form of the structures of the Newcastle area. Whereas this block model shows the general character of basement offset, it is oversimplified in at least two ways. First, the true inclination of the faults is unknown. Second, a single fault plane may not be adequate to form the broad, gentle folds such as the Black Hills monocline east of the junction and the Stockade Beaver Creek monocline. For these structures folding of the sedimentary units could result from draping across a series of faults of smaller offsets, from rotation due to shear along multiple close-spaced planes in the steeply dipped metamorphic basement, or from rotation of the basement between curving fault segments. In the first case, the draping of strata over multiple fault blocks, flow of ductile material is necessary in the basal part of the section if the faults are to die out upward in the section (Stearns, 1971, p. 130). As already noted flexural flow of the Opeche and Spearfish Formations is present, but this is at a higher stratigraphic level than that required. Thickness of the Cambrian Deadwood Formation and its shale content increase from south to north along the line of cross sections shown in Figure 15. In the area of sections C–C′ and D–D′ (Fig. 15) it is believed that sufficient shale is present to allow thickness change due to ductile flow during folding, and this is, in fact, shown in the sections. If the second case of shear across a broad zone with movement on multiple, discrete planes, or the third case, rotation of the basement rock, has occurred, such flowage of the shale is not necessary.

An additional difference between the surface geology and the block diagram arises from the fact that the block model would allow for draping of sedimentary strata only in the immediate fold area above a fault. In both the Stockade Beaver Creek and the Fanny Peak monocline south of the junction the downthrown side flattens for 1 to 2 km away from the monocline. For example, the regional dip of 2°W found throughout the eastern part of the Powder River Basin is seen in section A–A′ (Fig. 15) to begin at the west edge of the section approximately 2 km west of the fold. The intervening area, although structurally part of the basin, has the same westward tilt as the top of the uplift. This suggests either an early stage of rotation across a broad area prior to actual monocline formation or movement in the monoclinal zone accompanied by minor rotation of beds to either side.

SUMMARY OF LARAMIDE STRUCTURES

Gross Form of Uplift

The 290-km-long Black Hills uplift consists of two essentially flat-topped blocks that trend north and northwest and are separated by a west-facing monocline. The asymmetric uplift is bounded on the west by abrupt monoclines, with regional dip, and on the east by a broad, curving zone of increased, but gentle dips. The east flank and north and south ends of the east block are sites of numerous monoclines parallel to, with, and opposed to regional dip.

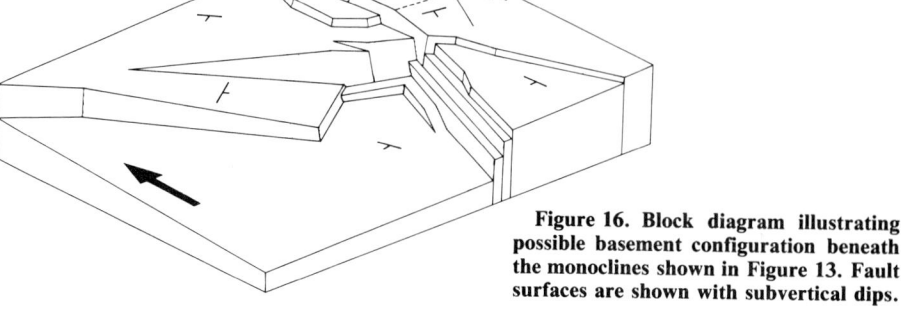

Figure 16. Block diagram illustrating possible basement configuration beneath the monoclines shown in Figure 13. Fault surfaces are shown with subvertical dips.

The basement consists of Precambrian igneous, metasedimentary, and metaigneous units that generally display anisotropic northwest-trending fabric in the exposures of the eastern block. The erosional surface of the basement displays very low relief. Although there is insufficient data to prepare a detailed structure contour map of this surface for the western block, the planar character of the overlying sedimentary units suggests that the basement surface is also planar. In the eastern block both a planar and, locally, a curved character is believed to occur. The planar character is clearly present in the immediate area of several small monoclines superimposed on the curved flank of the uplift and in the fold zone at the south end of the block.

The sedimentary cover consists of Paleozoic shelf facies carbonate and minor clastic units, of Mesozoic continental clastic units, and of Paleocene continental clastic units. During Laramide structural activity the sedimentary cover varied in thickness across the uplift from approximately 2,200 m on the southeast to 3,000 m on the northwest. Postuplift Eocene alkalic intrusive bodies disrupt the Laramide structures in the northern part of the uplift.

Monoclines on the uplift are the dominant structural features, and most trend north or northwest at scales ranging from a few kilometres to 100-km length and with structural relief of a few metres to 1,670 m.

Monoclines

As observed elsewhere in the Wyoming province (Prucha and others, 1965; Stearns, 1971), the monoclinal flexures grade from fold-fault couples at lower stratigraphic levels to folds at higher stratigraphic levels. Faults may extend to the massive units of Mississippian and Pennsylvanian carbonate rocks. In the single location where the basement is observed along such a fault in the Black Hills, a narrow zone of drag is found in the overlying sedimentary rocks. At higher stratigraphic levels, rotation of the sedimentary layers is broader and more pronounced on the downthrown block. In the Fanny Peak monocline both high-angle reverse and normal faults are found in the Paleozoic section.

Gravity data suggest that the faults in the basement are high angle, and that single faults occur beneath narrow, steeply dipping monoclinal segments and multiple faults beneath the broader flexures.

In the Paleozoic section at the Fanny Peak monocline, the rotated limb is divisible into five discrete blocks as recognized by Stearns (1971) for many monoclines of the Wyoming province. Whether similar geometry exists at higher stratigraphic levels is unknown, because the shales and siltstones either are removed by erosion or are not exposed. Locally, however, shale and siltstone of the Permian and Mesozoic section are known to have undergone flexural flow, which is inferred to cause a more arcuate character at higher stratigraphic levels.

The inclined limb of the monoclines shows evidence of stretching at the mesoscopic scale. This is recognized in abundant examples of conjugate faults, ubiquitous tension cracks, and thinning of ductile layers by flexural flow.

The plan view of the monoclines commonly shows linear segments that join at small, open, plunging cross folds. The combined segments usually have an arcuate trend.

Monoclines opposed to regional dip are expressed as asymmetric anticlines. Closure results at the join of segments in such structures if the trend of the anticlinal crest is down the regional dip in each segment.

The monoclines have subsidiary structures in the form of ramps, terraces, and adjoining anticlines on the uplifted block. Terraces are recessed into the uplifted block or protrude into the downthrown block. They are commonly located at, or near, the join of major segments of the monoclines. Ramps are also of two general types, those that descend from one structural level to another without interruption and those that are disrupted on their upper or lower ends by smaller monoclines. They form at the area of en echelon overlap of monocline segments. The anticlines at the upper bend of the monocline show structural relief slightly in excess of the surface of the uplifted block.

The maximum rotation found in the inclined limb of any fold is 90°. In Paleozoic rocks subvertical dips are seen in the fold where the structural relief is as low as 250 m. In the Mesozoic section vertical inclination is recorded where the structural relief is 1,100 m. In no case is the inclined limb overturned.

The monoclines terminate by intersecting other monoclines, by a gradual decrease in stratigraphic separation along strike, or by bifurcation into several folds that die out through either one or both of the above mechanisms.

Two zones of faulting and one major lineament are present near the southern end of the east block, and a third fault zone crosses the northern end of the west block. The fault zones of the eastern block affected Cretaceous sedimentation patterns, and the lineament lies along an apparent basement structure of Precambrian age. The fault zone of the west block lies along the hinge between the top of the uplift and a northwest-descending ramp structure. These features trend east-northeast, and individual faults have only a few metres vertical displacement.

CONCLUSIONS

The Laramide Black Hills uplift is not a simple dome rising in the Interior Lowlands province. It is of essentially the same character and origin as other uplifts of the Wyoming province although of lesser structural relief (Shapiro, 1971a). The two blocks of which it is composed contrast in structural style. The boundary of the western block with the Powder River Basin consists of abrupt monoclines across which the sedimentary units and, by inference, the top of the basement remain planar, although they may show a few degrees difference in dip. The boundary of the eastern block with the Interior Lowlands province is broad, and the eastern block appears to represent a partial dome along the curved eastern flank.

The monoclines of the Black Hills uplift appear to have formed by drape folding of sedimentary units over faulted basement blocks in the manner described by Stearns (1971) for uplifts of the Wyoming province. This model explains both the geometry of those folds that bound the uplift and the smaller folds that occur on its crest and flanks.

Subsidiary structures in the form of ramps and terraces are also explained by the process of drape folding. The origin of parallel anticlines at the upper bend of the monoclines is unclear, although they may result either from collapse of a keystone over reverse faults in the basement as described by Wise (1963) for the Owl Creek uplift of Wyoming, or through a subsequent settling of the uplifted block along the previously formed steeper part of a reverse fault in the basement (Sales, 1968). In either case the reverse fault would dip beneath the uplifted block.

During the draping process the rotating limb was placed in tension which resulted in mesoscopic faulting and jointing of the brittle layers, flexural flow of the ductile layers, and an overall thinning of the limb.

The planar character of the sedimentary units across the monoclines in the Powder River Basin and on the uplift is compatible with an origin by basement block rotation along curved faults as proposed by Stearns (1975) to explain the structure of the western Bighorn Basin and adjoining uplift. In the western Black Hills uplift this character is maintained even in the case of an anisotropic basement.

This model, however, does not adequately explain the broad zone of tilting along the eastern margin of the uplift. There is no evidence of major faults along the margins of the inclined segment, although monoclines are locally present within it. Therefore, another method of deformation is required in order to avoid room problems that would result at the margins of a rotating rigid block. Unlike the margins or other uplifts of the Wyoming province, the eastern flank of the Black Hills uplift is not bordered by a basin. In fact, uplift of the block was accompanied by lesser uplift in the adjoining Interior Lowlands province. This positive character for both appears to have resulted in the distribution of strain across a broad zone, perhaps by shear along individual schistosity planes in the basement, and the formation of a partial dome along the curved east flank of the uplift. On the west side of the uplift where there is both absolute uplift of the western block and subsidence in the basin, strain is taken up along individual faults in the basement.

The doming of the eastern block resulted in longitudinal and transverse faults in the basement and drape folds in the sedimentary section, which are seen as monoclines with, opposed to, and parallel to regional dip. At the north and south ends of the dome, tensional stress resulted in a series of drape folds plunging away from the uplift. Such faults are common on doubly plunging anticlines of the Wyoming province (Irwin, 1926, p. 113), and Wisser (1960, p. 53) noted that where anticlines plunge longitudinal faults may diverge. Such divergence is common in the folds at the ends of the dome. The arcuate plan of several folds may also result from local adjustment of the stress field on a rising dome.

Basement anisotropy appears to have locally controlled Laramide fault, and hence monoclinal, trends, but several large structures do not parallel the west-northwest trend of the basement fabric of the eastern block.

Monoclines with structural relief as much as 1,670 m lack any overturning in the inclined limb at the level of the Cretaceous. This suggests that at least for higher stratigraphic levels the "fold-thrust" model proposed by Berg (1962) for the margins of uplifts in the Wyoming province has not been initiated for such a vertical contrast between uplift and basin.

The model of faulting and rotation of the basement defined by Stearns (1975), based on the theoretical studies of Hafner (1951) and Sanford (1959), assumes a sinusoidal vertical movement of the upper crust. Such a sine curve can result from either buckling or bending and yields uplifts and basins. Buckling assumes lateral compression, bending a vertical movement only. It is beyond the scope of this paper to resolve the long-standing question of horizontal compression versus vertical tectonics in regard to the origin of stresses affecting the Wyoming province during Laramide time. I wish to comment, however, on the fact that to produce both absolute uplift and downwarp by horizontal compression and buckling requires a single mechanism operating over the entire region. To produce isostatic uplift and downwarp having the general wavelike character of the Wyoming province requires either positive and negative density changes of the lower crust and/or upper mantle in adjoining bands, or addition and removal of material from the base of the crust along generally parallel bands. The cause of such density changes or material transport in alternating bands across such a large region is not evident at present. The mechanism of lateral compression of the crust, as proposed in the plate tectonics model of Sales (1968), for instance, seems a less complicated choice. The generally sinusoidal uplifts and basins produced by such compression would be modified in the upper brittle zone, in the manner indicated by Stearns (1975).

ACKNOWLEDGMENTS

I express my appreciation to Jack Redden for his discussion of Black Hills geology and his review of the manuscript; to James Vandike, Harmon Heidt, Daniel Goodart, and Marilyn Hanson for their assistance; and to Vincent Matthews III for his patience. The Department of Geological Engineering at South Dakota School of Mines gave financial assistance for the preparation of the figures.

REFERENCES CITED

Berg, R. R., 1962, Mountain flank thrusting in the Rocky Mountain foreland, Wyoming and Colorado: Am. Assoc. Petroleum Geologists Bull., v. 46, p. 2019–2032.

Black, R. A., and Roller, J. C., 1961, Relationship between gravity and structure of part of the western flank of the Black Hills, South Dakota and Wyoming: U.S. Geol. Survey Prof. Paper 424-C, p. 260–262.

Blankennagel, R. K., Miller, W. R., Brown, D. L., and Cushing, E. M., 1977, Report on preliminary data for Madison Limestone test well no. 1, NE 1/4 SE 1/4 sec. 15, T. 57 N., R. 65 W., Crook County, Wyoming: U.S. Geol. Survey, Open-File Rept. 77-164, 97 p.

Brobst, D. A., and Epstein, J. B., 1963, Geology of the Fanny Peak quadrangle, Wyoming–South Dakota: U.S. Geol. Survey Bull. 1063-I, p. 323–377.

Cattermole, J. M., 1969, Geologic map of the Rapid City West quadrangle, Pennington County, South Dakota: U.S. Geol. Survey Map GQ-828, scale 1:24,000.

Darton, N. H., 1902, Description of the Oelrichs quadrangle, South Dakota: U.S. Geol. Survey Geol. Atlas, folio 85, 9 p.

——1904, Description of the Newcastle quadrangle, Wyoming–South Dakota: U.S. Geol. Survey Geol. Atlas, folio 107, 9 p.

——1905, Description of the Sundance quadrangle, Wyoming–South Dakota: U.S. Geol. Survey Geol. Atlas, folio 127, 12 p.

Darton, N. H., and O'Harra, C. C., 1907, Description of the Devils Tower quadrangle, Wyoming: U.S. Geol. Survey Geol. Atlas, folio 164, 9 p.

——1909, Description of Belle Fourche quadrangle, South Dakota: U.S. Geol. Survey Geol. Atlas, folio 164, 9 p.

Darton, N. H., and Paige, S., 1925, Description of the central Black Hills, South Dakota: U.S. Geol. Survey Geol. Atlas, folio 219, 34 p.

Darton, N. H., and Smith, W.S.T., 1904, Description of the Edgemont quadrangle, South Dakota: U.S. Geol. Survey Geol. Atlas, folio 108, 10 p.

Dobbin, C. E., Kramer, W. B., and Horn, G. H., 1957, Geologic and structure map of the southeastern part of the Powder River Basin: U.S. Geol. Survey Oil and Gas Inv. Map OM-185, scale 1:125,000.

Epstein, J. B., 1958, Geology of part of the Fanny Peak quadrangle, Wyoming–South Dakota [M.S. thesis]: Laramie, Univ. Wyoming, 90 p.

Gott, G. B., Wolcott, D. E., and Bowles, G. C., 1974, Stratigraphy of the Inyan Kara Group and localization of uranium deposits, southern Black Hills, South Dakota and Wyoming: U.S. Geol. Survey Prof. Paper 763, 57 p.

Hafner, W., 1951, Stress distributions and faulting: Geol. Soc. America Bull., v. 62, p. 373–398.

Hunt, C. B., 1972, Geology of soils: San Francisco, W. H. Freeman and Co., 344 p.

Irwin, J. S., 1926, Faulting in the Rocky Mountain region: Am. Assoc. Petroleum Geologists Bull., v. 10, p. 105–129.

Kelley, V. C., 1955, Monoclines of the Colorado Plateau: Geol. Soc. America Bull., v. 66, p. 789–804.

Kleindopf, M. D., and Redden, J. A., 1975, Bouguer gravity, aeromagnetic, and generalized geologic maps of part of the Black Hills of South Dakota and Wyoming: U.S. Geol. Survey Geophys. Inv. Map GP-903, scale 1:250,000.

Lisenbee, A. L., 1975, Structural geology: Black Hills, *in* Mineral and water resources of South Dakota: Washington, D.C., U.S. Govt. Printing Office Report to the Comm. on Interior and Insular Affairs, U.S. Senate, p. 52–56.

Love, J. D., 1960, Cenozoic sedimentation and crustal movement in Wyoming (Bradley volume): Am. Jour. Sci., p. 204–214.

Love, J. D., Weitz, J. L., and Hose, R. K., 1955, Geologic map of Wyoming: U.S. Geol. Survey, scale 1:500,000.

MacLachlan, M. M., 1972, Triassic System, *in* Geologic atlas of the Rocky Mountain region: Denver, Colo., Rocky Mtn. Assoc. Geologists, p. 166–176.

Mallory, W. M., 1972, Pennsylvanian System, *in* Geologic atlas of the Rocky Mountain region: Denver, Colo., Rocky Mtn. Assoc. Geologists, p. 111–127.

Mapel, W. J., and Pillmore, C. L., 1963, Geology of the Inyan Kara Mountain quadrangle, Crook and Weston Counties, Wyoming: U.S. Geol. Survey Bull. 1121-M, p. M1–M56.

——1963, Geology of the Newcastle area, Weston County, Wyoming: U.S. Geol. Survey Bull. 1141-N, p. N1–N85.

Mapel, W. J., Robinson, C. S., and Theobold, P. K., 1959, Geologic and structure contour map of the northern and western flanks of the Black Hills, Wyoming, Montana, and South Dakota: U.S. Geol. Survey Oil and Gas Inv. Map OM-191, scale 1:96,000.

McDowell, F. W., 1966, Potassium-argon dating of Cordilleran intrusives [Ph.D. thesis]: New York, Columbia Univ.

McGookey, D. P., Haun, J. D., Hale, L. A., Goodell, H. G., McCubbin, D. G., Weimer, R. J., and Wulf, G. R., 1972, Cretaceous System, *in* Geologic atlas of the Rocky Mountain region: Denver, Colo., Rocky Mtn. Assoc. Geologists, p. 190–228.

Meuschke, J. L., Johnson, R. W., and Kirby, J. R., 1963, Aeromagnetic map of the southwestern part of Custer County, South Dakota: U.S. Geol. Survey Geophys. Inv. Map GP-362, scale 1:62,500.

Mickelson, J. C., and Kulick, J. W., 1963, Pre-Minnelusa stratigraphy of the northern Black Hills: Northern Powder River Basin, Wyoming and Montana: Wyoming Geol. Assoc., 1st Joint Field Conf., and Billings Geol. Soc. Guidebook, p. 41–44.

Newton, H., and Jenney, W. P., 1880, Report of the geology and resources of the Black Hills of Dakota: U.S. Geog. and Geol. Survey, Rocky Mtn. Region (Powell), 566 p.

Noble, J. A., 1952, Structural features of the Black Hills and adjacent areas developed since Precambrian time: Billings Geol. Soc., 3rd Ann. Field Conf., Guidebook: p. 31–37.

Noble, J. A., and Harder, J. O., 1948, Stratigraphy and metamorphism in a part of the northern Black Hills and the Homestake Mine, Lead, South Dakota: Geol. Soc. America Bull., v. 59, p. 941–975.

Peterson, J. A., 1972, Jurassic System, *in* Geologic atlas of the Rocky Mountain Region: Denver, Colo., Rocky Mtn. Assoc. Geologists, p. 177–189.

Pillmore, C. L., and Mapel, W. J., 1963, Geology of the Nefsy quadrangle, Crook County, Wyoming: U.S. Geol. Survey Bull. 1121-E, 52 p.

Post, E. V., 1967, Geology of the Cascade Springs quadrangle, Fall River County, South Dakota: U.S. Geol. Survey Bull. 1063-L, p. 443–504.

Prucha, J. J., Graham, J. A., and Nickelsen, R. P., 1965, Basement-controlled deformation in Wyoming province of Rocky Mountains foreland: Am. Assoc. Petroleum Geologists Bull., v. 49, p. 966–992.

Rahn, P. H., 1978(?), Geology of the Rockerville quadrangle, Pennington County, South Dakota: South Dakota Geol. Survey Bull. (in press).

Rascoe, B., and Baars, D. C., 1972, Permian System, *in* Geologic atlas of the Rocky Mountain region: Denver, Colo., Rocky Mtn. Assoc. Geologists, p. 143–165.

Redden, J. A., 1975, Precambrian geology of the Black Hills, *in* Mineral and water resources of South Dakota: Washington, D.C., U.S. Govt. Printing Office, Rept. to the Comm. on Interior and Insular Affairs, U.S. Senate, p. 21–28.

Robinson, C. S., Mapel, W. J., and Bergnedahl, M. H., 1964, Stratigraphy and structure of the northern and western flanks of the Black Hills uplift, Wyoming, Montana, and South Dakota: U.S. Geol. Survey Prof. Paper 404, 134 p.

Sales, J. K., 1968, Regional tectonic setting and mechanics of origin of the Black Hills uplift: Wyoming Geol. Assoc., 20th Ann. Field Conf., Guidebook, p. 10–27.

Sanford, A. R., 1959, Analytical and experimental study of simple geologic structures: Geol. Soc. America Bull., v. 70, p. 19–52.

Shapiro, L. H., 1971a, Structural geology of the Black Hills region and implications for the origin of the uplifts of the Middle Rocky Mountain province [Ph.D. thesis]: Minneapolis, Univ. Minnesota, 213 p.

——1971b, Structural geology of the Fanny Peak lineament, Black Hills, Wyoming–South Dakota: Wyoming Geol. Assoc., 23rd Ann. Field Conf., Guidebook, p. 61–64.

Stearns, D. W., 1971, Mechanisms of drape folding in the Wyoming province: Wyoming Geol. Assoc., 23rd Ann. Field Conf., Guidebook, p. 125–143.

——1975, Laramide basement deformation in the Bighorn Basin—The controlling factor for structures in the layered rocks, Wyoming Geol. Assoc., 27th Ann. Field Conf., Guidebook: p. 149–158.

Steece, F. V., 1975, Precambrian rocks outside the Black Hills, *in* Mineral and water resources of South Dakota: Washington, D.C., U.S. Govt. Printing Office, Rept. to Comm. on Interior and Insular Affairs, U.S. Senate, p. 28–29.

Van Lieu, J. A., 1969, Geologic map of the Four Corners quadrangle, Wyoming and South Dakota: U.S. Geol. Survey Misc. Geol. Invest. Map I-581, scale 1:48,000.

Welch, C. M., 1974, A preliminary report on the geology of the Mineral Hill area, Crook County, Wyoming [M.S. thesis]: Rapid City, South Dakota School of Mines and Technology 83 p.

Wise, D. U., 1963, Keystone faulting and gravity sliding driven by basement uplift of Owl Creek Mountains, Wyoming: Am. Assoc. Petroleum Geologists Bull., v. 47, p. 586–598.

Wisser, E., 1960, Relation of ore deposition to doming in the North American Cordillera: Geol. Soc. America Mem. 77, 117 p.

Wolcott, D. E., 1967, Geology of the Hot Springs quadrangle, Fall River and Custer Counties, South Dakota: U.S. Geol. Survey Bull. 1063-K, p. 427–442.

Wulf, G. R., 1955, Geology of the Fanny Peak area, Weston County, Wyoming [M.S. thesis]: Rapid City, South Dakota School of Mines and Technology, 37 p.

MANUSCRIPT RECEIVED BY THE SOCIETY JUNE 27, 1977
MANUSCRIPT ACCEPTED AUGUST 25, 1977

Geological Society of America
Memoir 151

A relationship between strike-slip faults and the process of drape folding of layered rocks

ROBERT A. COOK
Exploration Resources, ARCO
1545 W. Mockingbird
Dallas, Texas 75235

ABSTRACT

The origin of strike-slip faults in the Rocky Mountains foreland is somewhat of an enigma; these faults occur in an area where the basement deformation has primarily involved vertical movements. Some of these faults may have originated not from lateral shifts within the crystalline basement, but from an interaction of detached and laterally shifting layers within the sedimentary pile and from the dynamics of drape folding.

The folds and related structures of Dinosaur National Monument of Colorado and Utah suggest such an origin for some strike-slip faults. These folds and related features evolved with the uplift and brittle deformation of the Uinta Mountains. Uplift, tilting, and extension along the south flank of this range ruptured the basement core into an array of rigid blocks. During this deformation, the Weber Formation became detached and slid between 1,000 and 1,500 m southward. Where the formation encountered rising steps (fault blocks) in the basement surface that were aligned normal to the sliding direction, it became draped and often thickened. Where steps did not exist, the lateral sliding was accommodated by buckle-type folding. In one region, however, the combination of lateral sliding, drape folding, and the effects of a low, inclined ramp structure caused the Weber Formation to tear in a small strike-slip fault with possibly 1,000 to 1,500 m of offset.

INTRODUCTION

Some of the most spectacular structural expressions of sedimentary rock ductility on a large scale are the drape folds of the Rocky Mountains. Throughout the Rockies are immense folds in which the layered rocks cascade thousands of metres, often unbroken by throughgoing ruptures or faults. Most of these folds are closely associated with basement faults over which the strata are draped. The process of draping, however, is far more complex than the simple bending or forced folding of layered rocks over some discrete, vertical displacement in the basement surface. The process includes a number of mechanisms that, in various combinations with the bending and the dynamics of the basement deformation (spatial and temporal changes), determine the final fold geometry. These mechanisms are lateral sliding along bedding surfaces, buckling, cataclasis, fracturing, and faulting.

Field studies conducted within Dinosaur National Monument of Colorado and Utah have defined and established some of the interrelationships between these mechanisms and the process of bending layered rocks (Cook and Stearns, 1975). One of the more interesting relationships of drape folding is that between folding, lateral sliding, tear faulting, and a geometry that I refer to as the "ramp structure": a low incline over which lateral sliding has occurred.

PROBLEM STATEMENT

The occurrence and origin of transcurrent or strike-slip faults in the Rockies is an interesting and somewhat controversial geologic problem. The Rocky Mountains foreland has a structural regime dominated by vertical crustal movements as opposed to strike-slip and horizontal tectonics. This does not exclude lateral movements during the basement deformation. Displacements such as crustal extension or shortening have taken place. The presence of the Absaroka volcanic rocks suggests some extension; likewise, the combined movements across the uplifts of the Uinta Mountains, Wind River Range, Rawlins uplift, and the Owl Creek Mountains necessitate some crustal shortening. It is apparent when one examines geologic

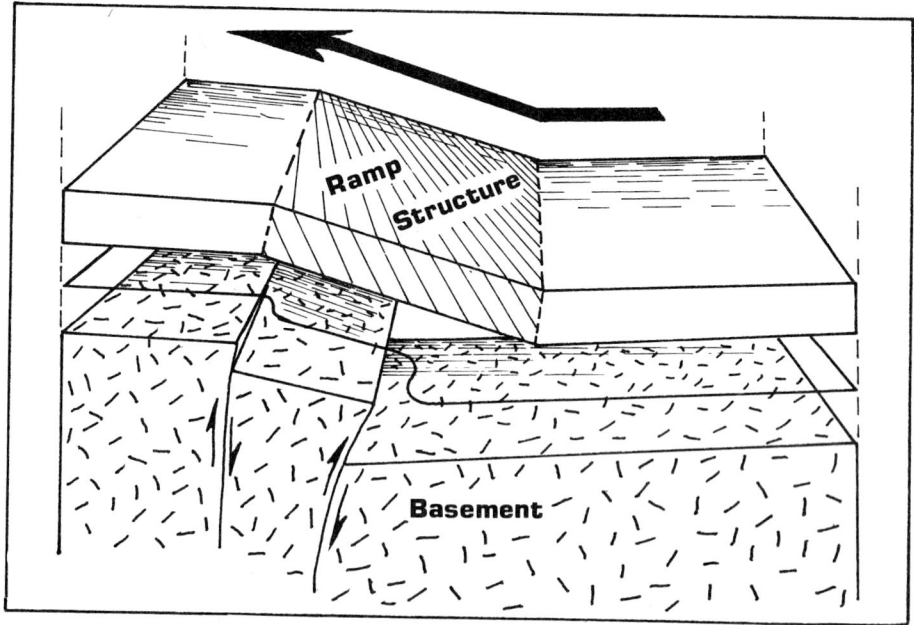

Figure 1. Idealization of a ramp structure over a reverse fault system prior to rupture and offset. This is just one of many possible geometries. Ideally, the term "ramp structure" should designate only steep, monoclinal flanks of drape folds that, because of their geometries, allowed lateral sliding over basement faults. In Dinosaur National Monument, the upper surface of the ramp is equivalent to the Weber detachment surface.

maps of the Rockies that transcurrent faults often exist in close proximity to vertical basement faults with, in many instances, only small components of horizontal movement. Thus, the origin of these transcurrent faults raises a basic question: Are they due to discrete lateral motions between blocks of the basement, or are they related in some manner to the drape-folding process?

The evidence from Dinosaur National Monument suggests that at least some of these transcurrent faults are the result of interaction between a detached, laterally sliding portion of the section, a ramp structure (Fig. 1) over which the rocks can pass, and the bending mechanisms of the draping process. The specific observations to be discussed in this paper further indicate that some of these faults arise from a tearing type of failure, and that the orientation of a fault closely defines the direction of lateral transport. Finally, the existence of such faults further specifies the spatial and temporal changes of the basement deformation—its dynamics.

To the exploration geologist, knowledge of the displacement field of a deformation is invaluable in predicting the position and geometry of potential structures and reservoirs and the paths and hydrodynamics of fluid migration. To the science of structural geology and its supporting disciplines, knowledge of the displacements and the interrelationships between the mechanisms of drape folding further clarifies the complexity of the folding process and its relationship to other geologic phenomena, both structural and stratigraphic.

REGIONAL GEOLOGY

The study area, Dinosaur National Monument, is situated on the southeastern flank of the Uinta Mountains of Utah and Colorado. The structural style of this mountain system and the adjacent depressions is typical of the Cordillera's foreland. The range strikes due east for some 250 km and is approximately 60 km at its greatest width. To the north and south and adjacent to the range are the immense sedimentary basins of Utah, Colorado, and Wyoming. On the north are the Bridger and Washakie Basins, separated by the smaller Rock Springs uplift. South of the range are the Uinta and Piceance Basins, and to the east is the Sand Wash Basin. These features, like the Uinta Mountains, are products of the Late Cretaceous and early Tertiary Laramide deformation.

The structural expression of the Uinta Mountains in cross section is that of a broad, symmetrical uplift bordered by two immense reverse faults (Fig. 2). These faults subparallel one another and have throws often in excess of 13,000 m. Between the two bounding faults are numerous faults of substantially lesser throw termed "strike faults," because their scarps parallel and subparallel the axis of the mountain range. Most of these secondary faults arose from the north-south extensional strain imposed upon the range as the result of its uplift. Locally, however, some faults have a reverse sense of throw and are not entirely related to the extensional phenomena. Still others originated in response to the late Tertiary collapse of the Uinta Mountains (Powell, 1876; Sears, 1924; Untermann and Untermann, 1954). It is the displacements across these strike faults that draped and deformed the Weber Formation and the other Paleozoic rocks within the study area.

The stratigraphy of the Uinta Mountains and the study area varies in age from Precambrian to Quaternary. Figure 3 is a generalized stratigraphic column summarizing the layered rocks cropping out within the monument and adjacent basins. The metasedimentary units of the Precambrian Uinta Mountain Group represent the oldest layered rocks exposed within the study area. Overlying them are the layered Paleozoic sequences. The Paleozoic rocks, in a regional sense, occur as blankets

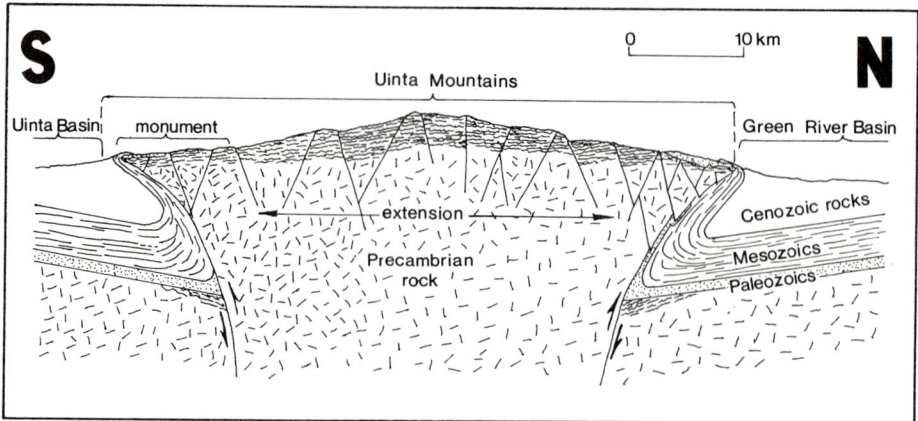

Figure 2. Generalized north-south cross section of the Uinta Mountains of Utah and Colorado. Strike faults across the range have accommodated the north-south extension imposed by the uplift.

or sheets which thicken to the west where the remainder of the Paleozoic section is present. The Cambrian Ladore Formation unconformably overlies the Uinta Mountain Group and is, itself, a transgressive sequence of calcareous shales and sands approximately 145 m thick. Overlying the Ladore Formation are Mississippian and Pennsylvanian carbonate rocks. The lower 180 m of this carbonate package in the study area is separated from the upper 140 m by 30 m of siliceous, fine-grained clastic rocks. The carbonate rocks of the upper portion become sandier toward the top and are interbedded with calcite-cemented, quartz sandstone layers near the top. This progressive lithologic change of the Morgan Formation is transitional upward into the massive sandstone sheet of the Pennsylvanian-Permian Weber Formation. Because of its excellent exposure across the drape folds in the monument, the Weber Formation is central to the forthcoming description and discussion. Within the study area, the Weber has an average thickness of about 300 m and is a sheet in overall form. Internally, however, it is a composite of at least eighteen internally cross-bedded units that are delineated by local truncation surfaces. Overlying the Weber Formation are the Mesozoic rocks. Most of these sedimentary units, however, are only partially preserved in the study area, and their principal importance to the structural problem at hand is their contribution to the overburden thickness and load at the time of the deformation. The best estimates of the overburden thickness are between 4,000 and 4,400 m (Untermann and Untermann, 1954).

LOCAL GEOLOGY

The most prominent structural features of Dinosaur National Monument are the large drape folds. The structural relationship of these folds and other structures is shown in Figure 4 and is a key to the understanding of the ramp structure and the tear fault.

The drape folding of the sedimentary sequences resulted from the Late Cretaceous and Tertiary breakup of the basement[1] complex into a series of rigid fault blocks.

---

[1] "Basement" is used here to refer to those rocks which behaved in a statistically homogenous, isotropic, and brittle manner to depths of at least 15,000 m (Stearns, 1971).

The nearly uniform, gently dipping flanks of the folds reflect the rotated upper surfaces of the basement blocks. Figure 5 is a model of the study area portraying the postdeformational configuration of the basement surface. Where the sedimentary rocks (in particular, the Weber Formation) are preserved across the upper surfaces of the basement blocks, they have normally experienced only rigid-body rotations

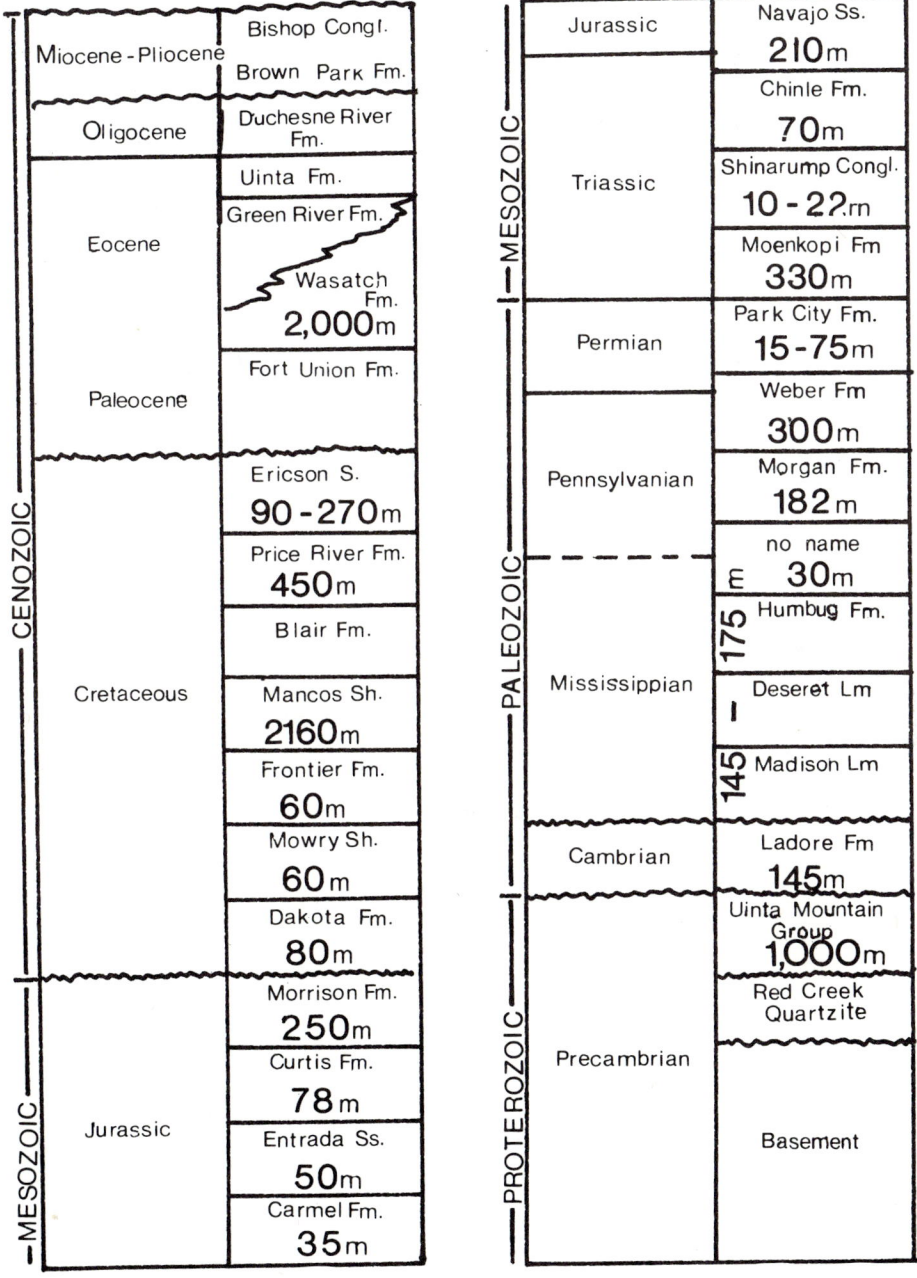

Figure 3. Stratigraphic column for the eastern Uinta Mountain region (after Untermann and Untermann, 1954).

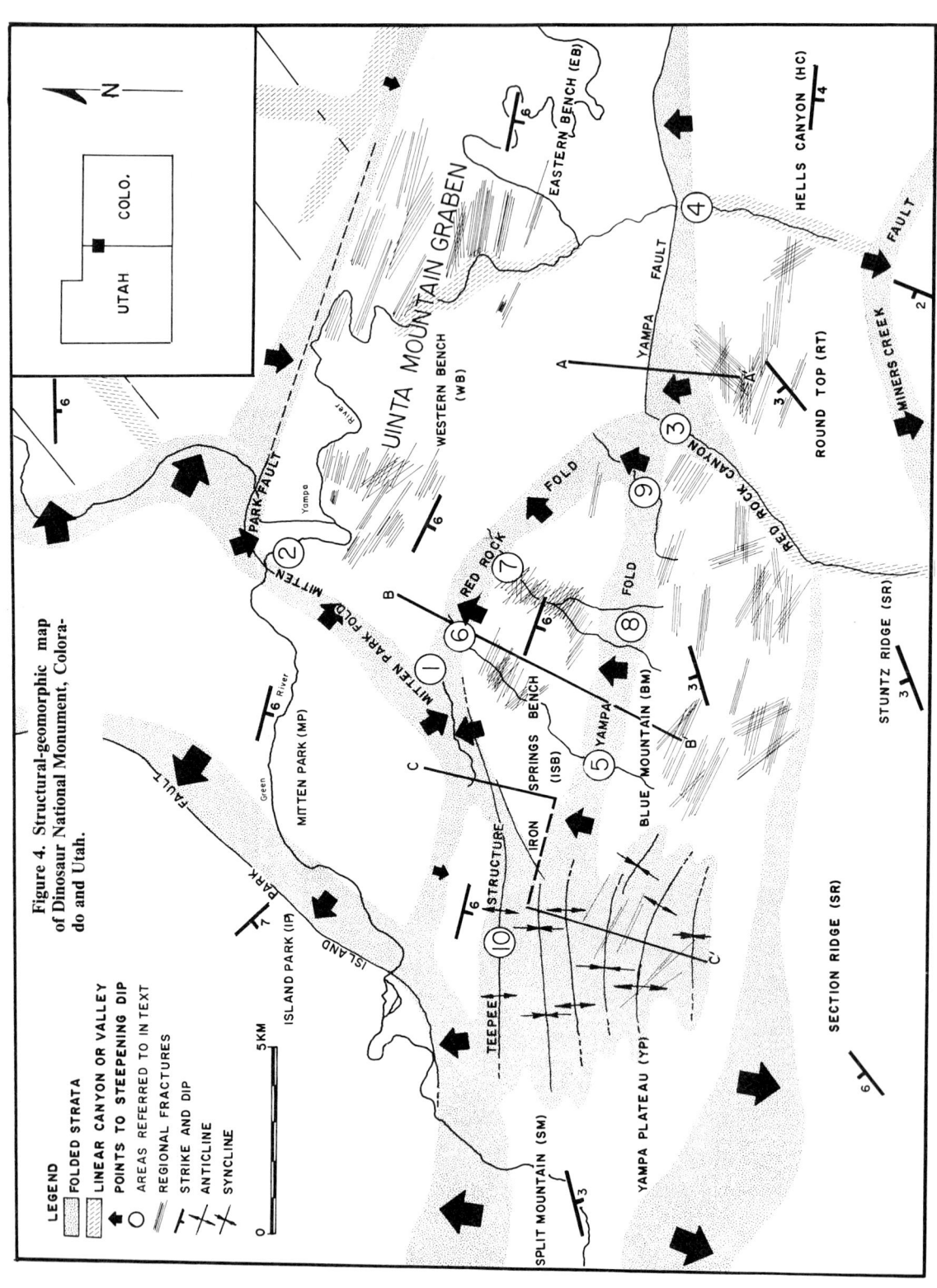

Figure 4. Structural-geomorphic map of Dinosaur National Monument, Colorado and Utah.

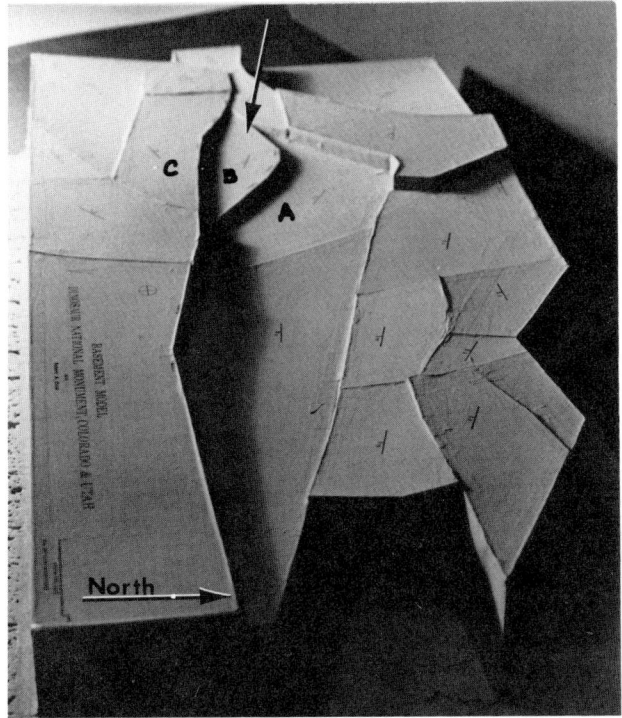

Figure 5. Basement model of study area. The Western bench (A), Iron Springs Bench (B), and Blue Mountain (C) form basement steps (arrow) which rise to the south. A structural reentrant or embayment lies directly to the right of the arrow's point. Model was constructed without vertical exaggeration.

and translations (designated by the clear areas in Fig. 4) and consequently display only minor internal disruptions. Such features as regional fractures and thin gouge zones along bedding surfaces are quite common.

In contrast to these nearly flat, undisturbed regions, the block edges show high distortional strains. Along the edges, the strata are draped and steeply inclined; moreover, they are often thickened, thinned, or ruptured internally by fabrics of small-scale faults that are not directly related to those that displace the basement. Across block boundaries of no apparent offset, only small strike and dip changes and preferential erosion (linear canyons, Fig. 4) indicate movement, fracturing, and intragranular cataclasis.

Briefly, the variations of the fold and fault geometries that are important to the problem here are as follows: Across the northern flank of Split Mountain (Fig. 4), the Weber Formation is draped and thinned to about 100 m, one-third of its original thickness. Farther to the east, the formation is ruptured and displaced by the Island Park fault against which the strata are upturned and attenuated. The throw across the fault averages about 1,300 m between the downdropped Island Park block to the northwest and Split Mountain.

Along the eastern flank of the high Mitten Park block, the Weber Formation is draped and is locally offset by the Mitten Park fault (Figs. 4 and 6). In the region of locality 1 in Figure 4, the Weber layer rolls unthinned across the 1,000 m of structural relief between the Mitten Park and Western Bench block. To the northeast along the strike of the fold, the Weber progressively thins, however, to about one-third of its original thickness, where it is locally offset and juxtaposed with sub-Morgan Formation carbonate rocks that are draped and unbroken by the fault (loc. 2 in Fig. 4). The fault curves to the southeast, rapidly loses throw, and becomes only a line of gentle warping in the layered sequences.

Figure 6. Aerial photograph looking southwest into Trail Draw. The Teepee structure (arrow) is the sharp chevron fold superimposed upon the Red Rock drape fold. The structure and adjacent benches are truncated by the Gilberts Peak (?) pediment surface. Benches shown are Blue Mountain (BM), Iron Springs bench, Split Mountain (SM), and Mitten Park block (MP).

Across the south-flank fault of Split Mountain and the Yampa Plateau, the sandstones of the Weber are also draped, but here they are continuous and unthinned. The maximum throw across the fault is not known precisely other than it equals or exceeds the 1,300 m of offset of the north flank. To the east, the throw uniformly decreases to zero at the southeast corner of Blue Mountain (near the trace of the Miners Creek fault, Fig. 4).

Figures 7 and 8 show the steeply inclined flanks of the drape folds defining the southern margin of the Uinta Mountain graben. Along this margin, east of the entrance to Red Rock Canyon (loc. 3 in Fig. 4), the Weber Formation and an undetermined thickness of underlying Paleozoic strata (possibly more than 500 m) are missing across the Yampa fault. The strata preserved on the high blocks (Fig. 7) south of the fault (Morgan Formation and older) are draped and steeply inclined as they roll into the fault and are in fault contact with the Mesozoic rocks remaining on the downdropped blocks. On the downdropped blocks of the graben, the Weber Formation and the other layered sequences maintain a low inclination (3° to 6° dip toward the south) into the Yampa basement fault, where they are sharply upturned and thinned. This is illustrated in cross section AA' (Fig. 9). Along the portion of the fault where the thinning is most conspicuous, the throw varies from a low of approximately 1,200 m, just east of the entrance to Red Rock Canyon, to a maximum of about 1,400 m at Hells Canyon (locs. 3 and 4, respectively, in Fig. 4). At Hells Canyon, the reverse character of the fault is clearly evident (Fig. 7), and this appears to have been an important element in the formation of a structural ramp in this region.

West of the entrance to Red Rock Canyon, the Weber Formation cascades across the Yampa and Red Rock folds (Figs. 4, 7, and 8). The rock in both folds is draped and segmented into smaller fold blocks in a manner similar to the classical fold studied by Stearns (1971) at Rattlesnake Mountain, Wyoming. The hinge lines between segments of each fold parallel and subparallel the main axis of the fold, but vary in form and continuity along the strike. In most regions the hinges are tight, narrow zones of small radius of curvature. In some areas, however, they

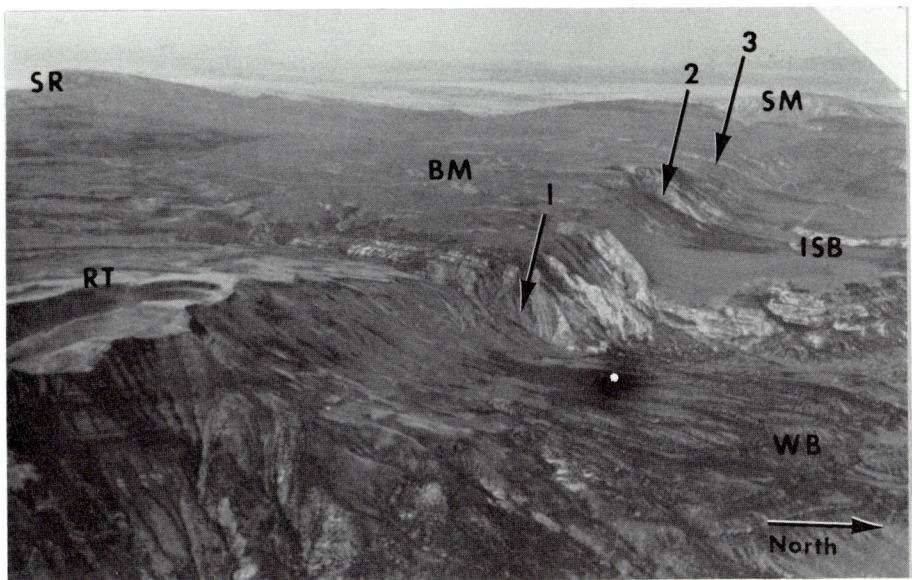

Figure 7. Aerial photograph of Yampa fault and fold trend. The Weber Formation is totally attenuated along the Yampa fault from Red Rock Canyon (arrow 1) to the east (lower left-hand corner); it is thickened at Canyon Overlook (arrow 2) and slightly thinned at Stuntz Draw (arrow 3). Benches shown here are Round Top (RT), Section Ridge (SR), Blue Mountain (BM), Split Mountain (SM), Iron Springs Bench (ISB), and Western Bench (WB).

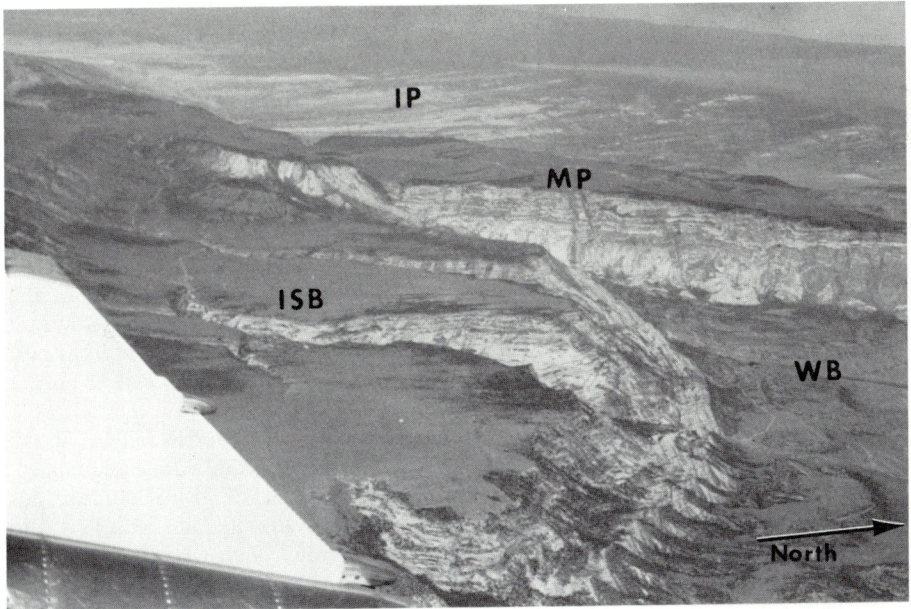

Figure 8. Aerial photograph looking west across the Iron Springs Bench (ISB), Western Bench (WB), Mitten Park block (MP), and Island Park block (IP). The Red Rock drape fold rolls around toward viewer from center of photograph to front center. Aircraft wing occupies lower left area.

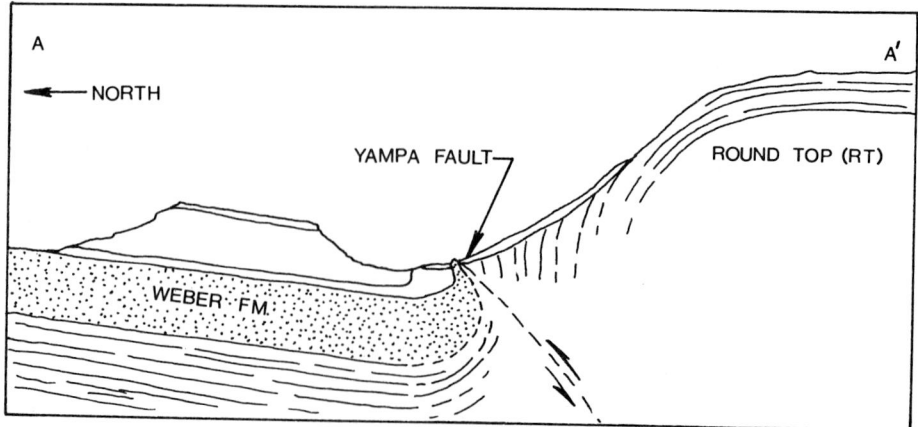

Figure 9. North-south structural cross section AA' across Yampa fault showing rapid dip reversal and attenuation of the Weber Formation against the Yampa fault. Section constructed with a minimum vertical distortion; location shown in Figure 4.

are marked by small reverse and thrust faults that necessitate shortening normal to the fold axis to allow for the sense of movement across the faults. In addition to the complexities of the internal geometries of these folds, the Weber layer is also thickened across each fold, except in one region of the Yampa fold (loc. 5 in Fig. 4). In this region it has been attenuated approximately 10%. The thinning occurred during the final stages of the deformation (Cook and Stearns, 1975). In the areas where the sandstone is thickened across the folds, the magnitude of change is significant. At localities 6 and 7 in Figure 4, for example, the Weber Formation is thickened by 25% and 45%, respectively, and at locality 8 it is approximately 40% thicker. At site 9 (Fig. 4) it is neither thinned nor thickened. Figure 10 (section BB') is a typical cross section showing the Weber sheet as it appears in the Yampa and Red Rock drape folds. Again, the basic form of the folds (continuity of section across the folds and segmentation into fold blocks) is a form closely resembling the Rattlesnake Mountain drape fold.

The structural relief across the Yampa fold, as well as the Red Rock fold, is in excess of 400 m along the eastern segment. To the west the vertical displacement of the Yampa fold decreases to a little more than 100 m (loc. 5 in Fig. 4). The displacement of the Red Rock fold, however, remains relatively constant. At the western end of the Red Rock fold (loc. 1 in Figs. 4 and 5), a structural re-entrant or embayment at the basement level is formed by the convergent geometries of the Iron Springs Bench and the Mitten Park blocks and the southwestern corner of the downdropped Western Bench block.

The deformation within the study area was not merely a brittle failure at the basement level accompanied by draping of the overlying sedimentary sequences. In the west-central map area, the Weber, as well as the other rock units, has been shortened laterally and buckled[2] into a wave train of folds (Figs. 4 and 11, cross section CC'). The axes of these folds parallel one another along their western lengths, but to the east the axes diverge and in two instances merge disharmonically in a superposed relationship with the draped layers of the Yampa and Red Rock drape folds.

---

[2] Buckled is used here in the traditional geologic sense, because the concept of buckling is currently subject to controversy (Fahy, 1976).

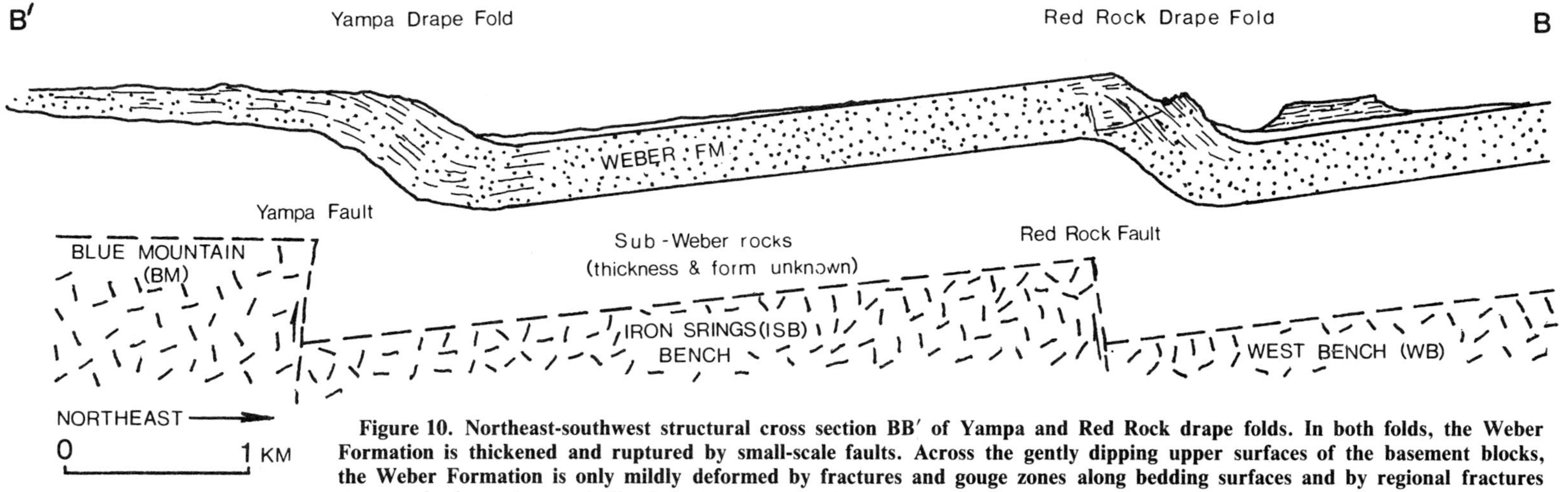

Figure 10. Northeast-southwest structural cross section BB' of Yampa and Red Rock drape folds. In both folds, the Weber Formation is thickened and ruptured by small-scale faults. Across the gently dipping upper surfaces of the basement blocks, the Weber Formation is only mildly deformed by fractures and gouge zones along bedding surfaces and by regional fractures that cut the formation vertically. Location shown in Figure 4.

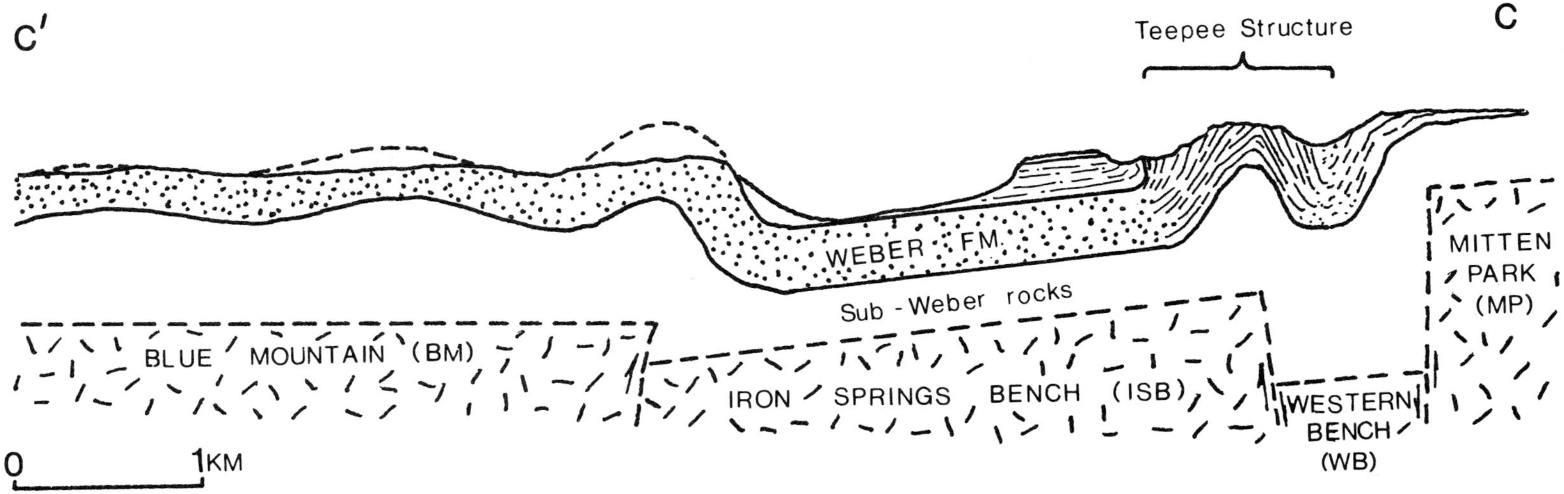

Figure 11. North-south cross section CC' with east-west offset showing the Teepee structure and related folds to the south. Location shown in Figure 4.

The northernmost fold of the wave train—dubbed the Teepee structure because of its cross-sectional appearance (loc. 10, Figs. 4 and 11)—is a sharp, chevron flexure (Figs. 6 and 8). For several kilometres in the region where the flexure is superimposed upon the Red Rock drape fold, the uppermost units of the Weber Formation, along the Teepee structure's southern flank, are slightly overturned (5°).

DISCUSSION

The thickened, steep flanks of the Yampa and Red Rock drape folds and the fixed hinges and fold segments of these flanks indicate excess Weber rock across the southwestern margin of the Uinta Mountain graben. Similarly, excess rock is also indicated to the west of the graben by the Teepee structure and related folds, as well as the lack of attenuation of the formation across the southern flank of Split Mountain and only a partial thinning over its northern flank. In order to resolve the origin of this excess rock and the fold geometries, the Weber Formation must have been detached across its lower boundary and transported laterally to the south during the basement deformation. In addition to the mechanism of lateral transport, the fixed hinges and fold segments of the drape folds restrict the sequence of basement faulting and block movement. Furthermore, the lateral movement indicated by the folds and the complete absence of Weber rock across the Yampa fault east of Red Rock Canyon (loc. 3, Fig. 4) over a strike distance of less than 50 m, necessitate the existence of a "ramp structure" (Fig. 1) over the Yampa fault during its initial stages of growth. Finally, the western boundary of the ramp structure is the tear fault at Red Rock Canyon, which has a right-lateral sense of slip.

Detachment and Lateral Transport

Prior to and during the breakup of the basement, the upper surface of the basement and the overlying sedimentary rocks were steeply tilted toward the south by 5° to 10° more than is now measurable. This situation resulted in a large component of the body forces inherent to the layered rock mass to be directed downdip. In view of this, it seems that gravity sliding toward the south was greatly enhanced and that updip movement of a detached mass was impossible. The thickened drape folds, the sense of displacement across small faults within these folds, and the sense of shortening of the wave train of folds does, indeed, support the southward sliding of the Weber Formation as a sheet (Cook and Stearns, 1975).

Along the Yampa and Red Rock drape folds, the lateral sliding of the Weber sheet was taken up in the folds as they grew. The thickening and small-scale thrusting and reverse faulting between fold segments indicate that the lateral movement, however, exceeded the amount of throw required to maintain a uniform, unthickened section of Weber rock across the folds. Best estimates place the amount of slip between 1,000 and 1,400 m.

In the west-central area, the lack of similar basement steps, such as those beneath the Red Rock and Yampa drape folds, resulted in the buckle folding of the Weber Formation. This behavior resolved any excess volume of rock created by the lateral sliding in this area, and the magnitude of movement indicated from amplitude and wavelength measurements of the folds suggests slip on the order of 800 to 1,500 m. Farther to the west the sliding was taken up across the immense bounding faults (mostly along the southern flank) of Split Mountain, and here the amount of lateral movement exceeds at least 1,300 m. In the area of the Teepee structure, the lateral movement, coupled with the development of the structural re-entrant (Fig. 5), locally created a severe space problem within the Weber Formation. The result was the sharp, chevron flexure of the Teepee structure that is disharmonically superimposed upon the pre-existing Red Rock drape fold.

Sequence of Basement Faulting

Within the immediate study area, the sequence of basement faulting and block movements is inferred from the above-mentioned relationships as follows. The Miners Creek fault (Fig. 4) was the first to experience vertical movement. Following closely were displacements along the south flank of Split Mountain and the eastward extension of this fault. Movements along the Yampa fault segment west of Red Rock Canyon followed, causing the Weber Formation to pile up and drape. East of the canyon, however, the displacement picture was different. Offset appears to have been delayed, or initially small, and of a different sort, as indicated by the reverse sense of movement. This difference not only suggests a different time of activity, but a somewhat different stress history for the area later in the deformation. At about the time the throw across the western segment of the Yampa fault reached 400 m at locality 9 in Figure 4, movement began along the basement fault beneath the Red Rock fold; by the time it reached a maximum of 450 m, the Mitten Park block began to move along its southeastern flank. This uplift resulted in the structural re-entrant and the chevron folding of the Teepee structure. Except for possibly the eastern extension of the Yampa fault, movement across the Island Park fault was initiated last with the downdropping of the Island Park block. Throughout this sequence of basement faulting, the regional tilt was in excess of present values, and the Weber Formation was sliding southward over the growing structural ramp.

Evidence for the sequence of basement disruption is found in the fixed-hinge geometries and fold segments of the drape folds. If, for example, the Mitten Park block had existed prior to either of the basement faults beneath the Yampa or Red Rock drape folds, then the Weber Formation would have had to move up and over the Mitten Park block without leaving a trace of thickening, thinning, folding, or unfolding. The same argument applies if the Red Rock basement fault and fold had pre-empted the existence of the Yampa fold. Again, the Weber Formation would have had to pass through the steep flanks of the Red Rock fold without leaving any evidence of the deformation. This seems improbable, because, as stated earlier, the Weber rock is essentially undeformed across the gentle flank of the Iron Springs Bench that lies downdip from the Red Rock fold (Fig. 10).

Ramp Structure and Tear Faulting

To the east of Red Rock Canyon (loc. 3, Fig. 4), the structural relationships and evolution are more problematic. This is particularly true when considered in terms of the 1,000 to 1,400 m of lateral sliding indicated for the Weber Formation just west of the canyon entrance. As stated previously, east of the canyon entrance, in less than 50 m, the entire thickness of the Weber Formation (300 m), plus an undetermined thickness of the sub-Weber rocks, are missing. As shown in the cross section across this eastern segment of the Yampa fault (Fig. 9), excess rock is not piled up against the fault. Furthermore, there is no strike-slip fault trending north from Red Rock Canyon that would have allowed for the lateral movement in the west, and at the same time indicated a fixed section of Weber rock to the east of the canyon entrance. On the contrary, all of the physical evidence points to 1,000 to 1,400 m of lateral sliding of the Weber Formation east of the canyon.

Although several solutions might explain this enigma, the one favored here involves a ramp structure. The ramp developed from the displacements along the eastern length of the Yampa fault and from the slight northward rotation—hence decrease in dip—of the Round Top and Hells Canyon blocks to something less than the regional tilt. Coupled with this ramp formation was a tearing of the Weber Formation at or near the entrance to Red Rock Canyon. Instead of forming a steep fault step in the basement surface, the delayed and minor movement across the fault and the tilting caused by the fault together gave rise to the ramp structure (Fig. 12). The ramp allowed the Weber sheet to slide southward across the increasing displacement on the Yampa fault. Once the fault ruptured the ramp, the Weber rapidly thinned. This rupturing appears to have occurred at a time when the southward movement of the formation had nearly ceased. If this were not the situation, the formation should either be piled up against the Yampa fault or have been preserved, in part, across the fault itself. Again, neither situation was observed.

The rapid thinning of the Weber Formation and the other sedimentary rocks across the entrance of Red Rock Canyon was accomplished by localized tearing. This tearing, the displacement field, and the specifics of the deformation are illustrated in the drawings of Figure 13. West of the canyon the sandstone piled up along the basement faults; this gave rise to the Yampa and Red Rock folds, respectively. Once the throw across the western extension of the Yampa fault reached some critical value, southward movement beyond the fault ceased; that is, the section became fixed (indicated by the wood screw in Fig. 13). To the east, however, southward sliding continued over the low ramp structure. The net result, therefore, was a situation of acute, localized shearing and tearing near the juncture of the Yampa and Red Rock drape folds at the entrance to Red Rock Canyon. The tearing at this juncture should have resulted in a small, right-lateral transcurrent fault trending toward the south. Measurable offset across the canyon is, however indeterminable because of poor exposure, but the magnitude of offset should nearly equal that (1,000 to 1,400 m) determined for the drape folds west of the canyon.

A fascinating aspect of this movement scheme is the effect of the ramp structure on the sandstone's constitutive properties as it passed southward over the ramp. Sliding up and over the ramp should have caused, at least in a kinematic sense, the sandstone to experience bending moments in one direction followed by a reversal of moments to the opposite direction. The outcome of such a deformation is not hard to visualize in light of the condition of the Weber Formation in the Yampa and Red Rock drape folds. The initial effects were probably predictable systems of fractures like those commonly found in folded strata (types 1 through 5, Stearns, 1968). Beyond a certain inclination the rock would probably have been totally pulverized in most regions to loose, unconsolidated sand. The absence of Weber rock across the Round Top and Hells Canyon blocks might just reflect such a strain history of intense distortion, fracturing, faulting, and cataclasis. The volume of rock affected in this manner, lying to the south of the ramp, should be a function of the amount of lateral transport over the ramp. Unfortunately, the complete picture is not preserved within the study area.

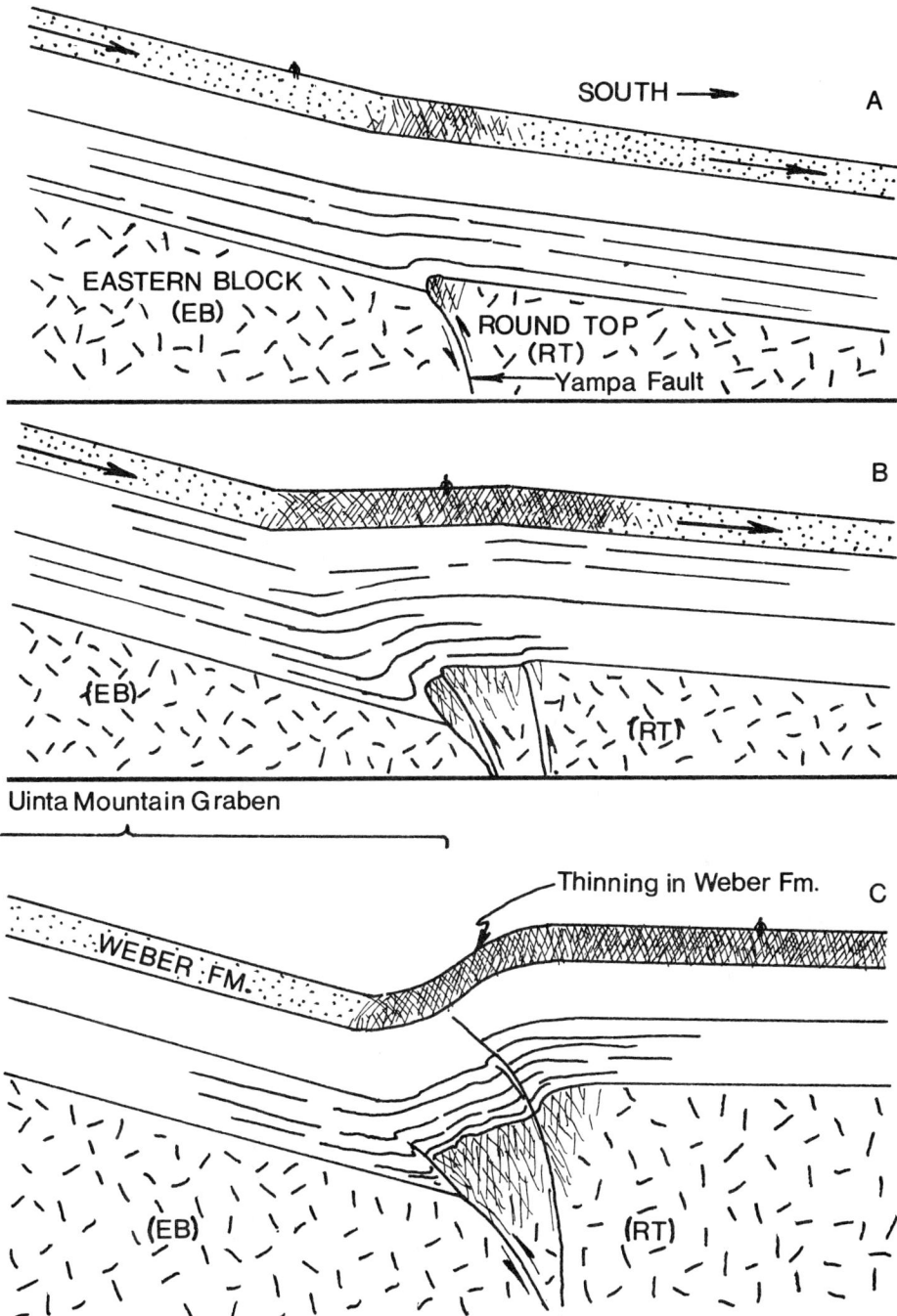

Figure 12. Diagrams illustrating development of a ramp structure. (A) Vertical movement across the basement fault results in a small incline over which the detached Weber Formation passes southward. (B) Displacement across the fault increases. This gives rise to a well-defined ramp over which southward movement of the Weber continues. The position of the human figure indicates the amount of sliding over the ramp. With this southward displacement, fractured and granulated rock is carried downdip, away from the ramp. (C) After lateral movement of the Weber sheet has ceased, continued movement on the Yampa fault rapidly attenuates the sandstone.

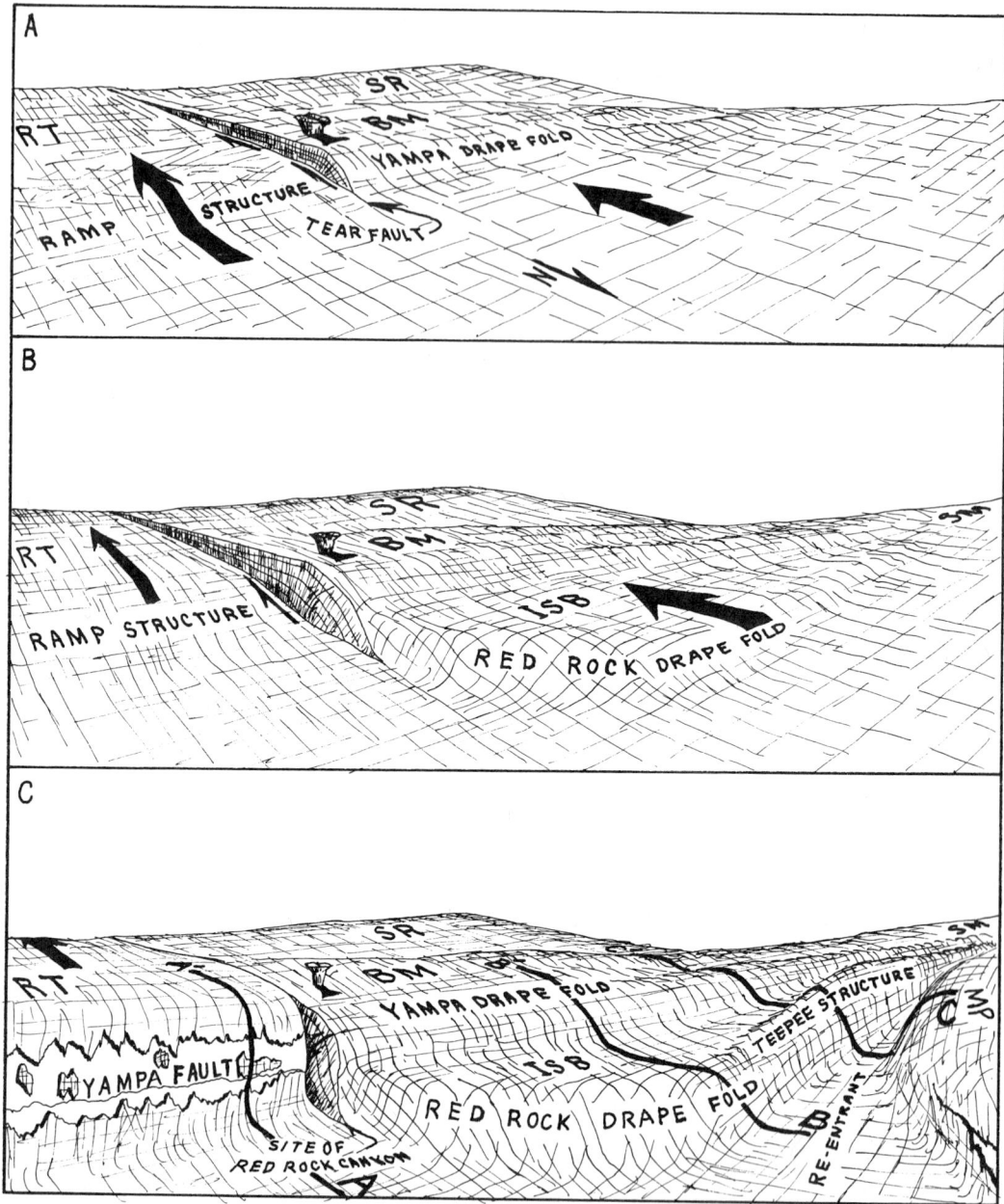

Figure 13. Drawings illustrating sequential movements and deformation envisioned for Dinosaur National Monument. Viewer is looking obliquely southwest across the western end of the Uinta Mountain graben; rocks above Weber Formation have been stripped away. In drawing (A), movements across the basement faults of Stuntz Ridge (SR) and Split Mountain (SM) blocks (south fault) have occurred, and movements are now occuring along the Yampa fault trend. The section on Blue Mountain (BM) is now fixed (denoted by wood screw), but tearing is occurring as the Weber sheet slides southward over the ramp. In (B), movement across ramp continues, but displacement beneath Yampa drape fold has nearly ceased, and the Red Rock fold is beginning to grow. The final stage in the deformation is shown in (C); the ramp structure is ruptured, the Red Rock fold is nearly complete, and the Teepee structure is taking its final form with vertical movements of the Mitten Park block. Other blocks shown are Round Top (RT), Mitten Park (MP), and the Iron Springs Bench (ISB).

SUMMARY

The thickened section of the Weber Formation across the steep flanks of the Yampa and Red Rock drape folds together with the volume of rock involved in the Teepee structure and related folds suggest that there was at least 1,000 and possibly 1,500 m of lateral sliding of the formation over the western end of the Uinta Mountain graben during its formation. This magnitude of movement, however, is not readily apparent east of Red Rock Canyon; neither is the lack of sliding. The only volume-consistent means of resolving the geologic relationships necessitates an equal amount of movement east of the entrance to Red Rock Canyon, a small, right-lateral tear fault along the canyon axis, and a ramp structure over the eastern extension of the Yampa fault. The evidence for the tear fault and ramp is indirect and entails the structural relationships throughout the entire expanse of Dinosaur National Monument.

CONCLUSIONS

The possible existence of tear faults and ramp structures and their relationships to the process of drape folding are important to the science of structural geology, its subdisciplines, and especially the petroleum industry for the following reasons:

1. The presence of transcurrent or strike-slip faults may reflect localized tearing as a result of lateral movements within the layered rocks and not strike-slip motions between elements of the underlying basement mass.

2. The trend of a strike-slip fault may not only reflect the existence of lateral movement, but also the direction of sliding; that is, the fault trend is probably parallel or subparallel to the direction of lateral transport.

3. Like the thickness variations and specific geometries found in the drape folds of Dinosaur National Monument, the existence of strike-slip faults (if related to lateral movements in the sedimentary pile) should allow for a more precise definition of the kinematics and strain history of the basement deformation (the deforming geometry) and the loading history experienced by the sedimentary rocks.

4. Finally, the existence of strike-slip faults may signal the presence of a ramp structure and, if so, the accompanying effects of intense fracturing and cataclasis in the mass of rock that passed over the ramp. Because the upward movements of a basement block could possibly cease or be reversed, evidence of any ramp associated with the block and even the ramp structure itself might be lost during the ensuing deformation. Consequently, large undetected volumes of fractured and granulated rock may occur off-structure or in an apparent, unrelated position to any obvious basement high. Such undisturbed regions might serve as potential petroleum traps because their porosity is increased relative to that of the surrounding rocks. This would be particularly true if they were situated in the migration path to a structural high. If such features exist, they more than likely represent major bulk-density variations within the sedimentary pile.

ACKNOWLEDGMENTS

I thank my close friend, David W. Stearns, for his support of my goals and his encouragement in this study. In addition, special gratitude is extended to John W. Handin, Dave Fahlquist, Bob Stanton, John Logan (all Texas A&M faculty), Pat Gilbert, Fred Meyers, Peter Barker (ARCO) and David Hite (ARCO). I also thank Frank Buono and the other members of the National Park Service of Dinosaur National Monument who assisted in the field investigations. Finally, thanks is extended to the National Science Foundation (Grant GA-3612-7X) and the Department of Geology of Texas A&M University for financial backing.

REFERENCES CITED

Cook, R. A., and Stearns, D. W., 1975, Mechanisms of sandstone deformation: A study of the drape folded Weber sandstone in Dinosaur National Monument, Colorado and Utah, in Bolyard, D W., ed., Deep drilling frontiers of the central Rocky Mountains—1975 Symposium: Denver, Rocky Mtn. Assoc. Geologists Guidebook, p. 21–31.

Fahy, M., 1976, Are any folds buckles?: Geol. Soc. America Abs. with Programs, v. 8, p. 19.

Powell, J. W., 1876, Report on the geology of the eastern portion of the Uinta Mountains and a region of country adjacent thereto: U.S. Geol. Survey Terr., v. 1.

Sears, J. D., 1924, Relations of the Browns Park Formation and the Bishop Conglomerate and their role in the origin of the Green and Yampa Rivers: Geol. Soc. America Bull., v. 35, p. 279–304.

Stearns, D. W., 1968, Fracture as a mechanism of flow in naturally deformed layered rocks, in Proc., Conf. Research Tectonics: Canada Geol. Survey Paper 68-52, p. 79–90.

——1971, Mechanism of drape folding in the Wyoming province: Wyoming Geol. Assoc., 23rd Ann. Field Conf., Wyoming Tectonics Symp. Guidebook, p. 125–143.

Untermann, G. E., and Untermann, B. R., 1954, Geology of Dinosaur National Monument and vicinity: Utah Geol. and Mineralog. Survey Bull., no. 42, 228 p.

MANUSCRIPT RECEIVED BY THE SOCIETY JUNE 27, 1977
MANUSCRIPT ACCEPTED AUGUST 25, 1977

Geological Society of America
Memoir 151

Monocline fold pattern of the Colorado Plateau

GEORGE H. DAVIS
Department of Geosciences
University of Arizona
Tucson, Arizona 85721

ABSTRACT

Geologic relationships in the Grand Canyon region demonstrate a spatial, genetic association of monoclines with reactivated, ancient, steeply dipping faults. If one assumes that these relationships hold for monoclines throughout the Colorado Plateau, the monocline pattern, in total, records the upper-crustal expression of many elements of the regional, basement-fault system. The map pattern of monoclines in northern Arizona seems to bear this out, for the curvilinear and branching nature of individual monoclines within an overall angular, orthogonal pattern implicitly favors a relationship of monoclines to basement faults. Analysis of the monocline pattern of the Plateau, in total, demands the additional but less obvious interpretation that many of the basement faults extend well beyond areas where their traces have clear-cut monoclinal expression. Specifically, the monocline (that is, basement-fault) segments can be fit to a systematic, interdependent network of lineaments, regional in extent, inferred to represent traces of basement-fracture zones, the loci of steeply dipping planar elements such as shear zones, faults, jointing, and lithologic contacts in basement rocks.

Two sets of major basement fracture zones are recognized in the monocline pattern, and these trend N20°W and N55°E. The fracture zones serve to subdivide the basement of the Plateau into a mosaic of crustal blocks, some of which moved differentially during the time(s) of monoclinal folding. The distribution of uplifts in the Colorado Plateau is seen to be very systematic when viewed in the context of a basement partitioned by basement-fracture zones.

Most of the monoclinal folding took place in the Laramide, a time of northeast-southwest compression in the Plateau. As a response to the regional compression, individual basement blocks within the mosaic were uplifted by reverse movements along segments of the high-angle fracture zones. Modeling of the deformation serves to emphasize that a very small amount of horizontal crustal shortening can produce significant structural relief when the shortening is accomplished along relatively widely spaced, steeply dipping faults.

INTRODUCTION

This analysis of monoclines of the Colorado Plateau tectonic province emphasizes the following:

1. The monoclines are upper-crustal expressions of near-vertical components of movements along reactivated, Precambrian, high-angle fault zones.
2. The systematic distribution and orientation pattern of the monoclines reflects attributes of a rejuvenated basement-block mosaic partitioned by ancient, deep-seated faults.
3. The reactivation of Precambrian fault and fracture zones to produce monoclines was caused by northeast-southwest regional compression.

RELATION OF MONOCLINES TO BASEMENT STRUCTURE

The literature is replete with a diversity of genetic models bearing on the origin of monoclines and monoclinal uplifts. The models vary considerably in the degree of emphasis of the relative roles of horizontal compressional tectonics and differential vertical tectonics. Powell (1873, 1876), Gilbert (1877), Dutton (1880), Walcott (1890), Nevin (1949), Prucha and other (1965), and Stearns (1970) have interpreted monoclines as products of the draping of near-surface strata over fault-bounded basement blocks; vertical forces were considered to be responsible for the movements. Baker (1935) postulated that the monoclines formed above deep-seated thrust faults in a stress system characterized by horizontal compression. Kelley (1955a, 1955b) also favored the concept that horizontal compression was the "dominating action" in the formation of the large monoclines. Noble (1914), Maxson (1961), Huntoon (1969, 1971, 1974), and Huntoon and Sears (1975), on the basis of their work in the Grand Canyon, observed that the monoclinal folds in that region commonly

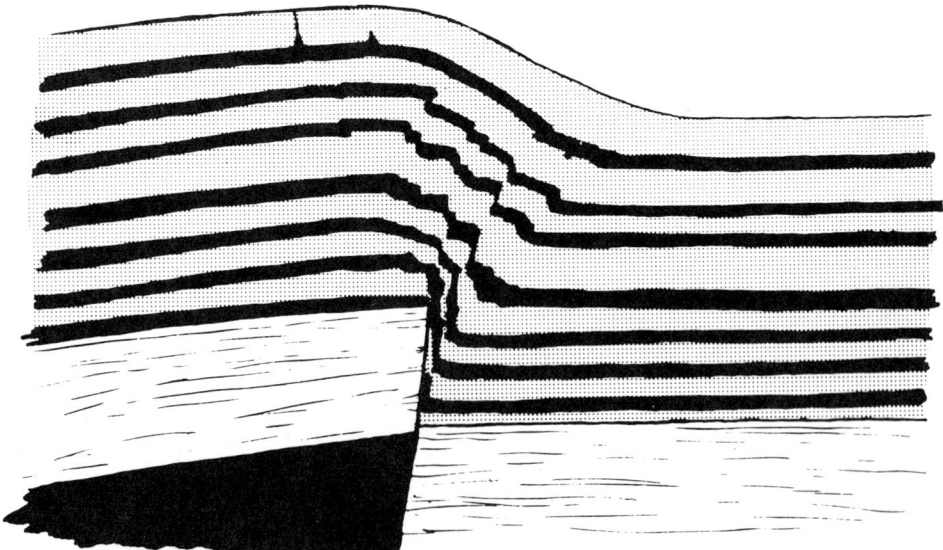

Figure 1. Photograph of experimental deformational model of "strata" of kaolinite and modeling clay resting on a rigid "basement" of pine board. High-angle faulting along pre-cut fault produces monoclinal fold in overlying thin layers.

are associated with reverse faulting along reactivated Precambrian high-angle faults. These workers then postulated that monoclinal folds are products of horizontal compression. Hodgson (1965) suggested that monoclines formed through differential vertical movements, without the influence of regional tangential compressive forces near the surface of the basement. Woodward (1973) attributed the development of monoclines to primary horizontal compression resulting in local secondary vertical components of stress. Barnes and Marshall (1974) envisioned the formation of monoclines to have involved a steplike faulting of the basement and the synchronous development of a curved principal stress trajectory in the supracrustal rocks, resulting in flexural folding of the sedimentary layers. The dynamic aspects of their model are based on the analytical and experimental work of Sanford (1959).

Despite differences in interpretation, most workers seem to concur that the folding occurred synchronously with near-vertical components of movement along ancient high-angle faults (Fig. 1) (Noble, 1914; Maxson, 1961; Hodgson, 1965; Huntoon, 1969, 1971, 1974; Lucchitta, 1974; Barnes and Marshall, 1974; and Huntoon and Sears, 1975). Huntoon (1974, p. 323) suggested that:

The sinuosity and branching that are characteristic of the monoclines in the eastern Grand Canyon result from pre-existing trends of the Precambrian faults that were rejuvenated as the monoclines developed.

Lucchitta (1974, p. 348) stated:

The association between monoclines and faults suggests that the monoclinal flexures are the surface expression of zones of weakness in the competent basement. . . . The antiquity and repeated reversal of movements on faults, as well as the association of monoclines with faults at depth, are well documented for the Grand Canyon.

Descriptive details of this association are reported by Huntoon and Sears (1975). For example, they point out that Laramide movements along the Bright Angel fault were superimposed on four phases of Precambrian movements and one phase of Phanerozoic movement. Further, two phases of normal faulting occurred along the Bright Angel fault zone in post-Laramide time.

In the discussions that follow, it is assumed that the association of monoclines with reactivated basement faults recognized in the classic Grand Canyon exposures holds for the Plateau province in general.

MONOCLINES IN NORTHERN ARIZONA

Pattern

Three distinct directional groupings of monoclines are recognizable in northern Arizona (Fig. 2, in pocket): North-northwest, northeast, and north to north-northeast. The north-northwest-trending monoclines are dominant and are represented by approximately 400 km (250 mi) of cumulative axial length. The north-northwest–trending monoclines include major segments of the East Kaibab, Echo Cliffs, and East Defiance monoclines as well as the Red Lake, Crazy Jug, and Oraibi monoclines. The north-northwest–trending set of monoclinal folds are asymmetric in an easterly direction (that is, the middle limbs dip east). The northeast-trending monoclines include segments of the Grandview, Coconino Point, Black Point, Cow Springs, Comb Ridge, and West Defiance monoclines. Their cumulative length is approximately 175 km (100 mi). With the exception of the segments of the West Defiance and Grandview monoclines, these northeast-trending monoclines are asymmetric to the southeast. The north to north-northeast-trending monocline segments constitute a minor set that includes portions of the West Kaibab, East Kaibab, Echo Cliffs, Organ Rock, West Defiance, and East Defiance monoclines (Fig. 3). With the exception of the West Kaibab and West Defiance structures, asymmetry of these north to north-northeast-trending structures is easterly.

The sinuosity of the monoclines in Arizona is conspicuous, as has been noted by Kelley (1955a, 1955b) and Kelley and Clinton (1960). The East Kaibab monocline is systematically curvilinear, a composite of north-northwest- and north-northeast-trending segments. The Echo Cliffs monocline comprises two north-northwest-trending segments connected by a relatively short north-northeast-trending segment. The East Defiance monocline displays the most extreme sinuosity; its form has been described in detail by Kelley (1967). Locally, the monoclines are seen to branch and split. At junctures of converging monoclines or monoclinal splays, relatively complex structural relationships are evident (Fig. 2). Such complications include (1) the structural terraces along the East Kaibab monocline (Babenroth and Strahler, 1945); (2) the fold interference patterns at Gray Mountain, the locus of convergence of the Grandview, East Kaibab, Coconino Point, and Additional Hill monoclines (Barnes, 1974); (3) the convergence of the Red Lake and Cow Springs monoclines near Tonalea; and (4) the convergence of the Comb Ridge, Organ Rock, and Cow Springs monoclines near Kayenta (Kiven, 1977).

The major monoclines in northern Arizona occupy critical structural positions with respect to the distribution of tectonic subprovinces within the Colorado Plateau. This is explicit in Kelley's discussion of tectonic subdivisions within the Colorado Plateau. Monoclines commonly mark the boundaries between adjacent, structurally distinctive subprovinces (Kelley, 1955a; Kelley and Clinton, 1960). Based on the magnitude of structural relief and total length, the major monoclines in northern Arizona, from west to east, include the West Kaibab, East Kaibab, Grandview-Coconino Point, Echo Cliffs, Red Lake, Cow Springs, Organ Rock, Comb Ridge, West Defiance, and East Defiance (Fig. 2). Of these, the West and East Kaibab monoclines mark the west and east margins, respectively, of the imposing north-trending Kaibab uplift (structural relief approximately 1,000 m [3,300 ft]). The East Kaibab and Echo Cliffs monoclines form the west and east boundaries, respectively, of the north-northwest-trending Echo Cliffs uplift. The polygonal Black Mesa basin is bounded on several sides by monoclines, most notably the Cow Springs, the southern extension of the Red Lake, and the disjointed West Defiance monocline. The convergence of the Organ Rock and Comb Ridge monoclines near Kayenta delimits the south-southwest terminus of the Monument upwarp. The Defiance uplift, whose structural relief is approximately 1,800 m (5,900 ft), is bounded by the West and East Defiance monoclines. The East Defiance monoclinal complex, in turn, marks the west boundary of the Gallup sag and a portion of the San Juan basin.

Interpretation

Given the probable relationship between individual monoclinal folds and underlying basement faults, the monocline pattern, in total, should record the position and trend of elements of the regional fault system in basement rocks underlying the Plateau province. This supposition seems to be born out by the pattern of monoclines in northern Arizona (Fig. 2). The abrupt changes in fold orientation that are evident in the map of monoclines are not characteristic of systems of flexural or buckle

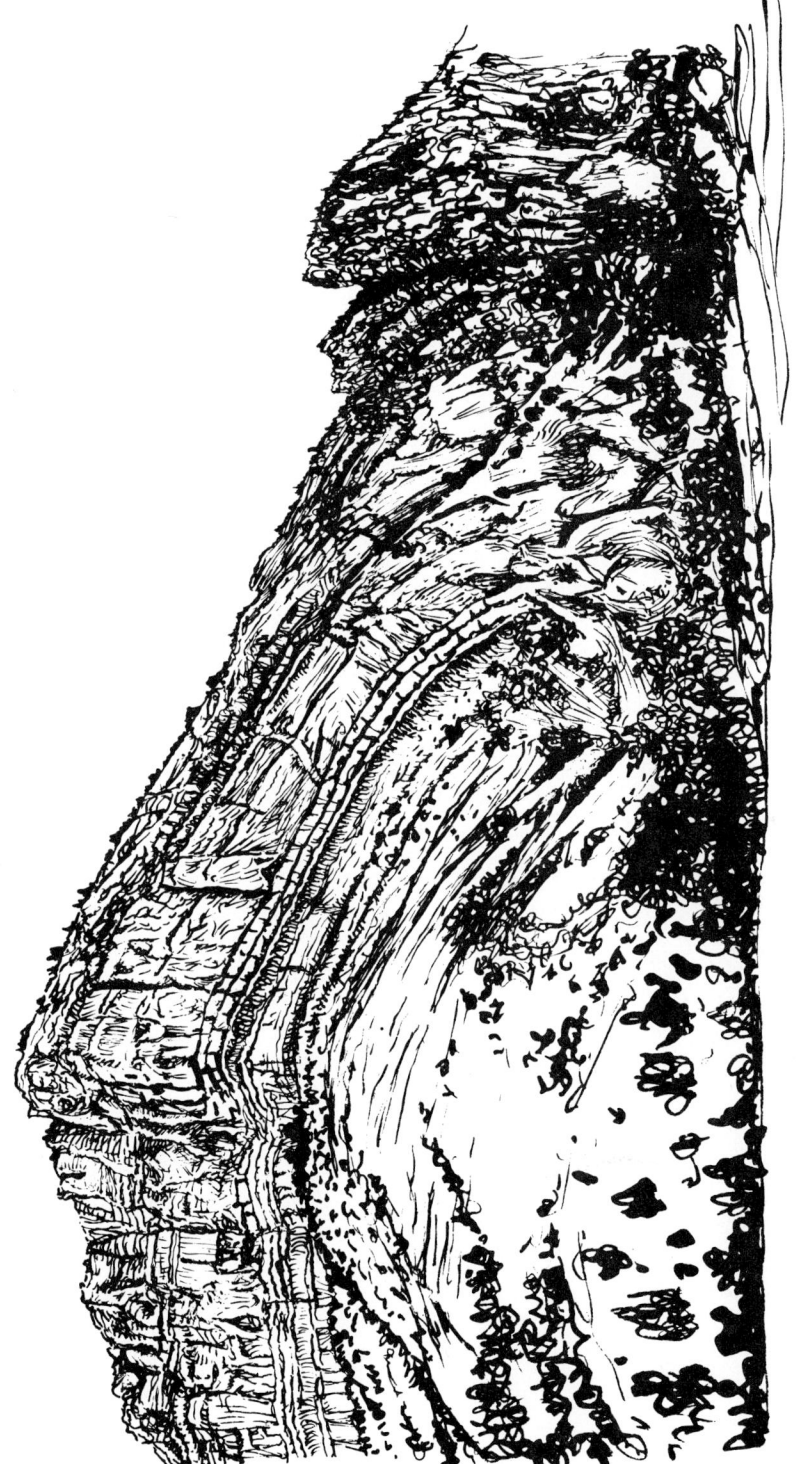

Figure 3. Tracing of a photograph of the Hunters Point segment of the East Defiance monocline; view to the northwest. Note vertical dip of middle-limb strata at far right of drawing.

folds in most sedimentary rock sequences. Rather, the network of curvilinear and branching folds implicitly favors a relationship of monoclines to major basement faults. The general influence of tectonic adjustments in the basement is also evident in the map of anticlines and synclines in northern Arizona (Fig. 2). Almost without exception, the anticlines and synclines are shallow-plunging, broad, open, upright folds with curvilinear traces. Most trend in a northwesterly direction, but many northeast-trending folds ignore the regional northwesterly orientation. The resultant orthogonal pattern, coupled with abrupt bends and convergences of traces, discloses the possible influence of basement block adjustments as a causal agent for some of the folding.

The locations and orientations of major basement faults in northern Arizona are interpreted to coincide with the steepened or "middle limbs" of major monoclines. The middle-limb segments of the monoclines shown in Figure 2 have been stippled in Figure 4 in order to emphasize the two-dimensional characteristics of part of the inferred basement fault system for northern Arizona. A major feature in the pattern is the Coconino lineament (Kelley, 1955b), the locus of segments of the Comb Ridge, Organ Rock, and Coconino Point monoclines (Fig. 2). Three junctions of converging monoclines occur along this "line," and many of the monoclines terminate and/or abruptly bend at it. The fold pattern suggests that the Coconino lineament served as a major partitioning element in the basement, separating different structural domains to the northwest and southeast. Notable in the pattern is the systematic spacing of north-northwest– and north-northeast–trending monoclines that lie northwest of the Coconino lineament. Where some of these intersect the Coconino lineament, branching, bifurcating patterns occur.

The patterns shown in Figures 2 and 4 are instructive because they emphasize a spatial and geometric interdependence not only of specific monoclines within the array of folds, but also of uplifts and basins within the Arizona portion of the Colorado Plateau. The most conspicuous example of the interdependence is

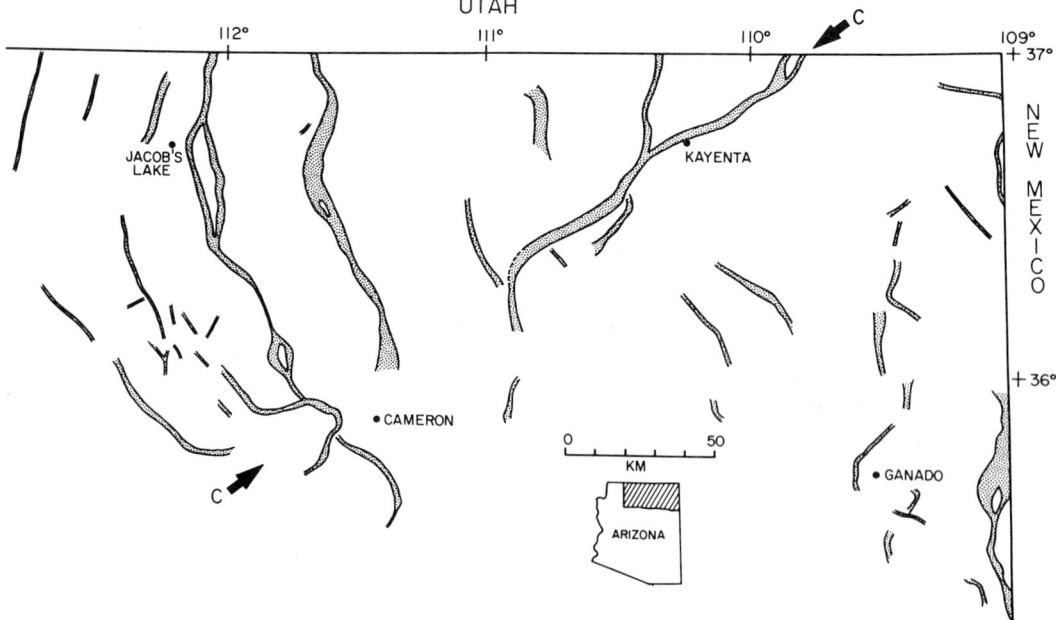

Figure 4. Map showing plan geometry and aerial extent of *middle-limb* segments (stippled) of monoclines in northern Arizona, based on Davis and Kiven (1975). Position of Coconino lineament noted (C-C).

exemplified, once again, in the Coconino lineament. The northeast-trending Cow Springs–Comb Ridge monocline segment and its southwest projection to Coconino Point mark the southern terminus of the Heather, Grandview, East Kaibab, Echo Cliffs, Red Lake, and Organ Rock monoclines (Fig. 2). The lineament also marks the south terminus of the Kaibab uplift, the Echo Cliffs uplift, and the Monument upwarp, as well as the northwestern terminus of Black Mesa basin.

MONOCLINES OF THE COLORADO PLATEAU AS A WHOLE

Pattern

The monoclines of the Colorado Plateau are enormous, ranging up to hundreds of kilometres in length and commonly displaying structural relief measurable in hundreds of metres. The pattern of monoclines for the Colorado Plateau, in total, has been described excellently by Kelley (1955a, 1955b) and Kelley and Clinton (1960). The basic properties of the pattern are identical to those described for northern Arizona. The map pattern of monoclines within the Colorado Plateau is multidirectional, sinuous, and branching, and has been described by Kelley in those terms (Fig. 5). The pattern is by no means characterized by constancy of fold trend. Rather distinct directional groupings of monoclines are evident, particularly north-northwest and northeast. North-northwest–trending monoclines include the Waterpocket and Nutria monoclines, as well as segments of the Comb Ridge and East Defiance. Northeast-trending monoclines include the San Rafael and segments of the Comb Ridge and Hogback monoclines. Northwest-trending monoclines are represented by the Uncompahgre, Grand, and Redlands monoclines. Sinuosity is conspicuous. Both the Comb Ridge and San Rafael monoclines, from north to south along their lengths, change in trend from north-northwest, through north, to northeast. The Hogback monocline, from west to east along its length, changes in trend from northeast to northwest. The East Defiance monocline displays its most extreme sinuosity in New Mexico, where it comprises a series of asymmetric Z-shaped folds (viewed in plan) that decrease in apparent wavelength and amplitude from north to south along its length. As in northern Arizona, monoclines in other parts of the Colorado Plateau are seen to branch and split. One of the best examples is at the north end of the Circle Cliffs uplift (Fig. 5).

Interpretation

Relationship of Monoclines to Basement Fracture Zones. Based on the characteristics of the monocline fold pattern in northern Arizona, described above, and the monocline–basement-fault relationships established in the Grand Canyon, it seems reasonable to interpret maps of monoclines in the Plateau province as maps of those basement faults that were active during monoclinal folding. Moreover, the map pattern of monoclines demands the less obvious interpretation that many of the basement faults extend well beyond areas of clear-cut monoclinal expression. The spatial-geometric interdependency of monoclines and uplifts, of the type already described for northern Arizona, permits some of the basement faults to be "tracked" through regions where the faults have no conspicuous expression in the Phanerozoic strata.

In analyzing the monocline pattern, it was assumed that each relatively straight-line monocline segment represents the approximate upward projection of a basement fault. By extending the numerous relatively straight-line monocline (that is, base-

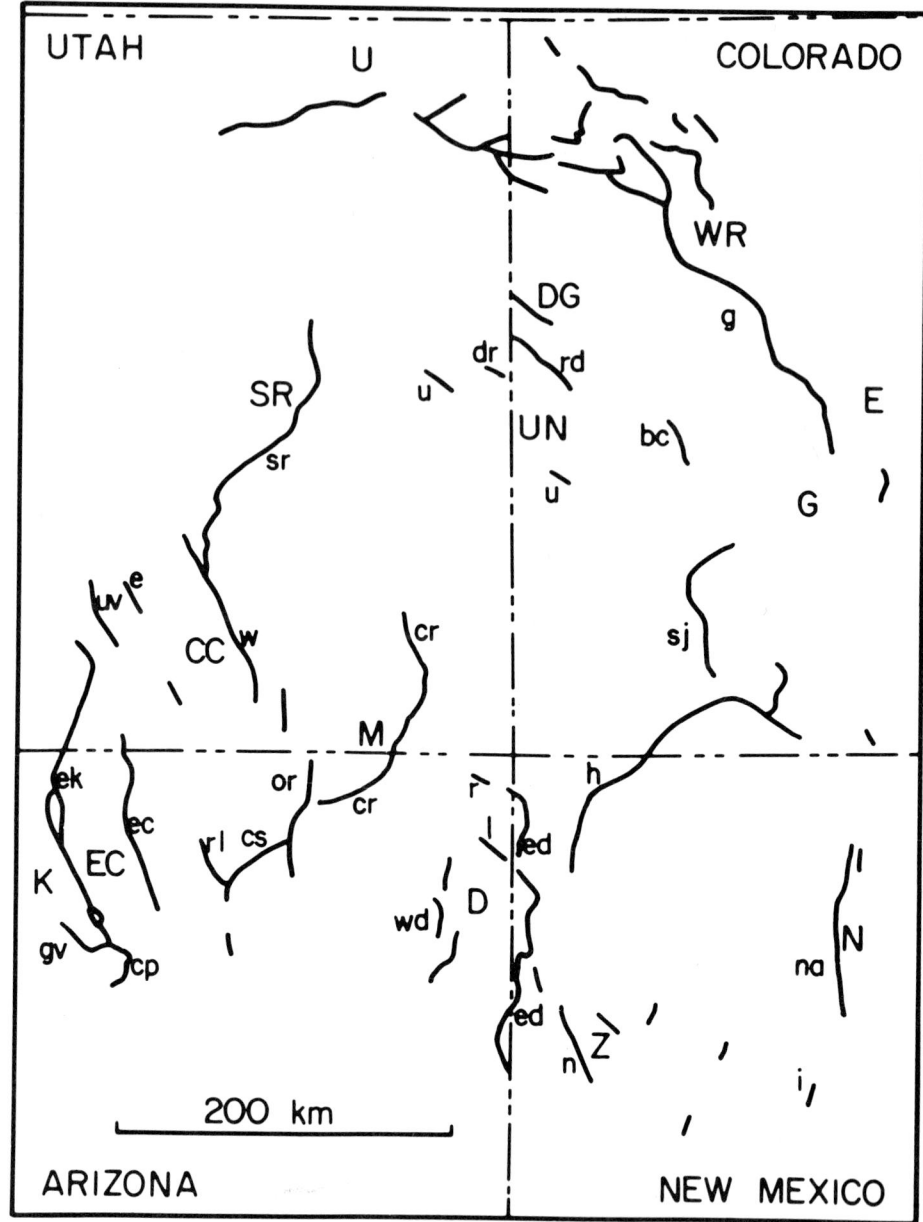

Figure 5. Map showing the distribution of monoclines and uplifts within the Colorado Plateau; from V. C. Kelley (1955b). *Monoclines:* bc = Book Cliffs; cp = Coconono Point; cr = Comb Ridge; cs = Cow Springs; dr = Davis Ranch; e = Escalante; ec = Echo Cliffs; ed = East Defiance; ek = East Kaibab; g = Grand; gv = Grandview; h = Hogback; i = Ignacio; l = Lukachukai; n = Nutria; na = Nacimiento; or = Organ Rock; r = Rattlesnake; rd = Redlands; rl = Red Lake; sj = San Juan; sr = San Rafael; u = Uncompahgre; uv = Upper Valley; w = Waterpocket; wd = West Defiance. *Uplifts:* CC = Circle Cliffs; D = Defiance; DG = Douglas; E = Elk; EC = Echo Cliffs; G = Gunnison; K = Kaibab; M = Monument; N = Nacimiento; SR = San Rafael; U = Uinta; UN = Uncompahgre; WR = White River; Z = Zuni.

ment-fault) segments, lineaments are disclosed that are identifiable as (1) loci of two or more monocline segments, (2) end points of monocline segments, (3) abrupt changes in trend of monoclines, and (4) zones of convergence of two or more monoclines (Fig. 6). Though it cannot be proved rigorously, these lineaments are inferred to coincide, at least in part, with basement-fracture zones (including faults), portions of which have been reactivated during Phanerozoic deformation. "Basement-fracture zones," as discussed by Spencer (1959), may be thought of as zones of crustal weakness controlled by the loci of steeply dipping planar elements such as shear zones, faults, jointing, and lithologic contacts in basement rocks. Field evidence for such an interpretation has been reported by Hodgson (1965) based on studies carried out in the Grand Canyon. He noted that the Bright Angel fault (which has monoclinal expression) is subparallel to northeast-trending, dominantly vertical foliation in the basement rock. Furthermore, the Bright Angel fault is partly localized at the contact of a large granite gneiss body enclosed in amphibolitic schist.

The inferred basement-fracture zones identified in this form of lineament analysis display a variety of orientations, but two major, distinct sets trend approximately N20°W and N55°E (Fig. 7). Although the inferred fracture zones are presented herein on a very small-scale map of monoclines, the actual analysis was carried out using Kelley's (1955b) 1:1,000,000 tectonic map of the Colorado Plateau.

The profound expression of the inferred north-northwest–striking fracture zones appears to be restricted to the southwest portion of the Colorado Plateau (Fig. 7). All are in part coincident with monoclinal segments (that is, zones of *demonstrable* vertical movement). The so-called Nutria fracture zone corresponds to a significant lineament cited by Kelley (1955b) as separating two tectonic subprovinces within the Plateau: a southwestern subprovince, dominated by major uplifts and characterized by northeast-facing monoclines, and a northeastern subprovince

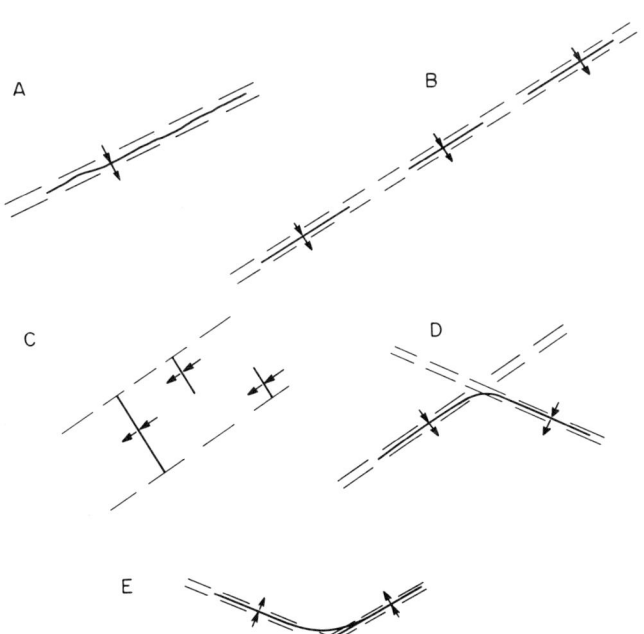

Figure 6. Examples of some of the criteria used to define the traces of inferred fracture zones. (A) monocline segment, (B) two or more monocline segments, (C) aligned end-points of monocline segments, (D) abrupt change in monocline trend, and (E) convergence of two or more monoclines.

containing the major basins and characterized by southwest-facing monoclines.

The N55°E–striking fracture zones are for the most part delineated on the basis of alignments of terminations of segments, and bends and branches of monoclines (Fig. 7). However, some coincide in part with relatively long monocline segments. The Coconino fracture zone, described above, is perhaps the best example of the inferred N55°E–striking fracture zones. The N55°E set parallels the trend of steeply dipping foliation in basement rocks within and immediately adjacent to the Colorado Plateau, as plotted by Kelley (1955b).

It is suggested that the entire array (Fig. 7) may reflect the characteristics (for example, trend, spacing, and length) of that portion of the system of basement-fracture zones of the Plateau that was, to some extent, active during monoclinal folding. The array of fracture zones represents a basement fracture "subsystem" for the Plateau in that it lacks regional basement-fracture elements that presumably exist but have no apparent expression in the monocline fold pattern.

Case and Joesting (1972) provide geophysical support for the reality of the inferred basement-fracture zones. They examined regional gravity and aeromagnetic patterns within a major portion of the Colorado Plateau and noted that the monoclinal uplifts are associated with gravity highs and high-amplitude magnetic anomalies bordered by relatively steep magnetic gradients. According to them (Case and Joesting, 1972, p. 10):

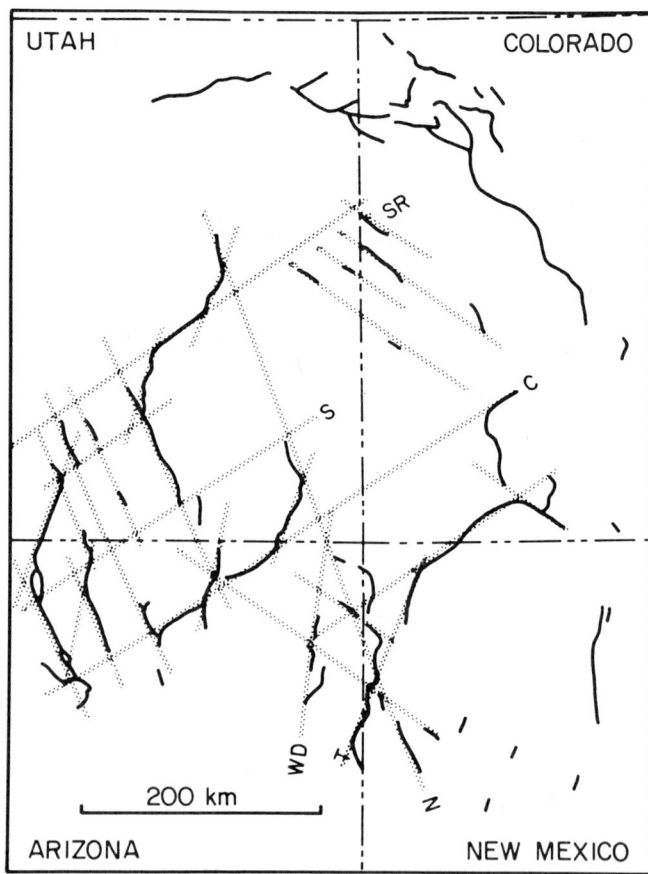

Figure 7. Map showing traces of the inferred basement-fracture zones in the Colorado Plateau. Base map of fold pattern from V. C. Kelley (1955b). SR = San Rafael; S = Snowmass; C = Coconino; WD = West Defiance; T = Todilto; N = Nutria.

The anomalies across the crests of the monoclinal uplifts are more heterogeneous than those over the adjacent structurally low areas, and this increased heterogeneity suggests a difference in Precambrian basement lithology, which may reflect the existence of Precambrian or late Paleozoic structural boundaries that were rejuvenated during the Laramide. . . . the near-linear zones of steepened magnetic gradients and lines of discontinuities of anomalies persist many miles in parts of the region. These zones reflect major lithologic discontinuities within the Precambrian basement and probably indicate a fundamental fracture pattern of Precambrian age. . . . dominant trends of these fracture zones are *northwest, northeast,* and *north* (emphasis added).

They concluded (p. 26):

Where the gradients and geophysical discontinuities fall along lines, they clearly indicate faults in the basement; so, we may generalize that the basement pattern is one of fault blocks, which correspond to zones of nearly straight steepened gravity and magnetic gradients.

Relationship of Monocline Pattern to Basement-Block Mosaic. The regional fracture zones thus defined serve to subdivide the basement of the Plateau into a mosaic of crustal blocks, some of which were activated during the time(s) of formation of the monoclines. The subdued, muted outlines of these blocks in Phanerozoic strata are characterized by loci of deformational features that are intense by Plateau standards. Abrupt changes in trend of the axial traces of the monoclines may be recognized as planview expressions of the "corners" of basement blocks. The blocks themselves may be viewed as complex polyhedra whose steeply dipping faces correspond to major fracture zones. High-angle movements have locally produced monoclinal folding of stratified rocks above the basement blocks.

The distribution of uplifts in the Colorado Plateau province is seen to be very systematic when viewed in the context of a basement partitioned by regionally extensive fracture zones. For example, in Figure 8, four of the longest and most significant of the inferred basement-fracture zones have been superimposed on a tectonic map showing the distribution of uplifts and monoclines in the Colorado Plateau province. In preparing Figure 8, the distribution and shapes of the uplifts, and the positions of the inferred basement-fracture zones, were plotted with great care. The shapes, aerial extent, and exact positioning of the uplifts were obtained from maps by Kelley (1955b), especially those showing structure contours. These data were transferred to 1:250,000 topographic maps of parts of Arizona, Utah, Colorado, and New Mexico, and then photographically reduced. As shown in Figure 8, the San Rafael zone marks the southeast border of the San Rafael swell and the northwest end of the Uncompahgre, Circle Cliffs, and Kaibab uplifts. The Coconino zone marks the southeast border of four major uplifts: the Kaibab, Echo Cliffs, Monument, and Uncompahgre. The Waterpocket zone projects south-southeast to demarcate the southwest end of the Monument upwarp. In fact, it may project further southeast and influence the south-southwest terminus of the Defiance uplift. The San Rafael swell, the Monument upwarp, Defiance uplift, and Zuni uplift all appear to share a common structural boundary, the Nutria zone. The number of "tie lines" in this system of uplifts seems sufficiently high to be regarded as significant. The distribution of the uplifts comprises an interdependent network interpreted to be related to a systematic arrangement of regional basement-fracture zones.

The kidney-bean shapes of many of the uplifts (Kelley, 1955a; 1955b) can be explained as the muted expression of underlying basement-block edges defined by two or more obliquely intersecting fracture-zone segments. The concave-east form of the Kaibab and Echo Cliffs uplifts differs from the convex-east form

226 George H. Davis

of the San Rafael swell and Monument upwarp, as has been pointed out by Kelley (1955a; 1955b). This difference exists despite the fact that the major displacement is consistently on the eastern flank of each of the uplifts. The difference in form of the two sets of uplifts may be attributed simply to dissimilar basement-block outlines at depth.

The most complex deformational patterns in Phanerozoic rocks within the Plateau appear to mark the border zones of adjoining basement blocks. Such structures become more readily interpretable in that light. For example, of all the monoclines, the East Defiance displays the most extreme sinuosity (Kelley, 1967). Woodward (1973, p. 97) has interpreted these cross folds as a zone of drag folding caused by a major shifting of the western part of the Colorado Plateau to the northeast. Another interpretation can be offered, recognizing that the inherent nature of the basement-fracture zones envisioned for the Colorado Plateau, coupled with the recurring nature of movements within the system, favors the phenomenon of cross folding in the vicinity of fracture-zone intersections. Anticlines or monoclines formed

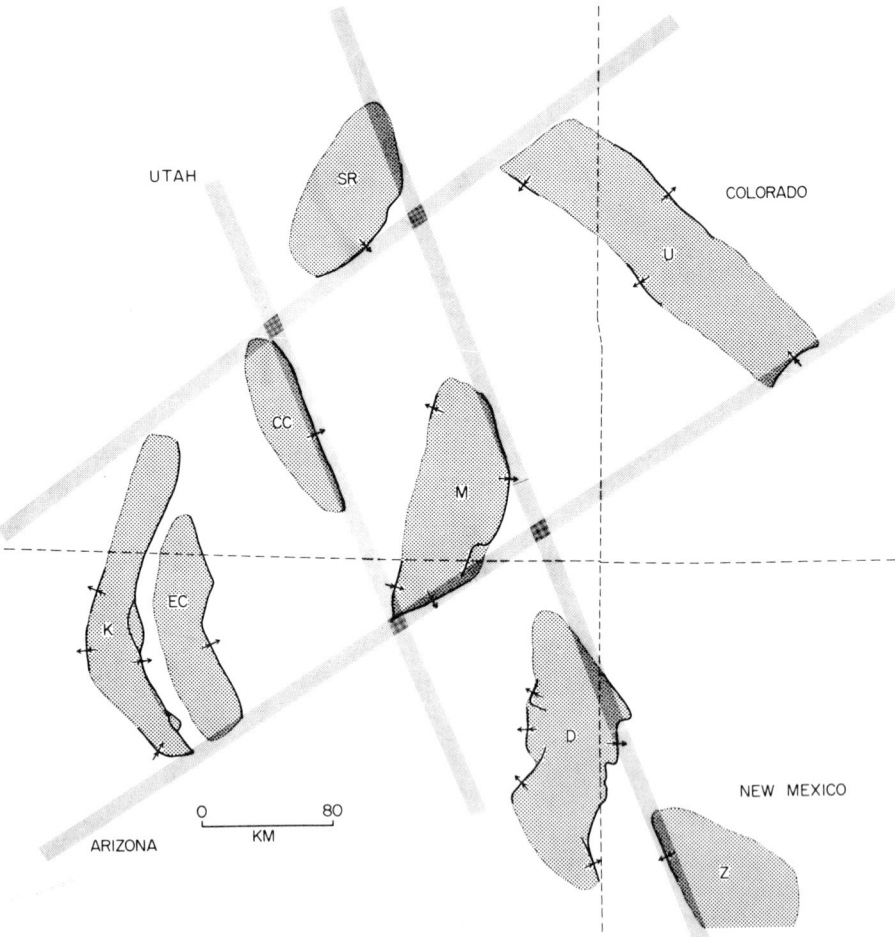

Figure 8. Map showing the distribution and shapes of the Kaibab (K), Echo Cliffs (EC), Monument (M), Circle Cliffs (CC), San Rafael (SR), Uncompahgre (U), Defiance (D), and Zuni (Z) uplifts, and the relation of these uplifts to four of the inferred basement-fracture zones (stippled). Only those monoclines associated with the uplifts are shown (see Fig. 7).

by movements along one fracture zone may be transformed into domes or doubly plunging anticlines by a superposed movement along a late, proximate fracture zone of different strike. The cross folds on the East Defiance monocline might be an interference product of southeast-directed, high-angle reverse faulting along the Todilto zone and southwest-directed, reverse faulting along the Nutria zone (Fig. 7). Southwest-directed, high-angle movement along the Nutria zone in this portion of the Plateau is consistent with the southwest-verging nature of the Nutria monocline. The largest cross fold lies just southwest of the trace of the Nutria zone; the cross folds systematically decrease in wavelength and amplitude to the south, *away* from the major locus of apparent interference.

The basic concept of a rejuvenated block mosaic is well established (Cloos, 1948; Spencer, 1959, 1969; Hoppin and Palmquist, 1965; Donath, 1962; Case and Joesting, 1972; Gay, 1972; Matthews and others, 1975; Saunders, 1975; and Thomas, 1974), even as applied to the Colorado Plateau tectonic province. Kelley (1955a, p. 802) commented that the distribution and form of the monoclines of the Colorado Plateau may be influenced by a mosaic of Precambrian nucleii whose shapes correspond to the tectonic subdivisions of the Plateau. Hodgson (1965, p. 935, 947–948) noted that in the Grand Canyon region:

Maximum deformation of the rocks occurs along narrow, linear zones which appear to follow elements of a primordial fracture pattern in the Precambrian basement. . . . The lack of uniform distribution of deformation in the rocks, coupled with the predominant vertical sense of displacements, suggest . . . differential vertical movements of discrete basement-blocks.

Shoemaker and others (1974, p. 383) concluded on the basis of their analysis of the Bright Angel and Mesa Butte fault system of northern Arizona:

Major (Precambrian) displacement probably occurred on a few main faults, which divide the crust into blocks tens of kilometers across. . . . Both the major and minor faults controlled later, dominantly vertical displacement in late Precambrian and Phanerozoic time.

ORIGIN OF MONOCLINES

Time of Formation

Monoclines are interpreted to have developed in Phanerozoic strata as natural responses to high-angle components of movements along reactivated basement faults and/or basement-fracture zones. The folding is regarded by most workers as Laramide in age (approximately 80 to 40 m.y.; Coney, 1972, 1976), but the evidence for this age assignment is not as strong as might be desired. The chief evidence cited by most workers consists of the angular unconformable stratigraphic relationships between "Late Cretaceous and Eocene strata" described by Gilbert (1877) and Gregory and Moore (1931) along the Waterpocket and East Kaibab monoclines, respectively. However, according to Huntoon (1974, p. 323), Bowers (1972) has reinterpreted at least a part of that contact as tectonic, the result of gravity sliding, so the date may be invalid. Kelley (1955a) concluded that most of the monoclines are Laramide products, and by way of support cites stratigraphic and structural relationships observed in the vicinity of the White River uplift, the Uinta uplift (Childs, 1950), the Defiance monocline, the Nacimiento uplift, and the Hogback monocline. Barnes (1974, p. 446) suggested, on the basis of detailed investigations in the Gray Mountain area, "that the major monoclinal uplifts, while presumably early Tertiary, are by no means synchronous." He thought that some monoclinal folding may be as young as Pliocene.

Compressional Dynamics

Coney (1976) has convincingly argued that the Laramide uplifts on the Colorado Plateau evolved as a response to regional compressive stresses transmitted northeast-southwest through the lithosphere. Burchfiel and Davis (1975) likewise have emphasized that during the Laramide the Colorado Plateau was affected by northeast-southwest, horizontal, compressive stresses. The compressive stresses were generated during the sustained convergence (collision) of North America and oceanic plates to the west (Coney, 1976; Burchfiel and Davis, 1975). That northeast-southwest compression was operative in the southern part of the Colorado Plateau during the Laramide is supported by coaxial, northwest-trending, upright folds in rocks as young as Late Cretaceous in the Colorado Plateau and neighboring Basin and Range tectonic province of Arizona. Many of those in the Basin and Range province of Arizona are tight folds with well-developed axial-plane cleavage. In some, deformed primary and secondary structure elements disclose 20% to 25% flattening.

Kelley (1955b) and Kelley and Clinton (1960) long ago recognized that monoclines in the Plateau represent a response of the upper crust to regional compressive stresses, an interpretation completely consistent with current plate tectonic models. Nonetheless, difficulties were encountered by them in explaining the diversity of trends of monoclines, anticlines, and synclines. They recognized two major monoclinal fold sets (northwest and northeast) and explained the development of these sets in a two-phased Laramide deformation (Kelley and Clinton, p. 96, 97):

> The northeasterly trending monoclines may have formed first under the influence of a dominant regional horizontal pressure from the Central Rockies. . . . the Plateau monoclines . . . of northwesterly trend, may have formed slightly later during a second major phase of Laramide orogeny. . . . Although the northerly trending monoclines might be the product of a separate diastrophic phase, it is preferable for the present to include them with the first phase. However, their relationship to the northeasterly trending monoclines is not clear.

The problem of reconciling the diversity of fold trends with an essentially uniform, northeast-southwest regional compressive stress is negated if the compression was impressed on a basement already partitioned by basement-fracture zones. It is suggested that as a response to strong regional compression, individual basement blocks within a block mosaic became relatively uplifted by reverse movements along segments of high-angle, reactivated, basement-fracture zones oriented northwest, approximately perpendicular to the direction of maximum compression. Although the northeast-trending monoclines and basement-fracture zones are not impressive in their structural relief, they effectively accommodated the movements of adjacent blocks, perhaps in part via minor strike-slip displacements.

Experimental Analog

In an effort to clarify kinematic details of this interpretation, the deformation was modeled physically in the laboratory. The experiments were not quantitatively scaled, and thus they provide mainly descriptive, not dynamic, insights. Nonetheless, the results may serve to clarify the concept being presented.

For purposes of the experiment, a 19 cm × 27 1/2 cm × 4 cm (7 1/2 in. × 10 3/4 in. ×

1½ in.) pine board was cut into 15 blocks along a series of steep-dipping saw cuts (70° and greater). The sides and bottom of each of the blocks were covered by aluminum foil and lightly coated with vaseline in order to reduce frictional resistance during the experimental modeling. The blocks were reassembled to form a block mosaic (Fig. 9) which was in turn covered by a thin clay layer simulating Phanerozoic strata. Then the block mosaic was subjected to an end-on, horizontal compression using a deformation table equipped with mechanical drive. As a response to the compression, blocks within the mosaic moved with respect to one another. Differential movement of the blocks under compression resulted in the development of monoclines in the clay layer (Fig. 10). Since the time these experiments were carried out, Friedman and others (1976) reported the results of elegant experimental folding and reverse faulting. Their work documented in detail the fabric and dynamic characteristics of deformed "cover" layers overlying faulted "basement" blocks.

The experimental results seem interesting in at least two respects. *First,* as the map of the experimentally induced array of monoclines shows (Fig. 11), diversely oriented, branching-converging monoclines did indeed develop when the block mosaic was rejuvenated by compressional stresses. The array of folds that developed in the clay displays an interdependency of position and geometry that is directly related to the existence and properties of the underlying system of saw cuts. The traces of most of the saw cuts can be tracked across the areas where the clay layer remained unfolded and flat-lying because of the systematic alignments of monocline segments with end points of monocline segments, monocline junctions, and bends in monocline trend (Fig. 11). Although the planarity of the saw cuts and thinness of the clay layer that recorded the block movements impart an angular, artificially obvious correspondence in form between "basement" and "cover" structures, similarities of details of the model with specific tectonic features of the Plateau are interesting. The monocline that marks the eastern edge of blocks 4 and 8 (Fig. 10) resembles the form of both the East Kaibab and the Echo Cliff monoclines (Fig. 2). The monocline at the upraised corner of block 9 (Fig. 10)

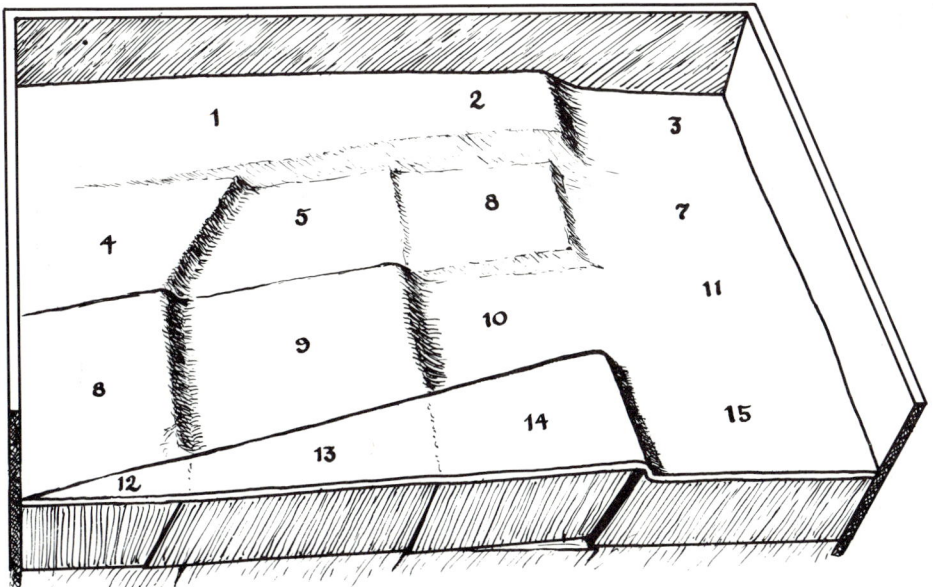

Figure 9. **Tracing of photograph of drape folds produced by compression of the block mosaic.**

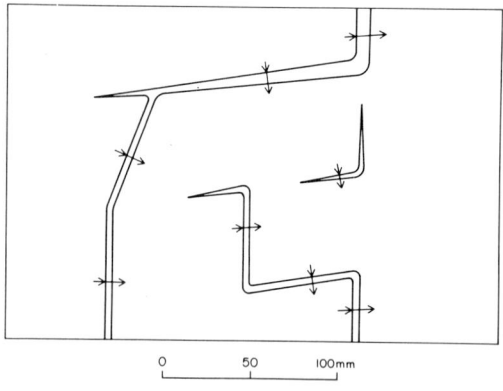

Figure 10. Map of monoclines in the experimentally deformed clay layer.

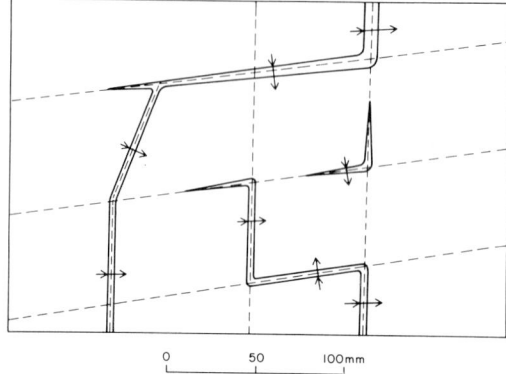

Figure 11. Relation of monocline pattern to the array of saw cuts.

is an analog for Coconino Point at Gray Mountain, Arizona (Fig. 2). The convergence of monoclines at the interface of blocks 1, 4, and 5 (Fig. 10) is equivalent to the convergence of the Organ Rock, Comb Ridge, and Cow Springs monoclines near Kayenta, Arizona (Fig. 2).

The *second* experimental result that seems important to discuss in the context of the origin of monoclines is this: The deformation served to emphasize how a tiny amount of horizontal crustal shortening produces "major," significant structural relief when the shortening is accomplished by movements along relatively widely spaced, steeply dipping faults. The monoclinal folding shown in Figure 10 was achieved through the shortening of the 275-mm-long block mosaic by a mere 3 mm (approximately 1% shortening). The structural relief displayed by individual monoclines in the clay layer reached a maximum of about 6 mm. Had the 275-mm-long block been cut by a single, 75°-dipping saw cut, the structural relief that would have developed during the 3-mm shortening would have been approximately 11 mm. [For an 80°-dipping cut, structural relief would have been approximately 15 mm.]. The sensitivity of the transfer of horizontal shortening to reverse-slip displacement is revealed in Figure 9. Structural relief along monoclines between blocks 2 and 3 is demonstrably greater than that of other monoclines in the model, simply because in the original experimental set-up (Fig. 11), the array of blocks 1–3 happened to be cut 1 to 2 mm longer than the arrays 4–7, 8–11, and 12–15. Had the blocks been cut in a perfectly regular manner, uniform shortening of the block mosaic would have produced identical cumulative displacements by monoclinal folding in each of the four panels of blocks (1–3, 4–7, 8–11, 12–15).

In comparing the experimental model to the real world, one gains the impression that the northeast-southwest shortening of the Plateau which produced the monoclines (not including anticlines and synclines) may not have been much more than 1%. Consider, for example, the 300-km-long, northeast-southwest segment of the Plateau that includes the Kaibab uplift, the Echo Cliffs uplift, and the Monument upwarp. Deformation within that "belt" has mainly been achieved by movements along the East Kaibab, Echo Cliffs, and Comb Ridge monoclines, with relatively minor movements along the West Kaibab, Organ Rock, and Red Lake monoclines (Fig. 2). Structural relief along the three major breaks does not exceed 2,500 m (8,250 ft). Using the model presented above and assuming that the basement faults underlying the major monoclines dip 75°, one finds that the net shortening to account for the monoclines need not have been greater than 0.87 km (total horizontal shortening of 0.3%). Perhaps the remarkable nature of this kind of transfer of shortening into lift is largely responsible for the decades of debate concerning the relative roles of vertical versus compressional tectonism in the formation of monoclines.

CONCLUSIONS

Previous workers have demonstrated that individual monoclines in the Grand Canyon region of the Colorado Plateau have resulted from high-angle movements along reactivated faults of Precambrian age. Applying this observation to monoclines in the rest of the Colorado Plateau, the monocline fold pattern is seen to be related to a reactivated basement-fracture system. Abrupt bends in the traces of individual monoclines, branching-converging monoclines, aligned segments of monoclines, and long monoclinal traces serve to outline some details of a basement-block mosaic partitioned by basement-fracture zones. Situated at the protrusion of the North American craton in Laramide time and converging directly with oceanic plates to the west (Coney, 1976), the Colorado Plateau, with its partitioned foundation, was subjected to a strong northeast-southwest compression. Under these conditions, the monoclines grew as a logical, inevitable accommodation to the need for a slight crustal shortening.

ACKNOWLEDGMENTS

Field investigations, structure mapping, and map compilation, interpretation, and drafting were largely funded by NASA Grant No. NGL 03-002-313 in cooperation with the Arizona Oil and Gas Conservation Commission, the Arizona Resources Information System, the Office of Arid Lands Studies, and the Department of Geosciences at the University of Arizona. Drafting of structural data on Figure 2 was by Cartographic Illustrators, Phoenix, Arizona. Art work for Figure 3 was done by David O'Day. I am indebted to V. C. Kelley and the thoroughness of his tectonic investigations of the Colorado Plateau. I am particularly grateful for the use of his tectonic maps and for constructive criticisms of the model presented herein. Peter Coney reviewed the manuscript; his input was very helpful to me.

REFERENCES CITED

Babenroth, D. L., and Strahler, A. N., 1945, Geomorphology and structure of the East Kaibab monocline, Arizona and Utah: Geol. Soc. America Bull., v. 56, p. 107–150.

Baker, A. A., 1935, Geologic structure of southeastern Utah: Am. Assoc. Petroleum Geologists Bull., v. 19, p. 1472–1507.

Barnes, C. W., 1974, Interference and gravity tectonics in the Gray Mountain area, Arizona, in Eastwood, R. L., and others, eds., Geology of northern Arizona, P. II—Area studies and field guides: Flagstaff, Ariz., Northern Ariz. Univ., p. 442–453.

Barnes, C. W., and Marshall, D. R., 1974, A dynamic model for monoclinal uplift and gravity gliding: Geol. Soc. America Abs. with Programs, v. 6, no. 5, p. 424.

Bowers, W. E., 1972, The Canaan Peak, Pine Hollow, and Wasatch formations in the Table Cliffs region, Garfield County, Utah: U.S. Geol. Survey Bull. 1331-B, 39 p.

Burchfiel, B. C., and Davis, G. A., 1975, Nature and controls of Cordilleran orogenesis, Western United States: Extensions of an earlier synthesis: Am. Jour. Sci., v. 275-A, p. 363–395.

Case, J. E., and Joesting, H. E., 1972, Regional geophysical investigations in the central Colorado Plateau: U.S. Geol. Survey Prof. Paper 736, 31 p.

Childs, O. E., 1950, Geologic history of the Uinta Basin, *in* Petroleum geology of the Uinta Basin: Intermountain Assoc. Petroleum Geologists, Guidebook to the geology of Utah, no. 3, Salt Lake City, p. 49-59.

Cloos, H., 1948, The ancient European basement blocks—preliminary note: Am. Geophys. Union Trans., v. 29, no. 1, p. 99-103.

Coney, P., 1972, Cordilleran tectonics and North America plate motions: Am. Jour. Sci., v. 272, p. 603-628.

——1976, Plate tectonics and the Laramide orogeny: New Mexico Geol. Soc. Spec. Publ. no. 6, p. 5-10.

Davis, G. H., and Kiven, C. W., 1975, Structure map of folds in Phanerozoic rocks, Colorado Plateau tectonic province of Arizona: Phoenix, Arizona Oil and Gas Conserv. Comm.

Donath, F. A., 1962, Analysis of basin-range structure, south-central Oregon: Geol. Soc. America Bull., v. 73, p. 1-16.

Dutton, C. E., 1880, Report of the geology of the High Plateaus of Utah with atlas: U.S. Geog. and Geol. Survey of Rocky Mountain Region (Powell), U.S. Dept. Interior, 307 p.

Friedman, M., Handin, J., Logan, J. M., Min, K. D., and Stearns, D. W., 1976, Experimental folding of rocks under confining pressure: P. III. Faulted drape folds in multilithologic layered specimens: Geol. Soc. America Bull., v. 87, p. 1049-1066.

Gay, S., Jr., 1972, Fundamental characteristics of aeromagnetic lineaments, their geologic significance, and their significance to geology: Salt Lake City, American Stereo Map Co., 94 p.

Gilbert, G. K., 1877, Report on the geology of the Henry Mountains: Washington, D.C., U.S. Geog. and Geol. Survey of Rocky Mountain Region, 170 p.

Gregory, H. E., and Moore, R. C., 1931, The Kaiparowits region, a geographic and geologic reconnaissance of parts of Utah and Arizona: U.S. Geol. Survey Prof. Paper 164, 161 p.

Hodgson, R. A., 1965, Genetic and geometric relations between structures in basement and overlying sedimentary rocks, with examples from Colorado Plateau and Wyoming: Am. Assoc. Petroleum Geologists Bull., v. 49, p. 935-949.

Hoppin, R. A., and Palmquist, J. C., 1965, Basement influence of later deformation: The problem, techniques of investigation, and examples from Big Horn Mountains, Wyoming: Am. Assoc. Petroleum Geologists Bull., v. 49, p. 993-1004.

Huntoon, P. W., 1969, Recurrent movements and contrary bending along the West Kaibab fault zone: Plateau, v. 42, p. 66-74.

——1971, The deep structure of the monoclines in eastern Grand Canyon, Arizona: Plateau, v. 43, p. 148-158.

——1974, Synopsis of Laramide and post-Laramide structural geology of the eastern Grand Canyon, Arizona, *in* Eastwood, R. L., and others, eds., Geology of northern Arizona, Pt. I—Regional studies: Flagstaff, Northern Arizona Univ., p. 317-335.

Huntoon, P. W., and Sears, J. W., 1975, Bright Angel and Eminence faults, eastern Grand Canyon, Arizona: Geol. Soc. America Bull., v. 86, p. 465-472.

Kelley, V. C., 1955a, Monoclines of the Colorado Plateau: Geol. Soc. America Bull., v. 66, p. 789-804.

——1955b, Regional tectonics of the Colorado Plateau and relationship to the origin and distribution of uranium: New Mexico Univ. Pubs. Geology, no. 5, 120 p.

——1967, Tectonics of the Zuni-Defiance region, New Mexico and Arizona, *in* Guidebook of the Defiance-Zuni-Mt. Taylor region, Arizona and New Mexico: New Mexico Geol. Soc. 18th Field Conf., p. 28-31.

Kelley, V. C., and Clinton, N. J., 1960, Fracture systems and tectonic elements of the Colorado Plateau: New Mexico Univ. Pubs. Geology, no. 6, 104 p.

Kiven, C. W., 1977, Kinematics of deformation at the southwest corner of the Monument uplift [M.S. thesis]: Tucson, Arizona Univ., 84 p.

Lucchitta, Ivo, 1974, Structural Evolution of northwest Arizona and its relation to adjacent Basin and Range province structures *in* Eastwood, R. L., and others, eds., Geology of northern Arizona, Pt. I—Regional studies: Flagstaff, Northern Ariz. Univ., p. 336-354.

Matthews, Vincent, Callahan, C. M., and Work, D. F., 1975, Vertical uplift versus lateral compression during the Laramide orogeny in the northern Front Range, Colorado: Geol. Soc. America Abs. with Programs, v. 7, no. 5, p. 627.

Maxson, J. H., 1961, Geologic map of the Bright Angel quadrangle, Grand Canyon National Park, Arizona: Grand Canyon, Ariz., Grand Canyon Nat. History Assoc.

Nevin, C. M., 1949, Principles of structural geology (4th ed.): New York, John Wiley & Sons, Inc., 410 p.

Noble, L. F., 1914, The Shinumo quadrangle, Grand Canyon district, Arizona: U.S. Geol. Survey Bull. 549, 100 p.

Powell, D. A., 1873, Some remarks on the geological structure of a district of country lying to the north of the Grand Canyon of the Colorado: Am. Jour. Sci., 3rd ser., v. 5, p. 456–465.

——1876, Report on the geology of the eastern portion of the Uinta Mountains and a region of country adjacent thereto: U.S. Geol. and Geog. Survey of Territories (Powell), 218 p.

Prucha, J. J., Graham, J. A., and Nickelsen, R. P., 1965, Basement-controlled deformation in Wyoming Province of Rocky Mountains Foreland: Am. Assoc. Petroleum Geologists Bull., v. 49, p. 966–992.

Sanford, A. R., 1959, Analytical and experimental study of simple geologic structures: Geol. Soc. America Bull., v. 70, no. 1, p. 19–52.

Saunders, D. F., 1975, Use of ERTS imagery in petroleum and minerals exploration (abs.): Wyoming Geol. Assoc. Newsletter, v. 21, no. 2, p. 5.

Shoemaker, E. M., Squires, R. L., and Abrams, M. J., 1974, The Bright Angel and Mesa Butte fault systems of northern Arizona, *in* Eastwood, R. L., and others, eds., Geology of northern Arizona, Pt. I—Regional studies: Flagstaff, Northern Arizona Univ., p. 355–391.

Spencer, E. W., 1959, Fracture patterns in the Beartooth Mountains, Montana and Wyoming: Geol. Soc. America Bull., v. 70, p. 467–508.

——1969, Introduction to the structure of the earth: New York, McGraw-Hill, 597 p.

Stearns, D. W., 1970, Drape folds over uplifted basement blocks with emphasis on the Wyoming Province [Ph.D. dissert.]: Texas A & M Univ.

Thomas, G. E., 1974, Lineament-block tectonics: Williston-Blood Creek basin: Am. Assoc. Petroleum Geologists Bull., v. 58, p. 1305–1322.

Walcott, C. D., 1890, Study of a line of displacement in the Grand Canyon of the Colorado in northern Arizona: Geol. Soc. America Bull., v. 1, p. 49–64.

Woodward, L. A., 1973, Structural framework and tectonic evolution of the Four Corners region of the Colorado Plateau, *in* Guidebook of Monument Valley and vicinity, Arizona and Utah: New Mexico Geol. Soc. 24th Field Conf., p. 94–98.

MANUSCRIPT RECEIVED BY THE SOCIETY JUNE 27, 1977
MANUSCRIPT ACCEPTED AUGUST 25, 1977

Printed in U.S.A.

Geological Society of America
Memoir 151

Development of monoclines: Part I. Structure of the Palisades Creek branch of the East Kaibab monocline, Grand Canyon, Arizona

ZE'EV RECHES
Weizmann Institute
Rehovot, Israel

ABSTRACT

The Palisades Creek branch of the East Kaibab monocline in Grand Canyon National Park, Arizona, can be divided into three structural levels with contrasting styles of deformation: a lower level composed of a vertical fault, the Palisades fault, and a sharp synclinal bend; an intermediate level characterized by a tight, steep monoclinal flexure; and an upper level dominated by an open monoclinal flexure. The three styles of deformation apparently were controlled by a combination of position relative to the fault and of structural behavior of various rock units. The profile of the monocline at depth might have been estimated by study of the profile of the monocline at the surface, but only if data concerning structural units were available.

Study of the internal strain and accurate mapping of the gross structure provide fundamental data for a theoretical model. Analysis of small-scale structures, including faults and folds as well as calcite twinning and thicknesses of rock units, indicates the internal strain of the rocks during monoclinal flexuring. Most strain indicators are consistent and imply shortening subparallel to layering at all levels within the monocline. An accurate cross section and local measurements of thicknesses of units indicate that layers in the monocline have changed thickness appreciably.

A general model of monocline formation must incorporate effects of layer-parallel shortening as well as differential vertical displacement along the underlying fault. Such a model is presented in Part II (this volume).

INTRODUCTION

About 100 years ago, John Wesley Powell (1873) coined the term "monoclinal folds" for some structures he observed during his famous expedition down the Colorado River, and since then the term has been applied to local steepening of uniformly gently dipping series of strata (Kelly, 1955). Powell inferred that faults and monoclinal flexures form a continuous transition from one to the other, so that faults can be transformed into monoclinal flexures in both the vertical and the horizontal directions. He did not explain the transformation; rather, he presented it as a fascinating puzzle. Today most geologists associate the term "monoclinal flexure" with a type of structure in which there is a fault below and a flexure above.

Since the time of Powell, monoclinal flexures have been recognized in a variety of structural settings throughout the world. They are perhaps most commonly known in the Colorado Plateau, where they are one of the dominant structural features (Powell, 1873; Dutton, 1882; Walcott, 1890; Baker, 1935; Babenroth and Strahler, 1945; Kelley, 1955; Wells, 1960; Lucchitta, 1974; Huntoon, 1974). However, they are well developed in orogenic belts such as the Rocky Mountains, the Jura Mountains, and the Appalachians (Stearns, 1971; Rohr and others, 1961; Goguel, 1962; Beloussov, 1962; Prucha and others, 1965) and in the Mideast (deSitter, 1962; Reches, 1976). In addition, sedimentary rocks at margins of some laccolithic intrusions in Utah and Montana have been deformed into monoclinal shapes (Johnson and Pollard, 1971, 1973; Johnson and Ellen, 1974). Kink bands in sedimentary and metamorphic rocks can be similar to monoclines (Ellen, 1971; Dewey, 1965; Weiss, 1968; Ramberg and Johnson, 1976; Reches and Johnson, 1976). Thus, monoclinal flexure is an important class of structure.

Most investigators assume that monoclinal flexuring is a passive response of a cover of sedimentary rocks to faulting in basement rocks, as the names "drape fold" and "passive fold" suggest. Monoclines, therefore, are considered by many to be secondary structures, much as drag folds, and the faults below are considered to be primary structures (Prucha and others, 1965; Stearns, 1971). Some investigators have found evidence for horizontal compression in monoclines; they used this information to infer that these monoclines formed by draping over reverse faults in the basement (Baker, 1935; Huntoon, 1974). Again, the implication is that monoclinal flexuring is produced solely by the faulting.

There have been a few theoretical and experimental studies of monoclines. Sanford (1959) theoretically analyzed the draping of a single homogeneous elastic layer over a vertical fault. Even though Sanford did not study monoclines specifically, his theoretical analysis and experiments are the basis for most interpretations of "drape folding" or "passive folding" (Prucha and others, 1965; Stearns, 1971). Monoclines have been produced experimentally in sand and clay layers (Sanford, 1959), rubber strips (Reches and Johnson, 1976), and rock layers (Friedman and others, 1976; Logan and others, this volume). Detailed maps of strain and stress within the experimental monoclines, produced by Friedman and others and Logan and others, indicate patterns similar to those predicted by the theoretical model derived by Sanford (1959).

Accurate maps of strain or stress distributions within large monoclines have not been published so far, even though such maps are fundamental to the development of sound explanations of monoclinal flexuring. Theoretical models of monoclines are presented as maps of strain and stress (Sanford, 1959), and therefore a theoretical model can be correlated with the field only through analogous maps of the large monoclines.

My main objective in this paper is to present a comprehensive and detailed field study of a monocline. The results are presented via accurate maps and cross sections as well as maps of strain and stress distributions. Theoretical analyses of monoclines as well as experimental results will be presented in Part II of this series (Reches and Johnson, this volume). In Part II we shall also correlate the theoretical models with the gross structure and strain and stress distributions reported here.

I briefly examined many monoclines on the Colorado Plateau as well as in the Bighorn Basin of Wyoming. Then I selected a monocline exposed in the Grand Canyon, Arizona, for detailed field study, because nearly perfect exposures there appeared to provide an opportunity to map internal structures of a monocline from the fault below to the open flexure above. Other reasons for selecting the monocline for detailed study are that the major fault is essentially vertical, so the extension associated with "draping" should be evident if it exists, and that the flexure has a relatively low amplitude, so early deformations should not be severely masked by subsequent deformations. I used various techniques of structural analysis to measure the state of deformation associated with the monocline.

Further, I made accurate topographic and geologic maps so that the gross geometry of the structure could be determined unambiguously.

This paper is an essential step toward development of a comprehensive model for monocline formation. A subsequent step is presented in a companion paper (Reches and Johnson, this volume) and is a theoretical and experimental study of possible mechanisms of monocline formation.

EAST KAIBAB MONOCLINE

One of the largest monoclines on the Colorado Plateau is the East Kaibab monocline, named by Powell after the Indian word "Kaibab," which means "mountain buried below." The East Kaibab structure is 240 km long, and is composed of flexures and faults. Its exposure changes laterally from a smooth flexure to a fault to a combination of fault and flexure. It trends generally north, but locally east, from the Bryce Canyon area, Utah, in the north, to the San Francisco Peaks area, Arizona, in the south (Fig. 1). The vertical displacement of the monocline ranges up to 1,200 m.

The history of tectonic activity along the East Kaibab monocline was outlined by Walcott (1890) in a study of the eastern Grand Canyon. According to Walcott, movement along the trend of the East Kaibab began in the region of the Grand Canyon as a Precambrian fault, downthrowing older, Algonkian strata on the west from 15 to 1,500 m. During the late Paleozoic, the sense of displacement on the basement fault reversed and an eastward-facing monoclinal fold was formed, displacing strata a few tens of metres. The same sense of movement resumed during the Tertiary, producing the present East Kaibab monocline and the accompanying faults. The net displacement aggregated more than 900 m in the vicinity of the Grand Canyon.

The Grand Canyon of the Colorado River cuts through the East Kaibab monocline and provides three-dimensional exposures for about 30 km along the structure. The Butte fault is intermittently exposed beneath the flexure for about 18 km, providing opportunities to study the fault-fold relations also. Figure 2 is a tectonic map of the eastern Grand Canyon which shows the main features of the East Kaibab monocline system. West of Chuar Lava Hill, the East Kaibab splits into two branches, one of which continues to trend southward, and the second of

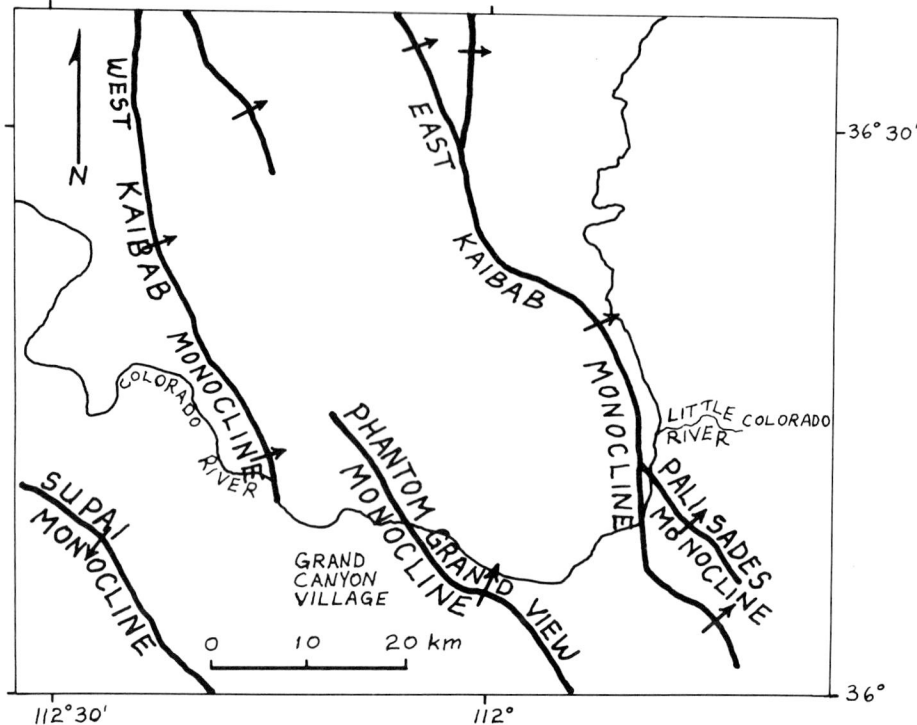

Figure 1. General structural features of the area of the East Kaibab monocline, Utah and Arizona.

which trends southeastward into Palisades Creek. About 4 km of the first branch appears as the Butte fault in Precambrian units. Some remnants of Paleozoic rocks (Fig. 2) indicate that this segment of the Butte fault was not active after Precambrian time (Walcott, 1890, p. 56). The second branch of the Butte fault continues into Palisades Creek, where one can observe the transition from fault to continuous flexure in the Paleozoic units (Fig. 2). The two branches of the East Kaibab monocline rejoin southeast of Desert View.

PALISADES MONOCLINE

The canyon of Palisades Creek, about 1,200 m deep, offers an excellent opportunity to explore a complete section of a monocline (Fig. 3). Several investigators mentioned the monocline in Palisades Creek (Dutton, 1882; Walcott, 1890; Huntoon, 1974); however, no detailed field work has been published for this area. In the beginning of the twentieth century, miners prospected for copper in the fault zone in Palisades Creek, and Tanner's mine drift is still open there (Billingsley, 1974). My field work in Palisades Creek included detailed mapping, collection of structural data, construction of a plane-table map of part of the area of the transition from a fault into a fold, and collection of samples for petrologic and petrofabric analyses and for age-dating. The various mapping methods are discussed in Appendix 1. An accurate vertical cross section of the Palisades monocline was constructed

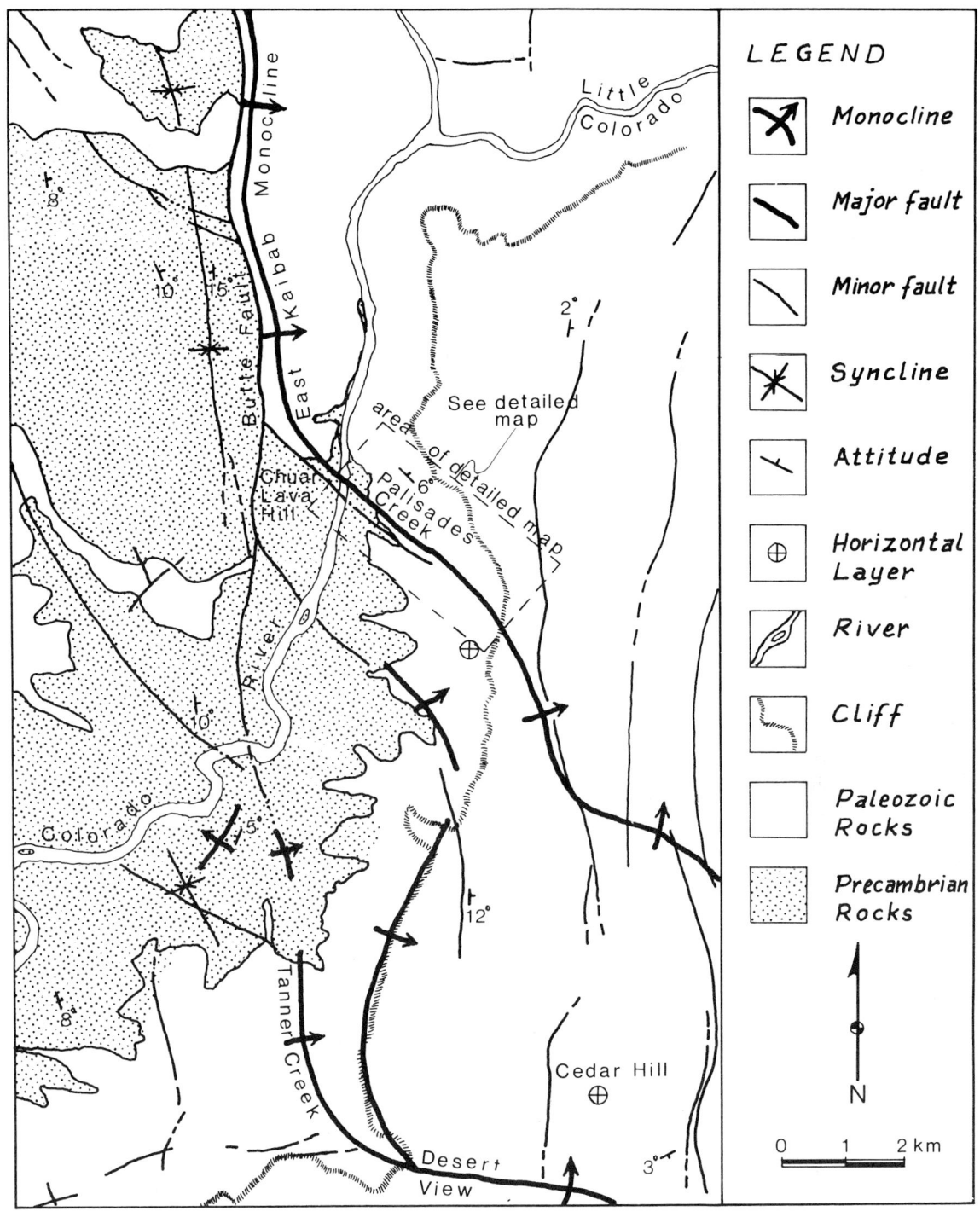

Figure 2. Structures of the eastern Grand Canyon National Park, Arizona (after Grand Canyon Natural History Association, 1976).

Figure 3. Oblique aerial photograph of the Palisades Creek area. View to the southeast.

by projection of stratigraphic contacts to a vertical section trending northeast (Fig. 5). Local variations, up to 10°, of the trend of the monocline were corrected by rotation around a vertical axis.

Stratigraphy

Rocks ranging in age from late Precambrian (Dox Fm) to Permian (Kaibab Fm) are exposed in Palisades Creek (Fig. 4). Most of the rocks are sandstone, shale, limestone, or dolomite. The only igneous rocks are Cardeñas lavas and basaltic dikes in the Dox Formation (both Precambrian). Igneous and metamorphic basement rocks lie at a depth of 1 km or more beneath the area. The uppermost unit, Kaibab Limestone, was once covered by at least 1 km of Mesozoic and Cenozoic rocks.

In structural analysis, one is primarily interested in mechanical rock units rather than in stratigraphic rock units. Therefore, map units such as Muav Limestone and Temple Butte Limestone, described by McKee and Resser (1945), were combined due to their apparent mechanical similarity. The Coconino Sandstone, Toroweep Formation, and Kaibab Formation were combined for the same reason.

General Structure

The geologic map of the Palisades Creek area shows three main structural blocks: the southern block, the northern block, and the monoclinal zone between the two (Fig. 6). Rocks within the southern and northern blocks dip up to 10° to the northeast and are locally disturbed by small structures. Attitudes of rocks within the northern block continue into the horizontal layers of the Marble Plateau, and the southern block merges into the Palisades-of-the-Desert area to the south (Fig. 2). The Palisades monocline is a transition zone between the structurally elevated southern block and the structurally depressed northern block. It contains a variety of styles of deformation: steeply to gently dipping layers, fault and breccia zones, altered rocks, tight folds, and overturned layers. The highly deformed and steeply dipping rocks are in the lower units, bounded by the Dox Formation and the Redwall Limestone, whereas the layers above the Redwall dip less than 20° (Fig. 5). The difference of structural elevation between the southern block and the northern block is about 250 m, which is accommodated by faulting, flexuring, and regional tilting (Fig. 7). In the lower part of the monocline, each of the three mechanisms accommodates about one-third of the throw. Higher in the monocline, the flexuring replaces the faulting, and at the level of the Bright Angel Shale, the flexure accommodates about two-thirds of the total throw, and the regional tilting accommodates one-third of it.

Palisades Fault

The major structure in the lower part of the monocline is the Palisades fault, which is a branch of the Butte fault. Fortunately, Palisades Creek flows in the synclinal zone of the monocline, and the fault underlies the anticlinal zone. Thus, the fault is completely exposed in the side of Palisades Creek. The Palisades fault is generally vertical and downthrown to the northeast. A zone of 5 to 60 m width and composed of breccia, leached sandstone, and altered basalt and sandstone occurs along the fault.

A cliff about 30 m high at the mouth of Palisades Creek is the best exposure of the Palisades fault zone (Fig. 8). A block of Shinamu Quartzite identified by D. Elston (1976, personal commun.) is bounded by two faults, the northern one of which has been the site of localization of copper mineralization. A white to purple sandstone, in which bedding planes cannot be recognized easily, appears on the northeastern side of the block of Shinamu Quartzite. Exposures of similar sandstone occur within the fault zone in Palisades Creek. Farther to the northeast in the cliff exposure, competent, brownish sandstone layers of the lower part of the Dox Formation are flexed sharply (Figs. 5, 8). In the southern block, the layers of the Dox Formation and the overlying Cardeñas lavas are inclined gently with the regional dip of 20°. The layers within the Dox are broken by small, closely spaced normal faults.

The Palisades fault zone has had several periods of activity since the late Precambrian; however, the monocline in the Paleozoic rocks formed during the Laramide orogeny. Even though our main interest is the Laramide deformation, we will briefly describe the various periods of activity. The first major displacement along the Palisades fault was probably during late Precambrian, when the northern

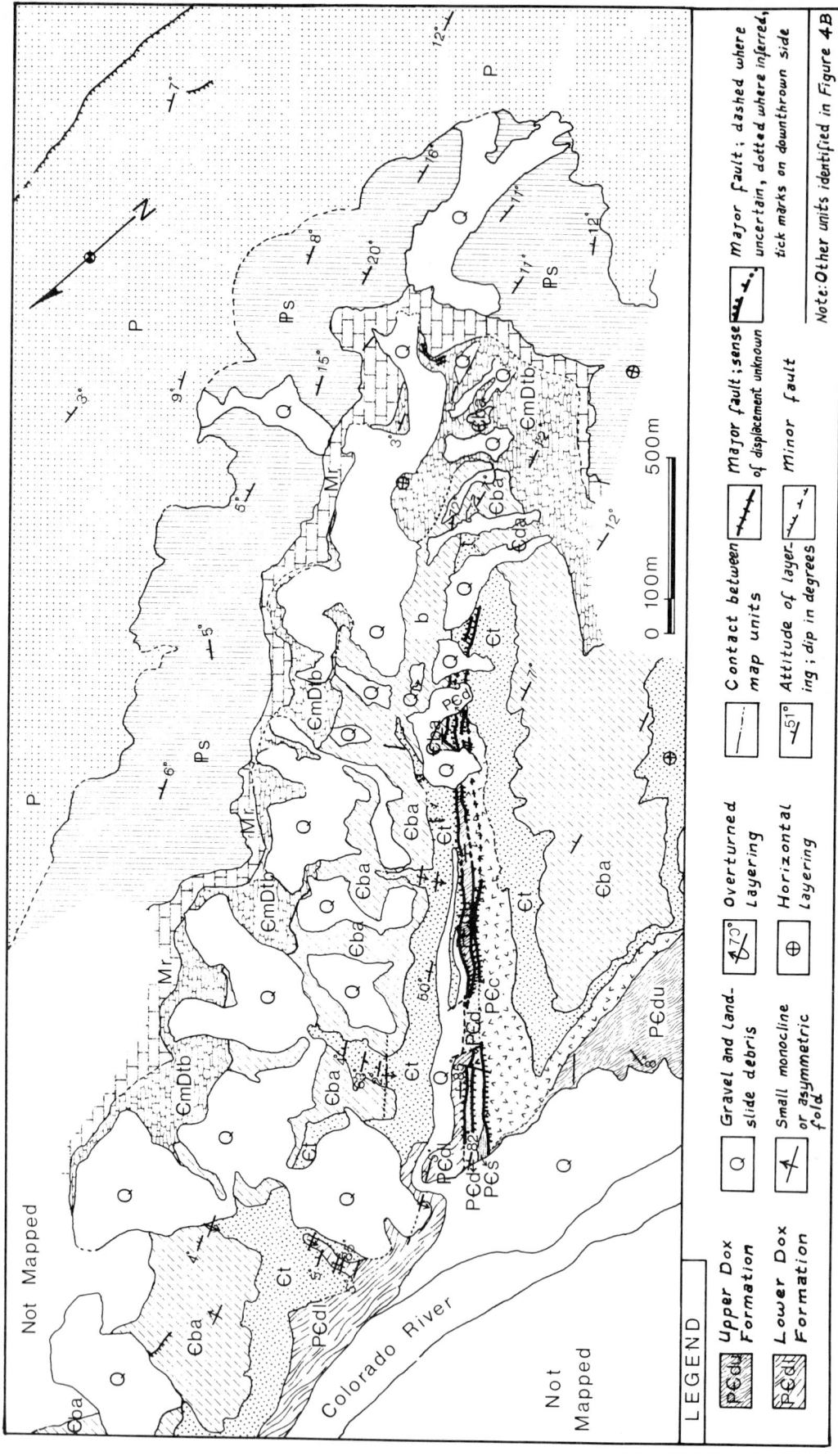

Figure 4A. Geology of the Palisades Creek area, eastern Grand Canyon National Park, Arizona.

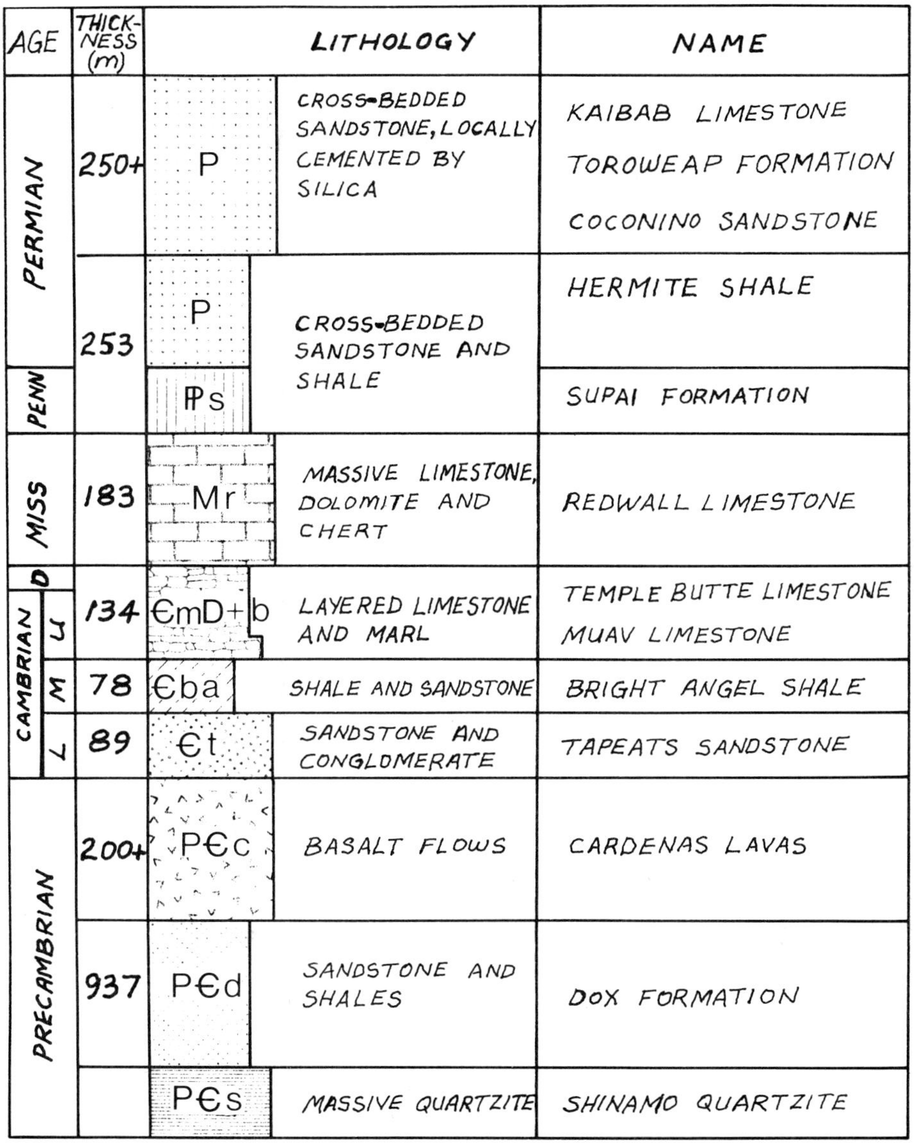

AGE	THICK-NESS (m)	LITHOLOGY		NAME
PERMIAN	250+	P	CROSS-BEDDED SANDSTONE, LOCALLY CEMENTED BY SILICA	KAIBAB LIMESTONE TOROWEAP FORMATION COCONINO SANDSTONE
PERMIAN	253	P	CROSS-BEDDED SANDSTONE AND SHALE	HERMITE SHALE
PENN	253	ℙs	CROSS-BEDDED SANDSTONE AND SHALE	SUPAI FORMATION
MISS	183	Mr	MASSIVE LIMESTONE, DOLOMITE AND CHERT	REDWALL LIMESTONE
CAMBRIAN D U	134	ЄmD+b	LAYERED LIMESTONE AND MARL	TEMPLE BUTTE LIMESTONE MUAV LIMESTONE
CAMBRIAN M	78	Єba	SHALE AND SANDSTONE	BRIGHT ANGEL SHALE
CAMBRIAN L	89	Єt	SANDSTONE AND CONGLOMERATE	TAPEATS SANDSTONE
PRECAMBRIAN	200+	PЄc	BASALT FLOWS	CARDENAS LAVAS
PRECAMBRIAN	937	PЄd	SANDSTONE AND SHALES	DOX FORMATION
PRECAMBRIAN		PЄs	MASSIVE QUARTZITE	SHINAMO QUARTZITE

Figure 4B. Stratigraphic section of the Palisades Creek area. Terminology after Hintze (1973).

block was upthrown relative to the southern block along two or three faults in the Palisades fault zone (Fig. 5). The three faults are subvertical; however, the central fault dips locally 60° to 80° to the southwest, which indicates normal faulting for the late Precambrian.

Hydrothermal alteration of layers of Precambrian basalt may be a manifestation of another period of faulting. Potassium-argon age determinations of altered basalt samples yielded 577 ± 23 m.y. for one sample and 525 + 21 m.y. for the second (Silberman and Reches, in prep.). If this alteration phase was associated with faulting activity, then the Palisades fault was active during Middle Cambrian. There

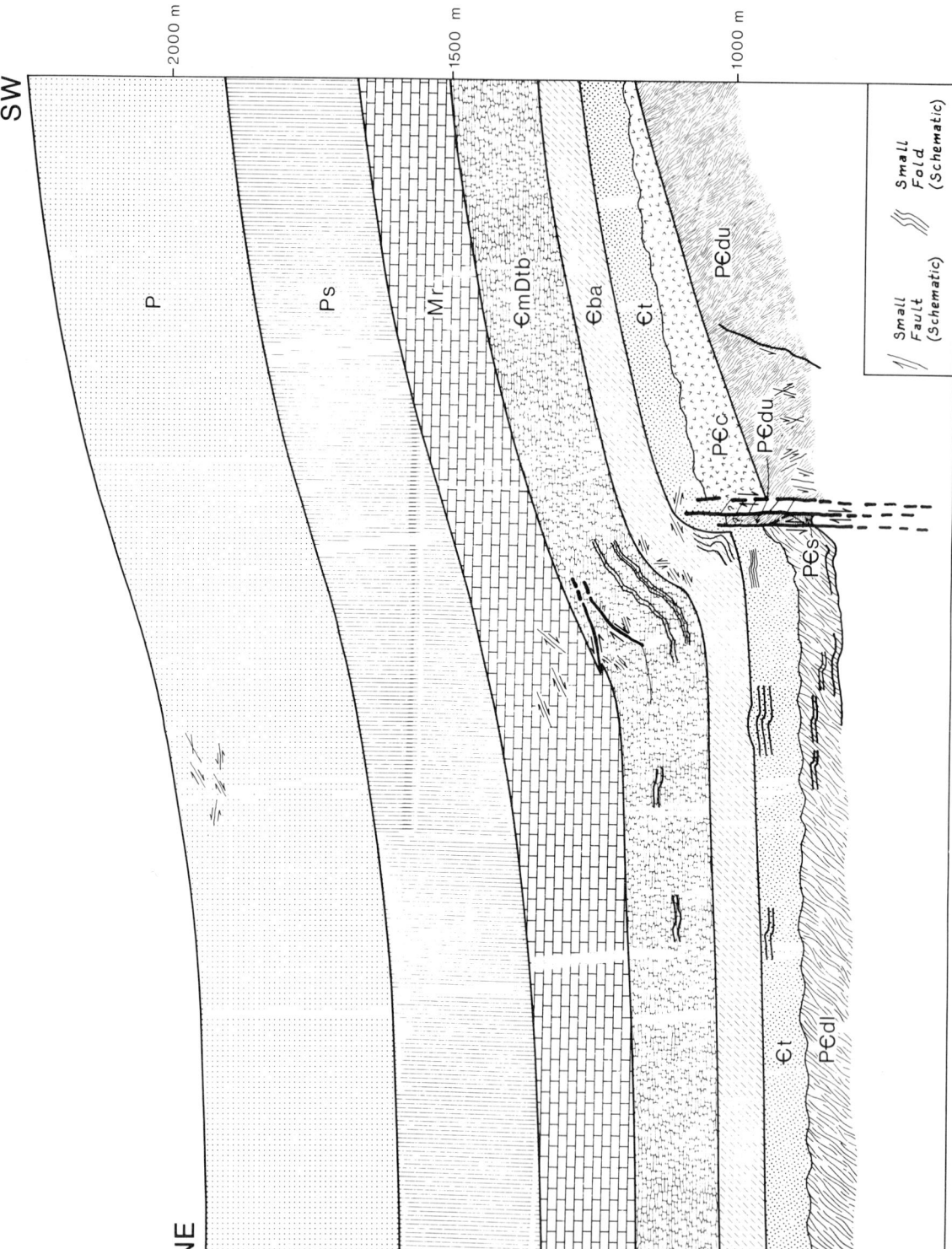

Figure 5. An accurate composite cross section of the Palisades monocline. Constructed by projection of all stratigraphic contacts and data stations to a vertical section striking northeast with accuracy of ±10 m. Small-scale structures are in correct position but drawn schematically. Legend in Figure 4.

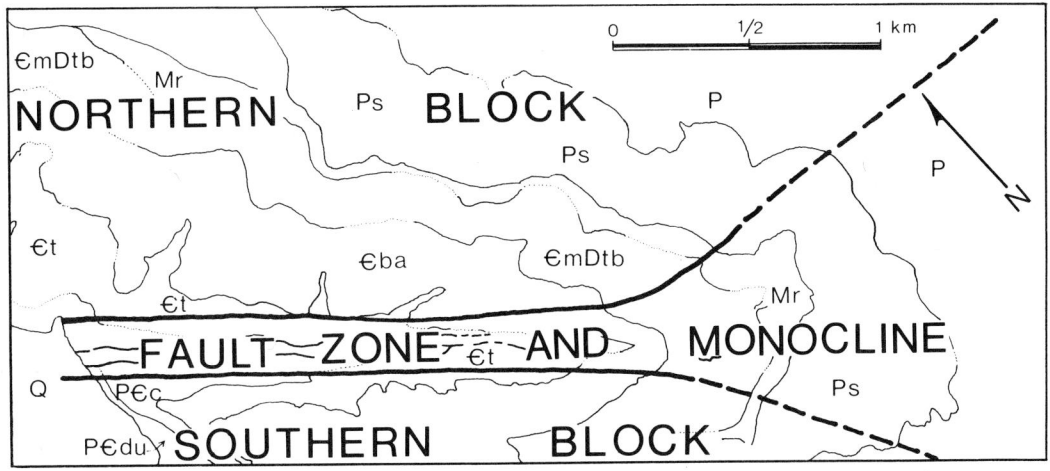

Figure 6. Simplified geologic map of the Palisades monocline showing the main structural units. Legend in Figure 4.

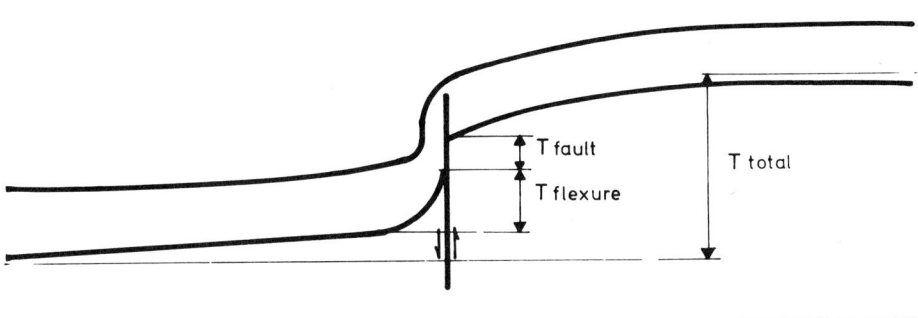

Figure 7. The difference in structural elevation, T_{total}, between the two sides of the Palisades monocline is a combination of throw along the fault, T_{fault}, amplitude of the flexure, $T_{flexure}$, and regional tilt, T_{tilt}. $T_{tilt} = T_{total} - (T_{fault} + T_{flexure})$.

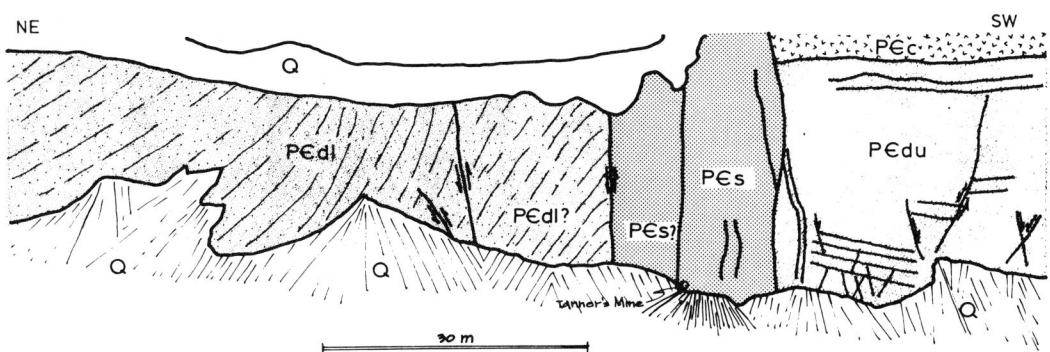

Figure 8. Cross section of the Palisades fault zone at the mouth of Palisades Creek (sect. a-a, Fig. 12). Legend in Figure 4.

is no direct field evidence for faulting during this time; however, thickness measurements indicate thinning of the Cambrian Bright Angel Shale in the anticlinal hinge zone of the monocline (App. 2).

Walcott (1890) suggested minor tectonic activity along the trend of the East Kaibab monocline during the Permian. Marshal (1972) found evidence for minor uplift during the Permian along the East Kaibab monocline at Gray Mountain, southeast of Palisades Creek, by means of a detailed study of the Toroweap Formation there.

The major deformation which resulted in the present monocline in Palisades Creek, however, was during the Laramide orogeny, starting at the end of the Mesozoic. The Laramide deformation uplifted the southern block, whereas the Precambrian deformation had uplifted the northern block. This reversal of slip direction has been reported in many other places on the Colorado Plateau (Walcott, 1890; Huntoon, 1974).

The net throw of the monocline is 250 m, but the net throw of the base of the Tapeats Formation across the fault is about 80 m (Fig. 5). The remainder of the throw was accommodated by flexuring of the strata within the Dox Formation and by regional tilting (Fig. 7). The new throw of 250 m during the Laramide deformation is much smaller than that which occurred in the Precambrian. Matching of units within and across the fault zone indicates that the stratigraphic throw of the Precambrian units across the fault is up to 700 m today. The net stratigraphic throw during the Precambrian, therefore, was at least 950 m, because the sense of throw was reversed between Precambrian and Laramide times. Actually, the throw during the Precambrian was probably larger, according to D. Elston (1976, personal commun.), because the slice of Shinamu caught up within the fault zone probably was dropped down 100 m during the Laramide deformation. Despite the long and complex tectonic history of this fault zone, the Laramide displacement which is associated with formation of the monocline can be traced accurately.

The Flexure

The zone of continuous flexuring within the Palisades monocline involves rock units from the Bright Angel Shale below through the Kaibab Limestone above, and the geometry of the flexure changes laterally within rock units and vertically from rock unit to rock unit (Fig. 5). The differences in geometry can be expressed in terms of differences of curvatures and shapes of contacts between layers and in terms of lateral changes in thickness of individual rock units.

Thickness variations of layers are useful for analyzing the gross strain within the layers. Therefore, I measured the thickness of several rock units both in the field and with a stereoplotter (App. 2) in 14 sections in Palisades Creek Canyon (Fig. 12, Table 2, App. 2). The accuracy of the thickness measurements is $\pm 8\%$ (Table 3, App. 2). Thickening or thinning was determined by comparing thicknesses of units within the monocline to thicknesses of units on either side of the monocline. Only the Redwall Limestone in section 5 and the Bright Angel Shale in section 9 show significant thickness changes (Table 2).

In addition to the direct thickness measurements presented in Table 2, one can observe thickness changes in the cross section of the Palisades monocline (Fig. 5). Furthermore, it is easier to visualize thickness variations in an accurate and continuous cross section (Fig. 5) than in a series of accurate but discontinuous stratigraphic sections (Table 2, App. 2).

Thickness variations of units in a profile of a fold are readily detected with the *isogon* technique proposed by Ramsay (1967). Because the cross section is accurate, the isogon pattern of the Palisades monocline will reflect actual thickness variations. Isogons are lines drawn on a profile of a fold and connecting points with equal dip on adjacent folded surfaces (Fig. 9). Ramsay proposed that folded layers may be divided into three classes:

1. Folded layers in which the curvature of the external surface is smaller than that of the internal surface. This class is divided into three types—1a, 1b, and 1c—in which type 1b is the concentric form (Fig. 9).
2. Folded layers with constant curvature (similar folds).
3. Folded layers in which the curvature of the external surface is larger than that of the internal surface. Each type of fold in Ramsay's classification is associated with certain thickness variations (Ramsay, 1967, p. 366).

Table 1 summarizes the fold types and the relative thickening of layers within the Palisades monocline (App. 2). The Palisades monocline is composed of three zones according to Table 1. Let us discuss the thickness variations in these zones. The thickening in the lower zone is probably within the accuracy of the measurements. The middle zone contains a synclinal bend and an anticlinal bend. All three rock units in the synclinal zone show significant thickening, whereas, two units in the anticlinal zone show constant thickness, and one unit, the Bright Angel Shale, is thinned (Table 1). The rock units in the upper zone maintain constant thickness.

These zones will be described in the following paragraphs.

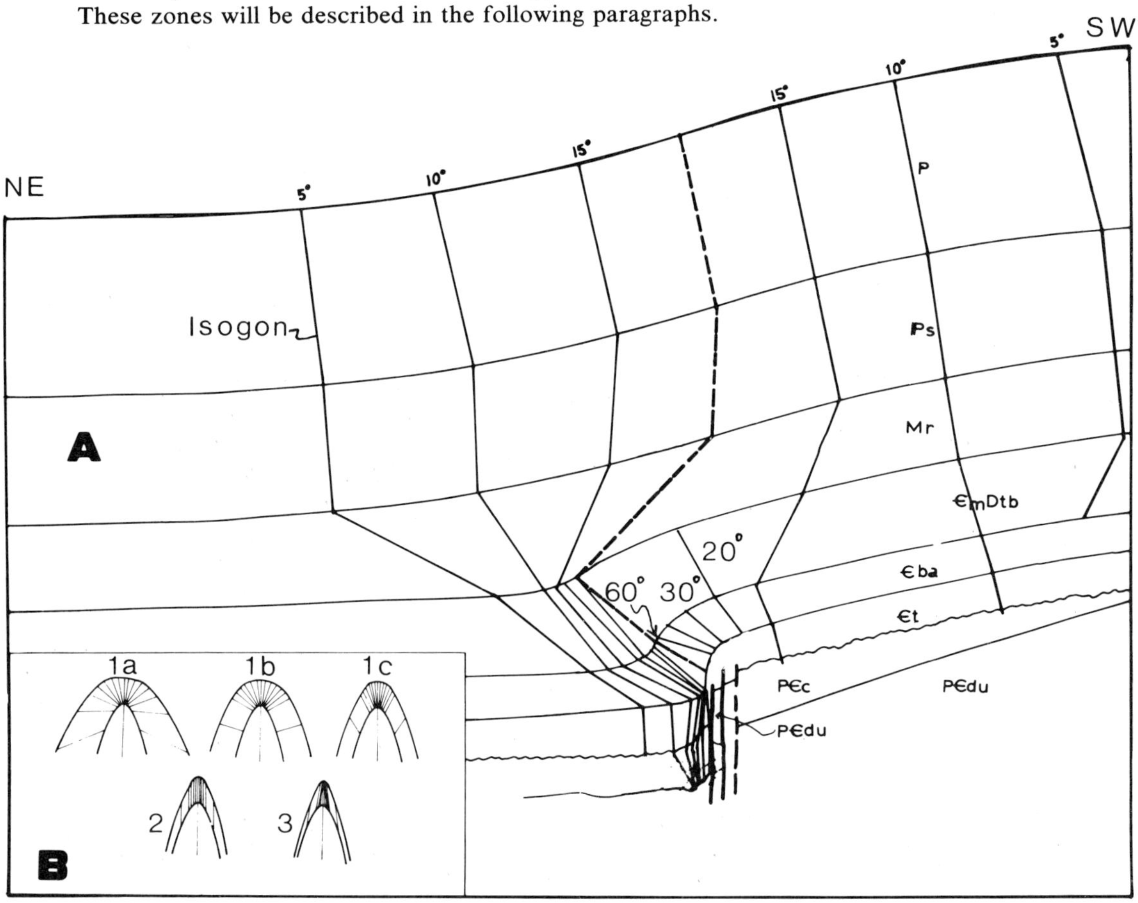

Figure 9. A. Isogon pattern of the Palisades monocline. B. Isogon patterns for five types of folds in the classification of folds by Ramsay (1967, Figs. 7-24).

Overturned Zone. The lowest structural level of the Palisades monocline is composed of both flexuring and faulting within the Dox Formation and the Tapeats Sandstone (Fig. 5). Only the synclinal zone of the monocline is developed there. The throw of the Palisades fault is about 50 m, and the throw of the flexure is about 110 m. Some of the layers are overturned within this lowest level, but they are nearly horizontal a few tens of metres from the Palisades fault. For example, Figure 10 shows the Tapeats Sandstone dipping about 20° northeastward on the upthrown side of the Palisades fault; altered basalt and leached sandstone occur within the fault zone. The Bright Angel Shale which overlies the Tapeats is overturned to about 150°, but it is nearly horizontal 20 to 30 m away. Despite the high curvature, the layers of the Bright Angel Shale are not faulted. The termination of the Palisades fault is about 2 km southeast of the Colorado River and 240 m above its level.

Continuous Flexure. A continuous flexure within the Bright Angel Shale completely replaces the Palisades fault within the middle structural level of the monocline (Fig. 5). The monocline includes a sharp anticlinal bend with a radius of curvature of about 10 m (Fig. 11) and a relatively open synclinal bend with a radius of curvature of about 150 m. The rocks in the anticlinal bend are intensely fractured, but the apparent displacements are small. The steeply dipping beds of the Bright Angel Shale and the overlying Muav Limestone are disturbed by small faults, by joints, and by small folds. The structural throw of the monocline is about 150 m.

Two relatively large faults occur in the Temple Butte Limestone which overlies the Muav Limestone. Both faults dip to the northeast and become bedding-plane faults toward the southwest (Fig. 5). The minimum dip slip along the upper fault is about 70 m. The layers above the fault dip 10° to 15°, whereas the layers below it dip 35° to 40°. Probably the steep layers under the fault were displaced to their current position from a zone of steeper dips to the northeast. Displacement along these faults accommodated local lateral shortening and vertical extension.

The Open Flexure. A striking contrast exists between the intense deformation of units below the Redwall Limestone and the open flexure of units above (Fig. 5). The Redwall Limestone was studied in the main gully of Palisades Creek,

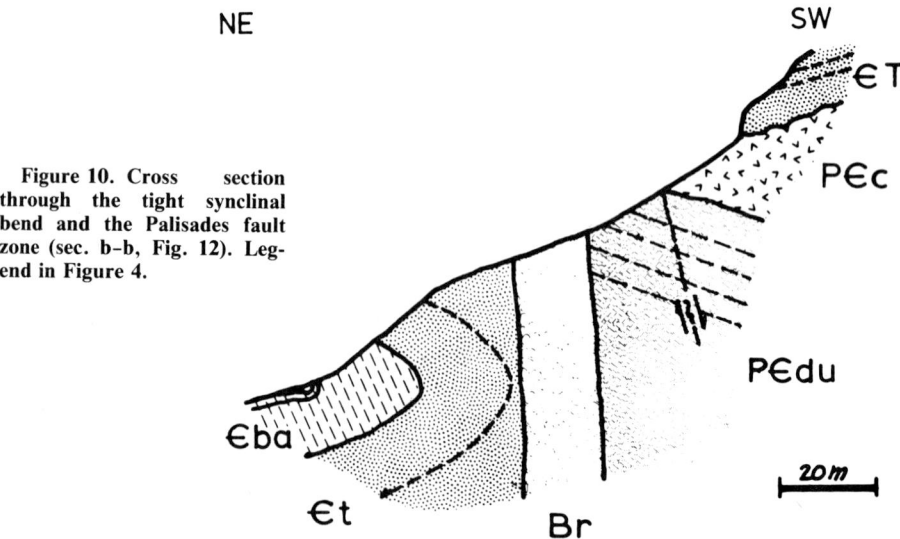

Figure 10. Cross section through the tight synclinal bend and the Palisades fault zone (sec. b-b, Fig. 12). Legend in Figure 4.

the only accessible trail to the top of the Redwall in the area (stations 334, 335, Fig. 12). Here, the Redwall Limestone is disrupted by many closely spaced fractures, some of which are slickensided. Some of the limestones and dolomites of the Redwall are essentially cemented breccia, with no clear bedding planes. The maximum dip at the top of the Redwall is 20°, whereas it is 39° at its base. The continuous flexure opens upward and, on the plateau, the Kaibab limestone dips up to 18°, with a radius of curvature of about 4 km (Fig. 5). The synclinal bend seems to be slightly sharper than the anticlinal bend. Vertical joints and minor faults are abundant, whereas faults with displacements of more than 10 cm and minor folds are scarce.

STRUCTURAL ANALYSIS

Analysis of Small Structures

Numerous small structures occur in the Palisades monocline, including faults, kink bands, folds and joints (Fig. 12). The slip along the faults or the amplitude of the folds ranges from a few millimetres to a few metres. The contribution of such small structures to the regional strain is limited to a small percentage (Reches, 1976); however, they seem to be useful indicators of strain orientation.

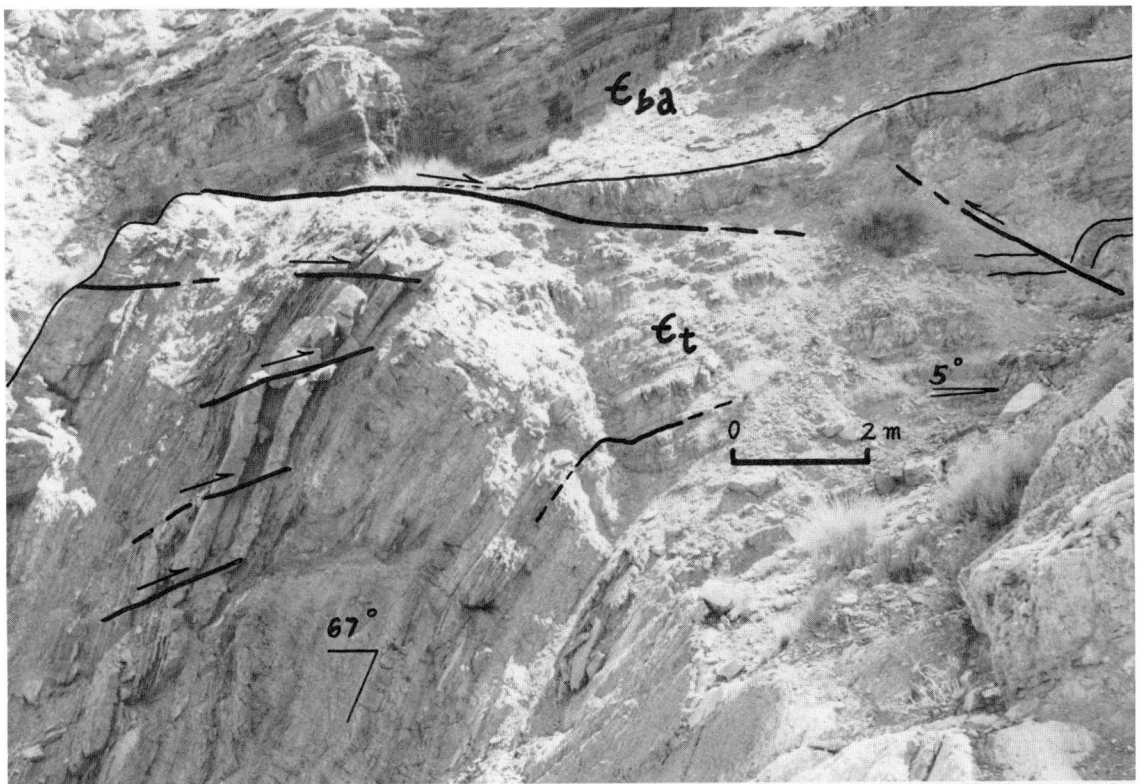

Figure 11. The continuous anticlinal bend in the sandstone layers of the upper member of Tapeats Sandstone (sec. b-b, Fig. 12). Radius of curvature is 10 to 15 m. The rocks are brecciated, but the layers are continuous. The termination of the Palisades fault is about 10 to 15 m below this exposure.

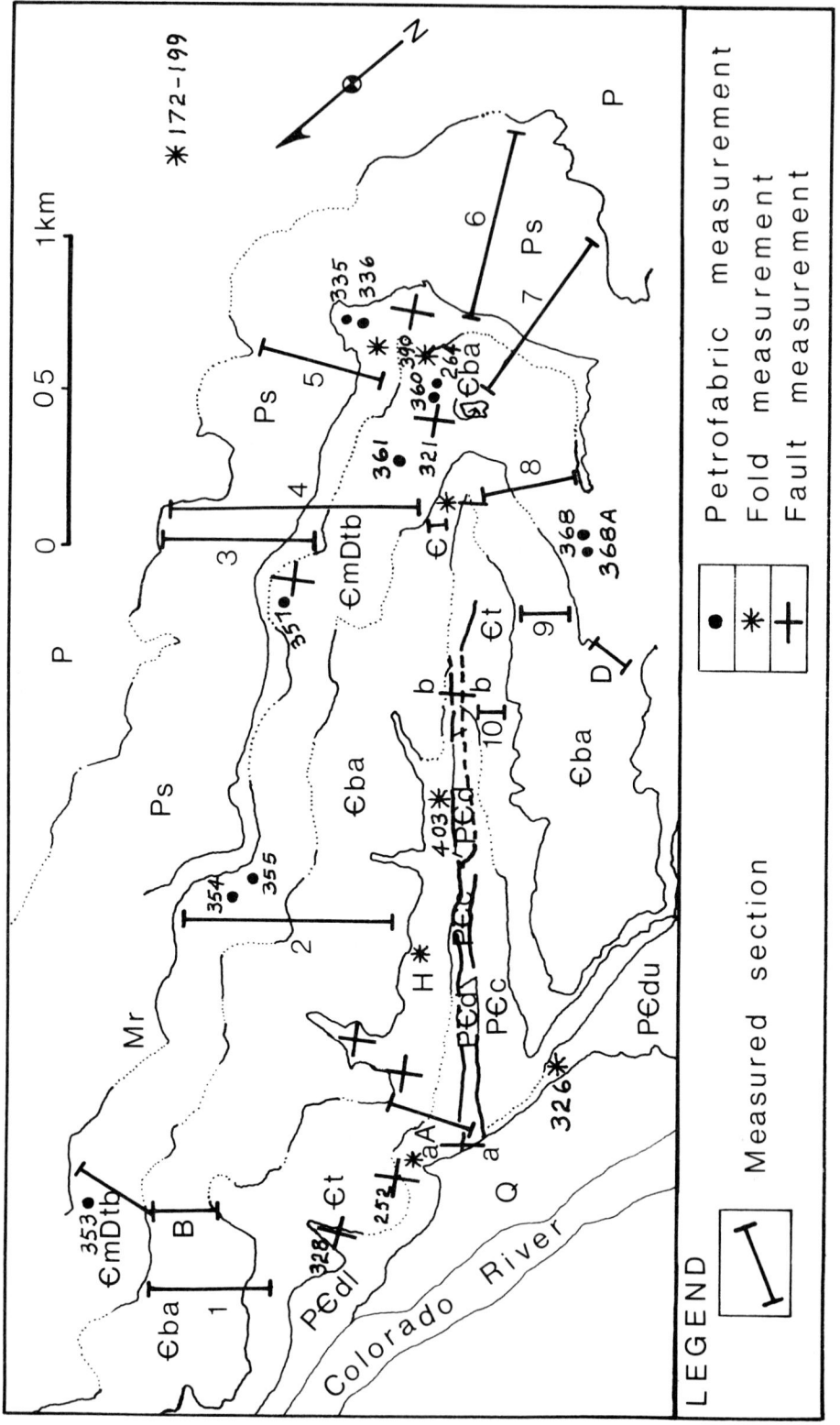

Figure 12. Locations of structural observations in the Palisades monocline. Each symbol of fault measurements represents grouped data for tens of faults. Each symbol of fold measurements represents either a single large fold or several small ones. Each symbol of petrofabric measurements represents a single sample. Thickness measurements marked by letters were measured on a stereoplotter.

Small structures were measured at many stations in the monocline and grouped according to their structural setting.

Small Faults. Normal and reverse faults as well as strike-slip faults are abundant in the Palisades monocline. Many of them carry slickenside striae. Faults are relatively scarce in the flat layers a few hundred metres on either side of the monocline (Fig. 5); however, they are abundant in flat layers closer to the monocline.

About 240 small faults were measured within the monocline, and several major faults were studied in the Palisades fault zone. These faults are considered to be strain indicators; for example, reverse faults indicate horizontal shortening and normal faults indicate extension. Two types of faults are characteristic of the synclinal bend of the monocline near the Palisades fault. The first are reverse, low-angle faults which are common in flat layers north of the synclinal axis; the second type are closely spaced faults which are abundant in the steep layers close to the Palisades fault.

Reverse faults, most of them with slickensides, were measured in subhorizontal sandstones of the lower Dox Formation, the Tapeats Sandstone, and the Bright Angel Shale close to the synclinal zone (stations 252, 403, H, Fig. 12). These faults are at low angles, generally 10° to 30°, to bedding planes, and in many cases they are continuous through bedding and cross-bedding surfaces. They are scarce in the shaley layers of Bright Angel Shale and the Muav Limestone. The patterns of both the attitudes of the faults and the orientations of the slickensides indicate subhorizontal shortening in the direction N65°E (Fig. 13A).

A different type of fault occurs southwest of the synclinal bend. In several places the steep layers are fractured by hundreds of closely spaced faults bearing small displacements. One can distinguish between two pairs of fault sets in which each pair has a common strike but different dip (Fig. 14). These faults resemble fractures produced by Oertel (1965) in claycake experiments. Following Oertel's analysis, these four sets of faults in the Palisades monocline indicate extension parallel to the dip of the layers and shortening normal to the bedding and parallel to the strike. The structural interpretation of faults in steeply dipping layers, however, is ambiguous (Fig. 15). If the faults were formed during the earliest stages of folding, they represent extension normal to the axis of the structure. If they were formed in the later stages, they indicate shortening normal to the axis of the structure.

Figure 13B is a stereoprojection of faults and slickensides measured at stations F, 390, and 245 in Bright Angel Shale, Muav Limestone, Temple Butte Limestone, and Redwall Limestone. Most of these faults have a component of oblique slip and imply shortening parallel to bedding. Despite the wide scatter of the orientations of these faults (Fig. 13B), the slickensides have a consistent direction with a mean plunge of 2° in the direction of 237°. These slickensides indicate layer-parallel shortening by displacement along faults.

Faults in the Kaibab Formation on the plateau generally have small displacements of 2 to 3 cm. A summary of the measurements of the faults at five stations is presented in Figure 13C. The sense of slip along some of the faults was deduced using the "steps" on the fault planes (Norris and Barron, 1968). Both reverse and strike-slip faults indicate subhorizontal shortening in the direction 071°.

All the faults described above show shortening in directions of 065° to 071° (Fig. 13A, 13B, 13C); however, study of normal faults in the sandstone layers of the upper Dox Formation south of the Palisades fault reveals a different pattern (Fig. 13D of station 326). Many of them are slickensided and show small displacement. The data are somewhat scattered (Fig. 13D), but two groups are distinct: steep faults dipping northeast or southwest. No clear slip direction can be deduced from

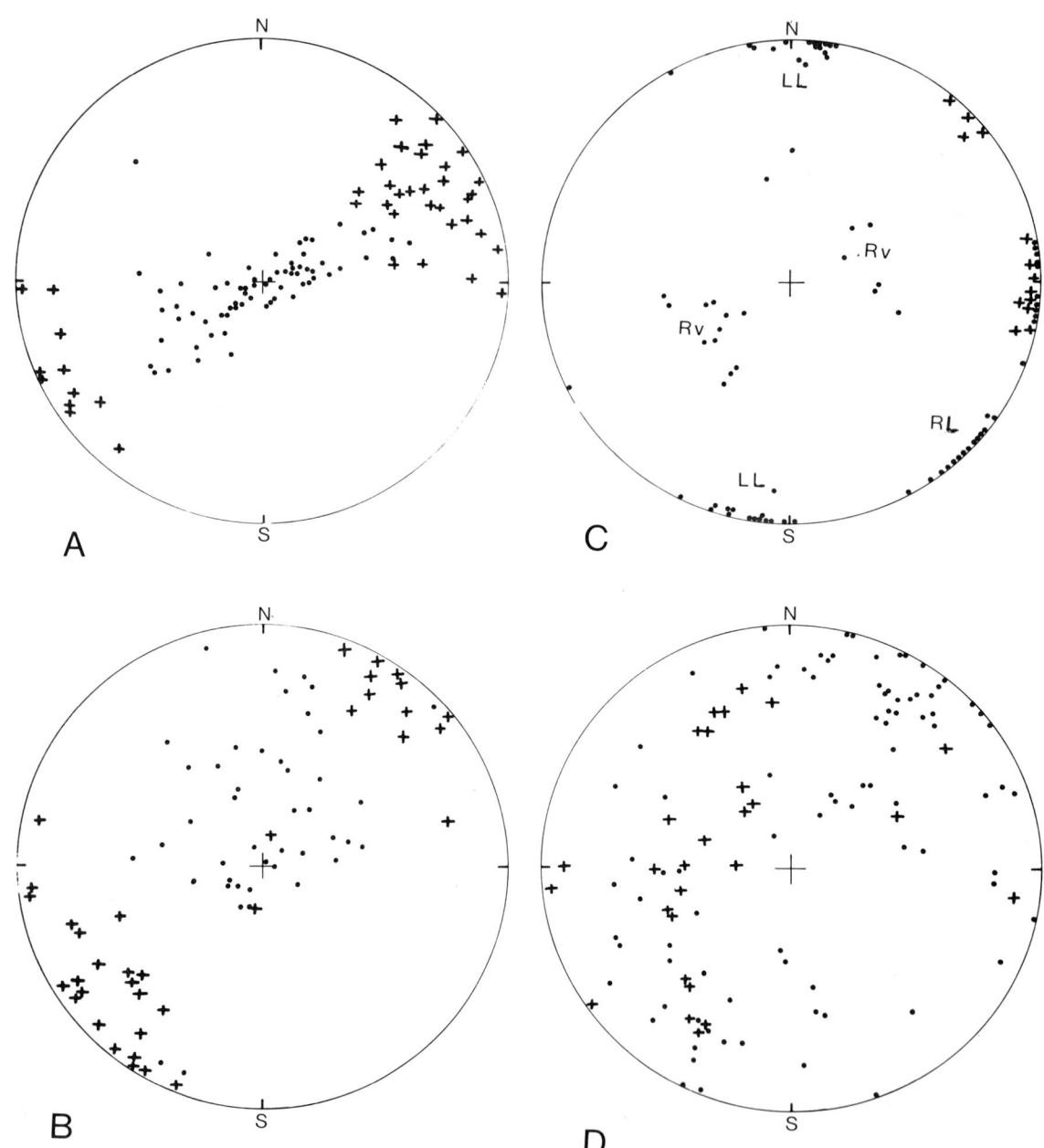

Figure 13. Projections of normals to small faults and of slickenside striations in the Palisades monocline. Equal-area projection, lower hemisphere. Solid dot represents normal to fault; cross represents slickenside striations in fault plane. A. Faults and slickensides close to the synclinal zone in flat layers of the Dox Formation, the Tapeats Sandstone, and the Bright Angel Shale (stations 252, 403, H, Fig. 12). Normals to 44 faults and 68 sets of slickensides. B. Faults and slickensides in the Bright Angel Shale and the Muav, Temple Butte, and Redwall Limestones in the monoclinal flexure (stations F, 390, 245, Fig. 12). Normals to 37 faults and 43 sets of slickensides. Rv is reverse. LL is left lateral. RL is right lateral. D. Faults and slickensides in the upper Dox Formation close to the Palisades fault zone (station 326, Fig. 12). Normals to 28 faults and 87 sets of slickensides.

the scattered slickenside data for these faults. The faults in the upper Dox Formation probably formed before the development of the monocline at the time of Precambrian normal faulting, so these faults need not be relevant to a study of the monocline.

Small Folds. The most striking small structures within the Palisades monocline are kink bands and asymmetric folds, the axes of which are roughly parallel to the axis of the monocline (Fig. 16). These folds occur in gently dipping layers of the lower Dox Formation and the Tapeats Sandstone and the lower part of the Bright Angel Shale in the downthrown block of the monocline. Most of the folds are asymmetric, facing the southwest, opposite to the Palisades monocline (Fig. 16). These folds range in form from sharp-hinged kink bands (Fig. 16A) through rounded-hinged kink bands (Fig. 16B) to smooth asymmetric folds (Fig. 16C); they range in amplitude from 1 to 20 m and in wavelength from 5 to about 50 m; sedimentary sections up to a 50 m thick are involved in these folds. The small folds occur in well-bedded sandstones locally interbedded with shale. They are closely fractured and faulted.

There are many kink bands in layered sandstones of the Dox Formation and the Tapeats Sandstone in the northern block of the Palisades monocline. The kink bands can be understood in terms of relatively high strength of contacts between layers within these units, of layer-parallel shortening, and of layer-parallel shear induced by flexuring of layers in the monocline. Relatively high strength of contacts is indicated by abundant, low-angle faults, inclined 10° to 30° to bedding within

Figure 14. Closely spaced faults in a steeply dipping layer of the Tapeats Sandstone near the Palisades fault (close to station H, Fig. 12). A. View of top of layer. B. Side view of same layer.

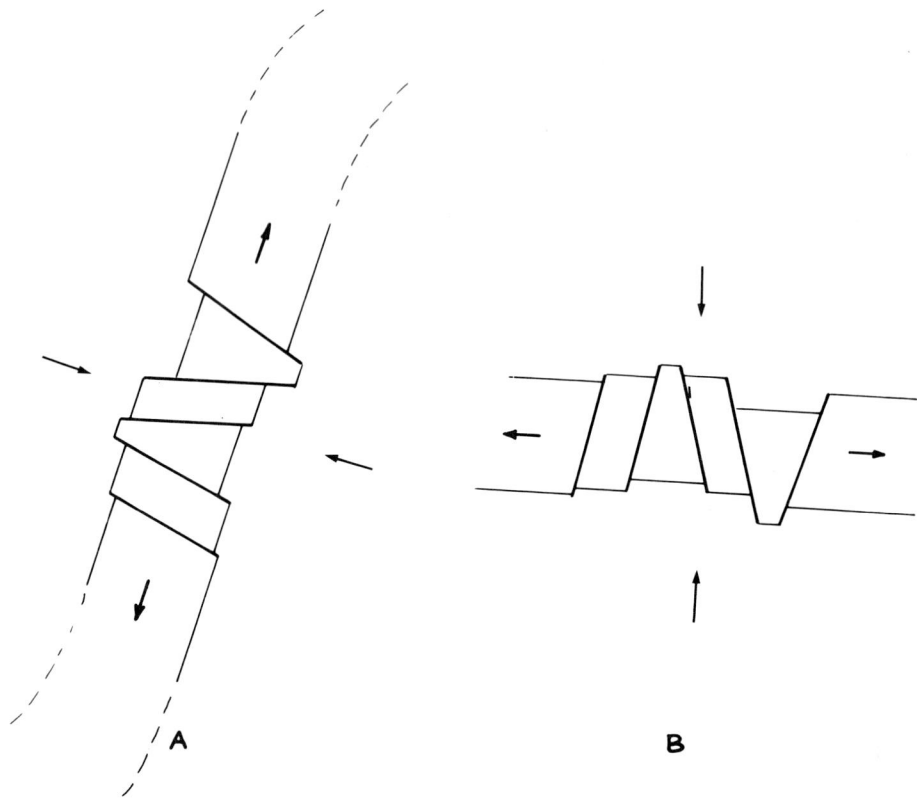

Figure 15. Sketch of faults in a steeply dipping layer. A. Cross section of the layer in present position. B. Cross section after retilting to horizontal position. The contribution of these faults to the gross strain depends on their age. If they formed during early stages of the folding, they represent horizontal extension (B); however, if they formed during late stages of folding, they indicate horizontal shortening (A).

these units. The existence of the low-angle faults demonstrates that the strength of contacts was almost as high as the strength of the rock within the layers themselves. Therefore, one can conclude that the Dox and Tapeats formations behaved much as multilayers with high contact strength between layers—a condition required for kink folding according to theories developed by Honea and Johnson (1976), Reches and Johnson (1976) and Johnson (1977). Further, Reches and Johnson (1976) showed that kink bands will have monoclinic symmetry, as those observed in the field, if there is a combination of layer-parallel shortening and layer-parallel shear. Finally, the layer-parallel shear is of the sense that would be induced by the folding of the synclinal bend of the Palisades monocline (Fig. 17). This layer-parallel shear is consistent with the layer-parallel shear found in the petrofabric analysis of the samples 353, 354, and 357 (see below).

Other small folds, some symmetric and others asymmetric, with amplitudes of 1 m or less, are at higher levels in the monocline, mainly in the Muav Limestone. They occur in limestone layers interbedded with shales, mostly steeply dipping (40° to 80°); they are particularly common in the steeply dipping Gateway member (Fig. 5). The axes of these folds are subparallel to the strike of the host layers, which trend about 145° to 150°. We consider these folds to imply layer-parallel shortening normal to the fold axes.

Petrofabric Study

Petrofabric analyses were made in order to determine orientations of stress in carbonate rocks in the Palisades monocline and adjacent areas. The analysis of calcitic rocks is based on the knowledge of low-temperature deformation of calcite crystals by twin gliding on $01\bar{1}2$ (e) crystalographic planes. The methods are described elsewhere (Turner and Weiss, 1962). The results presented below are of stress axes deduced by the dynamic method unless otherwise stated. Eleven samples from the Palisades monocline were studied, two from the Redwall Limestone (stations 335, 334, Fig. 12) and nine from the Muav Limestone. The samples of the Muav Limestone were collected from different structural locations within the upper part

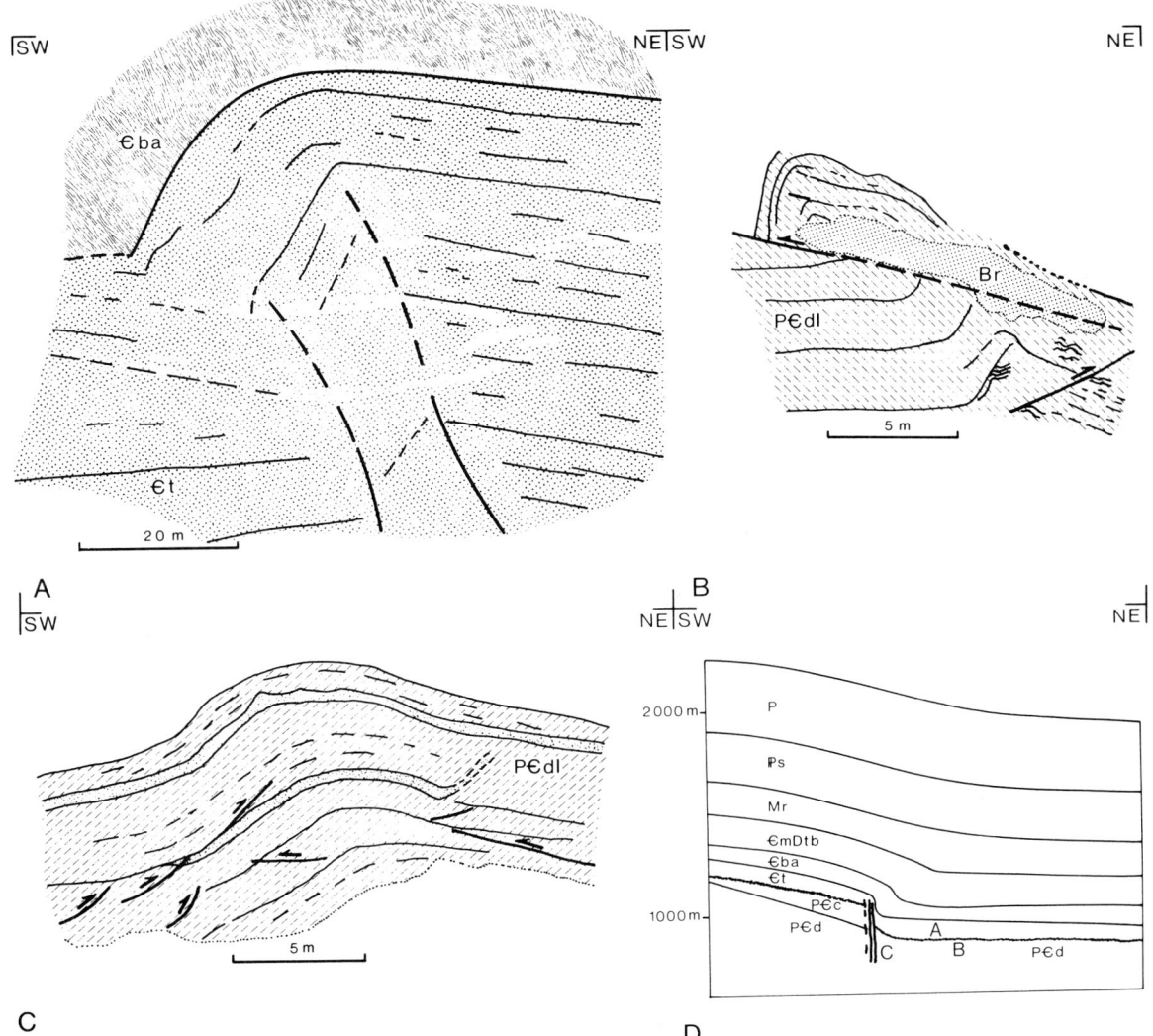

Figure 16. Small asymmetric folds in the Palisades monocline. A. A large kink band with faulted hinges developed in the upper Tapeats Sandstone (station 1, Fig. 12). B. A tight asymmetric fold associated with a thrust fault in the upper Dox Formation (station 38, Fig. 12). C. Open asymmetric fold in the upper Dox Formation (station 252, Fig. 12). D. Cross section of Palisades monocline showing locations of three asymmetric folds.

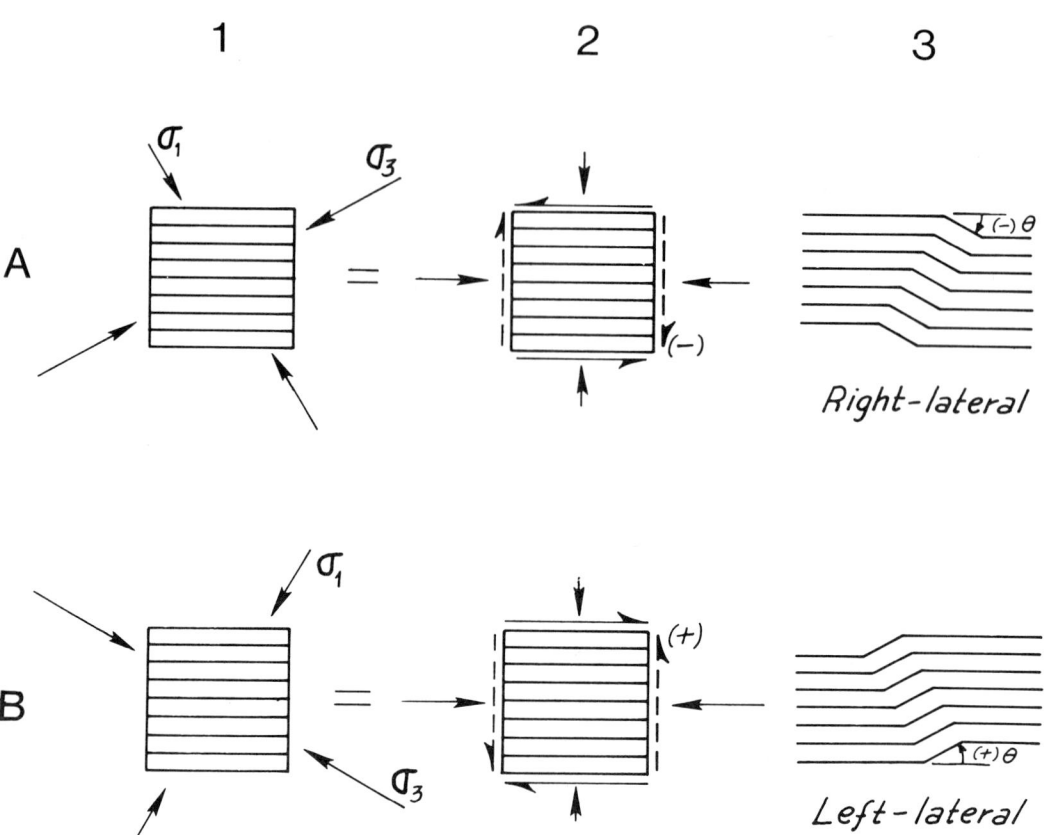

Figure 17. Orientation of the principal stress axes associated with the development of unidirectional kink bands (after Reches and Johnson, 1976). Maximum compression, σ_3.

of the Gateway member. Two are from the southern block (stations 368, 368A, Fig. 12), two from the monoclinal flexure (stations 264, 360, Fig. 12), and five from the northern block (stations 353–5, 357, 361, Fig. 12). Three pairs of samples were collected from the Muav Limestone to estimate local variations in strain orientations (samples 355 and 354, 368A and 368, 360 and 264).

Two mutually perpendicular thin sections were studied in eight samples, and one section was used for the other three. Highly twinned grains were selected for measurement. All the results were rotated to field orientation. The orientations of the optical C-axes were determined in two sections, and the results show a slight preferred orientation normal to bedding.

The samples of Muav Limestone have similar petrology. They are limestones composed of coarse intraclasts, pellets, fossil fragments, quartz grains, glauconite grains, dolomite crystals, and cement of fine- to medium-grained calcitic sparite. The intraclasts are up to 1 cm long and contain quartz and mica grains in a calcitic micrite. The sparite cement contains grains ranging generally from 0.01 to 0.10 mm in diameter in a mosaic texture. Few grains exceed 1 mm in length. Veins filled with large crystals of calcite and zones of recrystallization are common in several samples.

The two samples of Redwall Limestone also contain fossil fragments and intraclasts, but they show more features of brecciation and recrystallization. Dolomite crystals replacing the original calcite are common in the samples of the Redwall.

The percentage of twinned calcite crystals, the number of twins per millimetre, and the number of twin sets decrease with distance from the steep monoclinal flexures. Samples such as 264, 324, 335, and 360 show many twins in most of the grains and two sets of twins in some grains. On the other hand, in sample 353, far from the flexure, only 10 to 20% of the sparite crystals are twinned.

The results of the measurements of C-compression axes and T-extension axes for individual samples are shown in Figure 18, and the maximum point from all samples is shown in Figure 19. The pattern of the stress axes in the various samples seems to be consistent and can be characterized as follows:

1. The axes of extension lie within a plane striking north-northwest, which in general is normal to bedding. Therefore, the intermediate and maximum extensions are approximately equal.

2. The axes of compression generally are subparallel to bedding planes and trend east-northeast. They form a single maximum or several maxima close together.

3. Samples collected as pairs, such as 354–355, show similar results.

4. Samples with few measurements (for example, 353, 354) are consistent with other samples.

5. Samples from the monoclinal flexure (265, 335, 334, 360) have a more scattered pattern of strain axes. It seems that these samples reflect a component of shear parallel to the bedding. Sample 360 indicates compression normal to layering.

Figure 19 summarizes the orientations of the stress axes of the 11 samples in the Palisades monocline as well as other structural data. Due to result (2) above, only the mean direction of compression of each sample appears in Figure 19.

A strong concentration of maxima points indicates compression in mean direction 10°/073° (dip 10° in the direction 073°). A single maximum occurs at 72°/009°.

In order to determine whether horizontal compression is unique to the Palisades monocline or if it is common to the other branches of the East Kaibab monocline in the vicinity of the Grand Canyon, I collected a few samples from the Bright Angel monocline, Grandview monocline, Tanner monocline, and Coconino monocline. Ten samples of the Kaibab Limestone were collected in cooperation with Rick Groshong of Cities Service Oil Company, Tulsa, Oklahoma. Of the ten samples, two do not contain calcite crystals and five contain fine- to medium-grained calcite crystals with only a few twinned grains. Groshong analyzed the petrofabrics of two of the three suitable samples of Kaibab Limestone and one sample of Redwall Limestone. I analyzed two other samples. In total we analyzed five: one of Bass Limestone (analysis of twinning in dolomite), two of Redwall limestone and two of Kaibab Limestone. A single section was analyzed in every sample. Measurements in four samples indicate horizontal maximum compression (after retilting), whereas one sample indicates inclined maximum compression. The directions of maximum compression according to these measurements are shown in Figure 20.

The analysis of these few samples is inadequate to determine the regional pattern of principal stresses during the Laramide orogeny. However, the analyses do indicate subhorizontal compression over a wide area around the Palisades monocline and the other monoclines in the area of the Grand Canyon.

Summary of Structural Analysis

The orientations of the principal axes of compression deduced from the structural analyses of the Palisades monocline are plotted on a map (Fig. 21) and a section (Fig. 22). The map and section show high consistency of results among the various methods. Thus, the indicators of stress associated with the Palisades monocline,

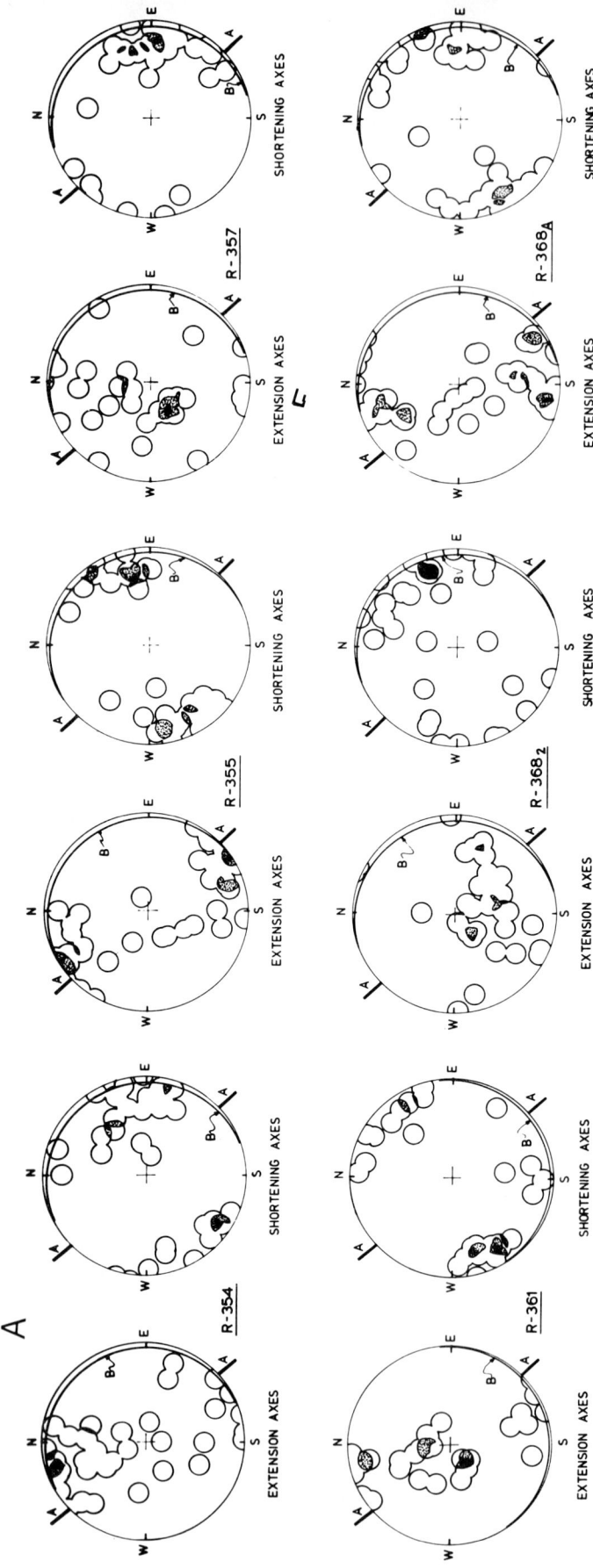

Figure 18. (facing pages). Stress axes deduced from analysis of calcite twinning in 11 samples from the Muav and Redwall Limestones. Projection on equal-area lower hemisphere. A. Samples from northern and southern blocks, where dips are small. B. Samples from monoclinal zone, where dips are large. For locations see Figure 12. Short lines A-A indicate the trend of the Palisades fault. The list below includes the following information in the given order: number of sample, number of thin sections analyzed, number of twinning sets analyzed, percentage of first contour, percentage of second contour, percentage of third contour:

R-264,2,58,1%,4%,10%
R-353,2,21,1%,8%,14%
R-357,2,28,1%,11%,14%
R-368,1,25,1%,12%

R-334,2,40,1%,8%,13%
R-354,2,30,1%,10%
R-360,1,52,1%,6%
R-368A,2,33,1%,9%,12%

R-335,2,48,1%,6%,10%
R-355,2,33,1%,9%
R-361,2,24,1%,13%,17%

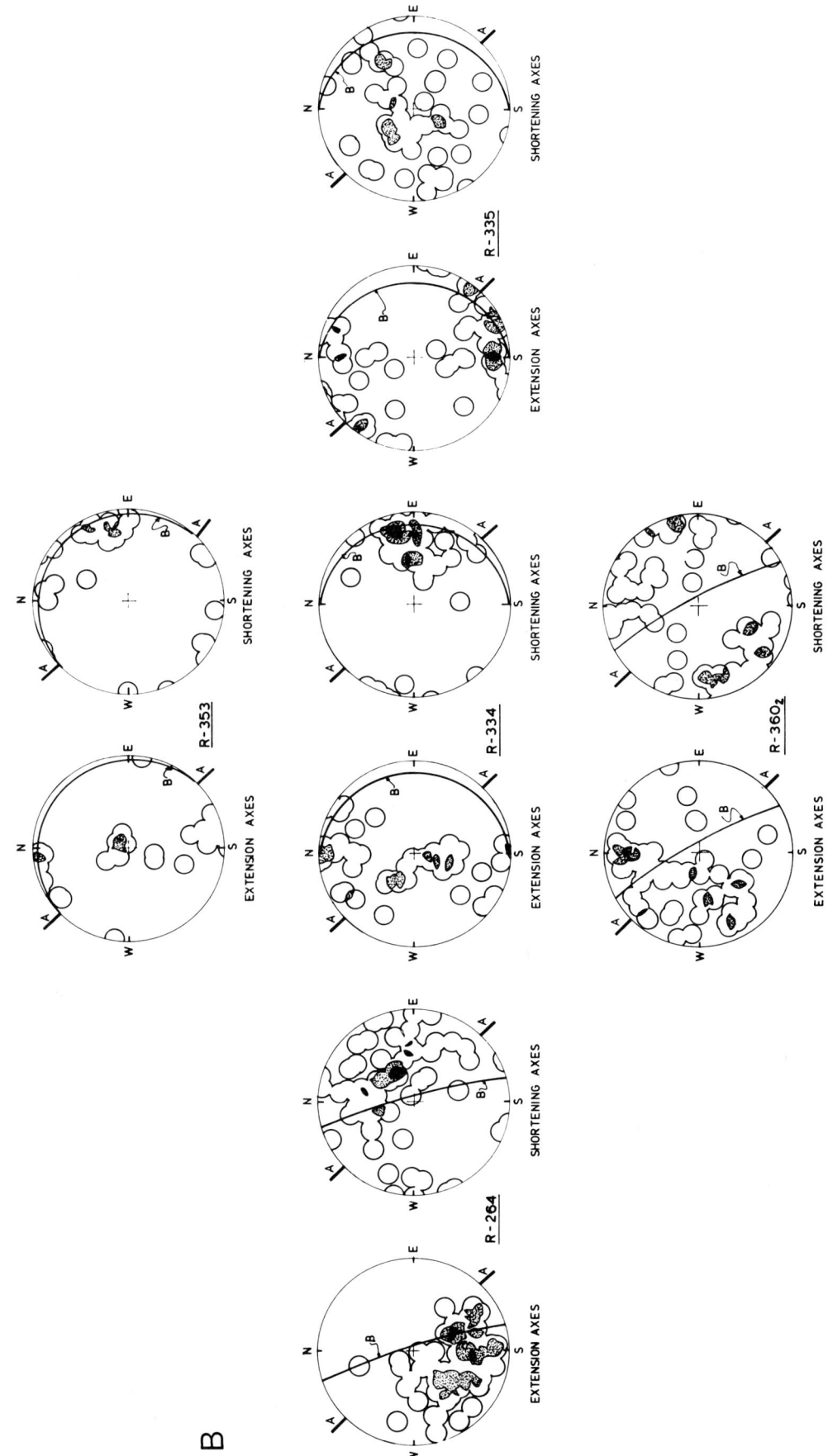

such as minor faults, kink bands and folds, variations of thicknesses of rock units, and petrofabrics of carbonate rocks, yield a coherent pattern through all scales of deformation, which can be summarized as follows:

1. Subhorizontal compression prevails in the two blocks bounding the monocline.

2. Layer-parallel compression is dominant in the steep zone of the monocline and in the synclinal bend; however, some deviations from the general result, such as layer-parallel shear (Fig. 18, station 264), the scattered fault pattern (Fig. 13B), and the four sets of small faults in the steep layers (Fig. 11), suggest that this steep zone underwent a complicated strain history. This complicated deformation could have resulted from rotation of layers in a constant strain field (Fig. 15).

3. The direction of maximum compression trends 8°/067°, the average trend of all stress indicators. This direction of maximum compression deviates 25° from the normal to the Palisades fault.

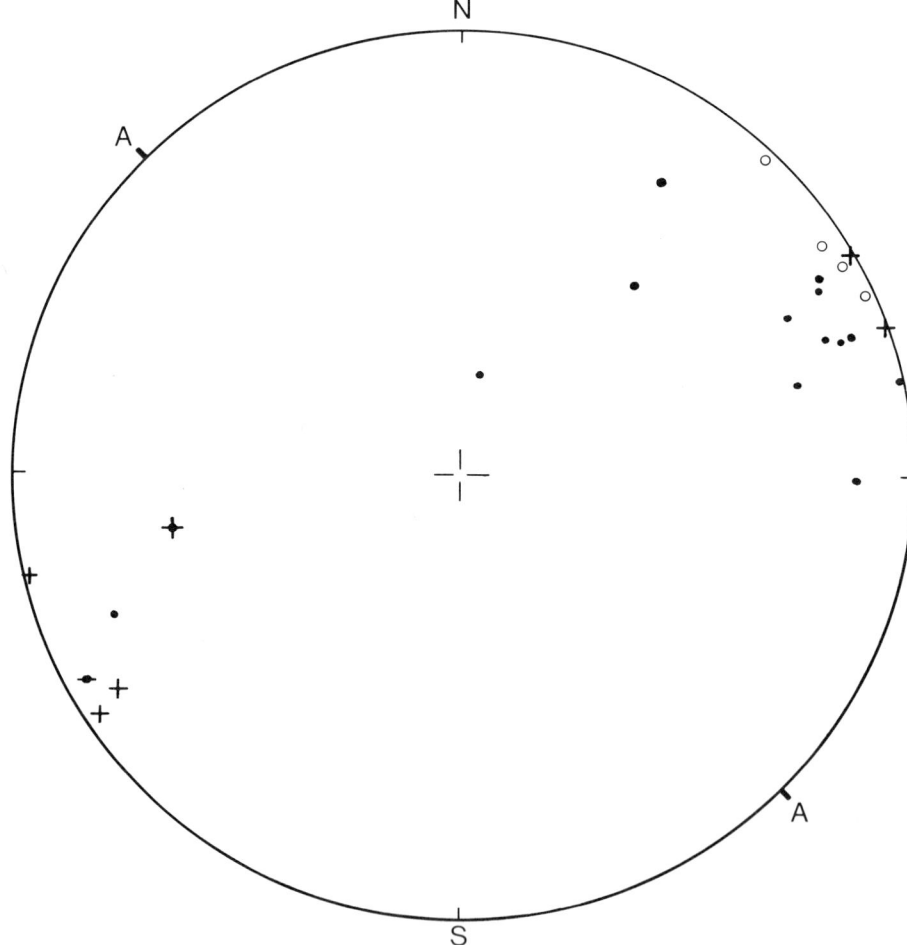

Figure 19. Synoptic diagram shows orientations of maxima of compression axes based on structural analysis of the Palisades monocline. Solid dot = maximum of compression axes determined by a petrofabric sample (summary of Fig. 18); open circle = shortening direction according to small folds; cross = shortening according to faults (summary of Fig. 13); line A-A = Palisades fault. Equal-area lower hemisphere projection.

4. The Palisades fault zone is subvertical; therefore, strain accommodated by displacement along the fault has no horizontal component.

5. The monocline can be divided into three levels. These structural levels differ in intensity of deformation, shape of flexure, and thickening of layers.

6. The normal faulting within the upper member of the Dox Formation (Precambrian age) in the southern block indicates horizontal extension subnormal to the trend of the monocline (Fig. 13D). This deformation probably resulted from Precambrian faulting and is therefore unrelated to the Palisades monocline.

DISCUSSION

The Colorado Plateau is disrupted by many monoclines, some with lengths of hundreds of kilometres and throws of hundreds of metres. Study of the Palisades monocline, a branch of the large East Kaibab monocline, provides several clues as to the origin of these fascinating structures. The Palisades monocline was chosen

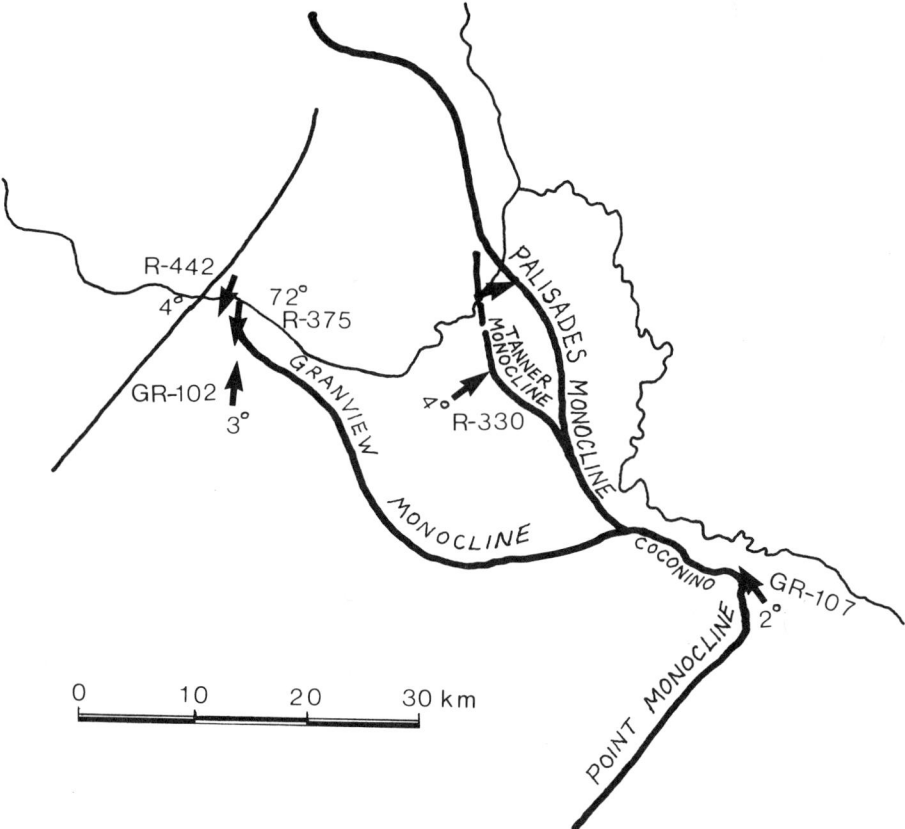

Figure 20. Compression axes in some branches of the East Kaibab monocline deduced from analysis of calcite twinning. Solid arrow indicates plunge of compression axis. Samples GR-102, GR-107, and R-330 were analyzed with the kinematic method (Groshong, 1972, 1974), and their shortening axes are drawn. One section of each sample was analyzed; 44, 47, 20, 38, and 19 twinning sets were measured in samples GR-102, GR-107, R-330, R-375, and R-442, respectively. The heavy arrow in the Palisades monocline area represents the mean shortening axis of all samples (Fig. 19). Samples GR-102, GR-107, and R-375 indicate strain axis, samples R-330 and R-442 indicate stress axis.

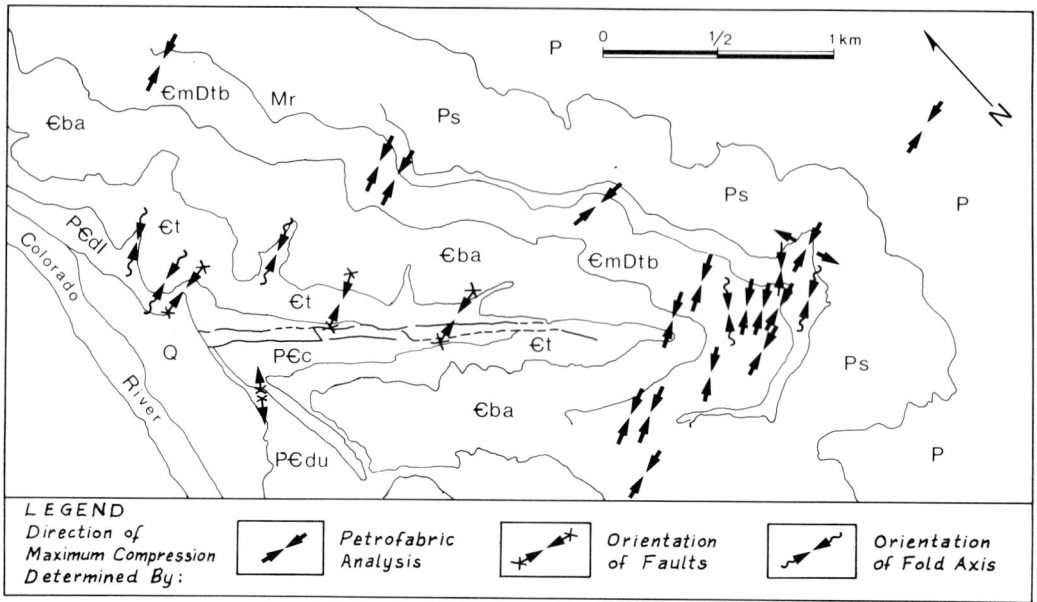

Figure 21. Directions of maximum compression in Palisades monocline (map view). Compression axes deduced from all structural elements are shown.

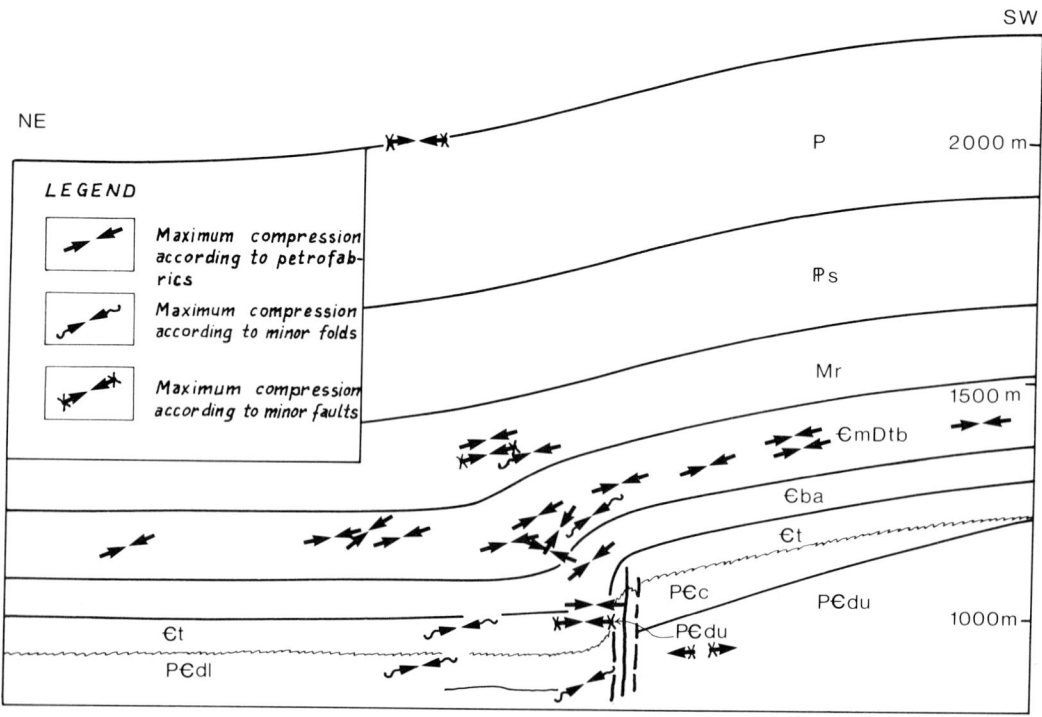

Figure 22. Directions of maximum compression in Palisades monocline (section view). Compression axes of all structural elements are shown.

for this study because of its three-dimensional exposure in the eastern Grand Canyon, its association with a basement fault, and its small size. The Palisades monocline is neither the ideal monocline that all monoclines should resemble, nor is it a special and unusual monocline. It is, rather, a monocline in which the shape, tectonic history, and internal strain have been documented relatively thoroughly. Therefore, any general theory of monocline formation must explain the features of the Palisades monocline. A general theory is studied in Part II (this volume).

In the following pages, the main results of the study of the Palisades monocline will be discussed and compared with results of field studies of other monoclines. I will discuss neither complex monoclines that cannot be compared with this monocline nor interpretative cross sections of monoclines that cannot be compared with the accurate cross section constructed during the course of my study.

Monocline as a Fold

The Palisades monocline is a double flexure, with an anticlinal bend and a synclinal bend, that connects layers at one level with the same layers at another level (Fig. 5). Throughout most of the section, the two flexures are continuous, merge gradually into the bounding blocks, and join each other in an inflection zone in the steepest part of the monocline. The shape of the flexures of the Palisades monocline varies according to both structural level and lithology.

Many investigators have described monoclines as continuous double flexures (Powell, 1873; Gilbert, 1875; Baker, 1935; Babenroth and Strahler, 1945; Collomb and Donzeau, 1974; Huntoon, 1974; Barnes, 1974); however, they have made no attempt to classify and analyze the geometry of a monocline as a fold. Kelley (1955, Fig. 6) presented a wide variety of profiles of monoclines, most of which are continuous flexed lines. On the other hand, Stearns (1971) indicated that the Rattlesnake monocline is composed of five relatively undeformed blocks separated by narrow hinge zones. He also suggested that fewer than five blocks are developed in most monoclines in Wyoming.

Recognition that the geometry of monoclines in general, and of the Palisades monocline in particular, is a continuous flexure probably is fundamental to the understanding and to the theoretical analysis of these structures. This is not to say that there can be no faulting within a monocline; indeed, it is well known that there is. Rather, this is to say that the main mode of deformation in *simple* monoclines is one of flexuring rather than of faulting.

Layer-Parallel Horizontal Compression

The study of a wide range of structural features and strain indicators in the Palisades monocline shows that subhorizontal compression in direction $8°/067°$ prevailed at the time of flexuring. This conclusion was deduced from petrofabric analysis of carbonate rocks and structural analysis of small faults, small folds, and kink bands. The directions of maximum compression deduced from all types of data are consistent (Figs. 21, 22). Moreover, the petrofabric analysis of carbonate rocks from other monoclines of the East Kaibab system reveals the existence of horizontal compression elsewhere (Fig. 20).

Evidence of horizontal or layer-parallel compression, such as small folds and faults, has been noted by investigators of the East Kaibab and of other monoclines. Walcott (1890) deduced significant shortening across the East Kaibab monocline from the profile of the monocline in the eastern Grand Canyon. Wells (1960) documented many small folds, small faults, and joints in the East Kaibab monocline in House Rock Valley, Arizona, about 80 km north of the Palisades monocline. He noted that axes of most small folds are subparallel to the monocline and that the folds are on the upthrown side of the monocline as well as in the flat area of the downthrown side. Wells also reported that there are asymmetric folds with their steep limb facing the major monocline in the downthrown side of that monocline. Such folds resemble the kink bands and other assymmetric folds in the northern downthrown block of the Palisades monocline (Fig. 16).

In a review of the monoclines of the Colorado Plateau, Kelley (1955) listed the following evidences of layer-parallel compression in monoclines: small *en echelon* folds above the anticlinal bend and below the synclinal bend, *en echelon* folds in the flat layers in uplifted blocks close to the monoclines, and small thrust faults. Small structures and thickness variations that indicate subhorizontal shortening across the axes of monoclines were reported by many other investigators (Van Gundy, 1946; Rohr and others, 1961; Prucha and others, 1965; Collomb and Donzeau, 1974; Cook and Stearns, 1975).

Local thinning of layers, which indicates layer-parallel extension, has been found in parts of several monoclines. For example, the Cambrian units in the Rattlesnake monocline of Wyoming are completely attenuated close to the fault in the basement, whereas the overlying units maintain constant thickness across the structure (Stearns, 1971). Slight thinning of the Bright Angel Shale was found in the anticlinal zone of the Palisades monocline immediately above the termination of the Palisades fault. It should be noted that thinning of rock units within simple monoclines occurs mainly in the immediate vicinity of the major faults associated with the monoclines.

Kelley summarized the evidence of shortening in monoclines (1955, p. 798): "There are several features about the total form of the monoclines, their general relationship to one another, uplifts, and certain thrusts which point to a complex of deforming actions involving *tangential compression* in most if not all of the monoclines" (italics mine).

This rather abundant evidence for compression during monocline formation is quite inconsistent with the simple passive folding model in which the overburden is "draped" over a displaced fault in the basement, as shown by the theoretical analysis by Sanford (1959). According to Sanford's analysis, there should be extension in the area of the anticlinal hinge of a monocline, whereas evidence for compression there is widespread.

Recognition of evidence of layer-parallel compression of rocks within many monoclines probably is critical to analyses of processes of formation of monoclines. Flexuring of multilayered rocks under conditions of layer-parallel compression will generally involve growth in amplitude by folding or buckling of the flexure as a result of the shortening. Therefore, any analysis of monocline formation that ignores the tendency for layered sedimentary rocks to fold or buckle is incomplete.

Major Fault in a Monocline

Probably the fundamental problem of the formation of monoclines is the relationship between the major fault below and the flexure above. We will discuss four aspects of the fault-fold relationship in simple monoclines which can be documented in the field.

Association of Vertical Fault and Horizontal Compression. The Palisades fault is vertical and has no significant component of horizontal displacement. Vertical displacement along the fault corresponds to an axis of maximum shortening trending N45°E and plunging about 45° southwestward. On the other hand, the analyses of strain and stress within and near the monocline indicate a subhorizontal axis of maximum compression trending N70°E (Fig. 21). These two results are incompatible with a single regional strain. The axis of horizontal compression trends obliquely to the fault, about 70°, and could cause some strike slip along the fault, but not vertical slip. On the other hand, this compression apparently was not large enough to lock the fault.

One can imagine a combination of deformations in which the Palisades Creek area was subjected to simultaneous horizontal compression and differential vertical uplift. The Palisades fault zone is a line of brecciated, weak gouge which would yield relatively easily in response to deep-seated vertical movement. The sedimentary layers above the fault deformed by a combination of the horizontal compression and the shear induced by movement on the fault to produce the internal strain found in them. Thus, one can imagine that the vertical displacement and the horizontal compression were results of independent processes. The source and implications of these regional tectonic processes are beyond the scope of my present study.

The study of the Palisades monocline shows that internal strain and type of major fault must be investigated separately, as they may be independent. If one suggested that the Palisades fault is a reverse fault in order to explain the subhorizontal compression, one would be wrong, because the fault is vertical. If, on the other hand, one suggested that the layers did not undergo horizontal compression because the Palisades fault is vertical, one would be wrong again.

Type of Fault. Monoclines have developed over vertical faults, including the Palisades fault (Walcott, 1890), over normal faults (Stearns, 1971), and over reverse faults (Walcott, 1890; Huntoon, 1974).

Baker (1935) postulated that most monoclines are underlain by reverse faults, to explain the lack of extension features in the monocline and the regional distribution of monoclines in the Colorado Plateau. However, reverse displacements along the major fault of a simple monocline are scarce and usually occur along high-angle faults, which would induce only minor shortening (Walcott, 1890; Prucha and others, 1965). It was shown that the attitude of the Palisades fault could not have been deduced from detailed study of the monocline above and, following Kelley (1955, p. 798), I conclude that the type of fault under a monocline cannot be deduced from the displacement pattern or the strain within a monocline.

Location of the Fault. The Palisades fault underlies the anticlinal bend of the flexure (Fig. 5). It propagated upward from the Dox Formation into the Tapeats Sandstone. A tight anticlinal bend with a radius of curvature of about 10 m replaces the fault above its termination (Fig. 11).

The fault associated with the Cremation monocline is in a different location. The Cremation monocline is another branch of the East Kaibab monocline in Grand Canyon National Park. There, the Tapeats Sandstone lies in part over the Precambrian crystalline basement. The vertical Cremation fault transects Paleozoic rocks in the *uplifted block* about 35 m southwest of the anticlinal bend, so the monocline is not directly above the fault.

The major fault underlying other monoclines was mapped in some localities. In the Rattlesnake monocline, Wyoming, Stearns (1971) showed that the fault in the basement is under the anticlinal bend, so it is similar to the Palisades monocline. Huntoon (1974), Prucha and others (1965), and deSitter (1962), showed the fault cutting through the inflection zone of the monocline. These investigators considered the monoclines to be "drag folds" developed along the major fault.

Probably there is no universal relation between the location of the fault and the location of the overlying monocline; it can be under the anticlinal bend, in the inflection zone, or even outside the monocline. It seems that the only way to locate the major fault under the monocline is to map it or to delineate it with geophysical field methods.

Pre-existence of Faults. Investigators of monoclines in the Grand Canyon area have demonstrated that the major fault zones are Precambrian in age (Walcott, 1890; Huntoon, 1974).

Most large monoclines in the Colorado Plateau probably are associated with major faults in the basement rocks. The faults are evident in the monoclines that are deeply eroded, and they can be inferred from the concentration of monoclines along margins of uplifts and basins. The pre-existing faults are probably weak zones in the basement of the Colorado Plateau; therefore, they tend to deform more easily than the intact blocks between them. At the beginning of the Laramide orogeny, Precambrian faults were reactivated, not because they were in the ideal orientation relative to the Laramide stress field within the overburden rocks, but because they were weaker than the continuous blocks. Thus, the faults presumably localized uplift, producing zones of inclined layering in the overburden which were amplified by the Laramide stress field.

Prediction of the Profile of a Monocline at Depth. Many monoclines are poorly exposed at depth, so it would be useful if we could derive the subsurface profiles of monoclines from surface geology. The detailed mapping of the Palisades monocline provides a good opportunity to examine the relationship between surface and subsurface geology in a monocline.

At the Plateau level, the Kaibab Limestone is flexed into an open gentle monocline with a maximum dip of 18°. The radius of curvature of the anticlinal and synclinal bends at this level is on the order of 4 to 5 km. The isogon pattern of the Palisades monocline (Fig. 9) shows that the open flexure maintains its general shape to a depth of 500 m below the Plateau surface. A similar open flexure probably extended upward into the Mesozoic rocks.

However, it seems impossible to deduce the structure of the deeper parts of the monocline from the structure of the Kaibab Limestone because of the existence of three structural levels in the Palisades monocline with transition zones between them. The three levels are: the lower, tight flexure adjacent to the Palisades fault; the central, tight, continuous monocline; and the upper, open, gentle monocline (Fig. 5, Table 1). The development of these three distinct zones presumably is not general to monoclines; rather, it is a result of pre-existing structural and lithological features. In the lower level, the slip along the Precambrian fault, coupled with the horizontal compression, probably controlled the shape of the tight flexure. The central level terminates at the Redwall Limestone. The base of the Redwall Limestone parallels the steep layers of the Temple Butte Limestone whereas the top of the Redwall dips gently at angles of up to 20°. The Redwall Limestone is 180 m thick and is composed of massive limestone, dolomite, and cherty layers. It lacks the thin layering of the unit above and below it. It deformed mainly by slip along small faults and fractures, rather than by buckling and bending. The slip along the discontinuities caused pseudo-continuous flow in which the Redwall thickened in the synclinal zone (Fig. 9). Therefore, the Redwall breaks the structural continuity of the Palisades monocline so that the structure below cannot be deduced from the surface geology above.

Changes of profiles of monoclines have been observed in other areas, particularly in deeply exposed monoclines (Walcott, 1890; Powell, 1873; Stearns, 1971; Prucha and others, 1965). Stearns (1971) suggested that the Heart Mountain detachment separated two structural levels in the Rattlesnake monocline, Wyoming. The lower level in that monocline is composed of attenuated, soft shaley units, and the upper level is composed of competent, calcareous rocks with constant thickness.

The study of the Palisades monocline enables us to provide some guidelines to the analysis of the subsurface geology of simple monoclines. The transition between structural levels probably is controlled by mechanical-lithological boundaries such as the Redwall above the layered limestone in the Palisades monocline or the massive carbonate rocks above the soft shale layers in the Rattlesnake monocline, Wyoming (Stearns, 1971). Therefore, thick, competent units should be suspected as loci of structural transitions, and the form of the flexure should be projected cautiously across them.

Another prediction usually made about monoclines is the attitude of the major fault. Kelley (1955, p. 798) stated: "It is not possible to determine from the surface or from the structural profile whether a monocline formed above (1) a high angle normal fault, (2) a vertical fault, (3) a high angle reverse fault, or (4) a deep seated buckle." The study of the Palisades monocline shows that Kelley's statement is probably correct—the type of fault at depth cannot be deduced simply from the shape of the monocline. It is suggested here that the observations in a monocline can be separated into two data sets: the first is the gross shape of the structure as presented in a profile; the second is the internal strain of the layers as obtained from analysis of small structures.

ACKNOWLEDGMENTS

I am especially indebted to Arvid Johnson of Stanford University for advice and helpful discussion since the beginning of this study. He also provided invaluable comments on earlier versions of this paper. Rick Groshong of Cities Service Oil Company kindly permitted the publication of some of his strain measurements. I thank Peter Huntoon of the University of Wyoming for suggesting Palisades Creek for my study.

My work benefited greatly from stimulating discussions with the following people: Atilla Aydin, Robert Compton, Paul Delaney, Ray Fletcher, Amos Nur, George Thompson, and Phil Salvador, all of Stanford University, Charles Barnes of the University of Northern Arizona, Nurit Hildebrand of the University of California, Los Angeles, Ivar Ramberg of the University of Oslo, and Rick Shimek of San Jose State University. I thank Robert Altenhofen and Richard Lugn of the U.S. Geological Survey for their help in the work on the stereoplotter, Ted Hatch of Hatch River Expeditions and Dave Ocsner of the Grand Canyon National Park for logistic support in the field.

This study was supported in part by Research Grants No. 1813-74 and 1998-75 from the Geological Society of America, by a grant from the Museum of Northern Arizona, and by a Sigma-Xi Grant-in-Aid. This research was mainly supported by the National Science Foundation Grant No. EAR 76-03273.

APPENDIX 1.—MAPPING TECHNIQUES

Knowing the accurate shape of the Palisades monocline is essential for an understanding of this structure. The steep gradient of Palisades Creek (1.2 km elevation difference over a horizontal distance of about 1 km), the inadequacy of the published topographic map (1:62500) for detailed study, and the inaccessibility of some parts of the Palisades Creek Canyon all required special mapping techniques in order to yield an accurate map and cross section. I chose two methods: plane-table mapping at a scale of 1:1200 of some parts of Palisades Creek and topographic, geologic, and structural mapping at a scale of 1:6000 on a Kern PG-2 stereoplotter (owned by the U.S. Geological Survey, Menlo Park, California).

The mapping was done on two pairs of vertical aerial photographs, with 60% overlap and an original scale of 1:24,000, which were especially flown in August of 1975. Specifications, model calculations, and corrections were made according to the standards of the Department of Photogrammetry of the U.S. Geological Survey.

Three maps at a scale of 1:6000 were made: a topographic map with a contour interval of 12.5 m, a geologic map of the Palisades Creek area, and a map of elevations of about 10 points per km along stratigraphic contacts. A cross section was prepared from the last two maps, incorporating results of the plane-table mapping.

There are three main sources for inaccuracies in the maps made on the stereoplotter: imperfect photographs, deviation of the Kern PG-2 stereoplotter, and operator error. The first source affected the northwestern corner of the map, as the northwestern photo pair could not be modeled accurately. An estimate of this inaccuracy is ± 5 m, but it varies from place to place. The second source causes less than one metre in elevation difference and is probably negligible relative to the errors introduced by the operator. I checked the reproducibility of elevation readings and found that in areas with moderate relief the results varied by ± 2 m, whereas in steep zones the results varied by ± 5 m. In very steep areas, the measurements are suspect. Also it was found that north-facing slopes yield more accurate results than south-facing slopes, probably due to enhancement of images of rock boulders on the north-facing slopes.

In spite of these inaccuracies, the map and section of the Palisades monocline are as detailed and as accurate as a map and section of this scale can be in such terrane.

APPENDIX 2. VARIATIONS OF THICKNESSES OF ROCK UNITS

The variation of thickness of rock units may be considered as an indicator of the gross strain of the layers. An attenuated layer, for example, indicates elongation parallel to bedding.

Stratigraphic sections in the Palisades monocline were measured in the field or computed from elevations of layers determined with stereoplotter. The variations of thickness of rock units presented in Table 2 can be attributed to several sources—inaccuracy of the measurements, variations due to sedimentary processes, and variations due to tectonic strain. The latter is of interest to us; however, it is difficult to separate it from the other two sources. The error of measurement was estimated as 8% by comparing thicknesses of a section in the northern block measured by different methods (Table 3). The evaluation of possible thinning or thickening effects by sedimentary processes involves careful stratigraphic analysis which is beyond the scope of my work. McKee and Resser (1945, Fig. 1) showed that regional

TABLE 1. FOLD TYPES AND THICKNESS VARIATIONS IN THREE STRUCTURAL LEVELS OF PALISADES MONOCLINE

Structural level	Rock unit	Synclinal bend		Anticlinal bend	
		Fold type	Relative thickness	Fold type	Relative thickness
Upper	Supai to Kaibab Fm	1c or 2	Slightly thickened*	1c or 2	Slightly thickened*
Middle	Redwall Limestone	3	Thickened	1b	Constant thickness
	Muav and Temple Butte Limestones	2	Thickened	1a or 1b	Constant thickness
	Bright Angel Shale	2	Slightly thickened	1a	Slightly thinned
Lower	Tapeats Sandstone	2	Slightly thickened*	Not folded	..
	Dox Formation	2	Slightly thickened*	Not folded	..

Note: Classified according to scheme of Ramsay (1967, chap. 7).
*Thickness variation probably insignificant; variation less than measurement error.

variations of thickness of Cambrian units are small and, therefore, it is assumed that the depositional thickness of the Paleozoic formations in the small area of the Palisades monocline is constant. However, possible activity of the Palisades fault during Middle Cambrian could have led to local erosion and changes of thickness in the Palisades Creek area. Therefore, it is suggested that any changes of thickness greater than ± 8% are the result of tectonic deformation (Table 1).

TABLE 2. THICKNESS (IN METRES) OF SOME ROCK UNITS IN PALISADES MONOCLINE

Rock unit	\multicolumn{13}{c}{Measured section}
	1	2	3	4	5	6	7	8	9	10	A	B	C	D
Supai Formation	..	..	266	252	256	237	256	..	..	..	..	..	..	..
Redwall Limestone	..	..	183	182	228	169		..	..	..	..	..	..	..
Temple Butte and Muav Limestones							248							
Upper	..	134	..	110	..	..		107	..	..	..	115	..	..
Lower	..		..	..	..	..	43	44	41	..	..	30	..	..
Bright Angel Shale	..	..	..	..	..	..	..	75	66	..	..	81	81	84
Tapeats Sandstone														
Upper	24	..	..	..	..	..	..	..	38	39	21	..	..	..
Lower	71	..	..	..	..	..	..	..	..	54	63	..	..	..

Note: Sections 1 to 10 measured on a stereoplotter, and sections A to D measured in the field. Locations of sections shown in Figure 10.

TABLE 3. ESTIMATION OF ACCURACY OF THICKNESS MEASUREMENTS OF STRATIGRAPHIC SECTIONS

Rock units	Thickness (m)			Mean and standard deviation of thickness	Standard deviation (%)
	McKee and Resser (1945) Close to sections A, B	This study			
		Field measurements sections A, B	Measurements on stereoplotter sections 1, 2		
Temple Butte Limestone					
Muav Limestone					
Kanab member	92	115	134	134 ± 10	8
Gateway member	32	30			
Bright Angel Shale	76	81	85	81 ± 5	6
Tapeats Sandstone					
Upper	27	21	24	24 ± 5	12
Lower	63	63	71	66 ± 5	7

REFERENCES CITED

Babenroth, D. L., and Strahler, A. M., 1945, Geomorphology and structure of the East Kaibab monocline, Arizona and Utah: Geol. Soc. America Bull., v. 56, p. 105–150.

Baker, A. A., 1935, Geological structure of southeastern Utah: Am. Assoc. Petroleum Geologists Bull., v. 19, p. 1472–1507.

Barnes, C. W., 1974, Interference and gravity tectonics in the Gray Mountain area, Arizona, *in* Karlstrom, N. V., and others, eds., Geology of northern Arizona, P. II: Flagstaff, Museum of Northern Arizona, p. 442–453.

Beloussov, V. V., 1962, Basic problems in geotectonics: New York, McGraw-Hill Book Co., 816 p.

Billingsley, G. H., 1974, Mining in Grand Canyon, *in* Breed, W. J., and Roat, E. C., eds., Geology of the Grand Canyon: Flagstaff, Museum of Northern Arizona.

Collomb, P., and Donzeau, M., 1974, Connections between decametric seal kink bands and basement faults in the Variscan aulacogen of Ougarta (western Sahara, Algeria): Tectonophysics, v. 24, p. 213–243.

Cook, R. A., and Stearns, D. W., 1975, Mechanisms of sandstone deformation: A study of the drape-folded Weber Sandstone in Dinosaur National Monument, Colorado and Utah, *in* Balyard, D. W., ed., Deep drilling frontiers: Denver, Rocky Mtn. Assoc. Geology, p. 21–32.

deSitter, L. U., 1962, Structural development of the Arabian Shield in Palestine: Geologie en Mijnbouw, v. 41, p. 116–124.

Dewey, J. F., 1965, Nature and origin of kink bands: Tectonophysics, v. 1, p. 459–494.

Dutton, C. E., 1882, Tertiary history of the Grand Canyon district: U.S. Geol. Survey Mon. 2, 264 p.

Ellen, S. D., 1971, The development of folds in layered chert of the Franciscan Assembage near San Francisco, California [Ph.D. dissert.]: Stanford Univ., 330 p.

Friedman, M., Harding, J., Logan, J. M., Min, K. D., and Stearns, D. W., 1976, Experimental folding of rocks under confining pressure: Pt. III, Faulted drape folds in multilithologic layered specimens: Geol. Soc. America Bull., v. 87, p. 1049–1066.

Gilbert, G. K., 1875, Report on the geology of portions of Nevada, Utah, California, and Arizona, *in* Wheeler, G. M., in charge, Report upon Geographical and Geological Explorations and Surveys west of the 100th Meridian: Washington, D.C., U.S. Govt. Printing Office v. 3, p. 21–63.

Goguel, 1962, Tectonics: San Francisco, W. H. Freeman & Co., 384 p.

Grand Canyon Natural History Association, 1976, Geologic map of the Grand Canyon National Park, Arizona: Grand Canyon Nat. History Assoc. Pub.

Groshong, R. H., Jr., 1972, Strain calculated from twinning in calcite: Geol. Soc. America Bull., v. 83, p. 2025–2038.

———1974, Experimental test of least-square strain gage calculation using twinned calcite: Geol. Soc. America Bull., v. 85, p. 1855–1864.

Hintze, L. F., 1973, Geologic history of Utah: Brigham Young Univ. Geology Studies, v. 20, pt. 3, p. 126.

Honea, E., and Johnson, A. M., 1976, Development of sinusoidal and kink folds in multilayers confined by rigid boundaries: Tectonophysics, v. 30, p. 197–239.

Huntoon, P. W., 1974, Synopsis of the Laramide and post-Laramide structural geology of the eastern Grand Canyon, Arizona, *in* Karlstrom, N. V., and others, eds., Geology of northern Arizona, Pt. I: Flagstaff, Museum of Northern Arizona, p. 317–335.

Johnson, A. M., 1977, Styles of folding: Elsevier Sci. Publishing Co., New York, 406 p.

Johnson, A. M., and Ellen, S. D., 1974, A theory of concentric, kink, and sinusoidal folding and of monoclinal flexuring of compressible, elastic multilayers, Pt. I, Introduction: Tectonophysics, v. 21, p. 301–339.

Johnson, A. M., and Pollard, D. D., 1971, Mechanics of intrusion of laccoliths: Final rept. to Natl. Sci. Foundation, Stanford, California, Branner Library, Stanford Univ., 173 p.

———1973, Mechanics of growth of some laccolithic intrusions in the Henry Mountains, Utah, Pt. I, Field observations, Gilbert's model, physical properties and flow of the magma: Tectonophysics, v. 18, p. 261–309.

Kelley, V. C., 1955, Monoclines of the Colorado Plateau: Geol. Soc. America Bull., v. 66, p. 789–804.

Logan, J. M., Friedman, M., and Stearns, M. T., 1978, Experimental folding of rocks under confining pressure: Pt. VI, Further studies of faulted drape folds, in Matthews, V., ed., Folding associated with the basement faulting in the Rocky Mountain region: Geol. Soc. America Mem. 151 (this volume).

Lucchitta, I., 1974, Structural evolution of northwest Arizona and its relation to adjacent Basin and Range province structures, in Karlstrom, N. V., and others, eds., Geology of northern Arizona, Pt. I: Flagstaff, Museum of Northern Arizona, p. 336–354.

Marshal, D. R., 1972, Gravity gliding at Gray Mountain, Coconino County, Arizona [M.S. thesis]: Flagstaff, Northern Arizona Univ., 83 p.

McKee, E. D., and Resser, C. E., 1945, Cambrian history of the Grand Canyon region: Carnegie Inst. Washington Pub. 563, 232 p.

Norris, D. K., and Barron, K., 1968, Structural analysis of features on natural and artificial faults, in Baer, A. J., and Norris, D. K., eds., Research in tectonics: Canada Geol. Survey Paper 68-52, p. 136–167.

Oertel, G., 1965, The mechanism of faulting in clay experiments: Tectonophysics, v. 2, p. 343–393.

Powell, J. W., 1873, Exploration of the Colorado River of the West and its tributaries explored in 1869–1872: Smithsonian Inst. Pub., 291 p.

Prucha, J. J., Graham, J. A., and Nickelson, R. P., 1965, Basement-controlled deformation in Wyoming province of Rocky Mountains foreland: Am. Assoc. Petroleum Geologists Bull., v. 49, p. 966–992.

Ramberg, I. B., and Johnson, A. M., 1976, A theory of concentric, kink, and sinusoidal folding and of monoclinal flexuring of compressible, elastic multilayers. Pt. VI, Asymmetric folding and unidirectional kinking: Tectonophysics, v. 35, p. 295–335.

Ramsay, J. G., 1967, Folding and fracturing of rocks: New York, McGraw-Hill Book Co., 568 p.

Reches, Z., 1976, Analysis of joints in two monoclines in Israel: Geol. Soc. America Bull., v. 87, p. 1654–1662.

Reches, Z., and Johnson, A. M., 1976, A theory of concentric, kink, and sinusoidal folding and of monoclinal flexuring of compressible, elastic multilayers, Pt. VI, Asymmetric folding and unidirectional kinking: Tectonophysics, v. 35, p. 295–335.

——1978, The development of monoclines: Pt. II, Theoretical analysis of monoclinal flexures, in Matthews, V., ed., Folding associated with the basement faulting in the Rocky Mountain region: Geol. Soc. America Mem. 151 (this volume).

Rohr, G. M., Rodenberg, K., and Tri-ger, J., 1961, Structural analysis of the Shell Canyon area, Bighorn Mountains, Wyoming: Iowa Acad. Sci. Proc., v. 71, p. 203–233.

Sanford, A. R., 1959, Analytical and experimental study of simple geologic structures: Geol. Soc. America Bull., v. 70, p. 19–52.

Stearns, D. W., 1971, Mechanisms of drape folding in the Wyoming province, in Wyoming Geol. Assoc. Guidebook 23rd Ann. Field Conf., Wyoming Tectonics Symposium: p. 125–143.

Turner, F. J., and Weiss, L. E., 1962, Structural analysis of metamorphic tectonics: New York, McGraw-Hill Book Co., 545 p.

Van Gundy, C. E., 1946, Faulting in the east part of Grand Canyon of Arizona: Am. Assoc. Petroleum Geologists Bull., v. 33, p. 1899–1909.

Walcott, C. D., 1890, Study of line of displacement in the Grand Canyon of the Colorado in northern Arizona: Geol. Soc. America Bull., v. 1, p. 49–64.

Weiss, L. E., 1968, Flexural slip folding of foliated model materials, in Baer, A. J., and Norris, D. K., eds., Research in tectonics: Canada Geol. Survey Paper 68-52, p. 294–358.

Wells, J. D., 1960, Stratigraphy and structure of the House Rock Valley area, Coconino County, Arizona: U.S. Geol. Survey Bull., 1081-D, p. 117–158.

MANUSCRIPT RECEIVED BY THE SOCIETY JUNE 27, 1977
MANUSCRIPT ACCEPTED AUGUST 25, 1977

Printed in U.S.A.

Geological Society of America
Memoir 151

Development of monoclines: Part II. Theoretical analysis of monoclines

ZE'EV RECHES
Weizmann Institute
Rehovot, Israel

ARVID M. JOHNSON
Department of Geology
University of Cincinnati
Cincinnati, Ohio 45221

ABSTRACT

Monoclinal flexures, which are isolated asymmetric flexures, range in scale from a few millimetres in kink bands to hundreds of metres in monoclines on the Colorado Plateau. A general model of monoclinal flexuring of multilayers is proposed here; the multilayers include layers with various rheologies, densities, thicknesses, and strengths of contacts between layers. The multilayers are subjected to displacements at their base, stresses at their edges, and a free surface at their tops. We study in detail three modes of this general model, assuming linear, incompressible elastic or viscous multilayers: *Drape folding*, in which a monoclinal flexure develops over a vertical fault; *buckling*, in which an initial monoclinal flexure is amplified by layer-parallel compression; and *kinking*, in which monoclinal kink bands develop unstably by compression inclined to the layering. Selected solutions are presented for the first two modes, and previous research is summarized for the kinking mode.

According to analyses of the three special cases of the general model, the profile of the monoclinal flexure, the displacement field, and the strain distribution within the flexure are useful criteria for distinguishing among the three modes of monoclinal flexuring. The Palisades monocline, described in detail in Part I (this volume), is interpreted to be a result of a combination of drape folding over a fault in Precambrian basement rocks and buckling, which together appear to account for most of the field observations. The Yampa monocline in Dinosaur National Park changes form along its length, but each form can be compared with characteristics of a combination of modes, including faulting at depth and layer-parallel compression. In some places it closely resembles a large kink band.

INTRODUCTION

Monoclinal flexures, which are isolated asymmetric flexures, range widely in scale and in tectonic setting. The smallest monoclinal flexures are kink bands, typically a few millimetres to a few hundred metres wide (Fig. 1A), which apparently form in response to a combination of layer-parallel shortening and shear in certain foliated materials; they are a buckling and shearing phenomenon (Reches and Johnson, 1976). Monoclinal flexures have developed at edges of several laccolithic intrusions also (Johnson and Pollard, 1971; Johnson and Ellen, 1974). Figure 1B shows a monoclinal flexure, with a width of about 100 m, at one edge of a laccolithic intrusion in the Henry Mountains of Utah. This flexure presumably formed in response to uplift of the sedimentary rocks by pressure in the magma of the intrusion. The largest of the monoclinal flexures are *monoclines* themselves, which occur in the Rocky Mountains, on the margin of the Arabo-Nubian massif, on the Colorado Plateau, and in other areas of the world. For example, Figure 1C shows a structural cross section of the East Kaibab monocline in Palisades Creek, Grand Canyon National Park (Reches, this volume). Although this paper is primarily concerned with mechanical analysis of large monoclines such as the East Kaibab, a complete mechanical analysis of monoclinal flexures must be capable of explaining other monoclinal flexures as well.

Monoclines normally are explained as drape folds, so that the flexure in the sedimentary sequence is assumed to have been the passive response of the sequence to displacement along a fault at its base (Prucha and others, 1965; Stearns, 1971). A necessary conclusion from this concept of drape folding is that the layers should be extended in the anticlinal part of the monocline if the dip and displacement of the fault below are vertical (Sanford, 1959). However, several investigators have noted evidence for significant shortening of layers within monoclines (Walcott, 1890; Baker, 1935; Kelley, 1955; Reches, this volume). Kelley (1955) examined most of the monoclines on the Colorado Plateau and reported widespread evidence of layer-parallel shortening. Detailed study of the East Kaibab monocline in Palisades Creek (Reches, this volume), indicates abundant evidence for layer-parallel shortening within and near the monocline. It is well known that layered sequences of rock subjected to layer-parallel shortening tend to buckle, so it is possible that many monoclines have developed partly by buckling, not merely by passive drape folding in response to faulting below.

The first part of this series (Reches, this volume) outlined an approach to the study of monoclines and presented the results of detailed field study. Theoretical analysis was avoided in the first part, and discussion was limited to interpretation of observations. Now we shall primarily discuss theories of monoclinal flexuring.

The purpose of this paper is to explore a relatively general theoretical model of monocline formation in multilayers, including effects of properties of contacts between layers, of dimensional and rheological properties of the layers, and of layer-parallel shortening and faulting displacement at the base of the multilayer. First, we shall define and describe the general model and define three special models that we have chosen to analyze. Then, we shall present and discuss the most important results of theoretical analyses of these three models; details of the theoretical analyses are presented in appendixes. And finally, we shall discuss the limitations of the theory and apply relevant results to several field and experimental examples.

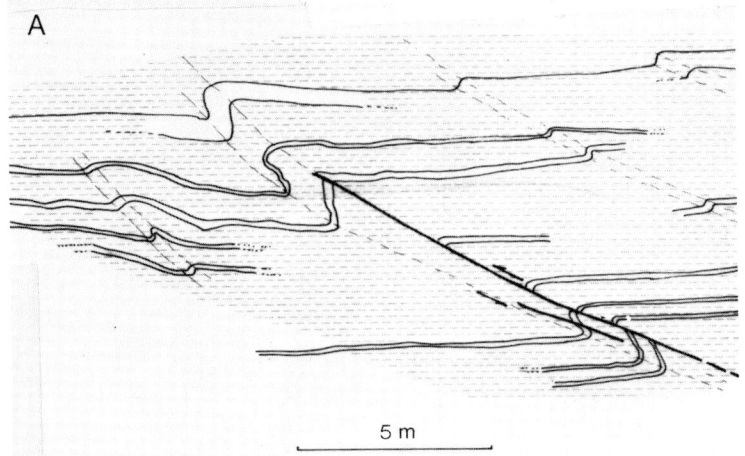

Figure 1. Three examples of monoclinal flexures. (A) Monoclinal kink bands in a sequence of shales and limestones, near Cody, Wyoming. (B) A monoclinal flexure above the periphery of a laccolithic intrusion, Henry Mountains, Utah (after Johnson and Pollard, 1971). (C) Vertical cross section of the Palisades monocline, Grand Canyon, Arizona (after Reches, this volume).

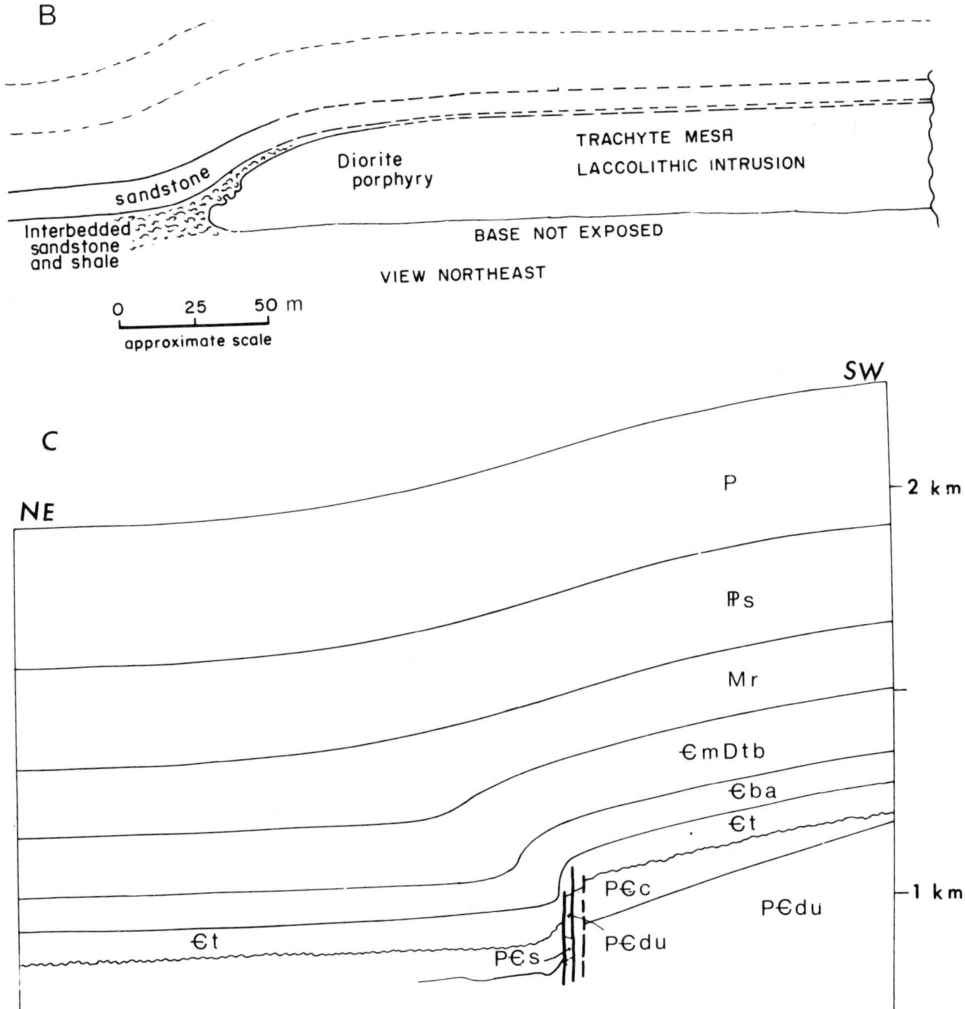

GENERAL MODEL OF MONOCLINAL FLEXURING

The occurrence of monoclinal flexures in a wide variety of rock types and tectonic settings and over a wide range of scales (Fig. 1) suggests that there are several mechanisms that can be responsible for the formation of monoclines. We would suggest a general model of monoclinal flexuring that incorporates all known mechanisms as well as a variety of material properties and boundary conditions (Fig. 2). The general model includes a multilayer with a free surface above and with a substratum below, subjected to various boundary conditions. The upper surface and the contact between the multilayer and the substratum might be planar or irregular. Displacement (U and V, Fig. 2) boundary conditions may be specified at the interface between the substratum and the multilayer. Also, the multilayer may be subjected to layer-parallel compression or extension (S_{xx}, Fig. 2). Further, the multilayer itself may have a wide range of dimensional and rheological properties. The layers might be elastic, viscous, or power-law materials. The contacts between layers might be frictionless, allowing free slip, or bonded, allowing no slip, or they might be described in terms of a yield strength, so that slippage between layers is possible if the shear stress equals some critical value. Finally, the multilayer might consist of interbedded stiff and soft layers with various thicknesses, or might even consist of a single layer. This is our general model of monocline formation. The theoretical problem is to determine conditions under which ideal monoclinal flexures might develop in such a model.

The general model of monocline formation is much too complicated to analyze completely; the number of combinations of variables is overwhelming. However, based on field observation, experimentation, and preliminary theoretical analysis, we can eliminate many combinations of variables. Thus, we know that layer-parallel extension tends to retard buckling, so we can specify that the layer-parallel stress, S_{xx}, is either zero or compressive. Similarly, we know that irregular interfaces of layers tend to induce irregular fold patterns, so we can simplify the analysis by assuming that the interfaces originally are planar or have certain simple forms, without the risk of losing fundamental insights into processes of monocline formation. The remaining variables in our general model are numerous, and in order to restrict them somewhat, we shall make rather arbitrary choices. Thus, we shall consider

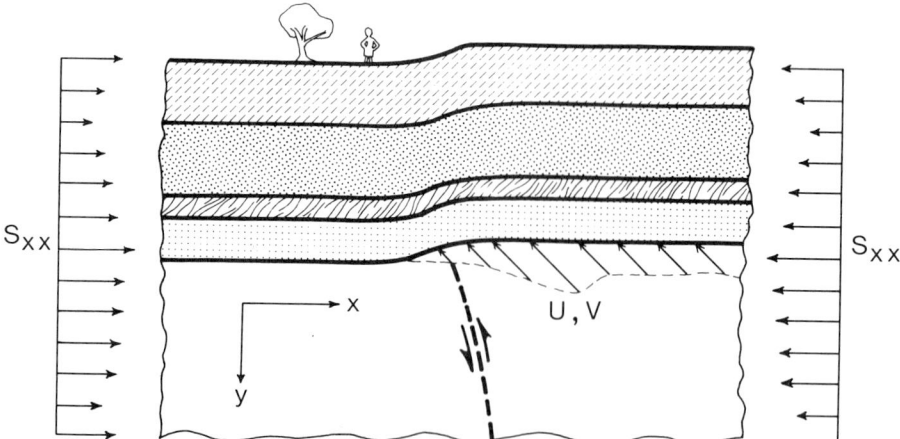

Figure 2. Schematic presentation of the general model of monoclinal flexuring. U and V are displacements of the base of the multilayer and S_{xx} is the horizontal stress. The layers may have various densities, rheologies, and thicknesses.

TABLE 1. SUMMARY OF ESSENTIAL FEATURES OF THE GENERAL MODEL OF
MONOCLINES AND OF THE THREE SPECIAL CASES

General model

Multilayer comprised of layers with elastic or viscous and linear or nonlinear rheologic properties. Finite contact strength. Finite strain. Includes effects of gravity. Displacement and stress boundary conditions.

Special models

Drape folding	Buckling	Kinking
Linear elastic or viscous layers. Infinite or zero contact strength. Vertical displacement at base. Infinitesimal plane strain.	Linear elastic or viscous layers. Infinite or zero contact strength. Layer-parallel shortening. Infinitesimal plane strain.	Elastic, or elastic-plastic layers. Finite contact strength. Layer-parallel shortening and shear. Finite plane strain. Large slopes.

multilayers with bonded or frictionless contacts as well as multilayers which have finite contact strengths. These three choices cover a wide range of properties of contacts. Finally, we shall assume that the multilayer is comprised of linear elastic or viscous layers. We shall not specifically study effects of compressibility, of nonlinear elastic, or of power-law behaviors here, but we know from other analyses of folding (Fletcher, 1974; Johnson, 1977, chap. 10) that general conclusions derived from one of these behaviors are valid for all the materials. Thus, we know that first-order analysis of nonlinear materials would not contribute fundamentally to our understanding of conditions of monoclinal flexuring (Johnson, 1977, chap. 10).

These considerations allow us to restrict our analysis to three special cases of our general model of monocline formation, and we have chosen to analyze these three models in some detail here (Table 1). We shall call them "drape folding," "buckling," and "kinking."

Drape Folding

The idea that monoclines are a response of a sedimentary sequence to faulting at its base was suggested by early investigators of monoclines, including Powell (1873) and Dutton (1882). This idea of drape folding of a multilayer is so simple and sound that it has been adopted by many geologists. Several theoretical analyses of drape folding have been published (Sanford, 1959; Howard, 1966; Min, 1974; Zanemonets and others, 1976).

According to the model of drape folding, the multilayer is deformed by displacement along its base (displacements U and V, Fig. 2). The multilayer might respond passively to these boundary displacements, where we use the term "passive" to imply that no mechanical instabilities are induced in multilayer, and in general the multilayer would deform much as a single layer, as analyzed by Sanford (1959). Several investigators have suggested that drape folding is a result of the contrast in mechanical properties between the faulted and stiff basement below and the layered and soft sedimentary cover above (Prucha and others, 1965; Stearns, 1971). However, our field observations (Fig. 1C), as well as our preliminary theoretical analysis of drape folding, indicate that the mechanism of deformation of the sedimentary cover is virtually independent of the source of displacement at the base. For example, a vertical fault in sedimentary rocks (Fig. 1C), a laccolithic intrusion of viscous magma (Fig. 1B), and a flexure of the basement surface would cause similar deformation in the sedimentary rocks above. Therefore, the development of monoclinal flexures by drape folding is not restricted to regions in which soft sediments are underlain by stiff basement.

It has been suggested that monoclinal flexures can form over both normal and reverse faults (Walcott, 1890; Baker, 1935; Stearns, 1971). However, analysis of drape folding should be restricted to boundary displacements normal (V, Fig. 2) to layering, such as those of a vertical fault, because displacements parallel (U, Fig. 2) to layering will cause the multilayer to shorten or lengthen. Any analysis that includes layer-parallel boundary displacements and ignores the layer-parallel deformation is incomplete. For example, layer-parallel shortening can lead to buckling.

We shall study the deformation of a horizontal multilayer due to displacement along a vertical fault at its base (Fig. 3B). The multilayer would accommodate the displacement of the base by draping over the fault to form a continuous flexure, by yielding, by faulting into several blocks, or by a combination of draping and faulting. We will analyze the draping mode only, because the faulting mode is beyond the limits of current theory.

Buckling

Walcott (1890) studied the East Kaibab monocline in the Grand Canyon area and suggested, apparently for the first time, that buckling processes are important in the development of monoclines. Other investigators (Baker, 1935; Kelley, 1955) recognized evidence of layer-parallel shortening within many monoclines, and they suggested that the monoclines are merely drape folds over reverse faults. They ignored possible effects of layer-parallel shortening on growth of the monocline itself; they imagined that layer-parallel shortening was required merely to account for compressional features such as small folds and reverse faults within the monoclines.

However, according to the general model of monoclinal flexuring, a possible

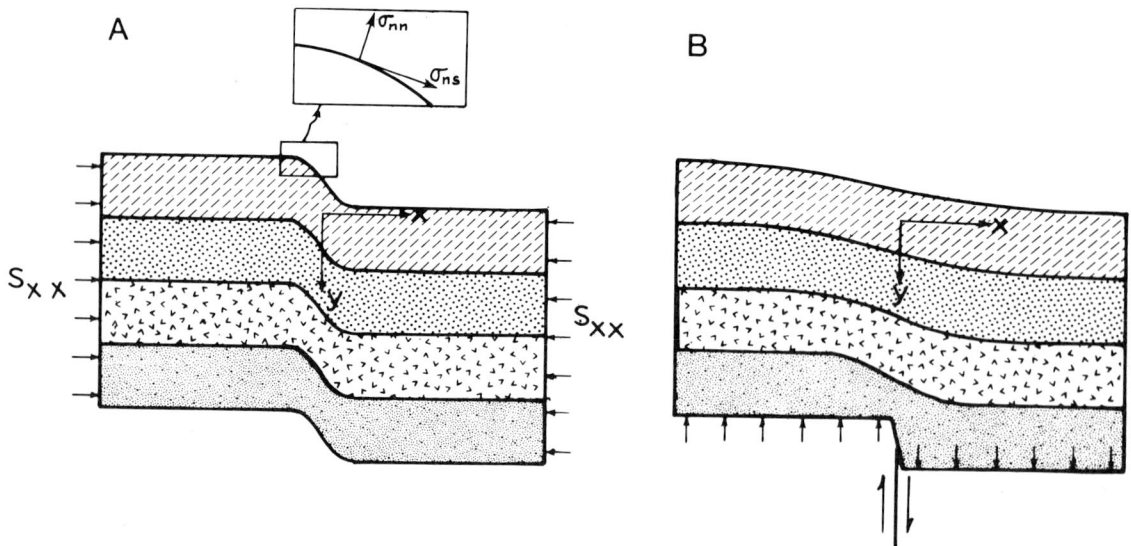

Figure 3. Boundary conditions of the buckling mode and the drape folding made of monoclinal flexuring. (A) The buckling is a result of shear, σ_{ns}, on inclined surfaces. (B) The drape folding is a result of displacement at base of flat multilayer.

way to generate a monocline is by buckling of the multilayer into an asymmetric fold. Folding theory indicates that a multilayer subjected to layer-parallel shortening tends to deflect into a train of folds, and that the amplification and shape of the folds depend on the properties of the multilayer (Ramberg, 1970; Johnson, 1970, 1977) and on the form of the initial deflection (Biot and others, 1961; Fletcher and Sherwin, 1978). On the other hand, it has been shown both experimentally and theoretically that asymmetric folds cannot initiate in a flat multilayer subjected to layer-parallel shortening and shear (Treagus, 1973; Reches and Johnson, 1976); rather, series of symmetric folds develop in such multilayers. Monoclinal flexures characteristically are isolated, asymmetric flexures and therefore cannot be simply related to repetitive, symmetric folds in a multilayer.

Even though a monoclinal flexure cannot develop spontaneously in a multilayer with initially flat contacts, it can develop in a multilayer with a low-amplitude, monoclinal initial deflection. A possible result of layer-parallel shortening of such a multilayer is amplification of the initial deflection into a large monoclinal flexure. The fundamental elements of this process of monocline formation can be elucidated by means of experimental and theoretical analyses of folding of a single stiff layer in a soft medium. Figure 4 shows results of three experiments with rubber layers of different thicknesses and properties embedded in thick gelatin media. All three rubber layers had the same initial monoclinal deflection (Fig. 4A1, B1, C1). Each rubber layer has a dominant wavelength defined approximately by the relation

$$W_d = 2\pi T \left[G_1/(6G_2)\right]^{1/3}, \tag{1}$$

where T is thickness and G_1 is shear modulus of the rubber layer and G_2 is shear modulus of the medium. The dominant wavelengths are shown in Figure 4A4, B4, C4; each corresponds with the wavelength that grows most rapidly with an increase in axial shortening.

In all three experiments the layer and gelatin were subjected to layer-parallel compression and layer-normal extension. In experiment A (Fig. 4A), the monocline grew in amplitude, and adjacent anticlines and synclines developed (Fig. 4A2). In experiment C (Fig. 4C), the monocline did not grow perceptibly; instead, many small folds developed along the rubber layer (Fig. 4C3).

Examination of the dominant wavelengths shown in Figure 4A4, B4, and C4 provides some insight into the differences among the three experiments. The dominant wavelength of the thick rubber layer in experiment A (Fig. 4A) is on the order of the width of the initial monoclinal deflection (Fig. 4A1), whereas the dominant wavelength of the thin rubber layer in experiment C (Fig. 4C4) is significantly shorter than the width of the initial monoclinal deflection (Fig. 4C). One may conclude that each rubber layer buckled according to its dominant wavelength, even though the initial deflections were the same. One can express this conclusion in a different form. The initial monoclinal shape can be described as the sum of an infinite number of sinusoidal waves (in the form of a Fourier series, equation 22a, App. 2). The dominant wavelength of the shortened rubber layer selectively amplifies more than the others (Johnson, 1970, p. 141). Accordingly, if the dominant wavelength of the layer is on the order of the width of the initial deflection (as in experiment A), the monocline will be strongly amplified, whereas if the dominant wavelength of the layer is significantly shorter than the initial deflection (as in experiment C), the monoclinal flexure will not be amplified significantly. If the dominant wavelength of the layer is much larger than the width of the initial deflection, the monoclinal flexure again will not be amplified significantly.

280 RECHES AND JOHNSON

The same general conclusion applies to growth of initial deflections in multilayers, but the results are slightly more complex for several reasons. For example, a multilayer can have more than one dominant wavelength (Ramberg and Strömgård, 1971; Johnson and Page, 1976), so that folds with a range of wavelengths may amplify significantly. In our analysis we will investigate conditions that favor

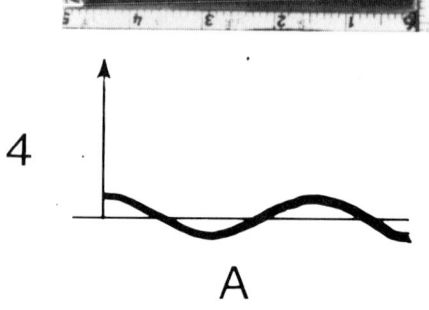

Figure 4. (facing pages). Three experiments of monoclinal flexuring of a rubber layer embedded in thick, soft gelatin. Plane strain, constant volume deformation was applied. The layers and the medium were both shortened horizontally. The initial monoclinal flexure is identical but the thickness of the rubber is different in the three experiments. Each unit on the frame of the folding machine is 3 cm. (1) The initial monoclinal flexure. The initial form was made by cutting the plate of gelatin to the desired curve. (2) The monoclinal flexure after small shortening. (3) The final monoclinal flexure. (4) The dominant wavelength for each experiment, calculated through equation (1).

amplification of initial monoclinal flexures in multilayers. We shall assume that small displacement along a vertical fault is amplified as a result of horizontal shortening. However, the initial deflection need not be restricted to a vertical fault; it could be the result of a local disturbance in the inclination of layers, such as a normal or reverse fault, or it could be the edge of an igneous intrusion

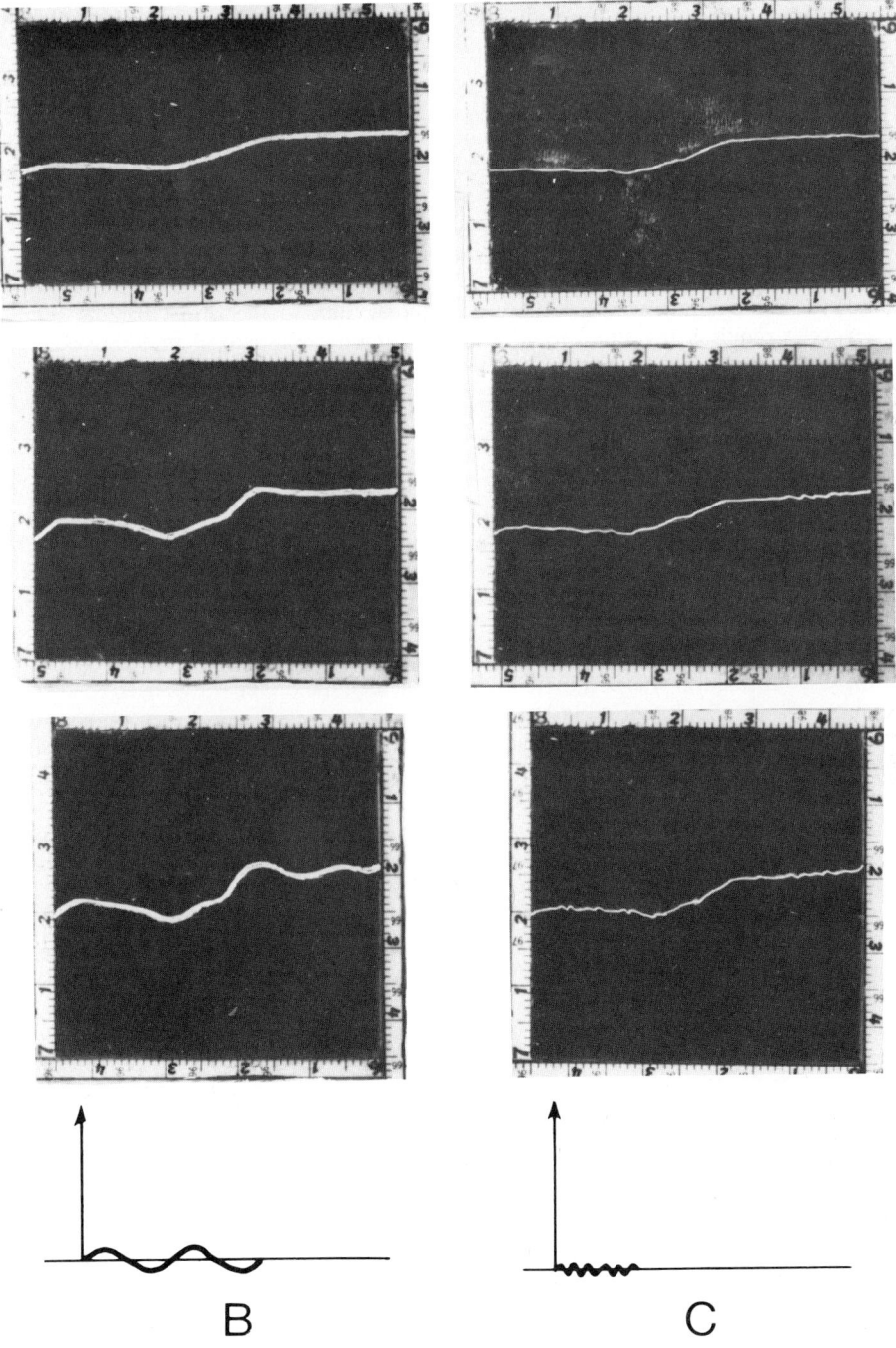

B   C

or an unconformity in the sedimentary rock sequence. Thus, according to the concept of monoclinal flexuring by buckling, the initial deflection serves primarily as a localizer or trigger for the monoclinal flexure rather than as the cause of monoclinal flexure as in the concept of drape folding.

Kinking

The third special case of the general model of monoclinal flexuring which we shall analyze is that of kinking, a process that depends upon contact strength between layers being overcome locally to produce a local flexure or kink band. We have analyzed conjugate and monoclinal kinking in some detail elsewhere (Honea and Johnson, 1976; Reches and Johnson, 1976), so we shall restrict the discussion here to general conclusions. Kink bands are characterized by relatively straight limbs and tight hinge zones (Fig. 1A) and resemble some monoclines geometrically. According to our analyses of kink folding, the essential processes are buckling and yielding instabilities. The kinking mode is more nearly akin to the buckling mode of monocline formation discussed above than to the drape folding mode, because it requires that the direction of maximum compression be subparallel to layering.

Several examples of monoclinal kink bands in various types of multilayers are shown elsewhere (Johnson, 1970, p. 319; Ramberg and Johnson, 1976, Figs. 5, 11; Reches and Johnson, 1976, Figs. 2, 3, 4, 7). In all cases the multilayers were subjected to a combination of layer-parallel shortening and shear. For example, Figure 5A shows monoclinal kink bands in thin rubber strips, the contacts between which were unlubricated, so that the contact strength was frictional. The kink band developed as the layer-parallel shear and shortening were increased. Conjugate kink bands, in contrast, developed where multilayers were subjected to layer-parallel shortening, without shear (Honea and Johnson, 1976, Figs. 2, 3, 4).

A series of experiments that are particularly relevant to the general model of monocline formation was conducted with the same type of apparatus used to subject multilayers to simultaneous layer-parallel shear and shortening but, in addition, one segment of one side of the apparatus could be faulted by retracting the segment (Fig. 5B). Each multilayer was first subjected to layer-parallel shortening and shear, and then to faulting. In one experiment the shear induced by faulting was of the same sense as the shear induced by the loading frame (Fig. 5B). A left-lateral monoclinal kink band developed over the fault. Subsequently, another parallel monoclinal kink band developed nearby. In another experiment the sense of shear induced by the faulting was opposite to that induced by the loading frame. Two distinct flexures developed, apparently simultaneously; a short, right-lateral kink band over the fault and a long, left-lateral kink band nearby (Fig. 5C). Kink bands could not be produced in the multilayers subjected solely to faulting nor to layer-parallel shear. The layer-parallel shortening was clearly required for kink

Figure 5. Monoclinal kink bands developed in thin rubber strips under plane-strain conditions. The rubber strips are unlubricated and therefore have frictional contacts. Solid arrows indicate layer-parallel shortening; half arrows indicate shear, and f indicates a fault introduced by displacing a short segment of the boundary of the apparatus. (A) Layers subjected to layer-parallel shortening and layer-parallel shear. A series of monoclinal kink bands developed. (B) Same conditions as in (A), except that a fault was introduced in the boundary. The shear produced by the fault, f, of same sense as the general layer-parallel shear. (C) Same conditions as in (B), except the shear along the fault was of opposite sense to the general layer-parallel shear. A local monocline induced by the fault triggered a through-going monocline of the opposite sense.

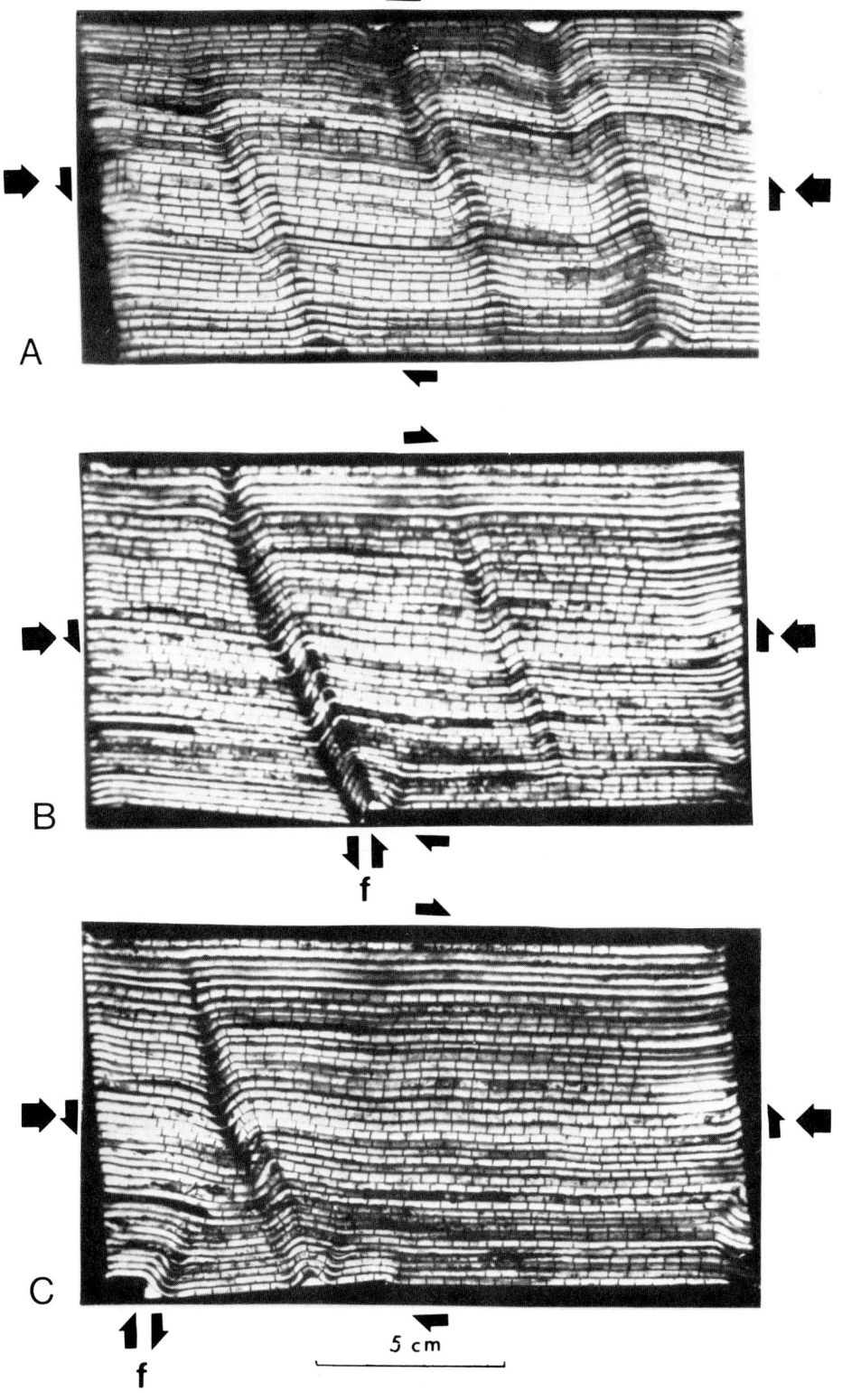

folding. Further, the fault acted as a trigger to the production of the monoclinal kink bands and as a source of shear stress within the multilayer, tending to favor monoclinal kinking.

Elsewhere we have presented analyses that distinguish conditions of repetitive, sinusoidal folding from conditions of localized, kink folding (Honea and Johnson, 1976; Reches and Johnson, 1976; Johnson, 1977). The same analyses distinguish between conditions that favor the buckling mode and conditions that favor the kinking mode of monocline formation. Briefly, the kinking mode of monocline formation is favored if the moduli of layers within a sedimentary sequence are similar, if the layers are thin, and if there is layer-parallel shear. If the contrast in moduli is high, or if the contacts between layers are virtually frictionless, the buckling mode of monoclinal kinking is favored. For example, if the contrast in moduli is high, the multilayer tends to buckle with relatively high amplitudes before local slippage can occur to produce the kink form.

The kinking mode in its pure form is characterized by hinge zones where layers are bent and by limbs along which unstable yielding of contacts is the primary process. The hinges may become unstable and yield plastically or fracture if bending becomes severe. Thus, the hinges might be loci of faulting as the kink band amplified (Fig. 1A). The width of the kink band is determined largely by the bending resistance of the layers so that the thicker the layers, the wider the kink band (Honea and Johnson, 1976).

METHODS AND RESULTS OF ANALYSES

Monoclinal flexures will be idealized by simple mechanical models in order to analyze conditions that could lead to the development of such structures. Nevertheless, the models are general enough to contain many of the mechanical characteristics of deforming sedimentary sequences.

Two of the special cases of the general model, drape folding and buckling, comprise a multilayer of homogeneous, incompressible, elastic or viscous layers with different rheological properties (Fig. 2). The layers may be bonded to each other or they may have frictionless contacts and may have the same or different rheological properties. The multilayer may be subjected to both displacement and stress boundary conditions (Fig. 2). Infinitesimal plane strain is assumed.

We analyze the deformation of multilayers using methods of continuum mechanics (Apps. 1, 2). These methods provide the necessary tools for analyzing low limb-dip folding of a wide variety of materials, including linear elastic and viscous, power-law elastic and viscous, and nonlinear elastic (Fletcher, 1974; Johnson, 1977, chap. 10). We will, however, present solutions only for simple multilayers comprised of homogeneous, linear incompressible elastic or viscous material. Effects of gravity have been excluded in the particular solutions presented here, but are included in the general solution (App. 1) and will be discussed below. The reasons for making these simplifying assumptions are brevity, clarity, and we do not know the rheological properties of the rocks involved in monocline formation. Nevertheless, results of previous studies of folding indicate that the results derived from the simple models studied here are qualitatively valid for many types of rheological behavior.

An important question is the nature of the contacts between the layers in the multilayer. In natural rock sequences, the contacts are probably frictional, so that slip between the layers occurs only when the shear stresses along the contact overcome the frictional resistance. However, the solution for a multilayer with frictional contacts is complex and not available, and therefore we solve two special cases: bonded contacts and frictionless contacts.

In the analysis of kinking we consider layered media with frictional contacts between layers so that slip along the contacts occurs only when the frictional resistance is overcome. We assume finite plane strain, and slopes of layers may exceed 60° or 70° (Reches and Johnson, 1976).

RESULTS OF THE THEORETICAL ANALYSES

Drape Folding

The displacements and strains within a multilayer, produced by displacement along a vertical fault at its base (Fig. 3B), can be calculated from the solutions presented in Appendix 2. We discuss two cases of drape folding:

1. Drape folding of a multilayer with bonded contacts between the layers. There are five layers of the same thickness, three stiff and two soft. The ratio of moduli of stiff to soft layers is five. The top surface of the multilayer is stress-free (Fig. 6).

2. Drape folding of a multilayer with frictionless contacts between the layers. There are five layers of the same thickness and the same modulus (Fig. 7).

The methods of calculation are outlined in Appendix 2.

The theoretical distributions of infinitesimal displacements and strains of a drape fold in a multilayer are presented in Figures 6 and 7. The displacements are exaggerated by three orders of magnitude. We shall examine four aspects of the distribution of displacements and strains in a drape fold: the profile, the variations of curvature, the variations of displacement, and the orientation of strain axes.

Bonded Contacts. Let us discuss these features in detail for a multilayer with bonded contacts (Fig. 6).

Profile. The profile of the drape fold is a simple monoclinal form with open anticlinal and synclinal bends.

Variations in Curvature. At low levels the displacements are concentrated in a narrow zone of high curvature, whereas at high levels the displacements occur in a wide zone of low curvature (Fig. 6A). In the extreme case of a very thick multilayer, the upper surface is deflected into an open curve, whereas the lower surface is displaced by a vertical fault (Sanford, 1959).

Variations of Displacements. The vertical displacement, V, of a point on the upper surface of a drape fold (Fig. 6A) is always equal to or smaller than the vertical displacement of a point with the same x-coordinate on the base. The decrease of vertical displacement up-section was demonstrated for a single layer by Sanford (1959, Fig. 8).

Variations of the Orientation of Strain Axes. The orientation of the axes of maximum shortening is plotted in Figure 6B. Continuous strain trajectories were not plotted, because the orientation of the maximum shortening changes at the contacts between layers. The strain axes in the lower layer or two are scattered; however, the strain axes above form a clear pattern. According to orientations of strain axes, one can divide the drape fold into three zones: downthrown zone in which shortening is subparallel to layering, upthrown zone in which extension is subparallel to layering, and central zone in which simple shear is subparallel to layering (Fig. 6B). Sanford (1959) derived similar results for a single layer.

Multilayers with Frictionless Contacts. The results of the analysis of a multilayer with five thick layers of the same rheology and frictionless contacts are presented in Figure 7. The main features of the drape folding of this multilayer are as follows (Fig. 7):

Profile. The profile of the flexure is of a simple monocline with open anticlinal and synclinal bends.

Variations of Curvature. The flexure at depth is narrower than the flexure at upper levels.

Zones of Layer-parallel Shortening and Layer-parallel Extension. These zones are evident in every layer (Fig. 7B). The corresponding neutral surfaces develop similarly to those in a bending plate.

The general profile and the variations of curvature are similar in drape folding of a multilayer with frictionless contacts and in the drape folding of a multilayer with bonded contacts (Figs. 6A, 7A). However, the patterns of strain axes differ in the two cases (Figs. 6B, 7B).

A drape fold in a multilayer with frictional contacts probably has the same general shape as the two cases described above, but it probably has a more complicated strain distribution. Where there has been slippage, one expects a neutral surface in many layers. Where there has been no slippage, one expects extension

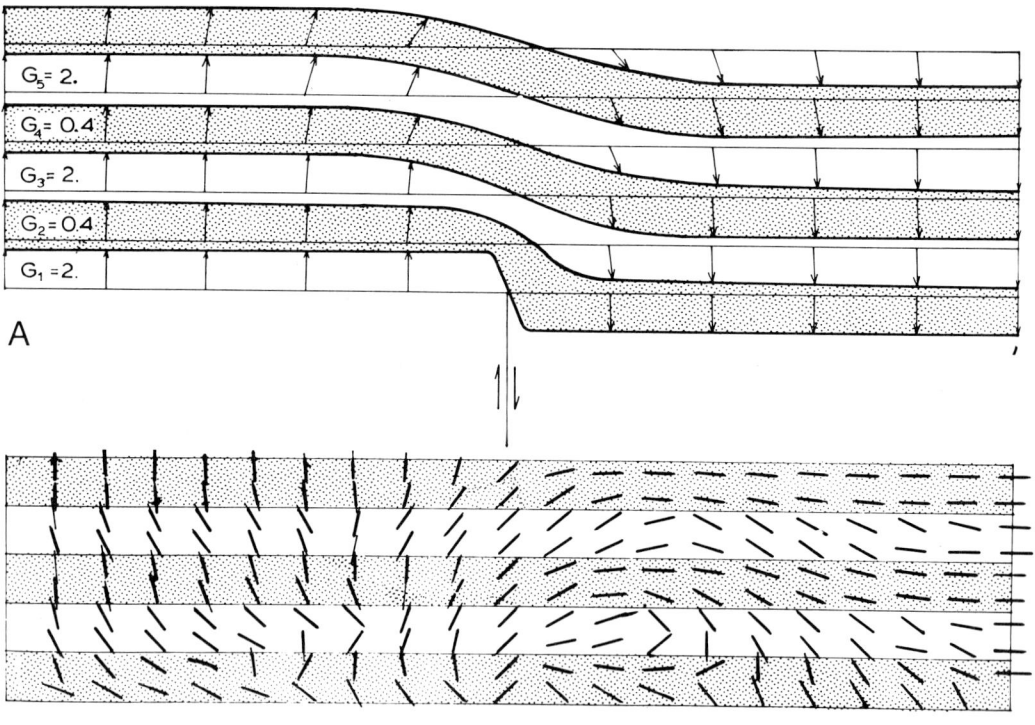

Figure 6. Drape folding of a multilayer composed of five layers, with bonded contacts, subjected to localized vertical displacements at its base. The displacements and strain orientations were calculated for an incompressible, linear multilayer (Apps. 1, 2). Shear moduli, $G_1, \ldots, G_5$, indicated on layers in their original positions. Stippled layers are the stiff layers in their deformed positions. (A) Displacements of contacts between layers. All displacements are exaggerated by three orders of magnitude. (B) Orientations of the axis of maximum shortening in the passively folded multilayer in the undeformed configuration. The orientations of shortening axes vary smoothly with position within layers, but change abruptly at contacts between layers with different properties.

or shortening within individual layers, depending upon their level within the large structure. Therefore, it is impossible to predict accurately the strain distribution within a multilayer in which contact strength has been overcome locally; however, one can conclude that zones of both layer-parallel extension and layer-parallel shortening will develop within all types of drape folds.

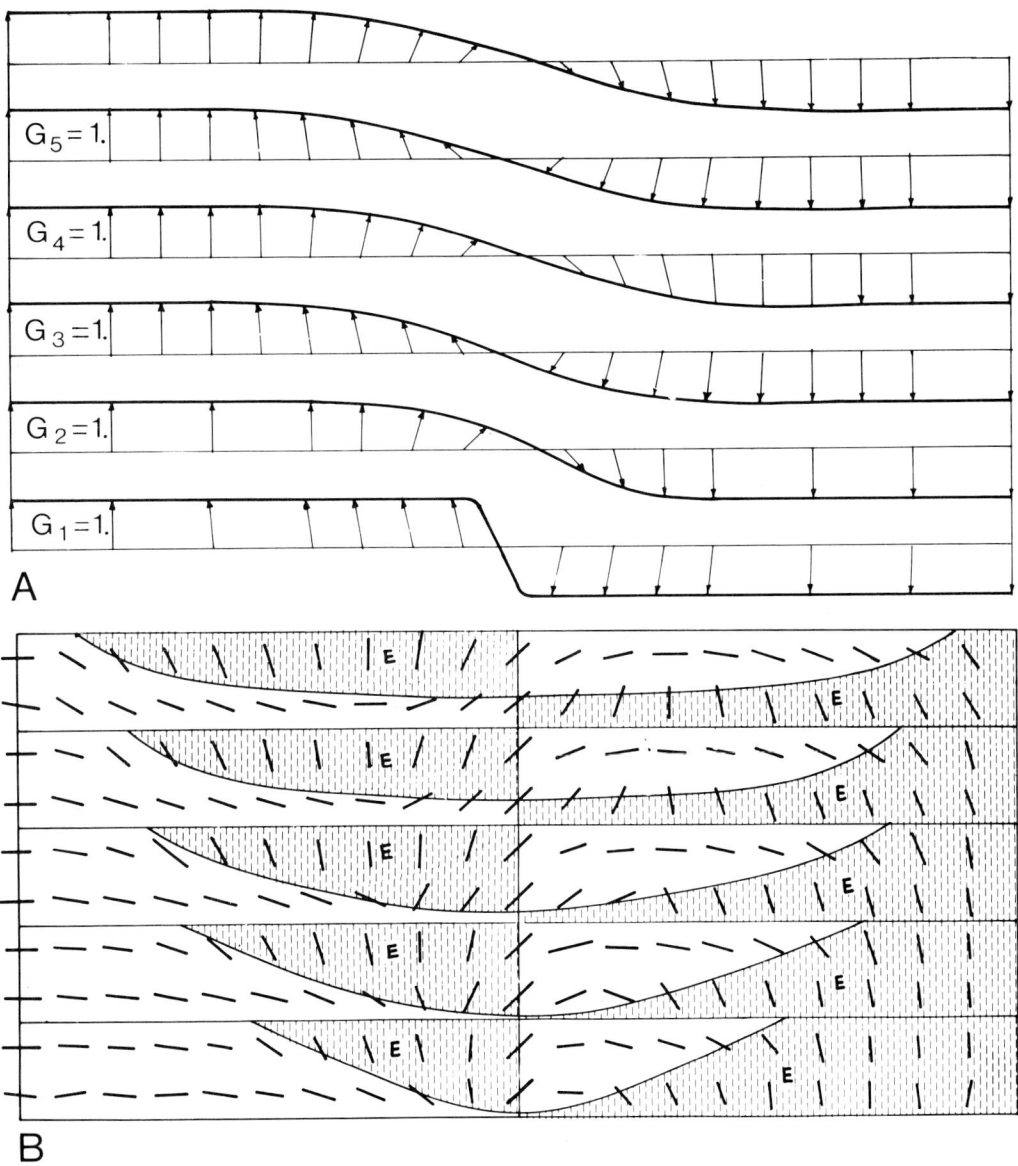

Figure 7. Drape folding of a multilayer composed of five layers with the same rheology, but with frictionless contacts. The multilayer is subjected to vertical displacement at its base. The calculations of the displacements and strain orientations are presented in Appendixes 1 and 2. (A) Displacements of contacts between layers. All displacements are exaggerated by three orders of magnitude. (B) Orientations of the axes of maximum shortening in the passively folded multilayer in the undeformed configuration. The principal strain axes are normal to all contacts between the layers. Stippled zones and the symbol E indicate areas of layer-parallel extension.

Combination of Buckling and Displacement of Lower Boundary of a Multilayer

The next step toward analysis of the general model of monoclinal folding is to consider both layer-parallel shortening and displacement of the base of the multilayer. The layer-parallel shortening can result in buckling in the sense that amplitudes of initial deflections become selectively amplified. In the general model we imagine the displacement of the base of the multilayer and the shortening to be simultaneous processes so that they interact with each other to produce the final monocline. However, our theory is linearized, so that we approximate the results of the simultaneous processes by applying the processes sequentially. First, we determine displacements within the multilayer as a result of drape folding

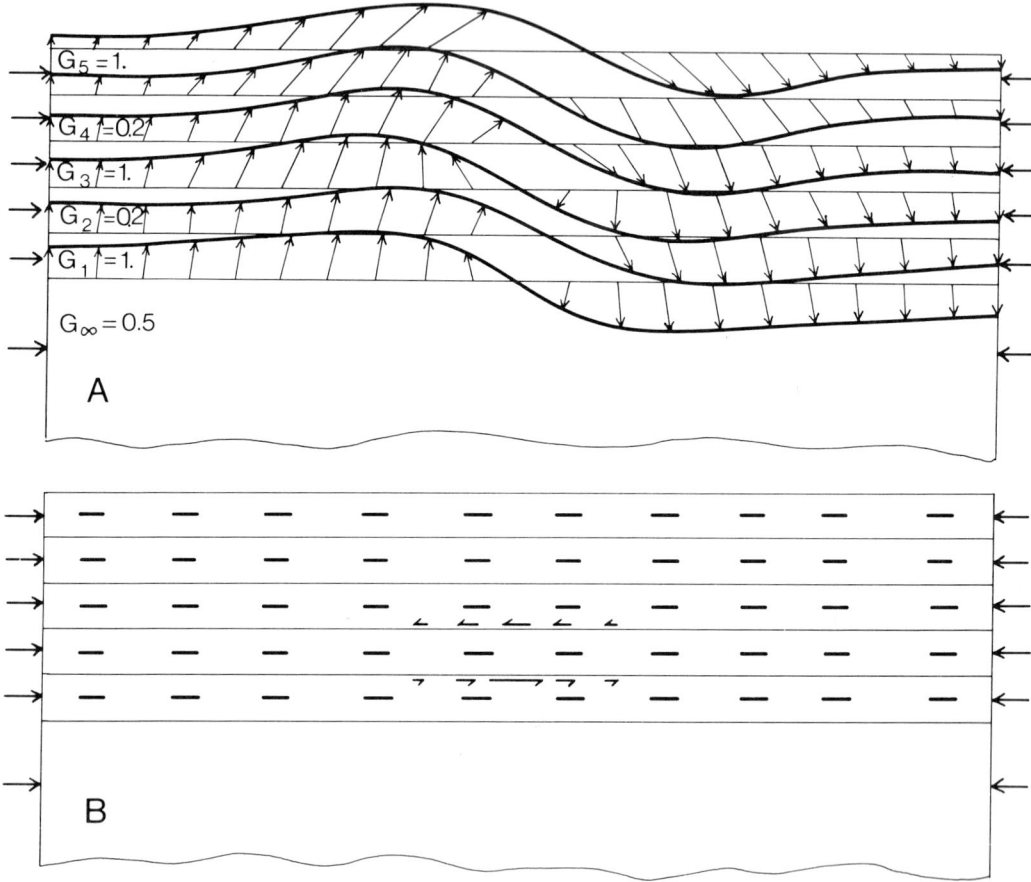

Figure 8. Buckling of a multilayer composed of five layers with bonded contacts. The lower medium is as stiff as the mean stiffness of the multilayer. The multilayer was first passively folded into a very low amplitude monoclinal flexure similar to that shown in Figure 5. Then the initial monoclinal flexure was buckled as a result of layer-parallel shortening of 0.001. Only the strains and displacements due to the buckling stage are plotted. (A) Displacements of contacts between layers. Displacements are exaggerated by three orders of magnitude. (B) Orientations of the axes of maximum shortening in the buckled monoclinal flexure in the undeformed configuration. The axes of the total strain, uniform and incremental (App. 1), are plotted. The small arrows marked above and below layer 2 indicate the relative magnitude and sense of the incremental layer-parallel shear (σ_{ns} in Apps. 1, 2).

in response to displacements along the base of the multilayer. Then we treat these displacements as *initial* displacements which become amplified as a result of layer-parallel shortening (App. 2). We should note that buckling is impossible in perfectly flat layers of viscous materials and virtually impossible in perfectly flat layers of elastic materials. Thus, we use the displacements due to drape folding to select the form of the initial displacements of interfaces between layers.

Choice of Multilayer. A single layer embedded in an infinite soft medium has one dominant wavelength which is determined approximately by equation (1). However, multilayers composed of many layers with arbitrary thicknesses and moduli can have several dominant wavelengths. We have chosen to analyze simple multilayers of alternating soft and stiff layers rather than complicated multilayers. The multilayer behaves much as the single layer described in earlier pages, in that the initial monocline will be amplified during shortening only in a multilayer which has a dominant wavelength on the order of the width of the initial monocline. Other multilayers also buckle, but into shorter or longer wave forms. The infinite lower medium (Fig. 8) below the multilayer has an important effect on the dominant wavelength. If the lower medium is stiff relative to the multilayer above, the dominant wavelengths will be short (Johnson, 1977, chap. 11) and may be less than the width of the monocline. On the other hand, if the lower medium is very soft, the dominant wavelengths may be larger than the width of the monocline. We arbitrarily chose a multilayer with alternating soft and stiff layers with bonded contacts overlying a lower medium with a shear modulus approximately equal to the mean shear modulus of the multilayer (Fig. 8).

The analysis of buckling of a multilayer with frictionless contacts indicates that the dominant wavelength depends mainly on the thickness of the layers and is virtually independent of the modulus of the lower medium if the modulus of the medium is greater than the modulus of the layers (Johnson and Page, 1976, Fig. 12).

We studied several multilayers with bonded contacts and with frictionless contacts and found that monoclines will be amplified in multilayers with a limited range of properties if contacts are bonded but will be amplified in most multilayers with frictionless contacts.

Results. The theoretical distribution of displacements and strains due to buckling of a multilayer are presented in Figures 8 and 9. In our calculations, the multilayer is first passively folded by a small displacement along a vertical fault at its base (Fig. 6). Then the initially deflected multilayer is subjected to layer-parallel shortening of magnitude 10^{-3}, which causes it to buckle. Methods of calculation and boundary conditions are presented in Appendix 2. We will again discuss the distribution of the deformation rather than the absolute magnitudes of the strains and displacements. We will discuss in detail the buckling of a multilayer with bonded contacts (Fig. 8), and we will discuss briefly the buckling of a multilayer with frictionless contacts (Fig. 9).

Bonded Contacts.

Profile. The profile of a monocline formed by buckling is characterized by a monoclinal flexure, with associated anticline and syncline (Fig. 8). The anticline and syncline replace the anticlinal bend and synclinal bend of the drape fold (Fig. 6). A similar profile was obtained in experiments involving a single rubber layer in gelatin (Fig. 4A).

Variation of Curvature. The profiles of the different layers in the buckled monocline are similar to each other (Fig. 8). One can, however, see a slight difference between the profile of the base of layer 3, for example, and the profile of the top of layer 3. This minor variation is the result of incremental shear which develops during the buckling (App. 1). In general, however, the layers maintain similar profiles at all levels.

Variations of Displacements. The vertical displacements of the layers in a buckled monocline increase up-section. Thus, the displacement, V, of a point at the top of the multilayer in the anticlinal bend is always equal to or larger than the displacement at the same distance, X, at the base of the multilayer (Fig. 8). This result is expected because the upper surface of the multilayer does not resist vertical displacement of the multilayer, whereas the lower surface in contact with the infinite medium does.

Variation of Orientation of the Strain Axes. The strains in the buckled monocline can be separated into uniform horizontal strain which causes the buckling and incremental strain which results from the buckling (App. 1). The total strain is the sum of the uniform strain and the incremental strain. However, the incremental strain is two to three orders of magnitude smaller than the uniform strain; therefore,

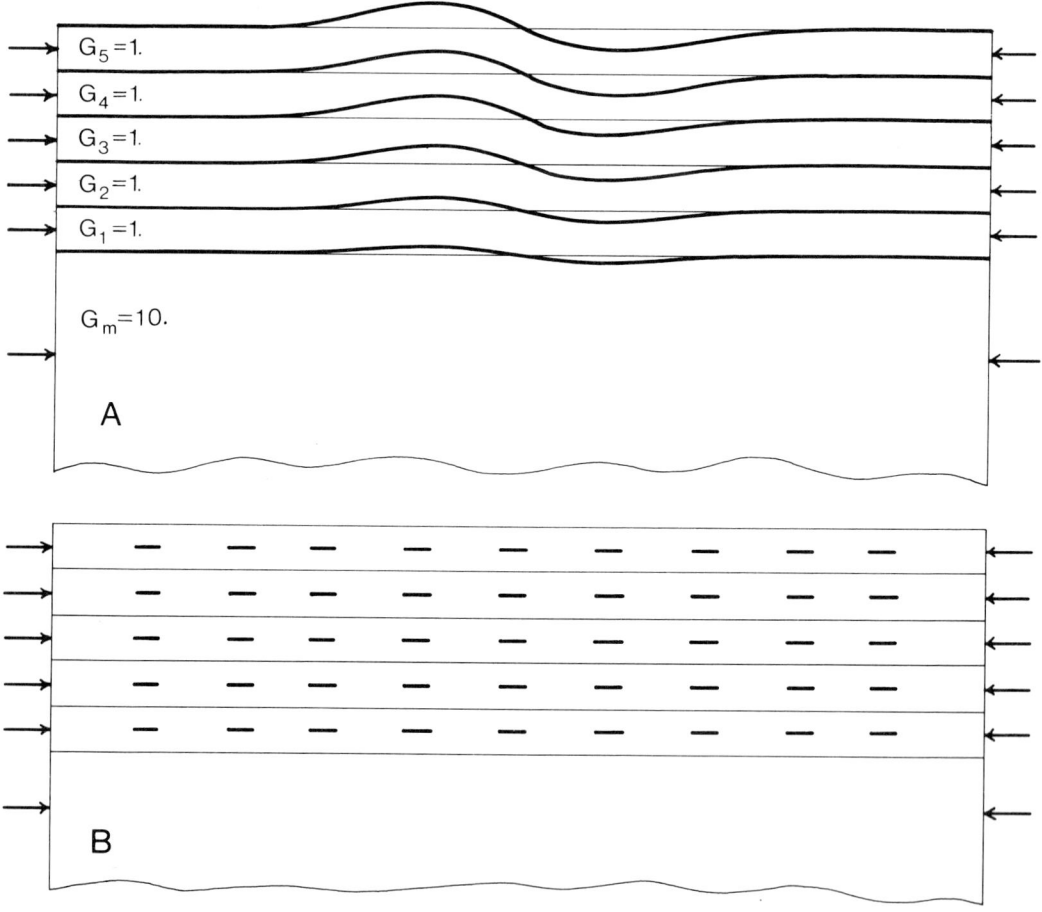

Figure 9. Buckling of a multilayer composed of five identical layers with frictionless contacts. The lower medium is ten times stiffer than the layers above. The multilayer was first passively folded as in Figure 6. Then the multilayer was buckled as a result of layer-parallel shortening of 0.001. Only the strains and displacements due to buckling processes are plotted. (A) Displacements of contacts between layers. Displacements are exaggerated by three orders of magnitude. (B) Orientations of the axes of maximum shortening in the buckled monoclinal flexure. The axes of the total strain, uniform and incremental (App. 1), are plotted.

the total strain in the monocline is essentially uniform layer-parallel shortening (Fig. 8B) for low-amplitude flexures.

Multilayer with Frictionless Contacts. The results of the buckling of a multilayer with frictionless contacts, due to layer-parallel shortening, are shown in Figure 9. The conditions assumed in the analysis differ from those for bonded layers in two respects: First, we assume the same properties for all five layers. Second, the lower medium is ten times stiffer than the layers above, whereas, for the multilayer with bonded contacts, the modulus of the lower medium was approximately equal to the average modulus of the layers above.

The most striking feature of the buckling of a multilayer with frictionless contacts is the strong amplification of the monoclinal flexure up-section. The vertical displacement at the top of the fold in Figure 9 is about ten times greater than the vertical displacement at the base of the multilayer. The cause of this difference in amplification is the stiff lower medium which is not able to buckle similarly to the multilayer, whereas the lack of confinement of the surface of the multilayer enables the monocline to be amplified up-section.

The monoclinal flexures we analyzed were initiated and localized by draping which deformed the multilayer before layer-parallel shortening. We presented orientations of strains, however, produced solely by the layer-parallel shortening and buckling. Zones of layer-parallel extension develop in the anticlinal bend of a drape fold (Figs. 6, 7), whereas layer-parallel compression prevails throughout the buckled monoclines. Therefore, it is important to superimpose the strains produced by the draping upon the strains produced by the folding. Unfortunately, there is no general relation between these strains because the amount of displacement along the fault during draping is unrelated to the amount of layer-parallel compression. We can, however, report the results of the superposition of strains for two multilayers analyzed here (Figs. 6, 8). The displacement of 1% of the total thickness of the bonded multilayer along a vertical fault (Fig. 6) produces maximum layer-parallel extension of about 0.7% in the upthrown block of the multilayer. Therefore, shortening of the multilayer by about 0.7% would eliminate the extension in the upthrown block due to draping.

Effects of Gravity

The general model of monoclinal flexuring includes the effects of gravity on stresses and displacements within a multilayer. We did not include effects of gravity in the solutions presented in earlier pages because the effects are negligible. We can demonstrate this by calculating the stresses and displacements in a homogeneous half space, the surface of which is deflected into a sinusoidal wave. For small deflections, where the amplitude-to-wavelength ratio (h/L) is much less than unity, the deformation of the half space can be evaluated with equations (21a–21h), Appendix 2. The displacements at the surface resulting from gravity tend to lower the elevated areas and to elevate the low areas, as would be expected.

Solution of equations (21a–21h) for an elastic half space shows that the surface displacements due to gravity are proportional to a parameter D,

$$D = \rho g / (4\pi G),$$

where ρ is density, g is acceleration of gravity, and G is the shear modulus. In order to estimate the maximum magnitude of the displacement, we use densities and shear moduli determined for rocks. A representative value for D for limestone, sandstone, shale, granite, and basalt, according to data presented by Clark (1966), is 10^{-9} cm^{-1}. Using this value and choosing a surface deflection of

$$h = 0.01\, L,$$

we compute the maximum displacement of the surface due to gravity

$$V_{\max} = 10^{-3} h.$$

That is, the correction to the displacements at the ground surface, where the correction is maximal, is about three orders of magnitude less than the displacement, h, of the surface due to draping or buckling. These corrections are clearly negligible for low-amplitude deformations such as those we consider in our linearized analyses.

Now let us consider the stresses. Stresses induced by gravity in a multilayer can be divided into two parts. One part is a result of the differences in altitude of different parts of the ground surface. The maximum difference of these stresses within the multilayer is proportional to the amplitude of the surface deflection and is therefore small if the deflection is small, as in our first-order analysis. The other part of the gravity stresses is hydrostatic and is directly proportional to the depth below the surface. This hydrostatic stress can be superimposed on the stresses produced by folding. Further, the hydrostatic stress produces strains in a compressible material, and these strains can be superimposed on the strains produced by folding. In incompressible materials, which we are considering here, hydrostatic pressure produces no strain, so there would be no correction to the strains computed by ignoring gravity.

Therefore, gravity has negligible effects on the strain and displacement distributions reported in previous pages, primarily because the deformations are small. An analysis of finite deformations, of course, must include effects of gravity.

DISCUSSION

The ultimate objective of our study is to identify general processes of monoclinal flexuring, including those responsible for the large monoclines on the Colorado Plateau and those responsible for monoclinal kink bands in foliated materials. A general model that we believe to include most of the processes is partly illustrated in Figure 2. The features of the general model not illustrated are reviewed in the discussion accompanying Figure 2. Not all possible conditions of the general model will lead to a flexure of monoclinal form, however, as has been indicated already by the analyses presented in earlier pages. Further analysis is required to specify those conditions that will; one based on finite-element techniques, using nonlinear, elastic-plastic materials, is in progress.

Perhaps the greatest value of the process of developing a general model is the selection of parameters that must be included in a comprehensive analysis. We have shown, through study of three special cases which we have loosely called "kinking," "drape folding," and "buckling," that a wide variety of conditions can be responsbile for monoclinal folding, even in simple linear materials. Clearly a notion that all monoclines are drape folds would be impossible to defend.

In our attempt to develop a general model of monoclinal flexuring, we have performed theoretical and experimental analyses of monoclinal kink bands in foliated materials (Reches and Johnson, 1976), we have made a detailed structural analysis of the Palisades monocline (Reches, this volume), and we have examined two special cases of our general model of monocline formation in previous pages.

A summary of the general results of the study of the theoretical models is presented in Table 2.

Limitations of the Theoretical Analyses

We will re-examine some of the field observations made in the Palisades monocline, as well as those made elsewhere by other investigators, and attempt to explain some of the observations in terms of the theoretical and experimental analyses. However, in order to compare the theoretical and field observations, we must know the limitations of the theory.

The theoretical analysis of the buckling mode of monocline formation is qualitatively valid for materials with a wide range of properties, including linear elastic or viscous, power-law elastic or viscous, and nonlinear elastic (Fletcher, 1974; Johnson, 1977, chap. 10). The analysis of drape folding is qualitatively valid for compressible or incompressible elastic or viscous materials. The specific results are valid only for incompressible elastic or viscous materials. However, the results

TABLE 2. SUMMARY OF TYPICAL FEATURES OF THREE SPECIAL MODELS OF MONOCLINAL FLEXURING ACCORDING TO ANALYSIS

Contacts between layers	Mode of monoclinal flexuring		
	Drape folding	Buckling	Kinking
Bonded contacts	1. Simple monoclinal flexure with anticlinal and synclinal bends 2. Decrease of curvature of profile up-section 3. Decrease of vertical displacement up-section 4. Zones dominated by layer-parallel shortening, extension, or shear	1. Monoclinal flexure with associated anticline and syncline 2. Increase, decrease, or constant curvature up-section, depending on properties of multilayer 3. Constant or increase of vertical displacement up-section in the anticlinal zone 4. Layer-parallel shortening dominates at all levels	Kinking not possible
Yielding contacts	Solution not available	Solution not available	1. Profile comprised of straight limb and tight hinge zones 2. Constant profile 3. Constant displacement 4. Yielding in hinges
Frictionless contacts	1. Simple flexure 2. Decrease of curvature up-section 3. Constant vertical displacement up-section 4. Every layer contains zones of shortening, shear and extension	1. Flexure with associated anticline and syncline 2. Curvature generally increases up-section 3. Increase of vertical displacement up-section in anticlinal zone 4. Layer-parallel shortening at all levels	Kinking not possible

Note: Summary of Figures 6, 7, 8, and 9 and Reches and Johnson (1976).

are presented in terms of nondimensional parameters such as relative displacements, strain or strain rates, and relative moduli or viscosity coefficients, so we consider the specific theoretical results presented here to be sufficiently general to elucidate the characteristics of the three special cases of the general model of monocline formation. There are, though, several important limitations in the analyses:

1. Rocks within certain parts of monoclines have been subjected to finite deformations and rotations, whereas the theories of drape folding and buckling discussed in previous pages assume infinitesimal strains and rotations. Thus, the theory is valid only for the inception of monocline development. The limitation of smallness of deformations is necessary in order to linearize the basic equations used to describe the deformations; also, it allows us to generalize the solutions to include certain aspects of power-law or other nonlinear rheological models.

The theory of kink folding, unlike the theories of drape folding and buckling, is not restricted to small deformations and rotations. As shown elsewhere (Reches and Johnson, 1976), we can investigate yielding instabilities of contacts between layers regardless of the slope angles of layers, and the yielding of contacts appears to control the high-amplitude growth of conjugate and monoclinal kink bands, including the orientations of kink bands and the locking angles of layering within kink bands. Thus, the theory of kinking incorporates certain nonlinearities that the other theories cannot incorporate, but the theory of kinking is so elementary that it cannot describe displacements and strain patterns in the way that the other theories can.

There is evidence derived from experiments and high-amplitude folding theory that sizes of folds are largely controlled by sizes of the first-formed folds, which we can study with the linearized theory presented here. The linearized theory apparently closely describes deformations within folding layers where maximum slopes of layering are less than 10° to 15° (Sherwin and Chapple, 1968; Dietrich and Carter, 1969; Hudleston and Stephansson, 1973). Therefore, it is quite likely that the linearized theories of monocline formation are valid for similar maximum slopes of layers, because the sources of error are the same in the theories. Hudleston (1973) has shown experimentally that wavelength selection and layer-parallel shortening are active primarily during early stages of growth of high-amplitude folds. Sherwin and Chapple (1968) and others have deduced useful information from high-amplitude folds by using infinitesimal strain theory and comparing theoretical wavelengths with arc lengths measured in the field.

On this basis we suggest that our theoretical analysis of drape folding and buckling is applicable to monoclinal flexures sloping up to 10° to 15° and may explain certain features of even steeper monoclinal flexures.

2. Idealized material properties of contacts between layers, such as bonded contacts, free-slip of contacts, or contacts with finite strength, were assumed in order to allow theoretical analysis. None of these idealized conditions corresponds with actual field conditions, of course, but they should allow us to understand some aspects of field conditions. Even though the cases of bonded contacts and free slip are end members of resistance to slip along layers, the two cases do not represent end members of deformations of contacts. Thus, if we derive the shape of a fold in a multilayer with bonded contacts and the shape of a fold in the same multilayer with frictionless contacts, we still cannot predict the shape of the fold in the same multilayer having finite contact strength. The reason is that localized slippage of contacts can fundamentally change the fold form. Indeed, it is precisely this condition that allowed us to develop a theoretical model for the isolated kink form (Honea and Johnson, 1976).

3. The multilayers we have selected to analyze are highly idealized, whereas

multilayers in the field are difficult to define. In our study of kinking we have assumed that each layer has the same rheological and dimensional properties, a condition closely approached in the experiments but never achieved in the field. In our study of buckling and passive folding, we have also selected relatively simple multilayers. It is no simple matter to identify structural units in the field (Chapple and Spang, 1974; Johnson and Page, 1976), so it is difficult to select realistic multilayers in terms of either rheological or dimensional properties. Detailed comparisons of theoretical results and field observations will require careful estimations of field conditions.

4. The analyses cannot account for some important behaviors associated with nonlinear rheologic properties or with faulting. We believe that the location and orientation of major faults within monoclines cannot be deduced from the analyses of buckling, drape folding, or kinking. One reason is that large deformations generally cannot be incorporated in the theory. Another is that the stress and strain distributions within multilayers may change considerably due to the existence of a large fault, so that, as soon as the fault begins to develop and propagate, the stresses become redistributed and must be recalculated. Analyses of this type are not yet available. Our analyses definitely are valid only for continuous and unfaulted monoclinal flexures. Small faults, which disrupt the stress distribution only locally, however, probably have minor effect on the gross deformation of a multilayer so that a multilayer containing small faults probably can be considered to be continuous.

Field Examples

Palisades Monocline. The Palisades monocline, which was described in a companion paper (Reches, this volume), will be analyzed by means of the general model of monoclinal flexuring. We shall review a few of its salient features and then compare these features with those of the idealized monoclines.

The Palisades monocline is a branch of the East Kaibab monocline in the eastern part of Grand Canyon National Park. It is nearly perfectly exposed in Palisades Creek, so that an accurate structural cross section could be prepared (Fig. 10). First let us consider the gross form of the monocline. The structure can be divided into three levels: a lower level that includes the vertical fault and steep layers in the adjacent synclinal zone; a central level of continuous but tight flexuring; and an upper level of open flexuring (Fig. 10). The lower, faulted level cannot be compared with our theoretical models because our models were derived for nonfaulted monoclines, so we are unable to discuss it.

The profiles of contacts between rock units within the upper two levels are characterized by a transition from a tight monocline, with layers dipping up to 90° in the Bright Angel Shale, Muav and Temple Butte limestones, into an open monocline with maximum dips of 20° to 25° in the Redwall Limestone and in units above (Reches, this volume). The anticlinal bend of the monocline has a radius of curvature of about 10 m, and the synclinal bend has a radius of curvature of about 150 m at the base of the Bright Angel Shale (Fig. 10), whereas both the anticlinal and synclinal bends have radii of curvature of about 4 km at the level of the Kaibab Limestone (unit P in Fig. 10).

The difference of structural levels across the base of the monocline is about 250 m, and the structural difference is constant or slightly decreasing up-section.

Thus, the profiles at various levels within the Palisades monocline are similar to those predicted by the drape folding case of our general model of monoclinal formation (Table 2). The relatively straight limb of the monocline in the central level, within the Bright Angel Shale and Muav Limestone, resembles a kink fold, but the proximity of this part of the structure to the large fault makes such an interpretation tenuous. We can state that there is no clear evidence for an anticline associated with the anticlinal bend or a syncline associated with the synclinal bend, as would be expected if much of the growth of the monocline were due to the buckling mode. Further, there is no evidence for increasing structural relief across the monocline up-section, as would be expected if buckling were strong.

On the other hand, the internal structures of the Palisades monocline do not correspond with those expected in a monocline developed as a drape flexure. The internal structures and measurements of strain orientations, using petrofabric analysis of carbonate rocks in the area, were described at length in a companion paper (Reches, this volume). Nearly all the measurements of small faults, small folds, twinning of calcite, and changes of thickness of units indicate maximum compression subparallel to layering at all levels within the monocline. These results are consistent with those expected in a monocline formed according to the idealized buckling or kinking mode but not the drape folding mode. If the Palisades monocline had formed via the drape folding mode, we would expect zones of layer-parallel extension in certain parts of the monocline, depending upon the properties of contacts between layers. For example, if the layers were bonded, we would expect marked layer-parallel extension in the anticlinal zone of the Palisades monocline, but we found evidence for compression there (Fig. 10).

Thus, analysis of the gross shape and the internal deformation patterns of the Palisades monocline leads us to the conclusion that the monocline formed as a

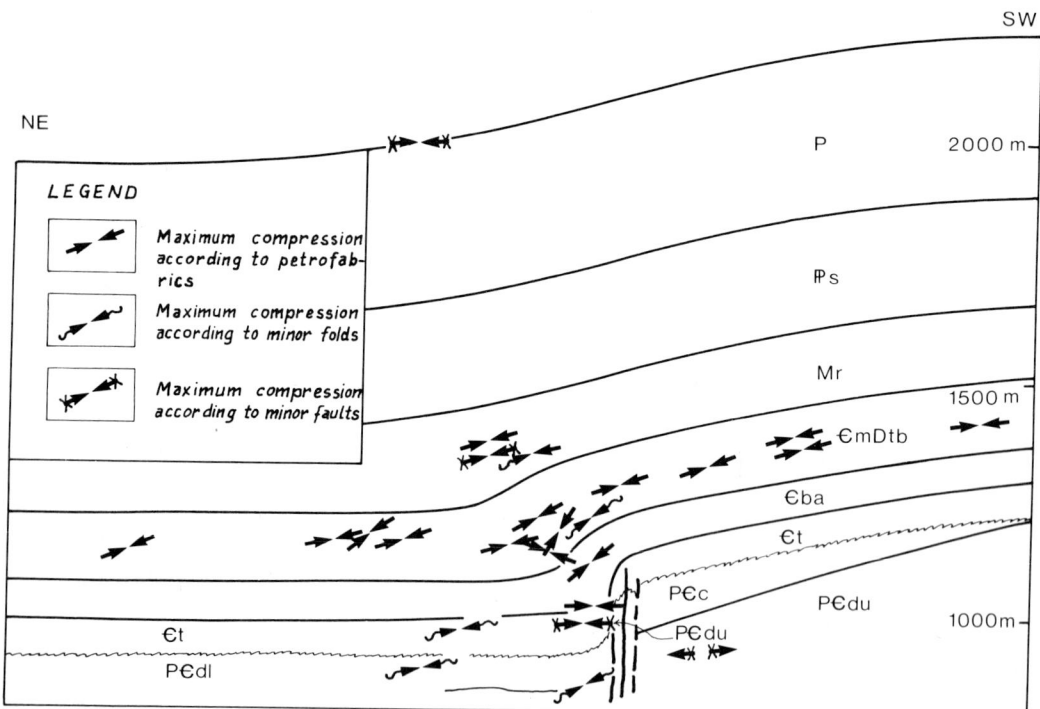

Figure 10. Vertical cross section of the Palisades monocline, Grand Canyon, Arizona. The solid arrows show the orientation of axes of maximum compression as interpreted from analyses of small structures and petrofabrics. Stratigraphic units range from Dox Formation of the Precambrian to Kaibab Limestone of the Permian (after Reches, this volume).

result of uplift along the Palisades Fault at its base, along with layer-parallel compression. The results do not correspond clearly with any of our three special cases of the general model of monocline formation. Without specific knowledge of the rheological and dimensional properties of structural units within the monocline, we are unable to determine the relative contributions of vertical displacements along the base of the monocline and buckling of layers within the monocline to the development of its present shape.

In a companion paper (Reches, this volume) we showed that the direction of maximum compression was oblique to the axis of the monocline, and one might well wonder if this is the reason there is no evidence for the buckling mode in the general form of the monocline. We think not. Analysis of three-dimensional folding based on plate theory, either for single layers or multilayers (Johnson and Page, 1976), indicates that the components of stress that are important to folding depend upon the shape of the perturbation to be amplified by buckling. Thus, the driving term in the differential equation describing the equilibrium of a layered system is (Johnson and Page, 1976, equation 24a)

$$[S_{xx}(\partial^2 v/\partial x^2) + S_{yy}(\partial^2 v/\partial y^2)], \qquad (2)$$

where x and y are horizontal axes, v is vertical displacement, and S_{xx} and S_{yy} are principal components of horizontal stresses. Let S_{xx} act normal to and S_{yy} parallel to the axis of a *cylindrical* monoclinal perturbation. The shape of the perturbation can be described in terms of a Fourier series, such as

$$v = a \sin(\ell x) + b \sin(2\ell x), \ldots, \qquad (3)$$

where $a, b, \ldots$, are coefficients and ℓ is a wave number defined as in Appendix 1. If equation (3) is substituted into equation (2), it is clear that the term involving the normal stress, S_{yy}, vanishes and only the term involving the stress, S_{xx}, normal to the axis remains. Thus, only the magnitude of the stress normal to the axis of the cylindrical perturbation is important in determining the amplification of the perturbation into a fold. Amplification of the perturbation will occur, therefore, if the axis-normal stress is compressive; it will occur even if the horizontal stresses are equal. This is merely one more example of the importance of the form of the initial perturbation on the shape of resulting folds, a concept introduced long ago by Willis (1894).

We suggest the following sequence of events during the development of the Palisades monocline. The strain field of the Laramide orogeny in the area of the Palisades monocline was dominated by subhorizontal, regional shortening. The regional trend of the shortening is unknown, but the areal trend in the vicinity of the eastern part of Grand Canyon National Park was about N65°E (Reches, this volume, Fig. 21). The branch of the Butte fault in Palisades Creek had developed during the Precambrian with a northwest trend. Its strike deviated about 70° from the axis of maximum shortening during the Laramide. The fault was unfavorably oriented relative to the directions of stresses in the sedimentary cover during the Laramide orogeny. However, the fault apparently provided a weak zone relative to blocks around it, and unknown, deep-seated processes caused large blocks of the Colorado Plateau to move vertically relative to each other. The sedimentary rocks above the fault became flexed into a monocline, apparently in response to a combination of horizontal shortening and differential vertical uplift.

Yampa Monocline. Cook and Stearns made a series of cross sections through the Yampa monocline in Dinosaur National Monument, Colorado and Utah, and mapped internal features of the monocline in some places (Cook, 1975; Cook and Stearns, 1975). We shall present some of the cross sections of the Yampa monocline and discuss its gross geometry and internal structures in terms of the theory developed here.

The Yampa monocline is about 20 km long, trends east, and has a maximum structural relief of about 500 m. A reverse fault is exposed along the eastern part of the monocline, whereas a continuous, unfaulted flexure occupies the western part of the monocline (Cook and Stearns, 1975, Fig. 1). Four cross sections of the western part of the monocline are shown in Figure 11. The most important characteristics of the monocline are

1. Small reverse and thrust faults are common within the monocline. They are particularly common in the anticlinal and synclinal bends (Fig. 11B, 11A) (Cook, 1975, Fig. 17). Intense fracturing with no clear displacement direction commonly occurs in the anticlinal and synclinal bends. Small normal faults were not observed in the Yampa monocline.

2. Thickening of the Weber Sandstone up to 50% prevails in the western part of the Yampa monocline (Fig. 11C) (Cook and Stearns, 1975, Fig. 6). At one location (Fig. 11B), "The Weber Sandstone is thinned approximately 10% normal to bedding; however, small thrust faults and shear fractures have shortened and thickened the formation perpendicular to the fold axis" (Cook and Stearns, 1975, Fig. 7).

3. The amplitude of the Yampa monocline and of neighboring monoclines within the Weber Sandstone increases up-section in some places (Cook and Stearns, 1975, Figs. 8, 9). For example, the structural relief across the monocline at the base is about half that at the top of the Weber Sandstone in one cross section (Fig. 11C).

4. The Yampa monocline is characterized in some places by blocks within which layers are straight but have different dips from layers in adjacent blocks (Fig. 11D). Fracturing is less intense within the blocks than within "hinges" between the blocks. In some places the "hinges" contain faults.

The Yampa monocline is fascinating for its variety of forms. However, all the observations seem to correlate with a mechanical model in which both vertical uplift and horizontal compression were important, as in the Palisades monocline.

Two cross sections of the Yampa monocline resemble large kink bands (Fig. 11A, 11D). In one, there is abundant evidence of layer-parallel shortening in the form of small reverse and thrust faults; the limb is quite straight and the hinges are rounded but narrow (Fig. 11A). In the other, there appear to be several limbs that are quite straight separated by very narrow hinge zones (Fig. 11D). According to our analyses, these forms require high, layer-parallel shortening and yielding of contacts between structural layers. A fault below the Weber Sandstone probably served as a trigger to the flexuring, much as in one of our experiments (Fig. 5B).

The third cross section (Fig. 11B) shows internal evidence of horizontal shortening, but the form of the flexure provides no information concerning the dominant mechanism of its formation. The anticlinal and synclinal bends and the limb are all curved. The fourth cross section (Fig. 11C) shows evidence of increase of amplitude of the flexure up-section, as in our idealized buckling mode of monoclinal flexuring.

Only the general model of monoclinal flexuring can account for the major features of the Yampa monocline that show evidence of buckling and kinking. No part of the Yampa monocline is consistent with a model of drape folding.

Experimental Monoclinal Flexuring. Some of the results of the theoretical analyses

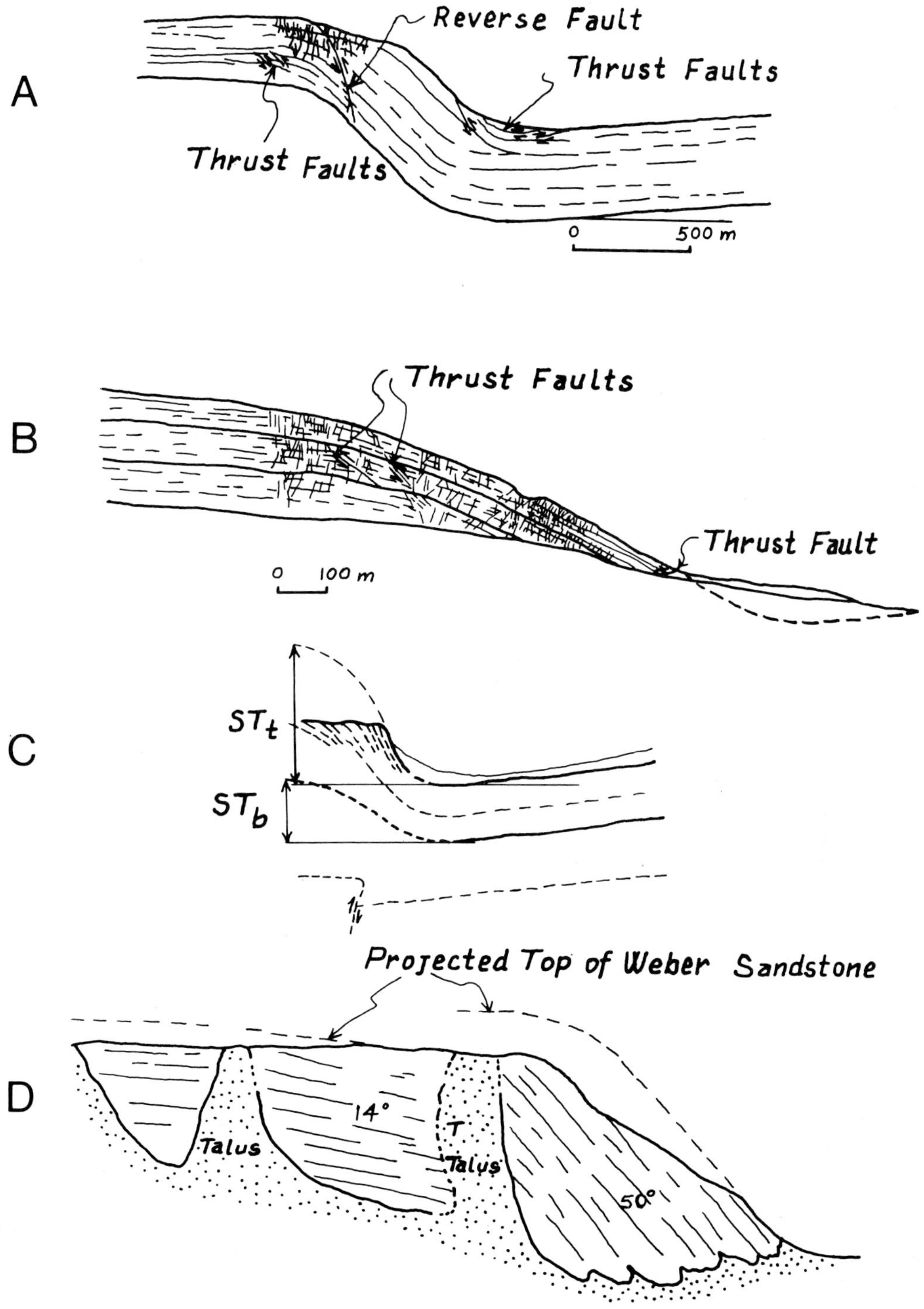

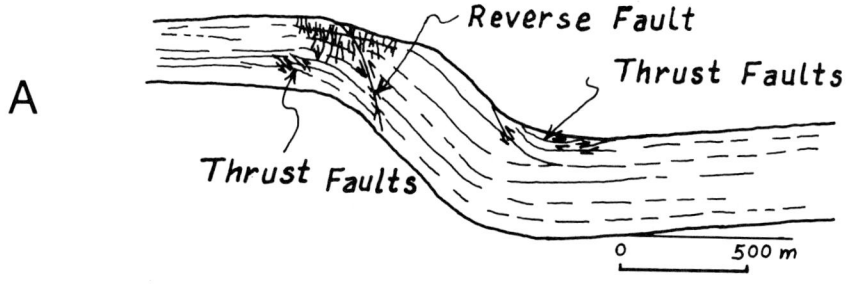

Figure 11. Vertical cross sections trending north, of the Weber Sandstone in the Yampa monocline, Dinosaur National Monument, Utah and Colorado. In C, ST_b and ST_t are the structural throw of the base and the top of the Weber Sandstone, respectively (after Cook, 1975; Cook and Stearns, 1975).

of monoclinal flexuring can be compared with experimental monoclines produced by Friedman, Logan, and others. Friedman and others (1976) and Logan and others (this volume) described experimental monoclinal flexures developed in multilayers comprised of sandstone and limestone, and presented distributions of strain and stress with the multilayers derived from analyses of microfractures, calcite twins, and thickening of layers. The multilayers were subjected both to confining pressure and to displacement along a fault at one side of the multilayer (Fig. 12). According to descriptions by Friedman and others (1976, p. 1053), bedding-plane slip within the multilayer was negligible. Thus, the conditions of the experiment shown in Figure 12 closely correlate with those assumed in the theoretical model of a multilayer with bonded contacts (Figs. 6, 8; Table 2). In both the experiment and the theoretical model the multilayer is simple, with alternating soft and stiff layers, and a small displacement occurs along a vertical fault.

According to Figure 12, the flexure is tight and narrow in the lower part, near the fault, and wide and open above. Also, layer-parallel extension is evident in the anticlinal bend of the experimental monocline, and layer-parallel compression is evident in the synclinal bend. The extension in the anticlinal bend is definitely inconsistent with the buckling mode of monocline formation (Fig. 8). Indeed, all the features are evident in the theoretical monoclinal flexure produced by drape folding of a multilayer with bonded contacts (Fig. 6).

CONCLUSIONS

Monoclines are generally simple-appearing structures with a single limb and smooth profile. Powell (1873) displayed usually sound physical insight when he recognized the close association of faults below and monoclinal flexures above. Further, Sanford (1959) provided a sound model of monoclinal flexuring in which the overburden is deformed passively as a result of faulting below. Thus, the model of drape folding is an attractive model of monocline formation. Yet, the simple drape-folding model cannot account for many observations made in detailed studies of monoclinal flexures. There is abundant evidence of layer-parallel shortening in anticlinal zones of many monoclines; therefore, shortening must be incorporated in a general model of monocline formation. Further, we have shown that monoclinal kink bands can develop in response to layer-parallel shortening and shear, even without the necessity of the marked local disturbance produced by displacement along a fault below (Reches and Johnson, 1976). We do not know

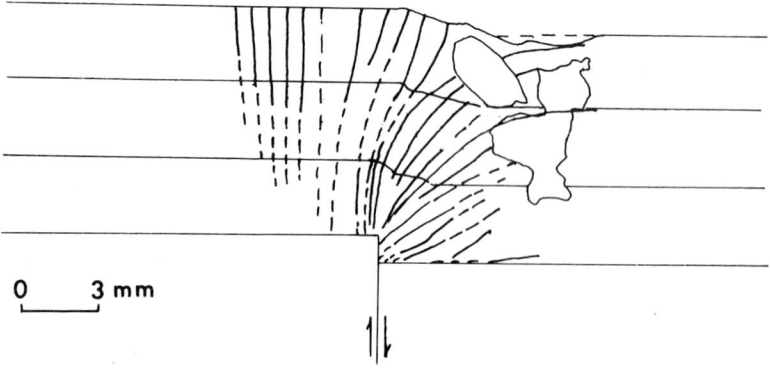

Figure 12. Map of trajectories of the maximum principal compressive stress as interpreted from orientations of faults and microfractures in an experimental passive fold. Stippled layer is sandstone; other layers are limestone (after Logan and others, this volume).

whether any large monoclines have been produced in this way, but it is possible.

We imagine that a monoclinal flexure generally initiates above a disturbance, such as a fault, an igneous intrusion, a local steepening of layers, or a buckle. Without such a disturbance, there generally is no theoretical basis for the initiation and localization of a monoclinal flexure. Our general model of monocline formation includes effects of layer-parallel shortening and shear, of vertical displacement of the base of a multilayer, and of a wide range of dimensional and rheological properties of layers as well as properties of contacts between layers. We have analyzed specific cases of the general model: drape folding, in which the disturbance at the base causes the overburden to deform passively; buckling, in which the layer-parallel shortening and the vertical uplift interact to amplify the perturbation at depth; and kinking, in which buckling and yielding instabilities result in a kink band in the overburden.

We solved some specific examples of multilayers subjected to drape folding and to buckling in order to determine the general characteristics of monoclines produced by pure forms of these processes. The characteristic features are summarized as follows:

Drape Folding

1. Monoclinal flexure is simple, with open anticlinal and synclinal bends.
2. Curvature increases downward so that monocline is tight below and open above.
3. Vertical displacement is constant or decreases up-section.
4. There are zones of layer-parallel extension and zones of layer-parallel shortening. The pattern of the zones can be complicated in irregular multilayers; however, the lack of zones of extension is incompatible with the drape-folding mechanism.

Buckling

1. Monoclinal flexure is associated with an anticline and a syncline.
2. Curvature is constant, increases or decreases up-section.
3. Vertical displacement is constant or increases up-section.
4. Layer-parallel shortening prevails at all levels (Figs. 7B, 8B).

Kinking

1. Straight limbs and distinct hinge zones.
2. Internal strain indicates layer-parallel shortening at all levels.
3. Yielding or faulting in tight hinge zones.

These features can be used as rules-of-thumb in order to recognize effects of the idealized processes in monoclines. However, it is clear from the analyses of a few field examples discussed in previous pages that none of the idealized processes adequately accounts for the field observations. In general, we must rely on the general model to interpret the field observations.

The general model accounts remarkably well for the main structural features of the Palisades monocline in the Grand Canyon and the Yampa monocline in Dinosaur National Park; none of the field observations contradicts the theoretical predictions. Further analysis of monoclinal flexuring will require consideration of high amplitudes and relatively complicated, realistic, rheological properties of layers and contacts between layers, and precise definition of structural units.

ACKNOWLEDGMENTS

We especially thank Raymond Fletcher, Stanford University, for advice and assistance with the theory, and David Pollard, U.S. Geological Survey, Menlo Park, California, for criticizing the manuscript. The research reported here was financed by the National Science Foundation Grant No. EAR 76-03273.

APPENDIX 1. DERIVATION OF BASIC EQUATIONS FOR ANALYSIS OF BUCKLING AND DRAPING

This appendix has two purposes. It presents several steps in the derivation of displacements or velocities as well as mean stress for two special cases, *drape folding* and *buckling*, of our general model of monoclinal flexuring. Also, it presents solutions for incompressible and compressible linear elastic or viscous materials. Analogous solutions for power-law elastic or viscous materials and for incompressible, nonlinear elastic materials have been derived in some detail by Johnson (1977, chap. 10). Solutions for linear viscous and power-law viscous materials were first derived by Fletcher (1974, 1977).

The general model of monoclinal flexuring is based on a multilayer with layers of various thicknesses and rheological properties, subjected to both stress and displacement boundary conditions. The solutions we present in the following pages include most aspects of the general model. They do not include second- or higher-order effects, including that of yielding of contacts, treated elsewhere (Honea and Johnson, 1976; Reches and Johnson, 1976), which are required to follow high-amplitude development of monoclinal flexures and to determine conditions required for localization as in kink bands. The solutions do represent a wide range of rheological properties, however, so they are relatively general first-order solutions to the general model of monoclinal flexuring.

One useful feature of the analysis is that the solutions presented here are valid for both drape folding and buckling, as well as for gravity instability. The differences between these modes become apparent only through boundary conditions; the general solutions are identical.

Thus, for drape folding, we specify displacements at the base of the multilayer. For buckling, we specify initial displacements throughout the multilayer as well as layer-parallel shortening. For gravity instability, we specify relative densities of various layers and initial displacements throughout the multilayer. We can, of course, combine all three of these. We do combine the first two in our analyses of buckling. For that model, we assume that the initial displacements throughout the multilayer are produced by differential vertical displacement on a fault or a local fold below. Then we determine the amplification of these displacements as a result of layer-parallel shortening.

Linear Elastic or Viscous Materials

The study of drape folding or buckling of linear elastic or viscous materials is especially simple. We solve the basic differential equations and specify boundary conditions at interfaces between layers. In order for buckling to be possible, the interfaces must be perturbed, but for drape folding they can be flat.

For drape folding, we specify the stress-strain relations for *incompressible* materials in plane strain,

$$s_{xx} = p + 2G(\partial u/\partial x) \tag{4a}$$

$$s_{yy} = p + 2G(\partial v/\partial y) \tag{4b}$$

$$s_{xy} = G(\partial v/\partial x + \partial u/\partial y) \tag{4c}$$

$$s_{zz} = p = (s_{xx} + s_{yy})/2 \tag{4d}$$

or *compressible* materials

$$s_{xx} = (2G + \lambda)(\partial u/\partial x) + \lambda\,(\partial v/\partial y) \tag{5a}$$

$$s_{yy} = (2G + \lambda)(\partial v/\partial y) + \lambda\,(\partial u/\partial x) \tag{5b}$$

$$s_{xy} = G(\partial v/\partial x + \partial u/\partial y) \tag{5c}$$

where s is stress, p is mean stress, u and v are displacements in the x- and y-directions, respectively, G is the shear modulus, and λ is a Lamé constant:

$$\lambda = 2G\nu/(1 - 2\nu). \tag{6}$$

Here ν is Poisson's ratio. For viscous materials, u and v are velocities, and G and λ are coefficients of viscosity (Johnson, 1970, p. 272–276).

Substituting the rheological equations (4) or (5) into the equilibrium equations, we derive, for *incompressible* materials,

$$\partial u/\partial x = -\partial v/\partial y \tag{7a}$$

$$G\left[(\partial^2 u/\partial x^2) + (\partial^2 u/\partial y^2)\right] + \partial p/\partial x = 0 \tag{7b}$$

$$G\left[(\partial^2 v/\partial x^2) + (\partial^2 v/\partial y^2)\right] + \partial p/\partial y = -\gamma \tag{7c}$$

Here equation (7a) is the condition of incompressibility and γ is the unit weight of the material (density times acceleration of gravity).

For *compressible materials*,

$$(2G + \lambda)(\partial^2 u/\partial x^2) + (G + \lambda)(\partial^2 v/\partial x \partial y) + G(\partial^2 u/\partial y^2) = 0 \tag{8a}$$

$$(2G + \lambda)(\partial^2 v/\partial y^2) + (G + \lambda)(\partial^2 u/\partial x \partial y) + G(\partial^2 v/\partial x^2) = -\gamma. \tag{8b}$$

We solve the equilibrium equations for *incompressible material* by eliminating the mean stress, p, between equations (7b) and (7c) and introducing a displacement function ψ such that

$$u = \partial\psi/\partial y \tag{9a}$$

$$v = -\partial\psi/\partial x \tag{9b}$$

which satisfy the condition of incompressibility, equation (7a). Thus, equations (7) and (9) provide the biharmonic equation in ψ:

$$\partial^4\psi/\partial x^4 + 2(\partial^4\psi/\partial x^2\,\partial y^2) + \partial^4\psi/\partial y^4 = 0. \tag{10}$$

The sinusoidal solution to equation (10) is

$$\psi = (1/\ell)\{[a + b(\ell y - 1)]\exp(\ell y) - [c + d(\ell y + 1)]\exp(-\ell y)\}\cos(\ell x) \tag{11}$$

where a, b, c, and d are arbitrary constants L is wavelength and ℓ is wave number, $2\pi/L$. Substituting equation (11) into equations (9),

$$u = \partial\psi/\partial y = [(a + b\ell y)\exp(\ell y) + (c + d\ell y)\exp(-\ell y)]\cos(\ell x) \tag{12a}$$

$$v = -\partial\psi/\partial x = \{[a + b(\ell y - 1)]\exp(\ell y) - [c + d(\ell y + 1)]\exp(-\ell y)\}\sin(\ell x). \tag{12b}$$

The mean stress is computed by substituting equations (12a) and (12b) into equations (7) and integrating,

$$p = -2G\ell[b\exp(\ell y) - d\exp(-\ell y)]\sin(\ell x) + p_0 - \gamma y \tag{12c}$$

where p_0 is a constant.

We solve equilibrium equations (8) for *compressible material* by determining displacements,

u and v, that satisfy them. First we eliminate u between the two equations and derive a biharmonic equation in v

$$\partial^4 v/\partial x^4 + 2(\partial^4 v/\partial x^2 \partial y^2) + \partial^4 v/\partial y^4 = 0. \tag{13a}$$

Then we eliminate v between the two equations to derive

$$\partial^4 u/\partial x^4 + 2(\partial^4 u/\partial x^2 \partial y^2) + \partial^4 u/\partial y^4 = 0. \tag{13b}$$

The sinusoidal solution to equation (13a) is

$$v = [(a + b\ell y)\exp(-\ell y) + (c + d\ell y)\exp(\ell y)]\sin(\ell x), \tag{14a}$$

where a, b, c, and d are constants. The differential equation for u, equation (13b), is of the same form as that for v, equation (13a), so the solutions for u and v should be similar. Also, the solutions must satisfy equilibriums equations (8), so we select the cosine solution to equation (13b)

$$u = [(e + f\ell y)\exp(-\ell y) + (g + h\ell y)\exp(\ell y)]\cos(\ell x),$$

where e, f, g, and h are constants. We can eliminate constants $e, \ldots, h$ by expressing them in terms of $a, \ldots, d$. We substitute the equations for the displacements into equations (8) and set the coefficients of terms in $\ell y \exp(-\ell y)$, $\exp(-\ell y)$, $\ell y \exp(\ell y)$ equal to zero. These terms provide four equations with which we can express constants $e, \ldots, h$ in terms of constants $a, \ldots, d$. The result is

$$u = -\{[a + b(\ell y + 4v - 3)]\exp(-\ell y) - [c + d(\ell y + 3 - 4v)]\exp(\ell y)\}\cos(\ell x). \tag{14b}$$

Here we have ignored effects of the weight of the material; that is, we have set γ in equation (8b) equal to zero. In order for equations (14) to satisfy equation (8b), we must add terms that account for displacements caused by the effects of gravity. We shall not do so.

Thus, we have derived the solution for displacements and mean stress for incompressible material, equations (12), and for compressible material, equations (14). The solutions are also valid for velocities for viscous materials, in which case u and v are velocities instead of displacements, and the material constants are viscosity coefficients instead of elasticity moduli. In order to solve problems where density contrasts and layer-parallel shortening are negligible; that is, in order to solve the *drape-folding* model, we merely specify boundary conditions and determine values for the arbitrary constants, a, b, c, and d. Details of this process are discussed in Appendix 2.

The solution of buckling problems requires a few more concepts which we shall discuss now. In buckling problems we consider the amplification of perturbations of flat surfaces of layers; that is, we determine conditions under which the perturbations grow in amplitude. This is the approach which Fletcher (1974, 1977) has introduced to study folding of viscous materials. According to this approach, we imagine that interfaces between layers are initially nonplanar and that the shapes of the interface surfaces can be expressed in terms of Fourier series. The initial shape, for example, might be a sine form,

$$v_i = \delta_i \sin(\ell x), \tag{15a}$$

where i refers to initial deflection, δ_i is the initial amplitude of the sinusoidal wave, and ℓ is the wave number

$$\ell = 2\pi/L, \tag{15b}$$

where L is wavelength. For viscous materials, v_i remains as the initial displacement of the interface, *not* the initial velocity.

Then, the layers are shortened and thickened uniformly as a result of layer-parallel shortening, producing displacements U and V and, nonuniformly, producing displacements u and v. The uniform stresses are designated S. They are, for *incompressible* materials,

$$S_{xx} = 2G(\partial U/\partial x) + P \tag{16a}$$

$$S_{yy} = 2G(\partial V/\partial y) + P \tag{16b}$$

$$P = (S_{xx} + S_{yy})/2 \tag{16c}$$

and, for *compressible* materials,

$$S_{xx} = (2G + \lambda)(\partial U/\partial x) + \lambda(\partial V/\partial y) \tag{17a}$$

$$S_{yy} = (2G + \lambda)(\partial V/\partial y) + \lambda(\partial U/\partial x). \tag{17b}$$

The nonuniform stresses are defined in equations (4) and (5). In the linear theory we ignore interactions between uniform and nonuniform states of stress and strain so they can be simply superimposed. Further, for the linear theory, the boundary conditions will be defined for the initial, undeformed boundaries. Thus, the displacements must be infinitesimal for the analysis to be valid.

Boundary Conditions

The boundary conditions are treated the same way for the linear elastic or viscous, incompressible or compressible materials. Figure 13 shows part of a boundary between two elastic materials or between an elastic material and a stress-free surface. The stresses acting on the surface can be expressed in terms of a normal stress, σ_{nn}, and a shear stress, σ_{ns}, which are parallel to local coordinates n and s. The slope angle of the surface is locally θ, the angle between the x-direction and the s-direction (Fig. 13). The stresses acting in the x- and y-directions are σ_{xx}, σ_{yy}, and σ_{xy}, where, according to derivations given earlier, the total stresses are the sums of the uniform and nonuniform stresses,

$$\sigma_{xx} = S_{xx} + s_{xx} \tag{18a}$$

$$\sigma_{yy} = S_{yy} + s_{yy} \tag{18b}$$

$$\sigma_{xy} = s_{xy}. \tag{18c}$$

We can derive the relation between the stresses acting on the surface and the stress components in the x- and y-directions by means of Mohr's circle, and they are

$$\sigma_{nn} = \sigma_{yy}\cos^2\theta + \sigma_{xx}\sin^2\theta - 2\sigma_{xy}\sin\theta\cos\theta \tag{19a}$$

$$\sigma_{ns} = (\sigma_{yy} - \sigma_{xx})\cos\theta\sin\theta + \sigma_{xy}(\cos^2\theta - \sin^2\theta). \tag{19b}$$

Now we introduce an approximation. We assume that the slope angle θ is so small that the sine is approximately equal to the tangent,

$$dv_i/dx = \tan\theta \simeq \sin\theta,$$

and that the cosine is nearly equal to unity. Further, we assume that products of the initial slope and the nonuniform stresses, such as $(s_{xx})(dv_i/dx)$, are negligible, but that products of the initial slope and the uniform stresses are significant. (We already have assumed that products of the uniform stresses and the nonuniform strains and rotations are negligible.) With these assumptions, equations (18) and (19) provide

$$\sigma_{nn} \simeq S_{yy} + s_{yy} = \sigma_{yy} \tag{20a}$$

$$\sigma_{ns} \simeq s_{xy} + (S_{yy} - S_{xx})(dv_i/dx). \tag{20b}$$

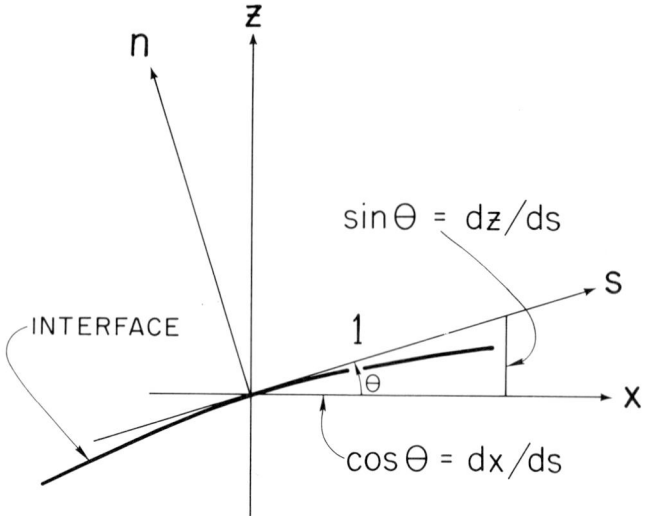

Figure 13. Boundary stresses at the contact between two layers with initial deflection.

One other aspect of boundary conditions must be considered for multilayers with contrast in densities. For such problems we specify that the boundary stresses are satisfied at the initial position of an interface. Thus, if a layer has a thickness of T, then the position of the interface is

$$y = T + v_i,$$

where v_i is the initial deflection of the interface. Then expressions for the mean stress, equations (12c) and (30), include terms such as γv_i, which might "drive" the folding (Ramberg, 1967; Johnson, 1977, p. 371).

Thus, we have presented all the basic equations required to solve problems involving either the drape-folding mode or the buckling mode of our general model of monoclinal flexuring for elastic and viscous materials. We shall discuss some of the details of solutions in Appendix 2.

APPENDIX 2. SOLUTIONS FOR MULTILAYERS

In this appendix we use the general solutions obtained in Appendix 1 to calculate the displacements and strains resulting from drape folding and buckling of a multilayer. We present solutions for incompressible, linear elastic materials, but the solutions are valid for viscous materials in which the displacements are replaced by velocities and the strains by strain rates. First we outline the general method used and then present the solutions for drape folding and for buckling of a multilayer with bonded contacts between layers.

The *sinusoidal solution* for an incompressible elastic material appears in equations (11). The stresses are derived by substituting equations (12) into equations (4),

$$s_{yy} = 2G\ell\{[a + b(\ell y - 1)]\exp(\ell y) + [c + d(\ell y + 1)]\exp(-\ell y)\}\sin(\ell x) - \gamma y + p \qquad (21a)$$

$$s_{xx} = -2G\ell\{[a + b(\ell y + 1)]\exp(\ell y) + [c + d(\ell y - 1)]\exp(-\ell y)\}\sin(\ell x) - \gamma y + p \qquad (21b)$$

$$s_{xy} = 2G\ell[(a + b\ell y)\exp(\ell y) - (c + d\ell y)\exp(-\ell y)]\cos(\ell x). \qquad (21c)$$

Let us rewrite equations (12a) and (12b),

$$u = [(a + b\ell y) \exp(\ell y) + (c + d\ell y) \exp(-\ell y)] \cos(\ell x) \qquad (21d)$$

$$v = \{[a + b(\ell y - 1)] \exp(\ell y) - [c + d(\ell y + 1)] \exp(-\ell y)\} \sin(\ell x). \qquad (21e)$$

The strains are

$$e_{xx} = \frac{\partial u}{\partial x} = -\ell[(a + b\ell y) \exp(\ell y) + (c + d\ell y) \exp(-\ell y)] \sin(\ell x) \qquad (21f)$$

$$e_{yy} = -e_{xx} \qquad (21g)$$

$$e_{xy} = \tfrac{1}{2}\left(\frac{\partial u}{\partial y} + \frac{\partial v}{\partial x}\right) = \ell[a + b\ell y) \exp(\ell y) - (c + d\ell y) \exp(-\ell y)] \cos(\ell x). \qquad (21h)$$

where e_{xx} and e_{yy} are the normal strains and e_{xy} is the shear strain. Other terms were defined in Appendix 1.

We shall assume that the density of each layer is the same, so that γ is the unit weight of the rock.

Equations (21) are the general solutions for each layer; thus, stresses, displacements, and strains in each layer are determined by four coefficients for each layer, a, b, c, and d. For n layers in the multilayer, there are n sets of equations similar to equations (21), but with different coefficients and shear moduli. One has to solve for all the coefficients to calculate the strains, stresses, and displacements in any layer. This solution, however, is for a single, sinusoidal wavelength, L, whereas a monoclinal flexure may be described in terms of the sum of many wavelengths in the form of a Fourier series. Therefore, we superimpose many (30 or more) wavelengths of the appropriate Fourier series to obtain solutions for monoclinal flexures.

The solutions for drape folding and buckling are essentially identical, only the boundary conditions differ. For drape folding, the initial configuration is a series of flat layers with zero uniform stresses, and the "driving" mechanism for the deformation is the displacement applied at the base. For buckling, the multilayer has an initial monoclinal flexure and is subjected to uniform layer-parallel shortening; the interaction between the uniform stresses and the initial slopes at interfaces of layers is the "driving" mechanism for the growth of the monoclinal flexure.

Drape Folding

In passive folding there are neither uniform stresses nor initial slopes (Fig. 3). Therefore, only the incremental strains and stresses are calculated. For the displacement, V, at the base, we choose the Fourier series (Fig. 14),

$$V = h \sum_{1}^{\infty} a_n \sin \frac{n\pi x}{L}, \qquad (22a)$$

where

$$a_n = \frac{2}{n\pi}\left[\cos(n\pi) - \cos\left(\frac{n\pi F}{L}\right)\right] + \frac{1}{\pi}\left[\frac{\sin(\pi/2 + n\pi)}{(N + n)} - \frac{\sin(\pi/2 - n\pi)}{(N - n)}\right] \qquad (22b)$$

for $N \neq n$, and

$$a_N = -F/L, \qquad (22c)$$

where F is defined in Figure 13, and $N = L/2F$.

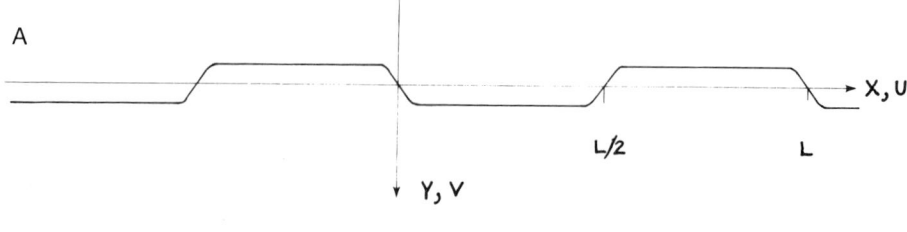

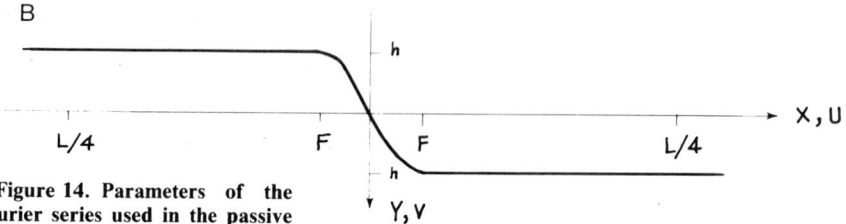

Figure 14. Parameters of the Fourier series used in the passive folding and buckling analyses of monoclinal flexures. (A) The periodic structure assumed. (B) Enlargement of the central part of the structure. $V(x)$ is the monoclinal wave.

$$V(X) = \begin{cases} h & F \leq X \leq L/4 \\ h \sin(\frac{\pi X}{2F}) & -F \leq X < F \\ -h & -L/4 < X < -F \end{cases}$$

One can vary the width of the flexure at the base from a vertical fault to a wide monoclinal flexure through the factor F (Fig. 14). Our calculations are for the range $F = 0.04 L_f$ to $F = 0.2 L_f$, where L_f is the longest wavelength of the Fourier series (Fig. 14).

Equations (21) and (22) should be substituted into the boundary conditions for drape folding. For example, the boundary conditions for the base of the multilayer with bonded contacts are at $y_1 = T_1/2$,

$$(u)_1 = 0 \qquad (23a)$$

$$(v)_1 = h \Sigma (a)_n \sin \frac{2n\pi x}{L_f}. \qquad (23b)$$

Substituting equations (21d), (21e), and (22) into equations (23),

$$a_1 \exp(k_1) + b_1 k_1 \exp(k_1) + c_1 \exp(-k_1) + d_1 k_1 \exp(-k_1) = 0 \qquad (24a)$$

$$a_1 \exp(k_1) + b_1(k_1 - 1) \exp(k_1) - c_1 \exp(-k_1) - d_1(k_1 + 1) \exp(-k_1) = h(a)_n, \qquad (24b)$$

where $k_1 = T_1 \ell/2$, $\ell = 2\pi n/L_f$, and $(a)_n$ is the coefficient of the Fourier series.

Similarly, by substituting equations (31) into the boundary conditions for drape folding of a bonded multilayer (Reches, 1977), a system of $4n$ linear equations is derived,

$$[A] \cdot [X] = [B], \qquad (25)$$

where A is the matrix of known constants, X is the vector of the unknown coefficients $a_1, b_1, c_1, d_1, a_2, \ldots, c_n, d_n$ and B is the vector of the "driving" terms (Reches, 1977). The "driving" terms are the displacements at the base in passive folding. The system of linear equations is solved by regular subroutines for simultaneous solutions which are available in any computation center.

After solving equations (24) for a single wavelength, we substitute the vector X of coefficients into equations (21) and calculate the strains and displacements within each layer. We repeat this procedure for at least thirty wavelengths defined by the Fourier series, equation (22).

Thereafter, we sum the strains and displacements to give the final results, which are summarized in several diagrams (Figs. 6, 7).

The procedure for drape folding of multilayer with frictionless contacts is essentially the same, but with different boundary conditions (Reches, 1977).

Buckling

We consider the buckling of a multilayer with initial monoclinal deflection over an infinite medium (Figs. 8, 9). Both the multilayer and the medium are subjected to horizontal shortening, $\partial U/\partial x$. Using equations (7a), (4a) (4b), and (15a), we rewrite equations (19),

$$\sigma_{nn} = S_{yy} + s_{yy} \tag{26a}$$

$$\sigma_{ns} = s_{xy} - 4G(\partial U/\partial x)\, dv_i/dx, \tag{26b}$$

where σ_{nn} is the normal stress and σ_{ns} is the shear stress at a contact, and dv_i/dx is the initial slope. The initial slope dv_i/dx is calculated for every layer by means of the drape-folding analysis described in earlier paragraphs and then substituted into the boundary conditions for buckling, equations (26).

Let us derive the boundary conditions for one contact as an example. We choose the contact between the lower medium and the multilayer, assuming bonded contacts. The coefficients, a_m and b_m, for the medium must vanish to satisfy the conditions of vanishing stresses and displacements for y_m approaching infinity, because these coefficients are associated with positive exponents, equations (21). Therefore, at the contact

$$y_m = \delta_i \sin(\ell x);\ y_1 = T_1/2 + \delta_i \sin(\ell x),$$

the boundary conditions are (Table 5)

$$(\sigma_{nn})_m - (\sigma_{nn})_1 = 0 \tag{27a}$$

$$(\sigma_{ns})_m - (\sigma_{ns})_1 = 0 \tag{27b}$$

$$(u)_m - (u)_1 = 0 \tag{27c}$$

$$(v)_m - (v)_1 = 0 \tag{27d}$$

Substituting equations (21) and (26) into equations (27) yields the following equations:

$$-c_m R_m + d_m R_m + a_1 \exp(k_1) + b_1(k_1 - 1)\exp(k_1) + c_1 \exp(-k_1) \\ + d_1(k_1 + 1)\exp(-k_1) = 0 \tag{28a}$$

$$c_m R_m + 0 + a_1 \exp(k_1) + b_1 k_1 \exp(k_1) - c_1 \exp(-k_1) \\ + d_1 \exp(-k_1) = 4(\partial U/\partial x)(1 - R_m)\partial V/\partial x) \tag{28b}$$

$$-c_m + 0 + a_1 \exp(k_1) + b_1 k_1 \exp(k_1) + c_1 \exp(-k_1) + d_1 \exp(-k_1) = 0 \tag{28c}$$

$$c_m + d_m + a_1 \exp(k_1) + b_1(k_1 - 1)\exp(k_1) - c_1 \exp(-k_1) \\ - d_1(k_1 + 1)\exp(-k_1) = 0 \tag{28d}$$

where $k_1 = T_i \ell/2$, $\ell = 2\pi n/L_f$, and $R_m = G_m/G_i$.

Similarly, by substituting equations (21) and (26) into the boundary conditions, we obtain a system of $4n$ linear equations for the bonded multilayer (Reches, 1977),

$$[A] \cdot [X] = [B] \tag{29}$$

where $\mathbf{A}$ is the matrix of the known constants, $\mathbf{X}$ is the vector of the unknown coefficients $c_m, d_m, a_1, b_1, \ldots, c_n, d_n$, and $\mathbf{B}$ is the vector of "driving" terms (Reches, 1977).

The "driving" terms in the buckling case are the products of the initial slopes, the uniform strain and a shear modulus, equations (26b) and (28b). The system of linear equations is solved by regular subroutines for simultaneous equations. In a manner similar to the calculations of drape folding, we repeat the computations for at least thirty wavelengths of the Fourier series, and we sum the displacements and strains (Figs. 8, 9). The procedure for analysis of buckling of a multilayer with frictionless contacts is essentially the same, but with different boundary conditions (Reches, 1977).

A term involving the unit weight of the rock would enter the equation for the normal stress at the free upper surface, but for convenience it has been deleted in our analysis. Specific consideration of its effect would require that another variable be specified. Its effect, however, is clearly that of reducing amplitudes of folding near the free surface, the amount depending upon the relative magnitudes of the uniform horizontal stress, S_{xx}, and the product, $\delta_i \gamma$.

REFERENCES CITED

Baker, A. A., 1935, Geological structure of southeastern Utah: Am. Assoc. Petroleum Geologists Bull., v. 19, p. 1472–1507.

Biot, M. A. Ode, H., and Roever, W. L., 1961, Experimental verification of the folding of stratified viscoelastic media: Geol. Soc. America Bull., v. 72, p. 1621–1630.

Chapple, W. M., and Spang, J. H., 1974, Significance of layer-parallel slip during folding of layered sedimentary rocks: Geol. Soc. America Bull., v. 85, p. 1523–1534.

Clark, S. P., Jr. (ed.), 1966, Handbook of physical constants: Geol. Soc. America Mem. 97, 587 p.

Cook, R. A., 1975, Mechanisms of sandstone deformation: A study of the drape-folded Weber sandstone in Dinosaur National Monument, Colorado and Utah [Ph.D. dissert.]: College Station, Texas A & M Univ.

Cook, R. A., and Stearns, D. W., 1975, Mechanisms of sandstone deformation: A study of the drape-folded Weber Sandstone in Dinosaur National Monument, Colorado and Utah: Rocky Mtn. Assoc. Geol., 1975 Symposium, p. 21–32.

Dietrich, J. H., and Carter, N. L., 1969, Stress-history of folding: Am. Jour. Sci., v. 267, p. 129–154.

Dutton, C. E., 1882, Tertiary history of the Grand Canyon District: U.S. Geol. Survey Mon. 2, 264 p.

Fletcher, R. C., 1974, Wavelength selection in the folding of a single layer with power-law rheology: Am. Jour. Sci., v. 274, p. 1029–1043.

—— 1977, Folding of a single viscous layer: Exact infinitesimal-amplitude solution: Tectonophysics, v. 39, p. 593–606.

Fletcher, R. C. and Sherwin, Jo-Ann, 1978, Relation between observed fold arc length and theoretical preferred wavelength in single-layer folding: Am. Jour. Sci. (in press).

Friedman, M., Handin, J., Logan, J. M., Min, K. D., and Stearns, D. W., 1976, Experimental folding of rocks under confining pressure: Pt. III. Faulted drape folds in multilithologic layered specimens: Geol. Soc. America Bull., v. 87, p. 1049–1066.

Honea, E., and Johnson, A. M., 1976, A theory of concentric, kink, and sinusoidal folding and of monoclinal flexuring of compressible, elastic multilayers. Pt. IV, Development of sinusoidal and kink folds in multilayers confined by rigid boundaries: Tectonophysics, v. 30, p. 197–239.

Howard, J. H., 1966, Structural development of the Williams Range thrust, Colorado: Geol. Soc. America Bull., v. 77, p. 1247–1264.

Hudleston, P. J., 1973, An analysis of "single-layer" folds developed experimentally in viscous media: Tectonophysics, v. 16, p. 189–214.

Hudleston, P. J., and Stephansson, O., 1973, Layer shortening and foldshape development in buckling of single layers: Tectonophysics, v. 17, p. 299–321.

Johnson, A. M., 1970, Physical processes in geology: San Francisco, Freeman, Cooper and Co., 577 p.

—— 1977, Styles of folding: New York, Elsevier Pub. Co., 406 p.

Johnson, A. M., and Ellen, S. D., 1974, A theory of concentric, kink, and sinusoidal folding and of monoclinal flexuring of compressible, elastic multilayers. Pt. I, Introduction: Tectonophysics, v. 21, p. 301–339.

Johnson, A. M., and Page, B. M., 1976, A theory of concentric, kink, and sinusoidal folding and of monoclinal flexuring of compressible, elastic multilayers. Pt. VII, Development of folds within Huasna syncline, San Luis Obispo County, California: Tectonophysics, v. 33, p. 97–143.

Johnson, A. M., and Pollard, D. D., 1971, Mechanics of intrusion of laccoliths: Final rept. to NSF, Branner Library, Stanford Univ., Calif. 173 p.

Kelley, V. C., 1955, Monoclines of the Colorado Plateau: Geol. Soc. America Bull., v. 66, p. 789–804.

Logan, J. M., Friedman, M., and Stearns, M. T., 1978, Experimental folding of rocks under confining pressure: Pt. VI. Further studies of faulted drape-folds, *in* Matthews, V., ed., Folding associated with the basement faulting in the Rocky Mountain region: Geol. Soc. America Mem. 151 (this volume).

Min, K. D., 1974, Analytical and petrofabric studies of experimental faulted drape folds in layered rock specimens Ph.D. dissert.: College Station, Texas A & M Univ., 90 p.

Powell, J. W., 1873, Exploration of the Colorado River of the West and its tributaries explored in 1869–1872: Washington D.C., Smithsonian Inst., 291 p.

Prucha, J. J., Graham, J. A., and Nickelsen, R. P., 1965, Basement-controlled deformation in Wyoming province of Rocky Mountains foreland: Am. Assoc. Petroleum Geologists Bull., v. 49, p. 966–992.

Ramberg, H., 1967, Gravity, deformation, and the earth's crust: New York, Academic Press, 214 p.

——1970, Folding of laterally compressed multilayers in the field of gravity: Physics Earth and Planetary Interiors, p. 203–232.

Ramberg, I. B., and Johnson, A. M., 1976, Asymmetric folding in interbedded chert and shale of the Franciscan Complex, San Francisco Bay area, California: Tectonophysics, v. 32, p. 295–320.

Ramberg, H., and Strömgård, K. E., 1971, Experimental tests of modern buckling theory applied on multilayered media: Tectonophysics, v. 11, p. 461–472.

Reches, Z., 1977, The development of monoclines [Ph.D. dissert.]: Dept. Geology, Stanford Univ., 234 p.

——1978, Structure of the Palisades Creek branch of the East Kaibab monocline, Grand Canyon, Arizona: *in* Matthews, V., ed., Folding associated with the basement faulting in the Rocky Mountain region: Geol. Soc. America Mem. 151 (this volume).

Reches, Z., and Johnson, A. M., 1976, A theory of concentric, kink, and sinusoidal folding and of monoclinal flexuring of compressible, elastic multilayers. Pt. VI, Asymmetric folding and monoclinal kinking: Tectonophysics, v. 35, p. 295–335.

Sanford, A. R., 1959, Analytical and experimental study of simple geologic structures: Geol. Soc. America Bull., v. 70, p. 19–52.

Sherwin, J. A., and Chapple, W. M., 1968, Wavelengths of single-layer folds: A comparison between theory and observation: Am. Jour. Sci., v. 266, p. 167–179.

Stearns, D. W., 1971, Mechanisms of drape folding in the Wyoming province: Wyoming Geol. Assoc., 23rd Ann. Field Conf. Guidebook, p. 125–143.

Treagus, S. H., 1973, Buckling stability of a viscous single-layer system, oblique to the principal compression: Tectonophysics, v. 19, p. 271–289.

Walcott, C. D., 1890, Study of line of displacement in the Grand Canyon of the Colorado in northern Arizona: Geol. Soc. America Bull., v. 1, p. 49–64.

Willis, B., 1894. Mechanics of Appalachian structure: 13th Ann. Rept. of the U.S. Geol. Survey, 1891–92, p. 213–281.

Zanemonets, V. B., Mikhaylov, V. O., and Myasnikov, V. P., 1976, A mechanical model of the deformation of block folding: Izvestia Physics of the Solid Earth, no. 10, p. 13–21.

Manuscript Received by the Society June 27, 1977
Manuscript Accepted August 25, 1977

Printed in U.S.A.

Geological Society of America
Memoir 151

Analytical solutions applied to structures of the Rocky Mountains foreland on local and regional scales

GARY COUPLES
DAVID W. STEARNS
Department of Geology and Center for Tectonophysics
Texas A&M University
College Station, Texas 77843

ABSTRACT

This paper presents conceptual models of basement configurations for several structural types in the Rocky Mountains foreland—upthrusts, rotated basement blocks, and plateau uplifts. These hypotheses of structural development are supported by calculated stress states and fracture patterns derived from them. The goodness-of-fit between field observations and the final geometries of the models leads to the suggestion of possible loading conditions for the initiation of Laramide structural development of uplifts in the foreland. These suggested loading conditions for individual features lead, in turn, to a hypothesized stress state consisting of a regional component of crustal horizontal compression superposed on the controlling stress fields caused by more local load variations at depth beneath the brittle upper crust.

INTRODUCTION

Although this volume is addressed to the folding of sedimentary rocks, this paper deals exclusively with the structural development of the fault-block geometries of the basement blocks that cause most of that folding. The empirical distinction between the mechanical response of the cratonic basement and that of the sedimentary veneer has been convincingly discussed by Stearns (1971, 1975) in his studies centered on the structural development of the Rocky Mountains foreland. For the type of basement uplift found there, he has proposed a conceptual model based on the folding (draping) of a ductile sedimentary section over the fault-bounded edge of a rigid block of uplifted basement rock.

His arguments relative to the mechanical decoupling of the sedimentary rocks from their underlying basement form an important philosophical construct, upon which this paper is allowed to stand separately. Those arguments also speak to the need to develop a comprehensive model to explain the variation of basement structural types across the foreland as components of a rational *system*.

In an attempt to explain the sequence of rotated basement blocks in just one cross section of the northern Bighorn Basin of Wyoming (where most basement faults intersect the basement/sedimentary rock contact at high angles), Stearns (1975) developed a hypothesis based on a solution in theoretical mechanics published earlier by Hafner (1951). However, the specific boundary-value solution employed by Stearns in that model does not lend itself to the explanation of another prominent structural style in the foreland: the upthrusts, especially those associated with rotated blocks. The upthrust structural type can be described as a mountain uplift with a margin characterized by a high-angle reverse fault at depth in the basement that flattens upward near the surface, with accompanying rigid body rotations and zones of extension (Stearns and others, 1975).

Opponents of the basement-uplift hypothesis (that is, those who say that loading conditions similar to those proposed by Stearns were not responsible for the deformations) apparently feel that the failure of this particular boundary-value solution to explain all the observed deformation types casts doubt on this type of model. Other objections to Stearns's assumed loading conditions arise from the widely accepted assumptions about crustal stress states derived from a generalized view of plate-tectonics theory—that crustal plates are in horizontal compression, with all structures necessarily related to thrusting and wrenching.

This paper will demonstrate that certain new boundary-value solutions, some of which incorporate aspects of the assumed loading conditions favored by plate tectonicists, explain the major features of the upthrust structural type and thereby support the general applicability of using analytical boundary-value solutions to study foreland structural development. The favorable goodness-of-fit of the structural configurations predicted from the theoretical solutions with cross sections constructed for the natural geometries, in conjunction with geometric arguments discounting the general applicability of the thrusting and/or wrenching hypotheses, supports the general form of the probable loading conditions leading to the observed structures of the Rocky Mountains foreland.

Previous Work

Within the geologic literature, there is a limited history of either developing or using analytical boundary-value solutions. Those publications significant in the study of the Laramide deformation of the basement in the Rocky Mountains foreland are discussed below.

Hafner (1951) was first to publish on this subject and in a sense may be considered a pioneer. His initial work has provided the basis on which other authors have founded their studies. Hafner was interested in the problem of crustal deformation. In dealing with this subject, he assumed that a segment of the crust could be modeled as an isolated beam with imposed boundary loads. Details of the mathematics are found in his paper, but the solutions were not applied to any specific geologic problem. Instead, he made some general speculations as to how his solutions might apply to a few large regions, for example, the Basin and Range and Rocky Mountains provinces.

One of the often-overlooked results of Hafner's study is that he was able to explain the existence of nonplanar (that is, curved in cross section) faults by means of an acceptable stress system that is also discussed in a companion paper by Hubbert (1951). Prior to that time, faults had been interpreted more simplistically, generally in terms of the supposed causative forces, but after Anderson's (1942) classic book on the dynamic interpretation of faults (which related faulting to the state of stress), they were generally assumed to be planar and representative of homogeneous stress fields. Hafner (1951) provided the first theoretical basis supporting curved faults and geologically reasonable patterns of mixed fault types.

Bengtson (1956) was the next contributor in sequence. His usage of boundary-value solutions was inspired by a field study on the Buffalo Fork area northeast of Jackson Hole, Wyoming. In this area, Bengtson mapped what we now would call drape folds over uplifted basement blocks. Associated with these he saw moderate-displacement thrust faults in the sedimentary section that steepened with depth ("upthrusts" in his terms). He presented solutions nearly identical to those of Hafner to explain the upthrusts *in the sediments* above a "deep-seated" vertical fault in the basement.

Sanford's (1959) solutions differed in character from those of Hafner (1951) in that he imposed displacements along the lower boundary of the beam instead of the stresses imposed by Hafner. Sanford did not specifically state a geologic model as inspiration, but he did discuss the *"folding" of sediments* above a displacing (forcing) member below (basement). Fracture-trajectory diagrams of the type presented by Hafner were not included by Sanford. He did, however, demonstrate that the potential fractures associated with a specific "piston-uplift" displacement boundary condition matched very closely with the shear zones developed in analog sand-box experiments.

In 1966, Howard used a technique nearly identical to that of Sanford (1959) to model the structural development of a specific geologic structure—the Williams Range thrust in north-central Colorado. Howard used the solutions to explain structures developed in the sedimentary rocks as a result of the motions of rigidly displacing basement blocks. Although we disagree with aspects of the *method* Howard used in employing the theoretical results, his solutions can stand as separate entities useful in constructing models of the type presented in this paper.

Couples (1977) published a suite of solutions for beams subjected to boundary stresses. The boundary conditions used in his study were more complicated than those in any of the previous publications. His investigation was initiated with the specific intent of explaining basement faulting in the foreland. In that paper, Couples also presented a model of the structural development of the basement blocks of the Wind River Mountains. However, as that model and others of its type are the principal subjects of this paper, Couples's model will not be discussed further at this time.

Gangi and others (1977) presented solutions for the problem of a uniform beam subjected to displacements at its base. In gross sense, their approach is similar to that of Sanford (1959) and Howard (1966), although their mathematical treatment is somewhat more elegant. Gangi and others (1977) directed their study toward the explanation of features observed in experimentally created drape folds. As such, their calculations did not consider the effects of body forces. Depending on the standard state (or gravitationally controlled stress field) assumed to exist in nature, body forces may or may not be important in influencing the orientation of faults (Couples, 1977). Therefore, the results of Gangi and others may not be useful directly. In any event, it is certainly more difficult to locate regions of the beam most likely to fracture when the influence on mean stress due to body forces is neglected, as in the case presented by Gangi and others. However, their procedure is more rigorous than previous studies and offers an exact solution to a sharp displacement discontinuity instead of to an approximate step.

The studies of Bengtson (1956), Sanford (1959), Howard (1966), and Gangi and others (1977) were not directed specifically toward explaining the generation of basement faults. Nevertheless, the mechanical assumptions they made in generating their solutions are essentially identical to those employed by Hafner (1951) and Couples (1977). Couples (this volume) shows that these assumptions render the solutions more suited to the study of faulting in basement materials. It is our opinion that their use should be so restricted, except in certain special cases such as that illustrated in the paper by Gangi and others (1977).

The remainder of this paper will be concerned with demonstrating our views concerning both the applications and limitations of these solutions in the study of certain geologic problems. Because of the restrictions inherent in the assumptions, the examples of their use that can be illustrated normally involve the deformation of basement rocks. Toward that end, we present four cases where the employment of boundary-value solutions has enhanced our understanding of the structural development of the basement in the Rocky Mountains foreland.

STATEMENT OF GEOLOGIC PROBLEMS

Geologic Constraints

1. Stearn's (1971) field observations (see also below) require that a rigid (that is, nonductile or nonfolding) condition apply to the basement portion of cross sections drawn for the foreland. Accordingly, "room" problems such as gaps or overlaps cannot be allowed to appear when the sections are restored. Similarly, models that contain holes or other room problems in the deformed state are also untenable. The plywood and clay model presented by Lowell (1974) violates this restriction.

2. The rigid nature of the basement blocks, coupled with the room-problem restrictions, makes it necessary to explain relative rotations between adjacent blocks by motion along curviplanar faults (see also below). Rotations of several different types are documented by Stearns (this volume).

3. The final major constraint on models is that they allow and/or explain *both* absolute up and absolute down motions during Laramide deformation (Stearns, 1971, 1975).

Stearns's Model

Stearns (1975) used boundary-value solutions in a study of basement-block faulting in the Rocky Mountains foreland. He did not generate any new solutions but instead employed those previously published by Hafner (1951). Primarily from subsurface information, Stearns constructed a profile of the top of the basement along a line drawn across the northern Bighorn Basin of Wyoming (Fig. 1). He was able to achieve a matching geometry by activating selected "faults" from one of Hafner's solutions (Fig. 2). In Figure 2 the top of the displaced rigid blocks conforms to the profile in Figure 1.

While the fit of the solution (in Fig. 2) is impressive, the fit in itself does not prove that the specific beam geometry, loading conditions, or actual fracture development are correct. It does, however, seem significant that the abstractions employed in the actual generation of the solution, as well as those invoked in the complex arguments leading to its application to the real world, result in a model that seems to explain so many interrelated portions of a total system. It is Stearns's model, and especially the logic of his approach, that form the basis upon which the remainder of this study depends.

Upthrust Problem

The term "upthrust" probably can be attributed to Link (1930), who defined it as a fault that is upwardly convex in cross section. There are a number of mountain ranges in the Rocky Mountains foreland that are probably bounded on at least one side by an upthrust fault in the basement. Prime examples are the southwestern margin of the Wind River Mountains and the southern boundary of the central Owl Creek Mountains, both of which are in Wyoming. Margins such as the eastern flank of the Beartooth Mountains might also be considered as upthrust, but there are some important differences noted below. Other upthrust ranges occur in the foreland, but those listed here are of particular interest to this study.

Interpretations of these structures have been diverse, ranging from simple thrusts, to fold thrusts, to upthrusts. The latter interpretation, which excludes the folding of basement rocks, is here considered to be generally applicable and is the one assumed. As used in this paper, the term "upthrust structure" will denote an uplift of basement rocks associated with a convex-upward fault. Empirical observations indicate that the rock mass on the uplifted block adjacent to the upthrust fault usually has experienced lateral extension, manifest and/or caused by normal faults within the basement rock mass itself (Stearns and others, 1975). Another important empirical observation is that the main portion of an upthrust mountain range normally is gently tilted. The basement surface normally dips about 10° to 15° away from the structurally highest part of the system that is almost universally immediately adjacent to the zone of extension.

It is important to note that these empirical "observations" are actually interpretive in part. The presence of the extension zone is inferred from either normal-faulted sedimentary rock sequences or intensely shattered basement rock. These features are determined from surface or near-surface information; subsurface data for these structures are very limited. The upthrust fault geometry itself is not well documented for most structures in the foreland; it only represents a prejudiced interpretation based on theoretical results, sand-box analog experiments, and a limited number of natural geometries. It is only one of several possible genetic interpretations that can honor the limited geometric data observed in the field, but it is the interpretation that most nearly honors *both* the behavioral properties of the basement rocks *and* the inferred structural configurations. The upthrust geometry is the working prejudice accepted here.

In summary, three generalizations can be made in describing the typical upthrust mountain range: (1) A convex-upward, upthrust fault in the basement rocks; (2) an extension region behind that fault; and (3) rigid rotations or plateau uplifts of the major basement blocks. Of course, the room problem restrictions and absolute up-and-down movement requirements discussed above apply to the upthrust structure as well.

Each of the authors discussed above (Hafner, 1951; Bengtson, 1956; Sanford, 1959; Howard, 1966; Couples, 1977; Gangi and others, 1977) derived solutions that contained potential fracture systems that included faults with upthrust geometries, and some of them (Sanford, 1959; Howard, 1966; and Couples, 1977) discussed the development of the extension zone behind the upthrust. Stearns and others (1975) have suggested a model (Fig. 3) for the development of an upthrust structure using a fault trajectory from Sanford (1959), but their model does not explain any rotation of the rigid blocks. The problem of rotations had not been addressed prior to the study of Couples (1977). None of the older solutions has proven useful in attempts to explain the attitudes of those basement blocks that are rotated away from the upthrust fault.

One conceptual model that (1) honors the three generalizations (upthrust fault, extension zone, rotations), (2) does not engender any room problems, and (3) allows for both absolute up and down motions is briefly described by Couples (1977). This model is based on two faults of circular cross-section concave toward one another. One fault has a "normal" sense of displacement at the surface (the extension zone), and the other exhibits "reverse" motion (the upthrust). The arcuate shapes of the faults allow relative rotations of the component blocks.

Stress-state calculations and potential shear-fracture orientations supporting the conceptual model just discussed also are presented by Couples (1977). A more comprehensive discussion of that model, and others developed subsequently, and of the use of analytical boundary-value solutions in structural modeling constitute the purpose of this paper. Discussions concerned with further details of the state-of-stress calculations and considerations related to the inexactness of the geometric match of the models are found in the paper by Couples (this volume).

APPLICATION OF SHEAR-FRACTURE TRAJECTORY DIAGRAMS ("HOW TO")

Stearns's (1975) application of shear-fracture diagrams (refer to Figs. 1, 2) is the approach on which this discussion is based. In simplest terms, the method employed in the application amounts to matching geometries inferred from nature to geometries predicted by the shear-fracture trajectory diagrams. Before describing the process involved in that matching procedure, it is necessary to discuss in some detail the kinematic constraints that arise from natural observations.

Kinematic Constraints

The rigidity of the basement rocks is deduced from the nonfolded, nonarched relationship between the Precambrian basement and the overlying sedimentary rocks in the long homoclinal (block 1) dip sections of the structures (Stearns, 1971). Although the exact basement–layered-rock contact is not always observable, other bedding surfaces higher in the stratigraphic section usually can be seen. It is assumed that these higher bedding surfaces, or at least those in the uniform Paleozoic part of the section, parallel the nonconformable planar surface of the Cambrian-Precambrian contact. This assumption is proven to be completely valid in any block 1 region where the sedimentary section is eroded deeply enough to expose the Precambrian-Cambrian contact.

But what does the rigidity mean? It means that individual blocks of basement behaved as rigid bodies, but the entire mass of basement in the foreland did not. In fact, if the entire basement had remained absolutely rigid, it now would be undeformed. Obviously, this is not the case.

Therefore, it should not be confusing to refer to "rigid basement blocks." It is understood that rigid basement blocks are finite (bounded) portions of the entire basement slab that behaved individually as rigid bodies at least to depths of several thousand metres. To a far-removed observer (changing the scale of observation), the deformation of the entire basement would appear cataclastic, with very large "grains" (rigid blocks) displaced past one another along sheared "grain boundaries" (faults). It is helpful to carry this conceptualization of cataclasis (and the implied kinematics) through the development of geometric models of basement-block configurations.

Consider the beam depicted in Figure 4A with precut surfaces I, II, and III.

If blocks R, S, U, and V are both rigid and laterally constrained, and if there is a small amount of slip on surfaces I, II, and III with the indicated senses of relative displacements, a geometry of the form shown in Figure 4B will result. Surfaces I and III are shaped so as to produce simple relationships between rigid blocks R and S and blocks U and V, respectively. However, the configuration of surface II results in point contacts M and N and gaps P and Q for the relationship of blocks S and U.

If the blocks shown in Figure 4B are viewed as rigid basement blocks in the crust and the sliding surfaces are taken to be faults, it is seen that faults I and III allow geologically reasonable motions to occur with no associated "room" problems. Fault II, with the gaps and point contacts associated with its motion, is geologically unacceptable. Therefore, we propose that only those faults that do not result in room problems (gaps, point contacts, overlaps) can undergo significant displacement without an alteration of the fault's geometry.

Given this hypothesis and a restriction that allows just plane deformations, only circular or planar faults (as seen in cross section) can be activated to cause block motion.[1] Planar faults (III in Fig. 4) present no problem, but neither can they produce rotations by themselves (the top surfaces of blocks U and V are parallel). Cylindrical faults (I in Fig. 4) allow both relative rotations and relative displacements of the adjacent blocks (top surfaces of R and S have different attitudes).

Stearns and Weinberg (1975, Fig. 11, p. 165) and Weinberg (this volume, Fig. 30) have illustrated one simple system of rigid blocks (Fig. 5A) that is "kinematically consistent"—there are no room problems. This is a system of rigid steel blocks designed so that, by driving the ends of the system together, the blocks are forced to differentially rotate and provide up and down movements (Fig. 5B). Layers of lead (Stearns and Weinberg, 1975) and multilayers of rock (Weinberg, this volume, and 1978) have been deformed under confining pressure in the laboratory using this system of blocks as "forcing members" (see Stearns, this volume). In these experiments, it is not the folding of the ductile layers that is important to the present discussion, but the fact that a system of rigid blocks is kinematically consistent. This system allows motions of the type interesting to the study of basement deformations of the foreland—rotations, up and down movements—yet no room problems are created during the deformation.

The Plexiglass Approach

On the basis of the preceding considerations regarding kinematic constraints, a general conceptual approach toward the geometric comparison of theoretical results and natural geometries can be outlined. The application of the shear-fracture trajectory diagrams amounts to a matching process whereby the deformation pattern "predicted" from the theoretical solution is "fitted" or compared to observed geometries in nature. The kinematic constraints discussed above lead to the following process, which is termed the "plexiglass approach."

Simply stated, the plexiglass approach consists of the following procedures. A thin sheet of plexiglass is laid over a copy of the theoretical fracture trajectories to be evaluated. The boundaries of the beam and the appropriate faults are cut into the plexiglass sheet. The free pieces of plexiglass are then displaced past

---

[1] In reality they do not have to be exact arcs of circles, but very near to it. This is because a single break that is not a perfect arc can develop a breccia zone so that the motion of the rigid blocks can occur as if the single break was a circular arc (see also the paper by Couples, this volume).

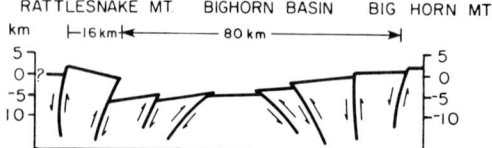

Figure 1. Cross section of basement surface across northern Bighorn Basin. Note absolute up and down movements and relative rotations of basement blocks. See section B in Figure 16 (after Stearns, 1975).

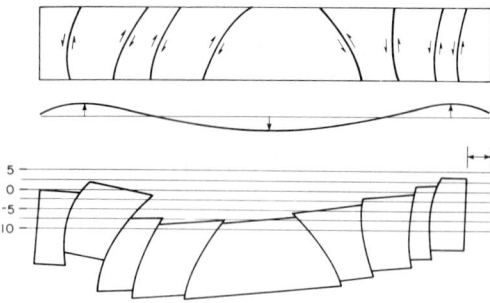

Figure 2. Selected faults from Hafner's (1951) case B cut through plastic sheet (upper diagram). Activating faults according to indicated senses of shear produces the geometry of Figure 1 but does not lead to room problems at depth (lower diagram). Scale to left in kilometres (after Stearns, 1975).

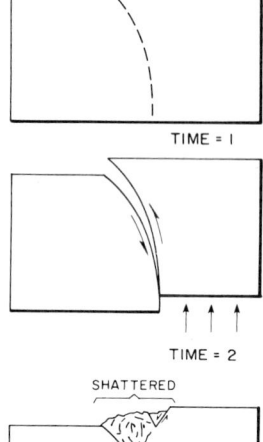

Figure 3. Proposed explanation for shattering of lobe rocks. Time = 1: initially underformed block with outline of potential fault (dashed). Time = 2: displacement along fault leads to unrealistic gap at depth. Time = 3: gravitational forces expected to break off lobe and produce shattering. Note lack of rotations (after Stearns and others, 1975).

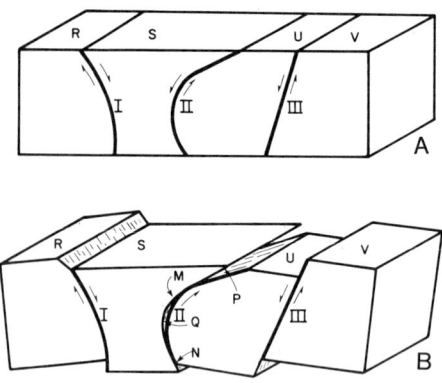

Figure 4. (A) Rectangular beam with precut surfaces I, II, and III. (B) Geometry resulting from activation of surfaces I, II, and III with the indicated senses of shear. Note gaps and point contacts.

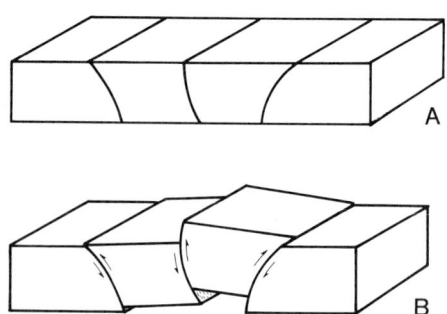

Figure 5. (A) Forcing-block assembly used by Weinberg (this volume). Precut surfaces are cylindrical. (B) Geometry resulting from driving the ends of the assembly together. Note that there are no room problems but relative rotations.

one another by observing the proper senses of movement while maintaining mutual contact. The final geometry resulting from this process is then visually compared to the present final geometry (in cross section) of the naturally deformed features.

There are two very critical points in this procedure that need to be discussed in greater detail. Which faults are "appropriate"? What are the "proper senses of movement"? As a first approximation, the proper sense of motion can be taken as that which initiated the fault in the theoretical solution. This is only a tentative suggestion based on the models constructed to date, for which the idea holds. But because the natural structures being modeled are *final* configurations caused by finite motions, this suggested method of selecting a proper sense of motion must be continually re-evaluated.

The means of selecting the appropriate faults are less readily communicated. It is perhaps easiest to describe the selection procedure by defining a key fault. A key fault is that fault in the natural cross section that is best controlled in terms of location, configuration, and sense of slip. By correlating the key fault in nature to the appropriate fault in the theoretical solution, the remainder of the selection procedure is essentially fixed.

The locations of the remaining appropriate faults in the theoretical solution are picked from the available real-world data that indicate the lateral positions of faults. One or the other of the conjugate faults intersecting the top of the beam at these lateral positions is then picked on the basis of which one will lead to the observed motions and rotations. If proper faults are not found in the solution, the match is either inexact or incorrect, depending on the degree of mismatch. Problems of mismatch are discussed in some detail by Couples (this volume).

Although the plexiglass approach is rarely performed exactly as described, the mental picture of rigid blocks is probably easier to carry forward after initially considering the motion restrictions imposed by the plexiglass. This mental picture is critical to the remainder of this study.

Viewpoint on the Geometric Comparisons

There are essentially two ways by which the fitting process described above can be initiated. Both are attempts at justification, but the goals of the justification are somewhat different.

Approach A. This is where a structural model, determined by field relations and behavioral constraints, suggests the investigation of a particular boundary-value problem as follows: Structural configuration X honors the field data. This geometry suggests that boundary conditions Y may have caused structure X. The deformations predicted by the theoretical solution derived from conditions Y fit the deformations observed for structure X. Therefore, boundary conditions Y may have been responsible for causing structure X. In approach A, a theoretical basis is being sought as justification or verification of a hypothesized structural model.

Approach B. This is where a theoretical solution suggests a deformation pattern to be sought in nature as follows: Theoretical solutions derived from boundary conditions P suggest possible deformations Q. In nature, there are examples of deformations with relations compatible with Q. Therefore, boundary conditions P may be geologically reasonable. Approach B is based on a desire to find geologic models that can be explained by a particular abstract solution. As such, it is a justification process whereby natural support is sought for a program of solution generating.

As a personal value judgement and opinion, we favor approach A. Ostensibly, A is the approach taken in generating the models discussed later in this paper.

However, there is nearly always the intrusion of parts of approach B. For the sequence of solutions reported earlier (Couples, 1977), approach A was initially followed, but the possible loading conditions suggested by natural features in themselves suggested that other related loading conditions be tried. This, of course, is approach B. It is our opinion that approach A should be the favored procedure, yet we admit that approach B often enters into the process and sometimes yields significant results. Thus, philosophically we prefer to uphold approach A, but we certainly do not deny that approach B is often usefully employed.

SPECIFIC EXAMPLES USING RECENT THEORETICAL SOLUTIONS

The recent solutions referred to are those presented by Couples (1977). That paper describes the techniques involved in generating the solutions and illustrates some of the basic concepts in the usage of this type of theoretical result. What follows here is a discussion centered on the usage of those solutions in structural modeling of the foreland.

Where the Solutions Come From

The recent theoretical solutions presented by Couples (1977) were derived by a procedure patterned after that used by Hafner (1951). Although the solutions in Couples's (1977) paper required the use of mathematical manipulations uncommon to the geologic literature (see also Sanford, 1959), the fundamental derivation of the equations is based on the concept of an Airy stress function as in Hafner's paper.

The important aspect to note in regard to the newer solutions is that their boundary conditions are more complicated than those used by Hafner or any other authors. For certain geologic settings, we feel that these more-complicated loading schemes may be more realistic. Hafner's solutions were based on the simple situation of a standard-state stress field (present even without imposed tectonic loads) plus a single tectonic load component, either laterally imposed or sinusoidally varying along the base. The more recent solutions of Couples (1977) contain combinations of a standard state, laterally imposed loads, and/or vertical loads on the base with "shapes" more complex than Hafner's sinusoidal one.

Figures 6, 7, and 8 present three of the solutions contained in the recent study. The boundary conditions for these three solutions are similar, consisting of (1) stresses due to an overlying sedimentary veneer, (2) a standard state, (3) a sawtooth- or step-shaped vertical normal stress on the base, and (4) laterally imposed loads (for Figs. 6, 7). The magnitudes of these normal stresses, along with the necessary boundary shear stresses, are illustrated diagrammatically in each of the figures. Note that the potential shear fractures derived from these loading conditions indicate concurrent extension and shortening (normal and reverse faults). Also note that even though there is a large lateral load (as much as 3 kb above the standard-state stresses), the deformation pattern is still dominantly vertical. The shear fracture patterns of these solutions are discussed in greater detail by Couples (1977).

The Wind River Mountains

Recall from Figures 1 and 2 and the accompanying text that Stearns (1975) initiated his modeling by constructing a cross section of the natural features. This procedure is followed here by first constructing a cross section of the Wind River

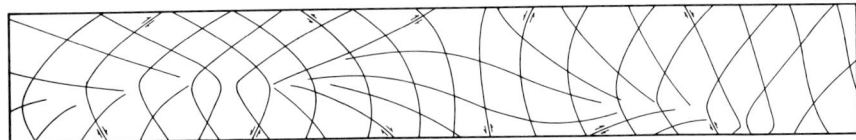
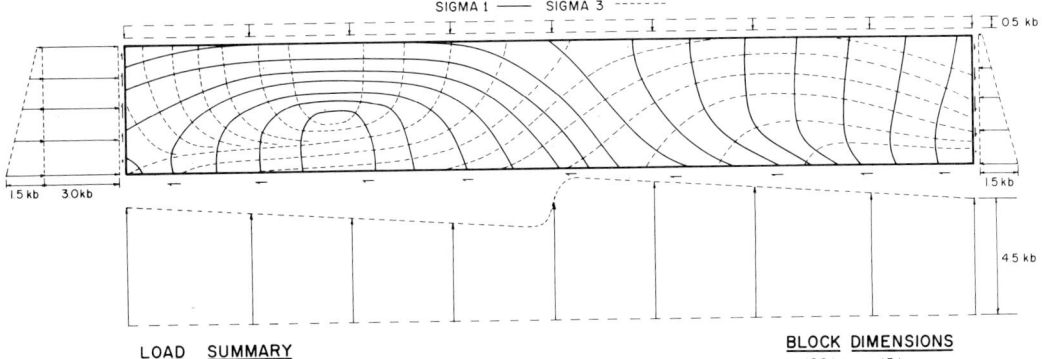

Figure 6. Theoretical solution from Couples (1977, Fig. 10). Boundary conditions indicated by scaled arrows; full arrows are normal stresses; half arrows are shear stresses; 0.5-kb load on top of beam represents weight of overlying sedimentary veneer. Potential shear fractures oriented ±30° to direction of maximum compressive stress.

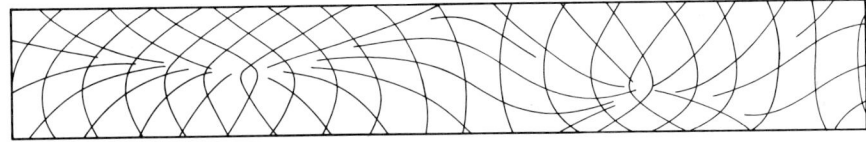
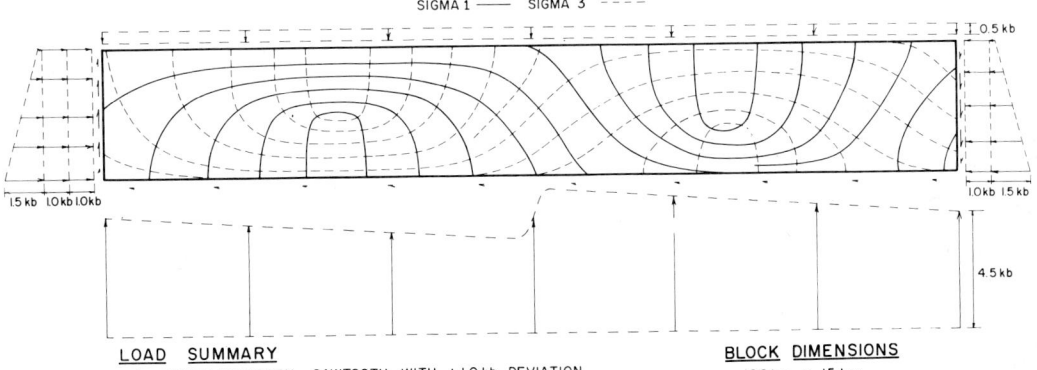

Figure 7. Theoretical solution from Couples (1977, Fig. 11). See caption to Figure 6.

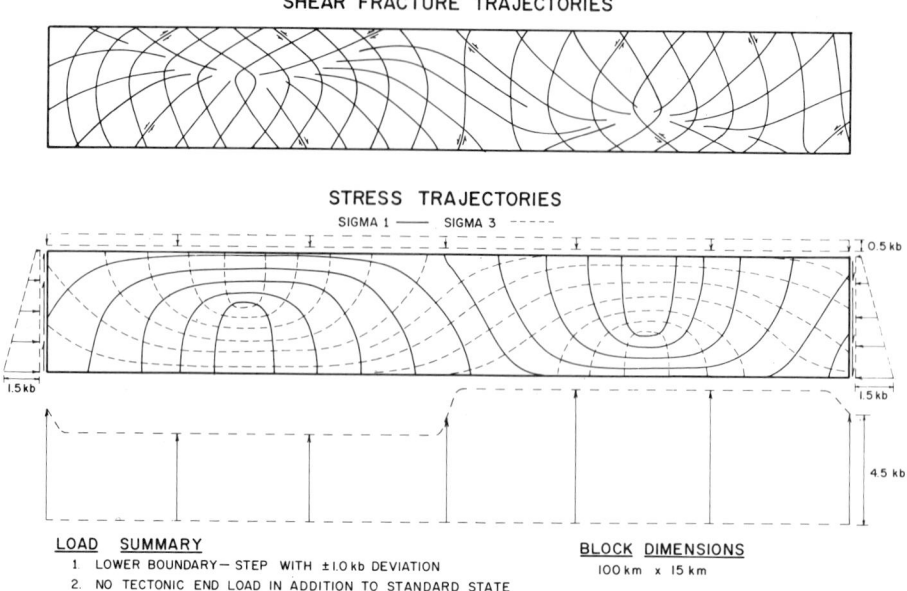

Figure 8. Theoretical solution from Couples (1977, Fig. 6). See caption to Figure 6.

Mountains (Fig. 9). In specific terms, the line of section is oriented N58°E, passing through the northwest corner of T. 32 N., R. 101 W. This line passes through Mount Arter, within 1 km of Wind River Peak, and within 4 km of Lander, Wyoming.

Data for structural control is minimal, being limited to (1) the attitude of the homoclinal sedimentary section on the northeast flank of the uplift (determined as 13° by us from several field measurements); (2) the location of the Cambrian-Precambrian contact (determined from the state geologic map by Love and others, 1955); (3) a generalized configuration for the Lander-Hudson anticline (J. J. Gallagher, 1977, oral commun.; also from a section by Keefer, 1970); and (4) the approximate location and character of the extension zone and frontal fault (observed and interpreted by us). The implied rotation of the "lobe" relative to the gently dipping block 1 of the northeast flank is not documented for this structure. However, O'Neill (1975) has documented a rotation of the same relative sense for the Never Summer Range in northern Colorado.

The cross section in Figure 9 is derived by utilizing these few control points and by accepting the assumption of rigid basement blocks and the restrictions on room problems. Only the near-surface portions of Figure 9 are well founded on facts, but we feel that several lines of reasoning all lead to this configuration.

Because detailed subsurface information for the deeper portions of Figure 9 is either nonexistent or proprietary, we have drawn on a general understanding of the nature and shape of the frontal fault as interpreted by Berg (1961, 1962). Although we use his geometric data, we disagree with Berg's mechanical interpretation of a fold-thrust origin for this structure. For the depth of the Green River Basin block, we have chosen to use the conservative value of −6.1 km (−20,000 ft). As just stated, the basement-fault configuration at depth is entirely schematic and represents, in large part, our working prejudices. However, it needs to be noted that Figure 9 depicts the only geometry we were able to construct that did not lead to room problems or behavioral impossibilities and yet explained the rotations and gross movement patterns. It was impossible to construct a cross

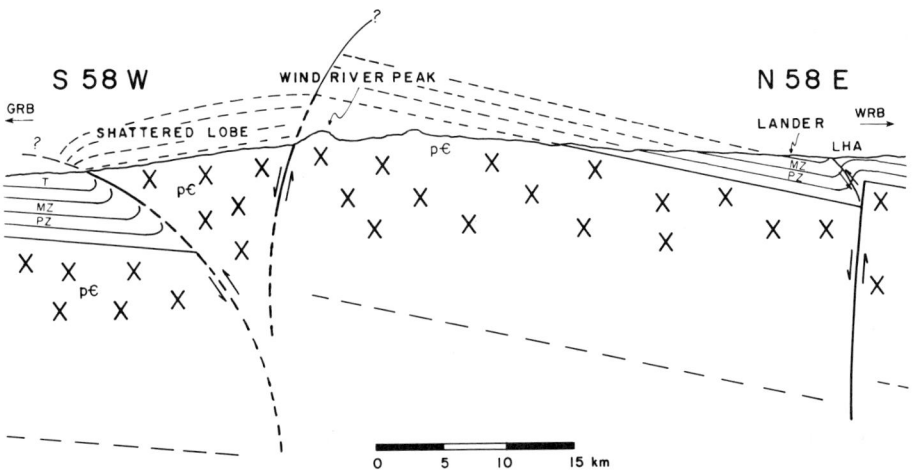

Figure 9. Schematic cross section of Wind River Mountains. Dashed line drawn at −15-km level in basement facilitates comparison to model. Stratigraphic units not accurately scaled. p€ = Precambrian basement, PZ = Paleozoic carbonate package, MZ = lower Mesozoic rocks, T = Tertiary basin fill, GRB = Green River Basin, WRB = Wind River Basin, LHA = Lander-Hudson anticline. See section D in Figure 16.

section where the frontal fault flattened under the range (a "sled-runner" thrust) unless room problems or unacceptable behaviors were allowed to occur. Because the configuration shown in Figure 9 does honor the observed as well as inferred geometries and does not lead to inconsistencies, it is deemed to be the acceptable model.

The upper diagram in Figure 10 is a reduced and simplified version of Figure 9, and the lower diagram is a simplified composite from Figure 6 with only two of the potential faults shown. If the plexiglass approach is applied to the lower diagram of Figure 10, it is possible to obtain a geometry very close to that of the upper diagram—a "fit" results. The basement fault beneath the Lander-Hudson anticline has no counterpart in this solution. This and other discrepancies of the fit are discussed in the paper by Couples (this volume).

There are several features to note in Figure 10. The two faults illustrated in the lower diagram allow (through the plexiglass approach) relative rotations of the component blocks; the senses of rotation correspond to those depicted in the cross section. It is also interesting to note that the model proposed here suggests that the lobe is a distinct and separate component, not merely a broken fragment of the rotated mountain block (refer to Fig. 3). The opposing curved faults that allow these motions in the model are not observed in any of Hafner's (1951) solutions. Individual faults with each of these configurations can be found in the solutions of Hafner and others, but the proximity and relative positioning of the two different faults was not seen in any of the earlier publications. Thus, the suggested cross section and the solution used in explaining it are indeed new, and the combination of the two is deemed highly significant.

The Owl Creek Mountains

Figure 11 is a cross section of the Owl Creek Mountains constructed in a fashion similar to that used for Figure 9. The line of section for this figure is north-south, passing through the townsite of Boysen and within a few kilometres of Thermopolis. The general character of the frontal lobe and the homoclinal dip of the sedimentary section on the north flank of the uplift are readily seen in Wind River Canyon. The configuration of the Thermopolis anticline is interpreted from our own field studies and from an interpretation provided by D. M. Weinberg (1976, oral commun.). The subsurface configuration of the basement is developed along the same lines used in constructing Figure 9 but also benefits from the adaptation of sections presented earlier by Fanshawe (1939) and Wise (1963). Likewise, this is the only geometry we could construct that honored the available data, yet did not violate behavioral limitations.

Figure 12 forms a comparison like that shown in Figure 10. The upper diagram is reduced and simplified from Figure 11, and the lower portion of the figure is simplified from Figure 7, with only two of the potential faults shown. The plexiglass approach yields a remarkably good fit with the cross section. Again, the basement fault underlying the Thermopolis anticline is not found in this solution. The discrepancies of this model are also discussed by Couples (this volume).

The preceding comments in regard to Figure 10 apply equally well to Figure 12. The prime difference to note is that deformation (displacement along the faults, particularly the Boysen fault) apparently has not "progressed" as far in the case of the Owl Creek Mountains as was the situation with the Wind Rivers. The extension zone does not seem as intensely deformed—the sedimentary rocks are present in coherent fault blocks, and the Boysen fault bounding the north side of the extension zone has a throw of only about 400 m. These differences may possibly be due to the apparently steeper orientations of the faults in the basement.

The Beartooth Plateau

Figure 13 is an east-west cross section drawn 2 km south of the Montana-Wyoming border. It was constructed from information in Foose and others (1961), Stewart (1975), Wellborn (1975), and Rea and Barlow (1975), along with our field observations of this flank of the Beartooth Plateau. The data contained in those sources were used as a base from which a basement configuration was constructed. The rigid nature of the basement, along with experimental data (Friedman and others, 1976), suggested a general sequence of development of the individual basement blocks, which was then used with the information on the dip of the Paleozoic rocks to construct the model. The drape subsidiary fold concept of Weinberg (this volume) was necessary in order to derive useful information concerning probable basement-block shape below the Terry Field structure (as shown in the section by Wellborn, 1975).

Figure 14 is constructed similarly to Figures 10 and 12. The upper diagram is simplified from Figure 13, and the lower portion is derived from Figure 8. In this case, however, unlike Figures 10 and 12, only one of the potential faults is illustrated.

What are the motions that must be explained by this particular conceptual model? In Figure 13, the mountain block has been uplifted uniformly with no rotational component. The basin block has been displaced downward and has been rotated slightly. The splinter blocks have moved differentially and their rotations are dissimilar to any of the others in the system. The plexiglass approach per se does not, in this case, lead to any obvious coincidence between model and cross section. Although this problem is covered in greater detail by Couples (this volume), we can briefly describe the requisite alterations of the fitting procedure.

The relative rotation between the main mountain and basin blocks is best explained by a fault concave to the east. The smaller indicated angle of rotation requires a radius of curvature that is somewhat larger than in either of the models illustrated

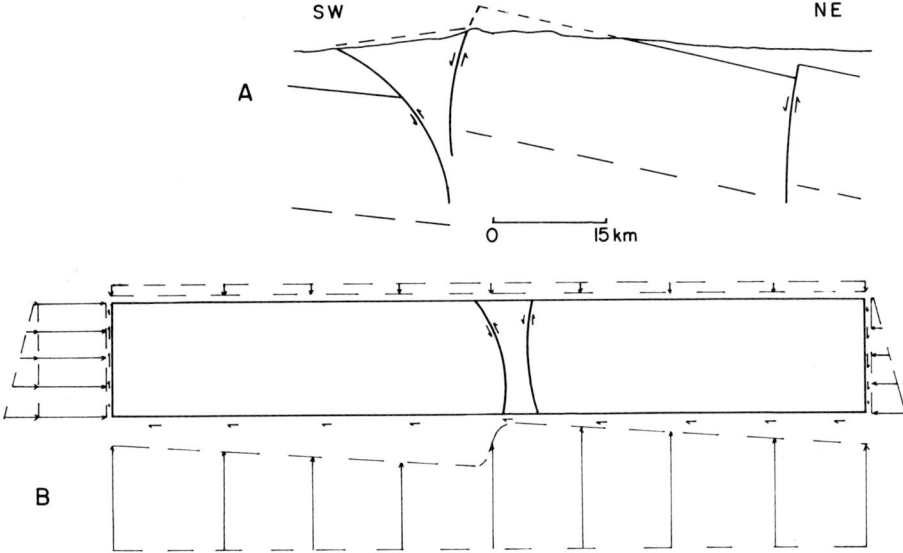

Figure 10. (A) Simplified cross section of Wind River Mountains from Figure 9. (B) Simplified composite diagram of theoretical solution from Figure 6.

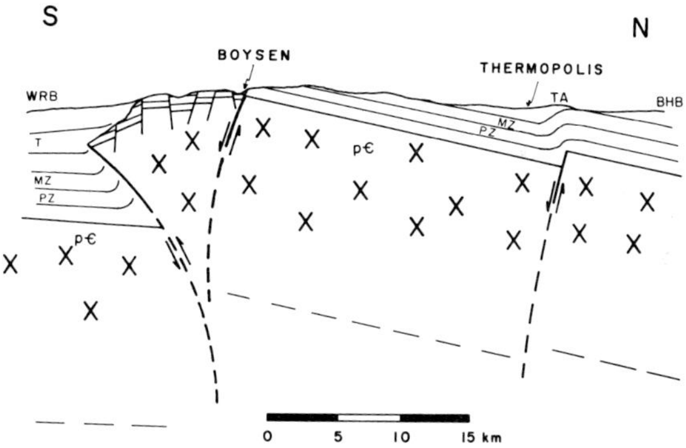

Figure 11. Schematic cross section of Owl Creek Mountains. See caption to Figure 9. TA = Thermopolis anticline, BHB = Bighorn Basin. See section C in Figure 16.

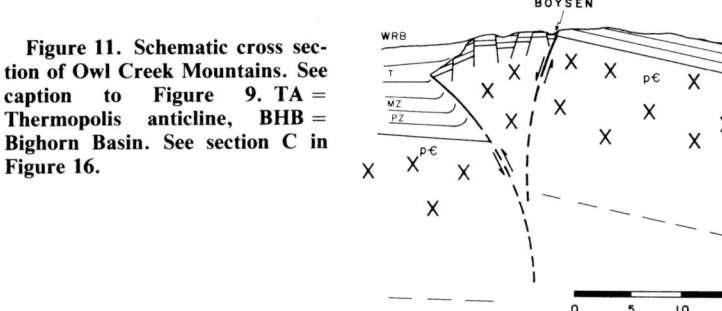

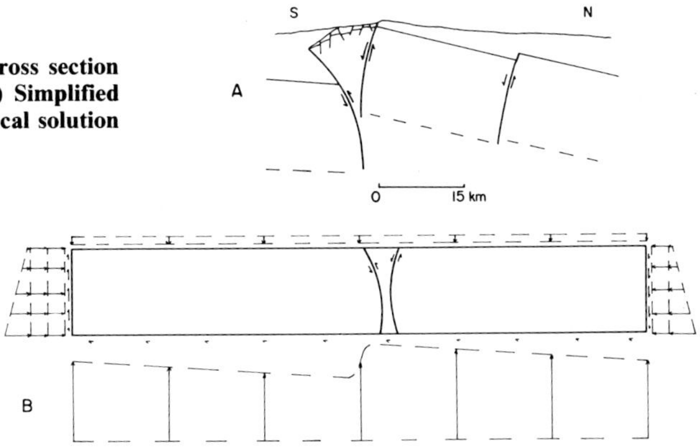

Figure 12. (A) Simplified cross section of Owl Creek Mountains. (B) Simplified composite diagram of theoretical solution from Figure 7.

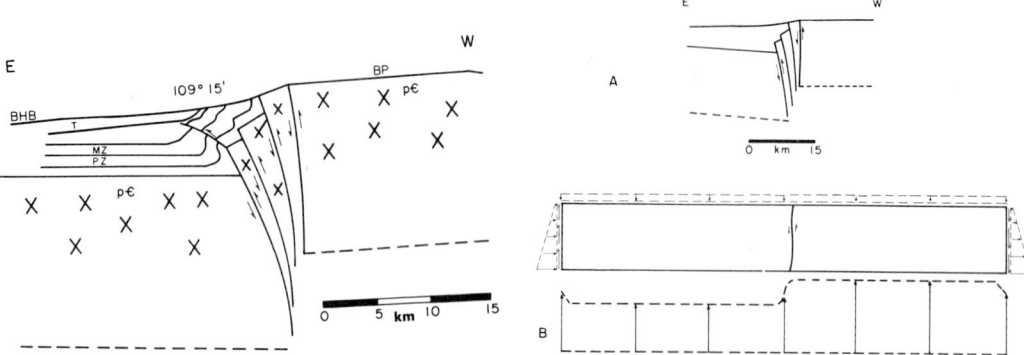

Figure 13. Schematic cross section of eastern flank of Beartooth Mountains. See caption to Figure 9. BP = Beartooth Plateau. See section A in Figure 16.

Figure 14. A. Simplified cross section of flank of Beartooth Plateau. (B) Simplified composite diagram of theoretical solution from Figure 8.

in Figures 10 and 12. The fault illustrated in Figure 14 corresponds to these requirements except at the top of the beam. We suggest that the real, finite displacements of the system occur in such a way as to modify the shape of the fault. This could be done by "breaking off" a small piece at the top to form a nearly circular shape for the main fault. Subsequent motion might then lead to the development of the splinter blocks by way of similar processes.

A complete theoretical model for this final basement configuration would require the solution of several boundary-value problems, each depending on the previous deformation history. Such solutions are both too complicated and too speculative to attempt at this time. However, it is interesting to point out that in the laboratory exactly such sequences have been observed (Friedman and others, 1976; Logan and others, this volume). These experiments are analog models where early deformation does influence final shape. Figure 15 illustrates one of the experiments where splinters did develop. Comparison of Figures 13 and 15 establishes that the postulations we have added to the plexiglass approach above are reasonable in that the theoretical solution presented here now fits with the natural cross section.

DISCUSSION

Hypotheses of Foreland Tectonics

Through time, a number of authors have proposed one or another model to explain the cause of the deformation of the Rocky Mountains foreland and its relationship to the remainder of the Cordillera. It seems that most authors have assumed that foreland tectonics and Cordilleran tectonics were causally related, with identical driving forces. Therefore, there have been as many theories (if not more) for foreland deformation as for orogenic processes in general.

In very simple terms, the various hypotheses can be classified into those that call upon laterally imposed loads (compressional tectonics) and those that are based on loads applied transversely to the crust (vertical tectonics). Within either of these categories, some authors have called upon either folding or faulting as a primary deformation mechanism. To complicate matters, there are also models based on a "mix" of horizontal and vertical loads (Lowell, 1974, among others), and models with a mixed style of deformation (the fold-thrust of Berg, 1962; Sales,

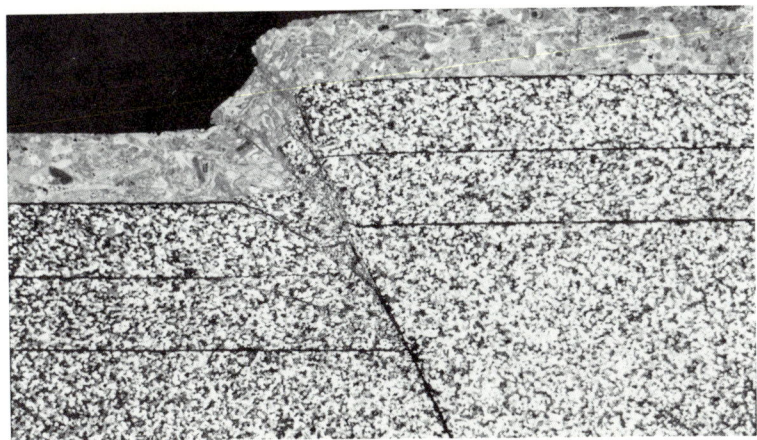

Figure 15. Photomicrograph of experiment 302 showing splinter blocks developed in brittle sandstone forcing member. Note rigid body rotations and zones of cataclasis. Upper limestone layer is 3 mm thick (after Friedman and others, 1976).

1968) have also been proposed. Within the compression classification, early models were based on thrusting. Theories based on the compression of a laterally inhomogenous crust (the wedge uplifts of Thom, 1955) and theories of wrenching (Stone, 1969; Thomas, 1971) have later been added. The concept of vertical tectonics was advanced by Osterwald (1961) and Prucha and others (1965).

It is not our purpose here to critically review the supporting data contained in the numerous publications that support one or another of these models: reviews of most of the literature can be found in Osterwald (1961), Blackstone (1971), and Lowell (1974). It *is* our purpose to discuss these theories in light of their internal consistency and applicability to foreland structural development.

Those hypotheses suggesting ductile folding of the basement have been discounted earlier in this paper and previously by Stearns (1971, 1975, and this volume). Such hypotheses will not be entertained further here.

The wide range of trends of the structures in the foreland (as discussed by Stearns, this volume) does not seem compatible with an origin based on uniform thrusting of the basement slab. Likewise the possibility of wrench faulting as a dominant mechanism in the near-surface rocks has been ruled out by Stearns (this volume) and will not be discussed further here. However, what should be pointed out is that the exclusion of these mechanisms (folding, thrusting, wrenching) applies only to the upper part of the basement. That is, they are not excluded deep within the crust and/or mantle where basement rocks become ductile. As a matter of fact, "folding" due to flow at depth may cause the lower boundary condition on the brittle basement. What is important is not to confuse stress conditions and flow at depth with stress conditions and faulting in the upper few thousand metres of the basement where the rocks behaved in a demonstrably brittle fashion.

The vertical motions proposed by Prucha and others (1965) consisted of distributed shear along vertical fractures in the basement. This proposal has been discounted as a general mechanism by Stearns (1971) on the basis of field evidence from Rattlesnake Mountain near Cody, Wyoming. Any other hypothesis of purely vertical motions for foreland structures can also be discounted owing to the inability of these motions to account for features such as the upthrusts. As pointed out by Stearns (1971), the probable reason for the large variety of hypotheses is that faults of all major types can be found at some place in the Rocky Mountain foreland. However, the relationships among the different fault types are not clear in the field, nor are they subject to direct observation in most cases. Nevertheless, certain nonunique associations, such as the upthrust-normal fault sequence, have been noted. It is the thesis of this paper that such *associations* are the norm and should be explained by any model proposed for foreland tectonics.

Mechanical Basis of Foreland Deformation

The various hypotheses discussed immediately above have little or no mechanical foundations. At best, they appear to have been formulated from vague, assumed, homogeneous stress fields. The selection of stress fields seems to have been based on each author's favorite hypothesis of orogenesis—either compressive or extensional.

The existence of the different fault types in nature suggests that models based on homogeneous stress states are unrealistic and inadequate. Hafner (1951) was the first to demonstrate that even simple load distributions lead to a variety of fault types in a single system. In our experience, we have found that the deformations observed for a system such as the simple sequence of basement blocks that make up the northern Bighorn Basin require the use of complicated stress fields in their explanation (Stearns, 1975).

Homogeneous stress fields (and their derived homogeneous deformations) do not help in attempting to understand even such simple systems as the northern Bighorn Basin. It is little wonder that the range of hypotheses described above (which are largely based on homogeneous conditions) fails to increase our understanding of the structural development of the foreland.

Some authors (among them Lowell, 1974) have suggested that a "mix" of crustal compression and vertical loads was responsible for the deformations of the foreland. However, stress states for loading schemes of this type had not been calculated until recently. Couples (1977) presented a suite of solutions for beams subjected to vertical loads shaped as sinusoids, steps, and sawtooth waves. In some cases, horizontal loads were also imposed on the beams. A sufficient variation of loading conditions exists in his suite of solutions to evaluate the efficacy of the perennial question: "How much vertical load as *compared* to the horizontal load?" Thus, there now exists a moderate variety of stress states—and derived fault trajectories—for use in attempting an understanding of foreland tectonics.

Loading Conditions for Foreland Structures

Because theoretical solutions are available for a range of the previously suggested loading conditions, we feel that it is now possible to better evaluate the different proposals. We propose to do this on the basis of the empirical usefulness of any hypothesis in terms of aiding our understanding of the structural types in the foreland. As was demonstrated previously (Stearns, 1975; Couples, 1977) and earlier in this paper, analytical boundary-value solutions have helped us tremendously in our desire to know possible loading conditions for the structural development of different uplifts in the Rocky Mountains foreland. They have provided support for conceptual models that explain the subsurface configurations of the major structural types—intermontane basins with simply rotated blocks, upthrusts, and plateau uplifts—while remaining consistent with the known behavioral limitations of the basement materials.

With such a background of "success," we feel it is appropriate to suggest that the types of boundary conditions used in solutions employed in this paper represent the normal range of conditions extant in the foreland at the initiation of Laramide deformation. We feel that the majority of the deformation can be interpreted on this basis. Thus, our view suggests that a good approximation to the probable loading conditions for the major structures of the province can be determined by evaluating the final configuration of the individual uplifts. This view leads us to speculate that the regional system of load distribution for the initiation of Laramide deformation is easiest to describe by describing its components.

It should be noted that there is a considerable horizontal load component in Figures 10 and 12 and none other than the standard state in Figures 2 and 14. This component decreases from a maximum in Figure 10 to none at all in Figure 14. Notice also the spatial distribution of the structures modeled in that sequence (Fig. 16). The horizontal-load models are in the southwestern part of the foreland and the no-horizontal-load models are in the more northeasterly portions. From these observations, we suggest that Laramide loading conditions contained a regional component of horizontal load (compression) directed approximately southwest-northeast and decreasing in magnitude to the northeast.

Along with the ever-present standard-state stress field due to gravity (see Couples, 1977) and the regional compressive field just discussed, vertical load distributions that were shaped much like the present structures were also present. It is our contention that the complicated stress field resulting from that combination was responsible for initiating the structures that we see today.

An interesting feature noted on many maps of the foreland is the presence of spurs on the flanks of the uplifts. The en echelon nature of these spurs has led some authors to suggest that they are evidence for strike-slip motion along the mountain flank. This hypothesis has apparently been derived from experiments of clay cakes on sliding boards and from data observed along major strike-slip faults like the San Andreas. The clay-cake models and the structures observed along the San Andreas are folds. It follows, therefore, that those who wish to fold the basement of the foreland could readily derive similar interpretations for the underlying motions.

However, Blanton (1975) has demonstrated that at least one of these spurs is not a basement fold but only a rotated basement block with marginal drape folds in the sedimentary rocks. We feel that this is a general interpretation applicable to most, if not all, spurs in the foreland. In our hypothesis the spurs may well represent complexities at the margins of the uplifts where the local uplift field has interacted dramatically with the regional compressive field.

As a possible source of the vertical load components, we are greatly impressed with the work of Sales (1968) on mud and plaster models. Sales deformed a layered, ductile sequence in a shear box oriented so as to represent the foreland. The deformation in his shear box might also be viewed as caused by a southwesterly oriented compression. Because he used a very ductile material, his deformations may be comparable to those expected to develop in a lower crustal upper mantle location subjected to loads of the general type we suggested above. If Sales's models are viewed as the expected lower crustal deformation type for our overall load configuration, we feel that he may have already provided the groundwork for studying another aspect of the problem—that of the source of the variations in the vertical load. Certainly one must be impressed by the fact that Sales's models produced many aspects of foreland geometry at the same time, not one model for one mountain.

Of course, these suggested load configurations are made without the benefit of calculated three-dimensional stress states; Couples (this volume) discusses some considerations that would enter into any such calculation. Nevertheless, we feel

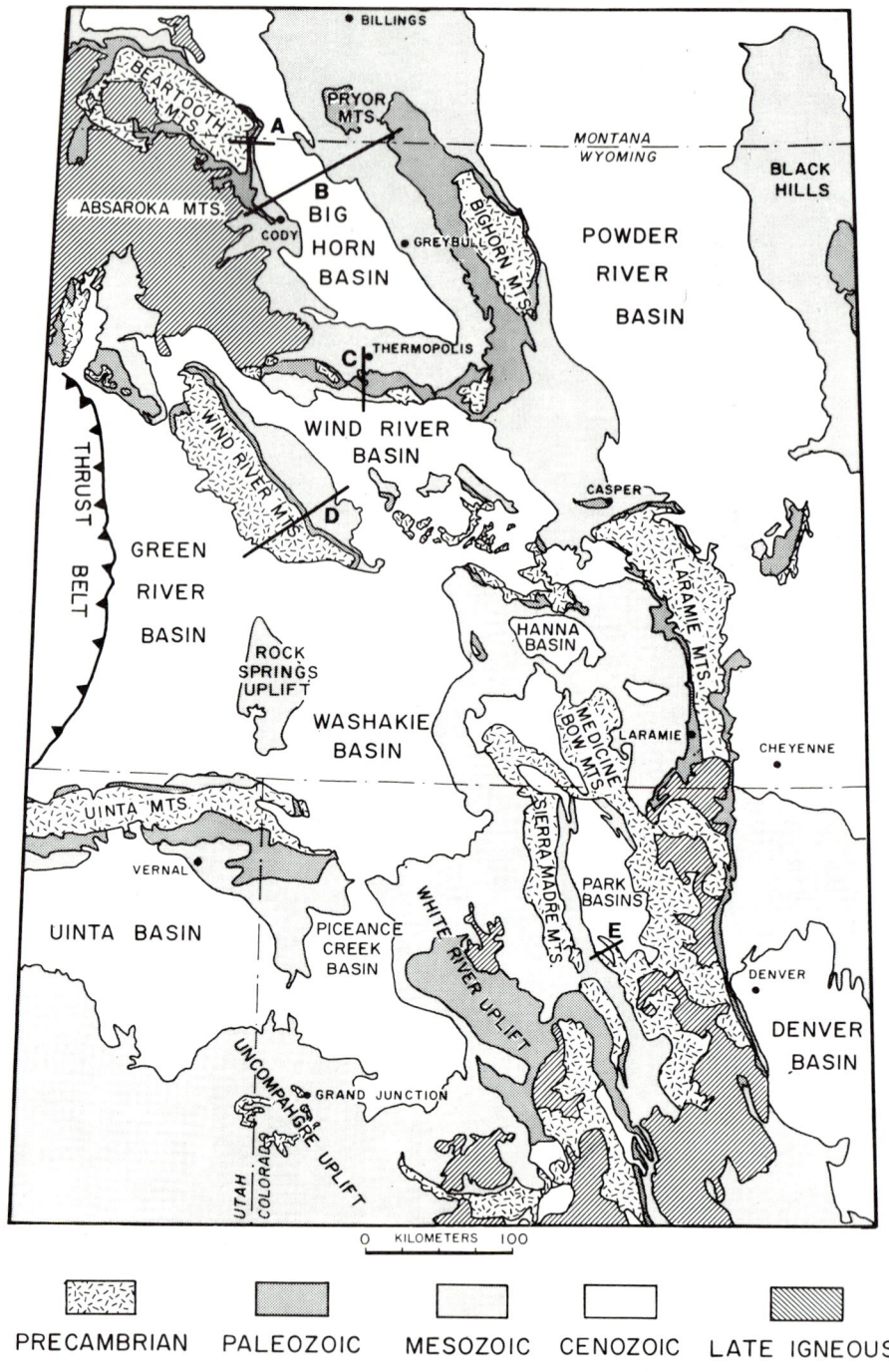

Figure 16. Generalized geologic map for part of the Rocky Mountains foreland. Cross sections are referenced as follows: section A, Beartooth Plateau (Fig. 13); section B, northern Bighorn Basin (Fig. 1); section C, Owl Creek Mountains (Fig. 11); section D, Wind River Mountains (Fig. 9); section E, Williams Fork Mountains, location of Howard's (1966) model study.

that our suggestion is sufficiently flexible so as to not be restrictive. Moreover, we feel that the concept we have advanced here provides a hypothesis to be critically tested by further work. Although portions of that critical test must await the appropriate stress-state calculations, other aspects of the general model may be tested by attempting to apply the other two-dimensional solutions to the explanation of other structures in the foreland.

CONCLUSIONS

Analytical boundary-value solutions have proven useful in understanding several aspects of Laramide basement faulting in the Rocky Mountains foreland. From field studies, three major basement structural types have been determined—rotated blocks, plateau uplifts, and upthrusts. Use of the boundary-value solutions has allowed us to suggest mechanically based subsurface configurations for these structures. The major features of each type—rotations, absolute up and down motions, fault zones—are explained by the solutions. This goodness-of-fit has allowed us to propose that the boundary conditions used to generate the solutions are representative of the loading conditions that initiated Laramide deformation of these foreland structures. Furthermore, we have been able to synthesize a possible regional variation in the loading conditions. This hypothesized regional system is composed of a regional horizontal component of crustal compression superposed with stress states related to local variations in the vertical loads. The eventual calculation of this regional stress state will allow better evaluation of some of the less-rigorously stated hypotheses of foreland tectonics.

ACKNOWLEDGMENTS

This paper is the outgrowth of ideas developed in extending an M.A. thesis submitted to Rice University by G. Couples. Funding for the field work has been provided by a Grant-in-Aid-of-Research from Sigma Xi, The Scientific Research Society of North America; by a research grant from the American Association of Petroleum Geologists Foundation; and by a Penrose Research Grant from the Geological Society of America. G. Couples was supported by a National Science Foundation Graduate Fellowship and by the Michel T. Halbouty Graduate Fellowship during the preparation of this manuscript.

We wish to thank Travis J. Parker and David M. Weinberg for their review of this manuscript and for their suggestions toward its improvement.

REFERENCES CITED

Anderson, E. M., 1942, The dynamics of faulting: London, Oliver and Boyd, 183 p.
Bengtson, C. A., 1956, Structural geology of the Buffalo Fork area, northwestern Wyoming, and its relation to the regional tectonic setting: Wyoming Geol. Assoc., 11th Ann. Field Conf., Guidebook, p. 158–168.
Berg, R. R., 1961, Laramide tectonics of the Wind River Mountains: Wyoming Geol. Assoc., 16th Ann. Field Conf., Guidebook, p. 70–80.
——1962, Mountain flank thrusting in Rocky Mountain foreland, Wyoming and Colorado: Am. Assoc. Petroleum Geologists Bull., v. 48, p. 2019–2032.
Blackstone, D. L., Jr., 1971, Plate tectonics and its possible role in the Rocky Mountains: Wyoming Geol. Assoc., 23rd Ann. Field Conf., Guidebook: p. 11–17.

Blanton, T. L., III, 1975, Fountain Creek flexure and basement deformation in the Manitou Spur, Colorado: Mtn. Geologist, v. 12, p. 119–126.

Couples, G., 1977, Stress and shear fracture (fault) trajectories resulting from a suite of complicated boundary conditions with applications to the Wind River Mountains: Pure and Appl. Geophysics, v. 115, p. 113–133.

——1978, Comments on applications of boundary-value analyses of structures of the Rocky Mountains foreland, *in* Matthews, V., III, ed., Laramide folding associated with basement block faulting in the Western United States: Geol. Soc. America Mem. 151 (this volume).

Fanshawe, J. R., 1939, Structural geology of Wind River Canyon area, Wyoming: Am. Assoc. Petroleum Geologists Bull., v. 23, p. 1439–1492.

Foose, R. M., Wise, D. U., and Garbarini, G. S., 1961, Structural geology of the Beartooth Mountains, Montana and Wyoming: Geol. Soc. America Bull., v. 72, p. 1143–1172.

Friedman, M., Handin, J., Logan, J. M., Min, K. D., and Stearns, D. W., 1976, Experimental folding of rocks under confining pressure: Pt. III. Faulted drape folds in multilithologic layered specimens: Geol. Soc. America Bull., v. 87, p. 1049–1066.

Gangi, A. F., Min, K. D., and Logan, J. M., 1977, Experimental folding of rocks under confining pressure: Pt. IV. Theoretical analysis of faulted drape folds: Tectonophysics, v. 42, p. 227–260.

Hafner, W., 1951, Stress distribution and faulting: Geol. Soc. America Bull., v. 62, p. 373–398.

Howard, J. H., 1966, Structural development of the Williams Range thrust, Colorado: Geol. Soc. America Bull., v. 77, p. 1247–1264.

Hubbert, M. K., 1951, Mechanical basis for certain familiar geologic structures: Geol. Soc. America Bull., v. 62, p. 355–372.

Keefer, W. R., 1970, Structural geology of the Wind River Basin, Wyoming: U.S. Geol. Survey Prof. Paper 495-D, 35 p.

Link, T. A., 1930, Experiments relating to salt-dome structures: Am. Assoc. Petroleum Geologists Bull., v. 14, p. 485–503.

Logan, J. M., Friedman, M., and Stearns, M. T., 1978, Experimental folding of rocks under confining pressure: Pt. VI. Further studies of faulted drape folds, *in* Matthews, V., III, ed., Laramide folding associated with basement block faulting in the Western United States: Geol. Soc. America Mem. 151 (this volume).

Love, J. D., Weitz, J. L., and Hose, R. K., 1955, Geologic map of Wyoming: U.S. Geol. Survey, scale 1:500,000.

Lowell, J. D., 1974, Plate tectonics and foreland basement deformation: Geology, v. 2, p. 275–278.

O'Neill, J. M., 1975, Rocky Mountain foreland deformation—An example from the Never Summer Range, north-central Colorado: Geol. Soc. America Abs. with Programs, v. 7, p. 1219–1220.

Osterwald, F. W., 1961, Critical review of some tectonic problems in Cordilleran foreland: Am. Assoc. Petroleum Geologists Bull., v. 45, p. 219–237.

Prucha, J. J., Graham, J. A., and Nickelsen, R. P., 1965, Basement controlled deformation in Wyoming province of Rocky Mountains foreland: Am. Assoc. Petroleum Geologists Bull., v. 49, p. 966–992.

Rea, B. D., and Barlow, J. A., Jr., 1975, Upper Cretaceous and Tertiary rocks, northern part of Bighorn Basin, Wyoming and Montana: Wyoming Geol. Assoc., 27th Ann. Field Conf., Guidebook, p. 63–71.

Sales, J. K., 1968, Crustal mechanics of Cordilleran foreland deformation: A regional and scale-model approach: Am. Assoc. Petroleum Geologists Bull., v. 52, p. 2016–2044.

Sanford, A. R., 1959, Analytical and experimental study of simple geologic structures: Geol. Soc. America Bull., v. 70, p. 19–51.

Stearns, D. W., 1971, Mechanisms of drape folding in the Wyoming province: Wyoming Geol. Assoc., 23rd Ann. Field Conf., Guidebook, p. 125–144.

——1975, Laramide basement deformation in the Bighorn Basin—The controlling factor for structures in the layered rocks: Wyoming Geol. Assoc., 27th Ann. Field Conf., Guidebook, p. 149–158.

——1978, Faulting and forced folding in the Rocky Mountains foreland, *in* Matthews, V., III, ed., Laramide folding associated with basement block faulting in the Western United

States: Geol. Soc. America Mem. 151 (this volume).

Stearns, D. W., and Weinberg, D. M., 1975, A comparison of experimentally created and naturally formed drape folds: Wyoming Geol. Assoc., 27th Ann. Field Conf., Guidebook, p. 159–166.

Stearns, D. W., Sacrison, W. R., and Hanson, R. C., 1975, Structural history of southwestern Wyoming as evidenced from outcrop and seismic: Rocky Mtn. Assoc. Geologists, Symposium on Deep Drilling Frontiers in the Central Rocky Mountains, p. 9–20.

Stewart, W. W., 1975, Recent drilling in the Line Creek area of Wyoming and Montana: Wyoming Geol. Assoc., 27th Ann. Field Conf., Guidebook, p. 203–208.

Stone, D. S., 1969, Wrench faulting and Rocky Mountain tectonics: Mtn. Geologist, v. 6, p. 67–79.

Thom, W. T., Jr., 1955, Wedge uplifts and their tectonic significance, in Poldervaart, A., ed., Crust of the Earth: Geol. Soc. America Spec. Paper 62, p. 369–376.

Thomas, G. E., 1971, Continental plate tectonics: Wyoming Geol. Assoc., 23rd Ann. Field Conf., Guidebook, p. 103–123.

Weinberg, D. M., 1978, Some two-dimensional kinematic analyses of the drape-fold concept, in Matthews, V., III, ed., Laramide folding associated with basement block faulting in the Western United States: Geol. Soc. America Mem. 151 (this volume).

Weinberg, D. W., 1978, Experimental folding of rocks under confining pressure: Pt. VII. Partially scaled models of drape folds: Tectonophysics (in press).

Wellborn, R. E., 1975, Structural interpretation of Terry Field, Park County, Wyoming: Wyoming Geol. Assoc., 27th Ann. Field Conf., Guidebook, p. 209–210.

Wise, D. U., 1963, Keystone faulting and gravity sliding driven by basement uplift of Owl Creek Mountains, Wyoming: Am. Assoc. Petroleum Geologists Bull., v. 47, p. 586–598.

Manuscript Received by the Society June 27, 1977
Manuscript Accepted August 25, 1977

Printed in U.S.A.

Geological Society of America
Memoir 151

Comments on applications of boundary-value analyses of structures of the Rocky Mountains foreland

GARY COUPLES
Department of Geology and Center for Tectonophysics
Texas A&M University
College Station, Texas 77843

ABSTRACT

The inexact geometric fits of the conceptual models of Couples and Stearns (this volume) are due largely to idealizations required by the theoretical analysis. The primary reasons for the mismatch are the assumptions of continuity and isotropy. Real rocks are not adequately described by these properties. Additional considerations of the state of stress—that is, stability index and principal-stress reorientation—are interesting in themselves but do not seem significantly to affect the construction of models using the theoretical solutions. On the basis of information currently available, the mechanical models of Couples and Stearns appear reasonably sound.

INTRODUCTION

The conceptual models of foreland structures presented in the paper by Couples and Stearns (this volume, Figs. 2, 10, 12, 14) are geometrically inexact in the sense that rigid-body motions derived *directly* from the theoretical solutions (that is, the plexiglass approach) do not lead to perfect coincidence with the cross sections constructed for natural structures. While exact matches are seldom, if ever, obtained in the modeling of any natural geologic structures, these particular models need further discussion regarding their goodness-of-fit.

There may be two reasons for these "minor" geometric discrepancies. One is that the stress-state solutions are inappropriate—that is, the boundary conditions inposed are not realistic and/or the dimensions selected are unreasonable. The other possibility is that the techniques for constructing the models *from* the stress-state information are faulty—that is, the shapes of real faults may be different from those derived from the analysis.

These questions comprise the subject matter of this paper. As such, this discussion is an extension of the paper by Couples and Stearns (this volume).

ASSUMPTIONS IN COMPARISON OF NATURAL AND THEORETICAL CROSS SECTIONS

Certain "knowns" are discussed by Couples and Stearns (this volume). These consist primarily of empirically observed attitudes and deformations of the upper portion of the Precambrian basement of the Rocky Mountains foreland. However, most of the information used to construct and interpret their models is itself inferred from a sequence of assumed relations that are based on generally accepted concepts but few "facts." The models can be better understood if the underlying assumptions, along with their corollaries, are more fully appreciated.

Of the many assumptions that enter into various phases of the type of modeling treated here (that is, the models of Couples and Stearns, this volume) two major groups have important geologic implications. One consists of those assumptions related to the calculation of the stresses in the unfractured material, as well as those made to convert the stress-state information into fracture-trajectory diagrams. The other group relates to the application of the fracture-pattern information to the study of natural deformations.

Assumptions of Stress-State Calculations and Construction of Shear-Fracture Trajectories

Determination of the State of Stress. The techniques for determining the state of stress in the model crust before failure occurs (Hafner, 1951; Bengtson, 1956; Sanford, 1959; Howard, 1966; Couples, 1977; Gangi and others, 1977) are all based on the same five assumptions. The model material (1) is a *continuum* (2) in *static equilibrium.* (3) The material is shaped as a *rectangular prismatic body* (a thick beam, see Fig. 1). (4) The behavior of the material is described as *elastic, isotropic,* and *homogeneous.* (5) Elastic distortions occur under a condition of *plane strain.* Because the solutions are used to model crustal deformations in the Rocky Mountains foreland, let us estimate the degree to which the natural material (the upper crust) meets the conditions so specified.

No rock mass in nature is strictly a continuum, especially if the volume treated is on the order of 1 m^3 or larger. The rock almost always contains one or more sets of fractures, which make the mass an aggregate, composed of fracture-bounded "grains." An analysis that accounts for all these material discontinuities would

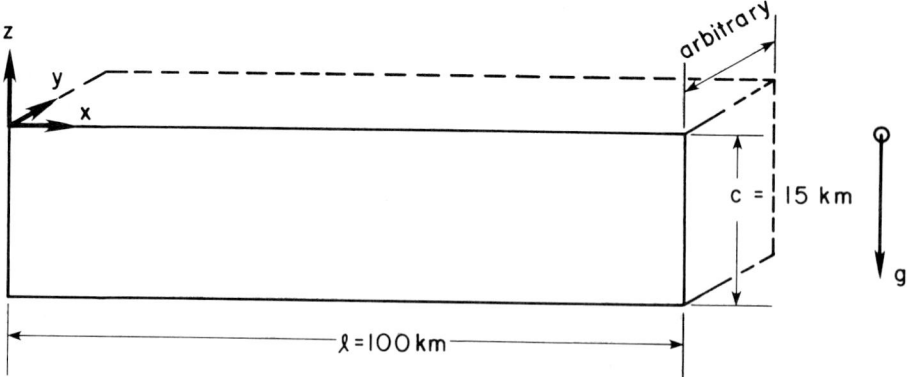

Figure 1. Geometry of rectangular beam showing coordinate directions and dimensions. Arrow labeled "g" represents the gravitational attraction acting parallel to z.

be difficult indeed. Therefore, a general procedure, followed by most previous workers, is to treat the rock mass as if it were, in fact, a continuum, or that at least its behavior can be described adequately by relations derived for strict continua. If continuum mechanics fails to describe the observed response of the rock mass, new approaches must be tried. The concept of the continuum is adopted here until the need for further refinement is demonstrated.

Static equilibrium is also assumed to hold. This eliminates the acceleration terms in the equations of motion (which are the real starting point for any mechanical analysis) and reduces the analysis to the solution of the equations of equilibrium. But, does this assumption seem geologically reasonable?

Cathles (1975) has recently compiled a large amount of data on recent crustal motions associated with postglacial mantle flow. In general, recent crustal uplifts (or downwarps) have rates on the order of 1 cm/yr. Chapin (1976) presented evidence that the western margin of the Sangre de Cristo Range in Colorado experienced 10^4 m of Miocene uplift (relative to the adjacent San Luis Valley) over a period of 4×10^6 yr, for an average displacement rate of 2.5 mm/yr. Tweto (1975) reported a minimum average displacement rate of 2 mm/yr for the Laramide deformation of the Front Range of Colorado. These minimum average displacement rates (2.5 and 2 mm/yr) for past tectonic processes are of the same order as the displacement rates (as much as 1 cm/yr) determined for current glacial-rebound phenomena and are somewhat smaller than present plate velocities ($\simeq$ 1 to 10 cm/yr).

The purpose of presenting these comparisons of rates is to point out that *if* the entire Laramide deformation of the foreland occurred at these rates, the motions would have been nearly imperceptible. With this in mind, the error of approximation made by assuming static conditions seems negligible. In one sense, however, the previous discussion was unnecessary; the analysis is strictly valid only up to the instant that the first fracture begins to form in the ideal beam. Such a problem is, by itself, quite uninteresting. Below, the restriction to initial deformations is relaxed; therefore, it is important to evaluate how well the assumption of static equilibrium holds for postinitiation deformations.

However, can the dynamic effects of the motions of earthquake-producing faults be ignored? The problem does not impose severe restrictions on my analysis, because it is used to predict fracturing, not the transient effects fault displacements have on the initial stress field.

The use of free-body diagrams is fairly old. Hafner's (1951) approach is only the logical extension of that concept. Probably the first to do so in the geologic literature, Hafner proposed the principle by which a portion of a half-space (that is, the Earth's crust from the free surface downward) could be modeled by mentally removing the element (in this case a beam) from the half-space and replacing the surrounding material with equivalent loads (or displacements) on the arbitrarily selected boundaries. In this manner, the intractable problem of modeling the entire crust is simplified to the problem of modeling a segment of it at any particular time and place, that is, to study the deformation of an equivalent isolated body—in this case, a thick prismatic beam (Fig. 1). Therefore, the large engineering literature on the mechanics of beams is opened to the structural geologist wishing to study crustal deformations.

The assumptions about mechanical response—elastic, isotropic, homogeneous— are probably the most difficult to reconcile with nature. Nearly all rocks respond to load elastically to some degree, but whether any rock mass in bulk behaves in a truly elastic fashion is moot. There is substantial evidence that rocks that fail in a brittle manner (fracture) can be treated as behaving elastically up to the point of failure (see, for example, Griggs and Handin, 1960). Nevertheless, the behavior of the then-fractured rock mass may obey some other law. Because testing rock masses at scales comparable to those of natural deformations is impractical, the validity of the assumption of elastic response can be checked only by the empirical agreement of models derived by assuming elastic response with natural features. Nevertheless, the study must begin at some point, and because the elastic-brittle correlation is reasonable, the assumption of elasticity is regarded as adequate for those rocks that have apparently behaved in a brittle fashion, especially for modeling the *initiation* of faulting in the beam.

The assumptions of isotropy and homogeneity must be treated in much the same manner as that for elasticity, that is, empirically. The assumptions concern the idealized *material response* and not the real texture, fabric, or composition. Let us consider the situations in which the assumptions are more likely to be met. Sedimentary-rock sequences composed of various lithologic types are expected to be inhomogeneous in their bulk response to load. On the other hand, "continental" or granitic basement rocks may well be substantially more homogeneous. Shales and other well-bedded, laminated, or schistose rocks tend to be anisotropic under experimental conditions (see, for example, Donath, 1961, 1964). Conversely, unjointed granite is all but statistically isotropic. Coarse, granitic-textured gneiss (composing the basement in much of the foreland) may also be nearly isotropic. My own field observations of granitic gneiss in Wyoming, along with those of Stearns (1971, 1976, personal commun., and this volume) and Weinberg (1976, personal commun.), among others, seem to support this view.

All these assumptions tend to limit the application of state-of-stress calculations to studies of the deformation of basement rocks (as defined by Stearns, this volume). Applications of analyses such as those of Couples (1977) to either layered-rock sequences or schistose metamorphic rocks are not justified.

The assumption that inital deformations (elastic distortions) occur under a condition of plane strain (that is, all displacements occur in the x–z plane of Fig. 1) is one that is intuitively acceptable, particularly when the deformations have taken place in the central regions of long structures. Experimental studies (Friedman and others, 1976; Logan and others, this volume) indicate that during *finite* deformations, material is transported in a direction longitudinal to the structure (y-direction in Fig. 1). Thus, finite deformations expected to occur under a condition of plane strain do not necessarily do so. This same antiplane strain condition may also apply to natural deformations (Weinberg and others, 1976), but its significance has not yet been studied analytically.

Note that the experimental conditions do not include lateral-displacement constraints—only the fluid-confining medium surrounds a specimen during its deformation. In nature, the "specimen" is surrounded by other masses of rock that can provide lateral constraint that might enforce the plane-strain condition. Until this assumption can be evaluated from further field, experimental, and theoretical studies, the plane-strain condition presently must be assumed to hold for deformations in a plane that is medial and transverse to the long dimension of the model beam.

Using these assumptions, expressions can be derived for calculating the state of stress within a beam for a given set of statically admissible boundary loads. These calculations usually involve the Airy stress function or some other method based on potentials. The derivations of the appropriate equations can be found in texts on the theory of elasticity (Love, 1927; Morse and Feshback, 1953; Timoshenko and Goodier, 1970), rock-mechanics books (Jaeger, 1969; Jaeger and Cook, 1970), and in scientific papers (Hafner, 1951; Sanford, 1959; Couples, 1977; Gangi and others, 1977).

For boundary-value problems, specific solutions depend on the selection of boundary conditions. The boundary conditions for my models can be specified as the "tectonic" and "standard-state" components. The reasoning behind the choice of the particular standard state (or gravitationally induced stress field) has been discussed in detail previously (Couples, 1977). The tectonic components are selected so as to yield structural configurations that approximate the real structures of concern here. This is done by imposing loads "shaped" like the final mountain systems (see Couples, 1977). There is no a priori basis for this procedure; at this time we can only choose a set of boundary conditions, calculate the results, and apply the plexiglass approach. The appropriateness of the boundary conditions can be tested only by the usefulness of the results (the goodness-of-fit).

What limitations do the assumptions behind the stress-state calculations impose? Sensu stricto, real rock materials seem untreatable by *any* theoretical modeling. The several approximations discussed above, however, allow us some leeway, and the stresses calculated for the ideal beam would seem most nearly to approximate those within certain relatively shallow parts of the granitic basement (say the upper 10 to 15 km). Experimental and field data strongly suggest that here deformations have been essentially elastic prior to rupture.

Representation of the Theoretical Results. Once the state of stress has been calculated, it remains to represent it diagrammatically to facilitate interpretation. The usual method is to draw stress trajectories, which are composed of two families of lines that are everywhere tangent to the orientations of their respective principal stresses; hence, the lines are everywhere orthogonal. Beyond this, additional information about the state of stress in plane-strain deformation can be displayed by sets of stress-magnitude or stress-difference contours, or some other means to show the intensity of the stresses, along with information on regions of the beam where stress reorientations have occurred.

Stress-trajectory diagrams alone do not adequately portray the expected deformation pattern, especially when faulting is expected. Stress-difference contours or some other measure of stress intensity may very well indicate those regions most susceptible to failure, but fault slip and orientation are not readily visualized from any stress map alone. Thus, as an aid to understanding the relation of the stress state to faulting, additional diagrams (shear-fracture trajectories) are desirable. To construct them one assumes that (1) the deformation is related to the calculated static stress state; (2) shear fracture is the primary macroscopic deformation mechanism; (3) the shear fractures that do develop are oriented at some fixed angle ($\theta \simeq 30°$, relative to the local orientation of the maximum principal compressive stress σ_1); (4) the other family of stress trajectories constructed in the plane of the diagram parallels the least principal stress σ_3; consequently, (5) the plane of the fracture is normal to the plane of the diagram—that is, parallel to the intermediate principal stress σ_2, which is itself parallel to the Y direction (Fig. 1; see below for modification of these last two statements), and (6) large stress differences and any possible associated deformations (fractures) do not influence the orientations of other fractures that might subsequently develop.

On the basis of the cumulative experimental results published during the past 25 yr, assumption 1 seems quite acceptable; assumption 2 is reasonable after considering Stearns's (1971; 1975; this volume) observations that faulting is the dominant deformation type for the basement rocks of the Rocky Mountains foreland. Assumption 3 is a good approximation to the large number of fault angles determined in experiments (Handin, 1966). Perhaps a more precise statement would be that θ is a function of σ_{ij} (Handin and others, 1967), but it is not greatly significant to this discussion. Statement 4 is not strictly true; however, the regions of the beam where this assumption fails are those least likely to experience fracture (see below). Most structural geologists agree that the intersection of conjugate shear fractures parallels the intermediate principal stress σ_2. Experimental work under true triaxial stress states by Mogi (1971) confirms this notion. Therefore, statement 5 is correct when assumption 4 holds.

One aspect of the evaluation of assumption 6 is discussed elsewhere (Couples, 1977). Briefly, this argument is based on the high degree of ordering observed for natural macrofractures (Stearns, 1968, 1972). It seems improbable that all fractures of a given ordered set could have formed simultaneously; they must have developed sequentially, so that the formation of each fracture could not have appreciably altered the causative stress field. Otherwise, the high degree of ordering could not obtain.

Macrofractures with little or no observable offsets (as in those cited by Stearns) do not provide sufficient basis for evaluating assumption 6 we must call upon the empirical agreement of experimentally created faults and those predicted by appropriate theoretical solutions. Sanford's (1959) experiments with sandboxes are the prime examples. He found analytical solutions for a thick beam subjected to an approximate step displacement on its lower boundary. The other boundary conditions included a stress-free upper surface and a gravitational standard state similar to that in my calculations. He also conducted analog experiments in which a piston at the bottom of a glass-fronted aquarium was forced into an overlying "beam" of sand.

Similar experiments have been conducted in the laboratories at Texas A&M University (Fig. 2A). Note the curving reverse-fault zone whose dip becomes shallower near the surface. Figure 2B shows the pertinent theoretical solution for this loading condition. Here the similar cross sections of the theoretical and experimentally created and natural faults provide striking support for assumption 6. The "upthrust" in the sandbox is actually composed of a sequence of faults that have coalesced into the final form of the fault zone, but this fact does not detract from the previous conclusion. In fact, Couples and Stearns (this volume) suggest that movement-related alteration of the geometry of the fault zone may very well be a normal phenomenon that occurs following the initiation of the fault geometry.

In summary, I conclude that the construction of the shear-fracture trajectories based on assumptions 1 through 6 is a valid operation. Correlations with experiments (Sanford, 1959) and usefulness in modeling (Couples and Stearns, this volume) suggest that at least some of the shear-fracture trajectories do, indeed, predict the geometry of faulting in materials subjected to loads like those assumed in generating the solutions.

Assumptions Involved in Using Shear-Fracture Trajectories to Model Natural Structures

Here the assumptions are not as readily justifiable as those discussed above. Two groups need consideration: those assumptions made in constructing the structural cross sections and those employed in applying the models—that is, the plexiglass approach.

Rules for Drawing Structure Sections. The construction of geologic cross sections is usually subjective. Consequently, the rules involved vary somewhat from case to case. In simplest terms, the rules are (1) honor available geometric data, (2) conserve volume, and (3) consider the probable mechanical behaviors of the rocks involved. The second rule may be restated: account for all displacements, including transport normal to the plane of the section. The third rule has a corollary: use knowledge of rock properties under the presumed environmental conditions to infer probable deformation mechanisms.

These rules were honored, insofar as possible, in constructing the cross sections of Couples and Stearns (this volume, Figs. 10, 12, 14). Geometric data were very limited for most of these locations, but they were honored wherever available. Behaviors for the basement rocks are everywhere observed to be "rigid" (Stearns, 1971, 1975, this volume). The assumption that this behavior extended to all relevant depths constrains the cross sections to be composed of components that fit together without invoking large distortions (see below and Couples and Stearns, this volume, for extended discussion of the meaning of "rigid"). Cylindrical faults between the basement blocks allow the required fault slips to be accomplished in a plane-strain condition. It is the ability to construct sections using *all* the rules that supports the validity of the subsurface predictions.

Plexiglass Approach. The plexiglass approach is described in the paper by Couples and Stearns (this volume). This name is applied to the technique used to construct the conceptual models. In these models, the natural structure section is compared to a geometric configuration that results from the motions of selected potential

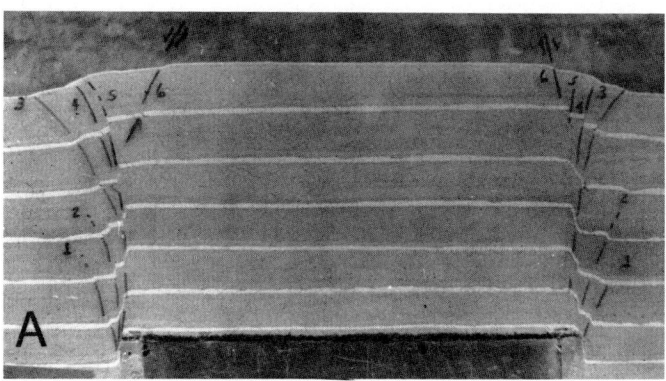

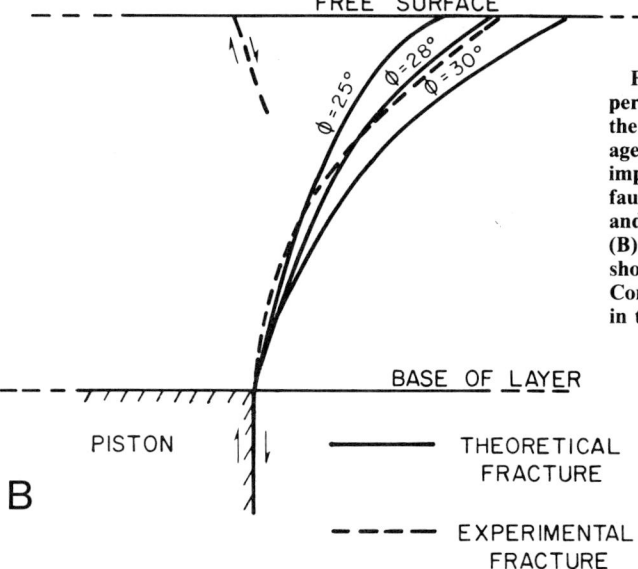

Figure 2. (A) Photograph of sandbox experiment. Piston was displaced vertically into the overlying thick, homogeneous sand package. White lines are passive markers and impart no mechanical effect to the sand. The faults in the sand are labeled 1, 2, 3, 4, 5, and 6, according to sequence of development. (B) Theoretical solution from Sanford (1959) showing geometry of the upthrust fault. Compare this geometry to that of the faults in the sandbox.

faults contained in the analysis. The name "plexiglass approach" arises from the visualization made possible by placing a thin sheet of plexiglass over the appropriate diagrammed analytical solution. The beam boundaries and the selected active faults are cut into the plexiglass. Then, these rigid pieces are displaced past one another, showing the geometric relations to be compared visually to the real structural cross sections previously prepared.

The assumptions inherent to this technique are that the component blocks of the system remain rigid and that no room problems are created. Both of these assumptions are discussed in some detail by Couples and Stearns (this volume), and they conclude that the plexiglass approach is, indeed, applicable to studies of basement faulting in the Rocky Mountains foreland.

FURTHER RESULTS DERIVED FROM THE STATE OF STRESS

At the time Hafner (1951) published his results, he recognized the need to evaluate the fracture trajectories in light of some sort of stability-instability criterion. On the basis of calculated stress states and a failure stress, he drew lines separating regions where the beam would not have failed (fractured) from those regions where the failure (fracture) stress was exceeded.

The general state of knowledge concerning rock behavior was limited then. Therefore, Hafner chose to use the Trescan criterion—that is, failure occurs when the stress difference at a point reaches some specified constant value, irrespective of normal pressure. Hafner drew stability boundaries at the arbitrarily selected stress differences of 1,000, 2,000, and 3,000 kg/cm^2 (approximately 1, 2, and 3 kb).

Bengtson's (1956) diagrams also contained areas labeled "stable," but he did not say how these were determined. Probably he merely followed Hafner.

Sanford (1959) approached the problem in a different way. He calculated and plotted distortional elastic strain energy for a grid of points in the beam. Although it is difficult to relate quantitatively distortional strain energy to regions of stability, it is probably reasonable to suppose that the regions of higher energy are the more likely to be unstable (to have previously failed).

Howard (1966) incorporated a specific failure criterion in his determination of stable regions. He used the Mohr representation of the Coulomb criterion to define a geometric comparison of stress states. He calculated the ratio of the radii of two Mohr circles representing (1) the actual stress state at each point in the beam, and (2) the stress state at failure. These ratios, scaled to 100, were contoured over the region of the beam. Unfortunately, the formula given in his paper does not perform the calculation just described, so Howard's results cannot be verified readily. Perhaps the formula is misprinted.

The Stability Index

Stability determinations have now been performed on my earlier results (Couples, 1977). The method of Howard (1966) was adopted, although his formula has been corrected. In Figure 3, the Mohr representation of the Coulomb failure criterion facilitates a comparison of the geometric relations used to derive the stability formula. In simple terms, the formula is designed to do the following: to *compare* the radius of the Mohr circle of the actual state of stress at each point to that of the circle tangent to the failure envelope with its center at the same mean stress, σ_m. The ratio is scaled so that a stress state just at the point of failure would have a stability index S of 100.

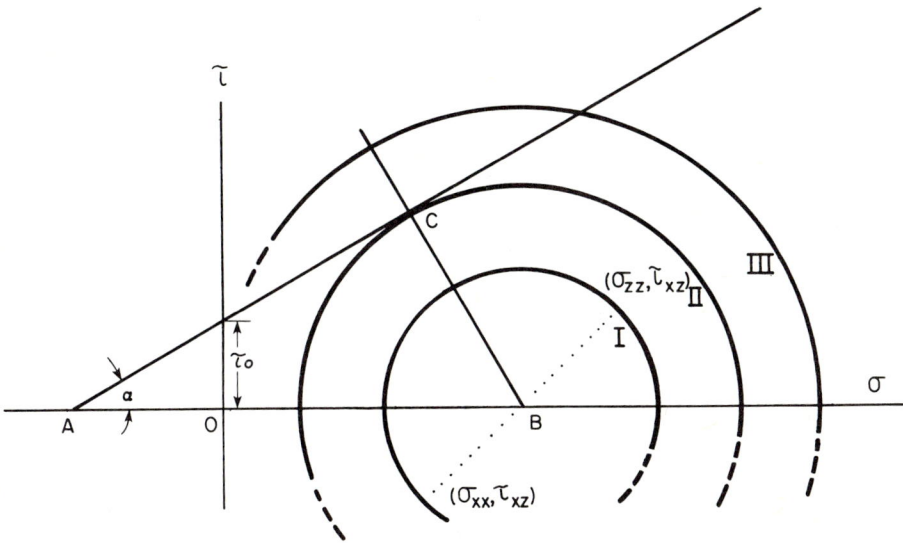

Figure 3. Mohr circle construction used in deriving stability index.

In Figure 3, the first length to be determined is $\overline{BC}$, which is completely determined from the segment $\overline{AB}$ and the angle ϕ:

$$\sin\phi = \overline{BC}/\overline{AB}.$$

$\overline{OB}$ represents the magnitude of the mean stress σ_m, defined here as

$$\sigma_m = \overline{OB} = (\sigma_{xx} + \sigma_{zz})/2,$$

with compressive stresses taken as positive. Another simple trigonometric formula is used to obtain the segment $\overline{AO}$:

$$\overline{AO} = \tau_o/\tan\phi.$$

Because $\overline{AB}$ is merely the sum of $\overline{AO}$ and $\overline{OB}$, the length $\overline{BC}$ is

$$\overline{BC} = \sin\phi\ \overline{AB} = \sin\phi\left(\frac{\sigma_{xx}+\sigma_{zz}}{2} + \frac{\tau_0}{\tan\phi}\right). \qquad (1)$$

The radius of the Mohr circle representing the actual state of stress is given by

$$R = \frac{\sigma_1 - \sigma_3}{2}. \qquad (2)$$

All quantities needed to calculate S have now been determined:

$$S = 100 \cdot \frac{R}{\overline{BC}} = 100 \cdot \left(\frac{\sigma_1-\sigma_3}{2}\right)\bigg/\left[\left(\frac{\sigma_{xx}+\sigma_{zz}}{2} + \frac{\tau_0}{\tan}\right)\cdot\sin\phi\right]. \qquad (3)$$

The circle labeled I in Figure 3 represents a stress state, the stability index of which is less than 100 when compared to the failure circle labeled II. The state of stress represented by circle III (in Fig. 3) results in a calculated stability index in excess of 100, an "unstable" condition. If the envelope (Fig. 3) represents a valid failure criterion, then stress states represented by circles such as III are impossible to achieve in that material; it must fail when the stability index reaches 100.

Since $S > 100$ cannot be attained, the presence of "unstable" regions within the beam (Figs. 4, 5, 6) needs explanation. The analysis requires that the beam remain continuous, hence it cannot fail and become discontinuous. Effectively, this restriction requires an idealized beam to have strength sufficient to sustain without failing the highest differential stresses caused by the boundary conditions. The beam is supposed to model a real material—in this case, continental crust that has a given strength of failure criterion. Therefore, whenever the stress state reaches the failure criterion for continental crust, the beam fails (fractures) at least at that point so that the real state of stress never exceeds those values within the stable field. The effect such failures have on the overall stress state in the remainder of the beam is unknown. As a first approximation, we have already assumed that the failure does not appreciably alter the stress state in regions beyond the immediate vicinity of any fracture (the "brittle" failure mechanism) that forms. Thus, it is possible to distinguish regions that have not yet failed under a given set of loads (stable) and those that have already failed (unstable), that is, failed at a lower differential load (Figs. 4, 5, 6).

Stress Reorientation

Another refinement of the earlier stress-state and fracture-trajectory diagrams (Couples, 1977) has also been completed. Sanford (1959) stated that "in a two-dimen-

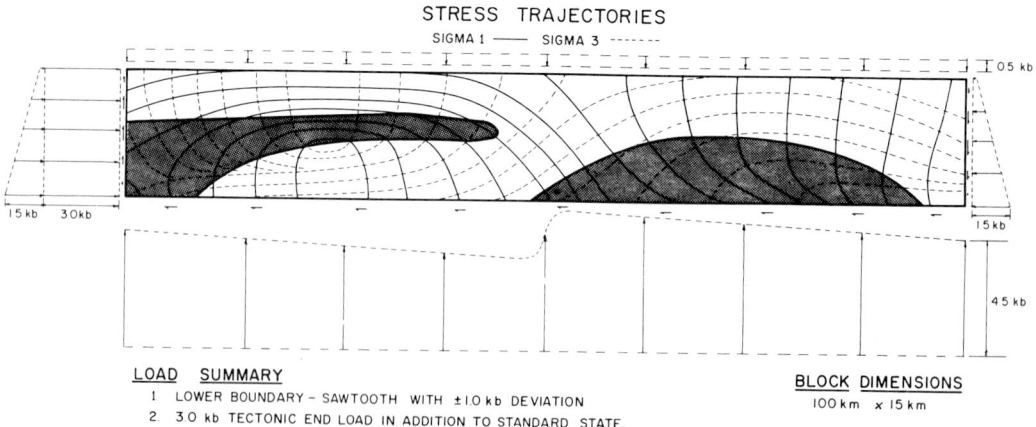

Figure 4. Updated version of Figure 6 of Couples and Stearns (this volume). In the upper diagram, the contoured lines represent 50%, 100%, 150%, and 200% stability indices. The areas of the beam in excess of 100% are unstable. In the lower diagram, shaded regions are locations of stress reorientation.

SHEAR FRACTURE TRAJECTORIES

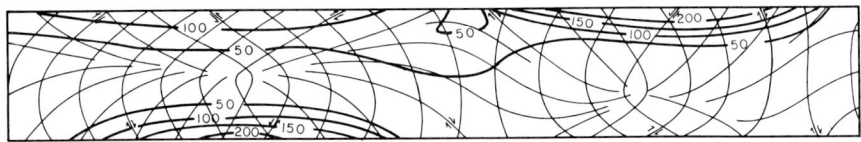

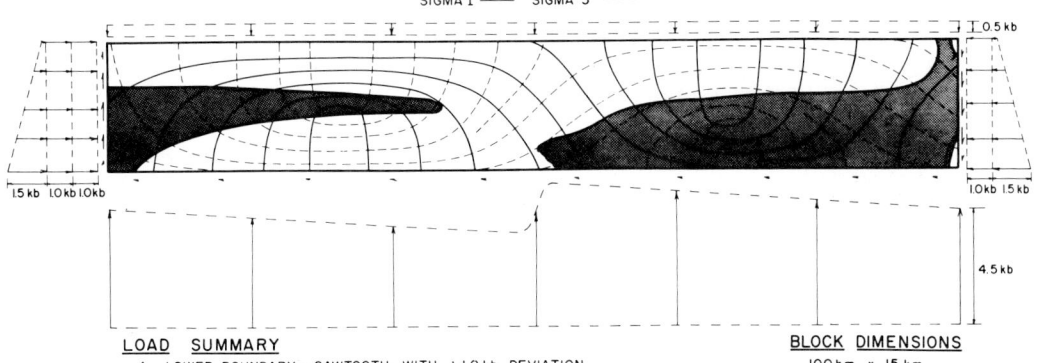

Figure 5. Updated version of Figure 7 of Couples and Stearns (this volume). Symbols same as Figure 4.

SHEAR FRACTURE TRAJECTORIES

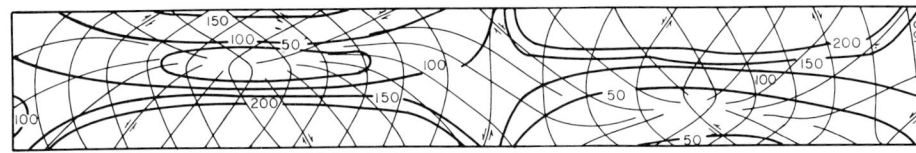

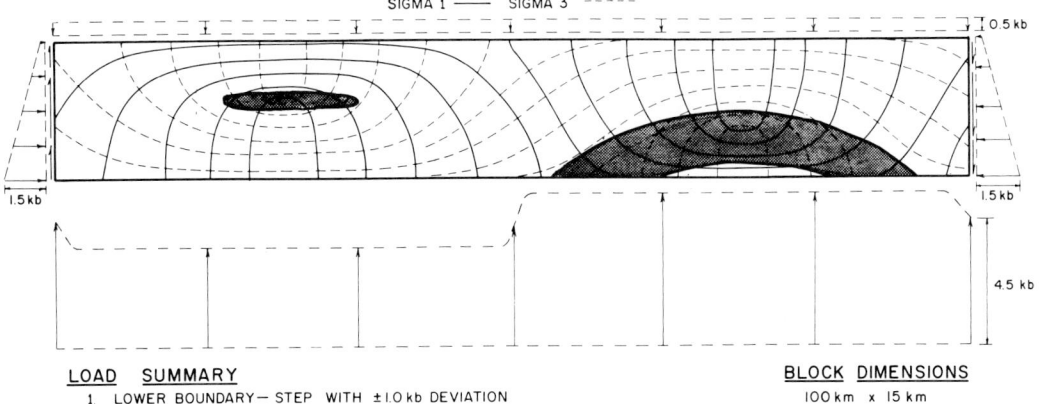

Figure 6. Updated version of Figure 8 of Couples and Stearns (this volume). Symbols same as Figure 4.

sional analysis, the intermediate principal stress (σ_2) parallels the z-axis." Hafner (1951) made a similar statement. In my coordinate system (Fig. 1), their statements would say that σ_2 parallels the y-axis, normal to the x-z plane. That statement is strictly true *only* when $v = 0.5$. For lesser values of Poisson's ratio, there are regions of the beam in which the σ_{yy} stress component has the value

$$\sigma_{yy} = v\,(\sigma_{xx} + \sigma_{zz}), \tag{4}$$

such that it becomes the least principal stress σ_3. The stress component σ_{yy} is merely required to maintain the plane-strain condition of the problem, expressed in terms of the other normal-stress components.

Here v is assumed to be 0.25 (compare Couples, 1977). Thus, there are regions where σ_{yy} is σ_3; they are located around the singular points (Figs. 4, 5, 6). The physical significance of this stress reorientation in unknown. To my knowledge, it has not been discussed in any previous analysis. If fracture orientation is dependent on the orientation of the principal stresses (as all previous workers have supposed), the fracturing represented in Figures 4, 5, and 6 is wrong. However, since these regions of potential stress reorientation are associated with low stress differences and low stability indices, the effects are probably not significant. Nevertheless, this problem should be studied further.

DISCUSSION

Explanations of the Geometric Mismatches

Although the models presented by Couples and Stearns (this volume) advance our knowledge of foreland structures, they leave many questions unanswered. Some features "predicted" from the theoretical analyses have not been observed in nature; other natural features are not explained by the theory. In other words, the plexiglass approach does not yield perfect coincidence.

Effects of Potential Anisotropies. Many observations of the basement rock of the Rocky Mountain foreland reveal a large and diverse variety of fractures. Only a few are easily associated with Laramide deformation; most have surely been inherited from pre-Laramide (probably Precambrian) events. Thus, these pervasive fractures imply the existence of wholesale material discontinuities in the rock mass *at the time* of Laramide deformation.

Pre-existing cracks are known to influence the mechanical response of rocks (Handin, 1969). Figure 7 illustrates how pre-existing surfaces can modify the allowable field of stress states. Lines $\overline{OB}$ and $\overline{BO}$ form the modified failure envelope of a now-cracked rock in σ-τ space. A stress state represented by circle I is expected to cause failure by sliding on the preformed surface. Circle II represents a stress state expected to cause a new fracture to form. Thus, the prefractured basement rocks may well have deformed non-ideally (that is, may not have fractured as expected), particularly in low mean-stress regions (circle I, Fig. 7).

Foliations are another feature often observed in the Precambrian rocks of the foreland (Prucha and others, 1965). However, as Stearns (this volume) has discussed, only in a few locations are the foliations strongly developed; these rocks are excluded a priori from consideration as basement. Weak foliation, as in coarse, granitic gneiss, does not seem to have influenced the deformation significantly.

Experiments have revealed that some foliations strongly affect the mechanical response of rocks, particularly at intermediate confining pressures (Donath, 1961).

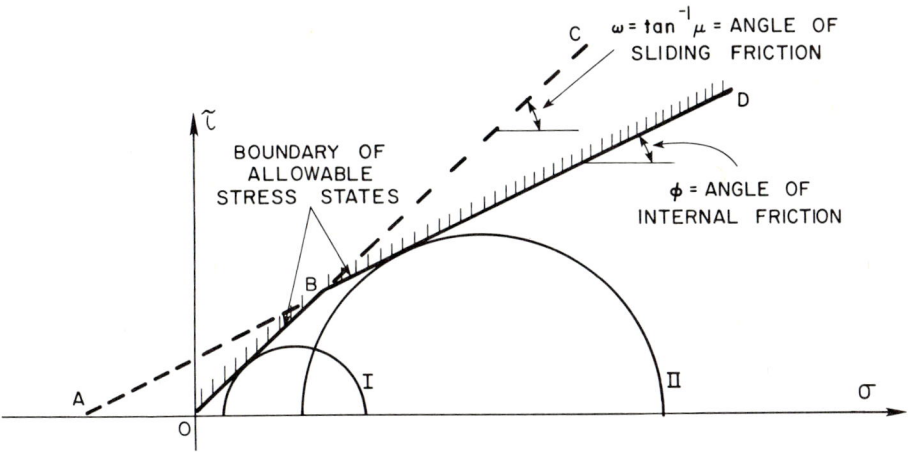

Figure 7. Diagram illustrating how the presence of pre-existing fractures can modify the allowable field of stress states. If the state of stress falls on the failure envelope along portion OB, failure is expected to occur as sliding on the fracture surface. If the Mohr circle of stress intersects line AD at *any* time, fracture occurs.

In some cases, this control is so pronounced as to exclude the possibility of treating the behavior as elastic up to some point of brittle failure. Borg and Handin (1966) showed that micaceous gneiss is also strongly anisotropic. The gneissic basement rocks of concern here *may not* have behaved as an ideally elastic, isotropic material.

Minor Deformations at the Top of the Basement. One of the results of the analysis for which evidence has not been observed in nature is the "early" extension and shortening at the top of the beam. Note that the contours of the stability index exceed 100 over large portions of the tops of the beams (Figs. 4, 5, 6). The analysis predicts that these regions should fracture early in the deformation, that is, at lower differential loads. Note that the right side of the top of the beam is expected to be in extension. In the field, no normal faulting that offsets the Precambrian-Cambrian contact is observed. On the left, where thrust-fault fractures are shown, the basement blocks usually are downdropped, and this contact is not often observable. Therefore, comparison of nature and model is not feasible here.

The lack of coincidence between natural and predicted faulting requires some form of explanation. The most reasonable guess follows from the recognition of pre-existing fractures. If a rock mass is required to extend, it can do so in at least two ways: (1) the creation of and motion along shear fractures (small normal faults), or (2) the creation and opening of extension fractures. It has already been shown that suggestion 1 is not supported by field studies. However, a modification of suggestion 2 is compatible with the observed geology. If the pre-existing (mostly steeply dipping) cracks are opened slightly, the displacement predicted by the analysis is accommodated without the need of additional fracturing. The shortening required at the opposite end of the beam could be accommodated by the closing of the pre-existing cracks, perhaps without further fracturing. These suggestions lack proof, but they can explain away certain of the inconsistencies while requiring minimal additional deformations.

Mismatch in Dip of the Basement Faults. Yet another of the disparities between the mathematical models and observed relationships is basement-fault dips. In nature, the attitudes of basement faults must usually be determined by indirect methods such as fault-parallel fractures. The attitudes of high-angle faults are often inferred by this method to be greater than 75° relative to the Precambrian-Cambrian contact (Weinberg, 1977, this volume; Stearns, this volume). The attitudes of the low-angle basement faults are seldon known at all. Their orientations can be determined with precision only by three-point methods when drill holes are available. Berg (1962) reported a dip of 30° for the frontal fault of the Wind River Mountains.

The theoretical analyses indicate that faults should intersect the upper surface of the basement at angles of 30° (for reverse faults) or 60° (for normal faults). The former cannot be checked in the field. However, the latter suggests that the analytical solutions *do not* completely describe the fault orientations.

An explanation for this discrepancy can be based on the existence of pre-Laramide fractures or foliations. The influence of these elements is most pronounced at low mean stress. It is in the shallow regions of the beam where the solutions appear to disagree most with geologic observations. Perhaps, then, pre-existing cracks or foliations have caused faulting to be steeper than it would have been in homogeneous, isotropic rock.

On the other hand, the solutions themselves may be "incorrect"—that is, some of the assumptions may be unjustified. Perhaps fractures do not really follow trajectories oriented at 30° to an initial σ_1 trajectory. This possibility seems unlikely. If the solutions are incorrect, it is more likely that the assumed boundary conditions are wrong—for example, the top of the beam should not be treated as a principal plane of no shear stresses. A different condition would allow for other fault attitudes, particularly near the top of the beam. This suggestion is the more likely.

Possible Inappropriate Assumptions in the Plexiglass Approach. Several factors contribute to the inexact fit between the projected geometry of a model and the configuration observed in nature. One is the possibility of changes in shape and/or volume resulting from finite deformations. Models based on infinitesimal theory simply may be inadequate. The shattered lobe of the Wind River Mountains is an example. Here the rocks must clearly have behaved cataclastically at least during the later stages of deformation. *Exact* geometric coincidence between elastic models and real cataclastic structures should not be expected in that case.

Cataclastic flowage in the lobes may have allowed the thrust portion of the basement faults to flatten even beyond the configuration depicted by Couples and Stearns (this volume, Figs. 9, 11). Flow could also lead to exaggerated horizontal motions. If it were possible to remove the flow component of the deformation, one might better model its structural development.

Faults in nature also may change shape with deformation. In the plexiglass approach, this possibility is not allowed. Motion is easiest along smooth circular arcs. Energetically, more work must be expended to move two rigid blocks past one another on noncircular surfaces or surfaces with asperities. A noncircular fault might prove to be an energy barrier that favors adjustment of the shape of the fault by ancillary fracturing and development of a zone of faulting. Such may well be the case in the model for the Beartooth Mountains discussed in Couples and Stearns (this volume).

In the models presented by Couples and Stearns (this volume), only circular faults were allowed. This is the only volume-conserving scheme that is possible in the plexiglass approach. However, a small averaging of the fault geometry is always presumed in developing the models. To be sure, S-shaped potential faults are not "averaged" into circular shapes, but nearly circular faults are. This procedure is not considered cheating; it is merely a useful means of initiating the modeling with a configuration that is reasonably close to the shapes expected to result from the continued deformation.

One more discrepancy needs clarification. Large structures occur on the "back" sides of many of the uplifts—the Lander-Hudson dome and the Thermopolis anticline, respectively (Couples and Stearns, this volume, Figs. 9, 11). The basement faults controlling these structures have not yet been modeled successfully because either (1) the solutions are wrong—perhaps the lower-boundary loading conditions should have additional "bumps," or (2) these structures are purely secondary features related to excessive rotations of the crustal slab of the upthrust system not accounted for by the analysis.

Usefulness and Limitations

At least a few geologic structures are entirely visible to the field geologist. In principle, then, it is fully possible to determine their structural histories from careful, detailed field observations alone (including petrofabric studies). Theoretical modeling of these structures serves primarily to direct attention to particular locations in a search for deformation features of specific interest. However, for large-scale structures (for example, the Wind River, Owl Creek, or Beartooth Mountains), modeling plays a somewhat different role. To be sure, field studies are undertaken as quite necessary, but most of the geometric data are limited in scope by size and outcrop conditions or are inferential in character (seismic records and a few data from boreholes). Much of the structure at depth is speculative. Here modeling can be useful indeed. The models provide well-educated guesses about subsurface structures that are not directly observable.

In brief, models can serve us well in two ways. They aid location of the features critical for the determination of structural history. They also provide the initial state from which to work out structural development. An immediate use of these solutions is to suggest geometries for yet unstudied uplifts in the region.

As a possible long-term application of the method, consider the following concept. If the structure of the Wind River Mountains is close to that suggested by Couples and Stearns (this volume, Fig. 9), then it may be feasible to select a location favorable for study of the lower crust. Notice the dashed line at -15 km, drawn only to facilitate comparison of the real cross section to the model. This line may represent a significant level, provided that no flowage has occurred. There are suggestions that 15 km is about the lower limit of granitic rocks in the continental crust.[1] If so, then basaltic rocks of the lower crust should be relatively near the surface beneath uplifted mountains. The Wind River Mountains might be a good site to drill into lower crustal rocks.

Regardless of the potential benefits of modeling, there are limitations that cannot be ignored. Existing solutions do not inform us about which of all the potential faults will actually form, the sense of net slip to expect, or the magnitude of displacement. Stearns (1975) used one of Hafner's (1951) solutions to construct his model (Couples and Stearns, this volume, Figs. 1, 2). Hafner's original diagram shows regions of stability and instability. Although the boundaries of these regions are not exactly correct (see above), the theory does predict that some of the faults are more likely to form than are others. Stearns (1975) ignored these distinctions. He treated all potential faults as equally likely to form.

The three new models of Couples and Stearns (this volume) suffer likewise.

---

[1] Lamping (1970, Fig. 4.2) presented crustal columns for locations within the continental interior of the United States. Separation of the crystalline rocks into Precambrian (= granitic) and lower crust (= basaltic) was accomplished from heat-flow modeling and seismic refraction. The sections illustrate granitic crust ranging in thickness from about 10 to 20 km, averaging about 15 km thick. Warren (1968) indicated similar results.

The analysis is used only to "document" a required fault shape at a particular location. Comparison of Figures 4, 5, and 6 of this paper to Figures 10, 12, and 14 of Couples and Sterns (this volume) reveals that these faults may not be the *least* likely to experience significant movement, but neither are they the *most* likely.

FUTURE WORK

Experience to date suggests that chances for improving our modeling techniques are good. So far the mathematical treatment has required some constraints that are not geologically reasonable. End effects can be minimized by increasing the length of the beam and restricting interpretation to the central region far from the ends. Further variation of loading conditions can be tried readily as I (Couples, 1977) have pointed out. Numerical methods (finite element) are now in wide use by structural geologists. Even so, problems of finite deformations, including the perturbations of an initial state of stress by faulting, need time-sequence analyses. Much more must be known about fracture initiation and propagation, slip along faults, and nonlinear rheological properties before numerical modeling can be adequate. Finally, three-dimensional models must eventually be developed. Aspects such as the stress reorientation will certainly enter into these more complicated models.

CONCLUSIONS

1. Couples (1977) described the techniques for generating elastic solutions for a wide variety of boundary-loading conditions and concluded that the resulting deformation patterns remain essentially "vertical," even under large components of lateral load. Potential-fault trajectories are fairly sensitive to the magnitudes of the loads applied.

2. Couples and Stearns (this volume), on the basis of my analyses (Couples, 1977), present new conceptual models for two tilted, upthrust mountain systems of the Rocky Mountains foreland—the Wind River and Owl Creek Mountains, and a plateau, "splinter-block" uplift, the Beartooth Mountains.

3. Couples and Stearns (this volume) suggest a possible simplified stress system for the initial deformation of the entire foreland—a regional, northeastward-decreasing horizontal compression superposed on local variations due to vertical loading.

4. Here, I have tried to show that the formal assumptions behind the analyses are not so restrictive as to preclude applications to rocks in nature. However, the assumptions of continuity and isotropy may not always be valid, and this is probably why the models of Couples and Stearns (this volume) do not fit exactly.

5. The concepts of stability index and stress reorientation do not seem to have important implications for modeling at this time. However, they deserve further investigation.

ACKNOWLEDGMENTS

My thinking has been stimulated by trying to answer the critical questions posed to me by fellow students and faculty at Texas A&M; I wish to thank all who, knowingly or not, sparked my interest.

David W. Stearns and David K. Parrish read early versions of this manuscript and suggested several important modifications.

I have been supported by a National Science Foundation Graduate Fellowship and by the Michel T. Halbouty Graduate Fellowship at Texas A&M University.

This manuscript has been reviewed by John Handin and David K. Parrish.

REFERENCES CITED

Bengtson, C. A., 1956, Structural geology of the Buffalo Fork area, northwestern Wyoming, and its relation to the regional tectonic setting: Wyoming Geol. Assoc., 11th Ann. Field Conf., Guidebook, p. 158–168.

Berg, R. R., 1962, Mountain flank thrusting in Rocky Mountain foreland, Wyoming and Colorado: Am. Assoc. Petroleum Geologists Bull., v. 48, p. 2019–2032.

Borg, I., and Handin, J., 1966, Experimental deformation of crystalline rocks: Tectonophysics, v. 3, p. 249–368.

Cathles, L. M., III, 1975, The viscosity of the Earth's mantle: Princeton, N.J., Princeton Univ. Press, 386 p.

Chapin, C. E., 1976, Evolution of the Rio Grande rift: Geol. Soc. America Abs. with Programs, v. 8, p. 808–809.

Couples, G., 1977, Stress and shear fracture (fault) trajectories resulting from a suite of complicated boundary conditions with applications to the Wind River Mountains: Pure and Appl. Geophys., v. 115, p. 113–133.

Couples, G., and Stearns, D. W., 1978, Analytical solutions applied to structures of the Rocky Mountains foreland on local and regional scales, in Matthews, V., III, ed., Laramide folding associated with basement block faulting in the Western United States: Geol. Soc. America Mem. 151 (this volume).

Donath, F. A., 1961, Experimental study of shear failure in anisotropic rocks: Geol. Soc. America Bull., v. 72, p. 985–990.

——— 1964, Strength variation and deformational behavior in anisotropic rock, in Judd, W. R., ed., State of stress in the Earth's crust: New York, Elsevier, p. 281–297.

Friedman, M., Handin, J., Logan, J. M., Min, K. D., and Stearns, D. W., 1976, Experimental folding rocks under confining pressure: Pt. III. Faulted drape folds in multilithologic layered specimens: Geol. Soc. America Bull., v. 87, p. 1049–1066.

Gangi, A. F., Min, K. D., and Logan, J. M., 1977, Experimental folding of rocks under confining pressure: Pt. IV. Theoretical analysis of faulted drape folds: Tectonophysics, v. 42, p. 227–260.

Griggs, D., and Handin, J., 1960, Observations on fracture and a hypothesis of earthquakes, in Griggs, D., and Handin, J., eds., Rock deformation: Geol. Soc. America Mem. 79, p. 347–364.

Hafner, W., 1951, Stress distribution and faulting: Geol. Soc. America Bull., v. 62, p. 373–398.

Handin, J., 1966, Strength and ductility, in Clark, S. P., ed., Handbook of physical constants: Geol. Soc. America Mem. 97, p. 223–289.

——— 1969, On the Coulomb-Mohr failure criterion: Jour. Geophys. Research, v. 74, p. 5343–5348.

Handin, J., Heard, H. C., and Mcgouirk, J. N., 1967, The effect of the intermediate principal stress on the failure of limestone, dolomite, and glass at different temperatures and strain rates: Jour. Geophys. Research, v. 72, p. 611–640.

Howard, J. H., 1966, Structural development of the Williams Range thrust, Colorado: Geol. Soc. America Bull., v. 77, p. 1247–1264.

Jaeger, J. C., 1969, Elasticity, fracture and flow: With engineering and geological applications: London, Methuen and Co., Ltd., 268 p.

Jaeger, J. C., and Cook, N.G.W., 1970, Fundamentals of rock mechanics: London, Chapmen and Hall, 515 p.

Lamping, N. E., 1970, The Mohorovičić discontinuity as a phase transition [Ph.D. dissert.]: College Station, Texas A&M Univ., 190 p.

Logan, J. M., Friedman, M., and Stearns, M. T., 1978, Experimental folding of rocks under confining pressure: Pt. VI. Further studies of faulted drape folds, *in* Matthews, V., III, ed., Laramide folding associated with basement block faulting in the Western United States: Geol. Soc. America Mem. 151 (this volume).

Love, A.E.H., 1927, A treatise on the mathematical theory of elasticity: New York, Dover, 634 p.

Mogi, K., 1971, Effect of the triaxial stress system on the failure of dolomite and limestone: Tectonophysics, v. 11, p. 111–127.

Morse, P. M., and Feshback, H., 1953, Methods of theoretical physics: New York, McGraw-Hill Book Co., 1978 p.

Prucha, J. J., Graham, J. A., and Nickelson, R. P., 1965, Basement controlled deformation in the Wyoming province of Rocky Mountains foreland: Am. Assoc. Petroleum Geologists Bull., v. 49, p. 966–992.

Sanford, A. R., 1959, Analytical and experimental study of simple geologic structures: Geol. Soc. America Bull., v. 70, p. 19–51.

Stearns, D. W., 1968, Certain aspects of fracture in naturally deformed rocks, *in* Reicker, R. E., ed., Rock mechanics seminar: U.S. Air Force Cambridge Research Labs., Terrestrial Sci. Lab., Contr. AD 669375, v. I, p. 97–118.

——1971, Mechanisms of drape folding in the Wyoming province: Wyoming Geol. Assoc., 23rd Ann. Field Conf., Guidebook, p. 125–144.

——1972, Structural interpretation of fractures associated with the Bonita fault: New Mexico Geol. Soc., 23rd Ann. Field Conf., Guidebook, p. 161–164.

——1975, Laramide basement deformation in the Bighorn Basin—The controlling factor for structures in the layered rocks: Wyoming Geol. Assoc., 27th Ann. Field Conf., Guidebook, p. 149–158.

——1978, Faulting and forced folding in the Rocky Mountains foreland province, *in* Matthews, Vincent, III, ed., Laramide folding associated with basement block faulting in the Western United States: Geol. Soc. America Mem. 151 (this volume).

Timoshenko, S. P., and Goodier, J. N., 1970, Theory of elasticity: New York, McGraw-Hill Book Co., 566 p.

Tweto, O, 1975, Laramide (Late Cretaceous–early Tertiary) orogeny in the Southern Rocky Mountains, *in* Curtis, B. F., ed., Cenozoic history of the Southern Rocky Mountains: Geol. Soc. America Mem. 144, p. 1–44.

Warren, D. H., 1968, Transcontinental geophysical survey (35°–39°N.)—Seismic refraction profiles of the crust and upper mantle from 100° to 112°W. longitude: U.S. Geol. Survey Misc. Geol. Inv. Map I-533-D.

Weinberg, D. M., 1977, Two-dimensional kinematic analyses of selected aspects of folding in the Rocky Mountain foreland, and their geologic applications [Ph.D. dissert.]: College Station, Texas A&M Univ., 110 p.

——1978, Some two-dimensional kinematic analyses of the drape-fold concept, *in* Matthews, V., III, ed., 1978, Laramide folding associated with basement block faulting in the Western United States: Geol. Soc. America Mem. 151 (this volume).

Weinberg, D. M., Holyfield, P. E., and Couples, G., 1976, Fractures associated with folds: An addendum: Geol. Soc. America Abs. with Programs, v. 8, p. 71.

MANUSCRIPT RECEIVED BY THE SOCIETY JUNE 27, 1977
MANUSCRIPT ACCEPTED AUGUST 25, 1977

Geological Society of America
Memoir 151

Plate tectonics of the Laramide orogeny

WILLIAM R. DICKINSON
Geology Department
Stanford University
Stanford, California 94305

WALTER S. SNYDER
Lamont-Doherty Geological Observatory
Columbia University
Palisades, New York 10964

ABSTRACT

In terms of plate tectonics, most orogenic belts are arc, collision, or transform orogens marked by regional batholiths, overthrust nappes, and en echelon fold trains, respectively. None of these models fits the crustal buckling of the classic Laramide orogeny, marked in the central Rocky Mountains by fault-bounded, basement-cored uplifts separated by intervening sediment-filled basins. Reported patterns of current seismicity, volcanism, and deformation in the modern central Andes document two modes of subduction; one involves plate descent at an abnormally shallow angle and may simulate Laramide conditions. In the more familiar mode, a plate descending steeply into the asthenosphere beneath the continental margin generates standard arc morphology with an active volcanic chain; crustal seismicity outside the subduction zone near the trench is confined mainly to a back-arc fold-thrust belt. In the unfamiliar mode, the descending slab of lithosphere slides along under the overriding plate of lithosphere, with which contact is maintained; crustal earthquakes are widespread across the dormant arc massif, within which local block uplifts bounded by reverse faults are prominent, and magmatism is meanwhile suppressed because the asthenosphere is never penetrated by the descending slab. The largely amagmatic Laramide style of deformation can be ascribed to the dynamic effects of an overlapped plate scraping beneath the Cordillera. That inference is strongly supported by the close correlation, in both space and time, between a prominent magmatic null or gap in the western Cordillera and the classic Laramide orogeny in the eastern Cordillera.

INTRODUCTION

Plate tectonics explains orogenesis as a product of plate interactions (Dewey and Bird, 1970). Three common types of orogenic belt can be identified:
1. Arc orogens, where (a) continued consumption of oceanic lithosphere at a

subduction zone builds an accretionary prism or subduction complex of deformed oceanic materials along one flank of the orogen and (b) magmatism induced by plate descent into the mantle constructs a volcanic chain capping a linear belt of injected batholiths along the spine of the orogen.

2. Collision orogens, where (a) continued consumption of oceanic lithosphere at a subduction zone eventually welds two once-separate continental blocks together and (b) the subduction complex of deformed oceanic materials is trapped along a linear suture belt marking the line of final ocean closure.

3. Transform orogens, where a component of contraction across an imperfect transform combines with the dominant translation to cause transpression (Harland, 1971), which gives rise to a train of en echelon wrench folds in an elongate belt parallel to the transform (Lowell, 1972; Wilcox and others, 1973).

The classic Laramide orogeny of latest Cretaceous and Paleogene age in the central Rocky Mountains—centered on Wyoming and Colorado—was unlike any of these three standard orogen types. It was marked especially by crustal buckling and associated fracturing to form giant fault-bounded, basement-cored uplifts separated by intervening basins in which sediment accumulated while deformation was in progress. There was minor magmatism similar to that common in arc orogens, but no continuous batholith belt and metavolcanic terrane or volcanic cover. There are no internal ophiolites indicative of oceanic closure, and no regionally integrated system of nappes like those common for collision orogens. Although some broad wrench deformation occurred, especially near the margins of the classic Laramide tract (Sales, 1968), no throughgoing transform structures showing major strike slip were present.

An adequate plate-tectonics model for the Laramide orogeny thus must differ from all three standard orogen models to some degree. We believe that a modified arc model is implicit in the conclusions of a number of recent workers. Our purpose here is to specify such a general qualitative model for the Laramide plate setting to serve as a basis for quantitative analysis of the plate interactions involved in the Laramide orogeny. Whereas we focus here on relative plate motions, Cross and Pilger (1978) elsewhere have speculated about absolute plate motions. Coney (1976, 1978) has discussed patterns of plate boundaries and motions throughout surrounding regions during Laramide time. Recently, Noble and McKee (1977) also discussed topics related to ours here. On the basis of the presence of eclogitic xenoliths in diatremes of the Colorado Plateau, Helmstaedt (1974) has previously proposed for Laramide orogenic movements a causative model closely similar to the one that we develop here.

LARAMIDE SETTING

The classic Laramide orogeny (Coney, 1976, 1978) occupied a particular time span and place within the Cordillera (Fig. 1), and we do not use the term here as a generic one for all deformation at the end of the Mesozoic Era. The characteristic signature of Laramide deformation in latest Cretaceous and Paleogene time was wholesale buckling and shear of the continental crust to produce asymmetric fractured uplifts and depressions oriented crudely parallel to the continental margin, but with notable local departures from that trend (Sales, 1968). In the central Rocky Mountains, the rumpling and rupture of the basement disrupted the previously integrated foreland basin with its typical wedge-shaped profile (Dickinson, 1976). The structural style contrasts markedly with older thin-skinned tectonic patterns within the fold-thrust belt of the late Mesozoic Sevier orogenic belt farther west

(Armstrong, 1968). To some extent, however, this older regime continued into Paleocene time and thus overlapped partly in time with formation of the classic Laramide structures (Dorr and others, 1977). Continuation of thin-skinned tectonics into Paleogene time was confined to the Idaho-Wyoming segment (Armstrong and Oriel, 1965) of the thrust belt near the juncture between Sevier and Laramide trends; thrusting did not persist along the Nevada-Utah segment (Armstrong, 1968) farther south (see Fig. 1).

The segment of the Cordillera that underwent classic Laramide tectonism along its eastern side was the same segment that experienced a pronounced Paleogene lull in magmatism farther west (Armstrong, 1974). This null in igneous activity within the Sevier-Laramide segment of the Cordillera can be interpreted as a temporary gap in the continuity of the magmatic arc that was evolving within the Cordillera in response to subduction along the Pacific margin of the continent (Snyder and others, 1976). We now have firm paleontologic data (Evitt and Pierce, 1975) from the coastal belt of the Franciscan subduction complex in the Northern Coast Ranges of California to confirm that coastal subduction persisted beyond Mesozoic and through much of Paleogene time (Travers, 1972). Both north and south of the Sevier-Laramide segment of the Cordillera, thin-skinned Sevier-like deformation apparently persisted into Paleogene time while Laramide-style deformation was in progress in the intervening area (Burchfiel and Davis, 1975). Significantly, the magmatic null or gap did not extend far into Canada or Mexico (see below).

The Paleogene lull in arc magmatism in the western Cordillera was accompanied by pronounced migration of the inland limit of arc activity far into the eastern Cordillera where Laramide deformation was simultaneously in progress. Thus, significant reduction of arc activity along its accustomed trend—marked by the great Mesozoic batholiths lying well west of the Sevier belt—was accompanied by a spread of desultory arc-related magmatism into the classic Laramide region well to the east of the Sevier belt. Lipman and others (1971) first suggested that

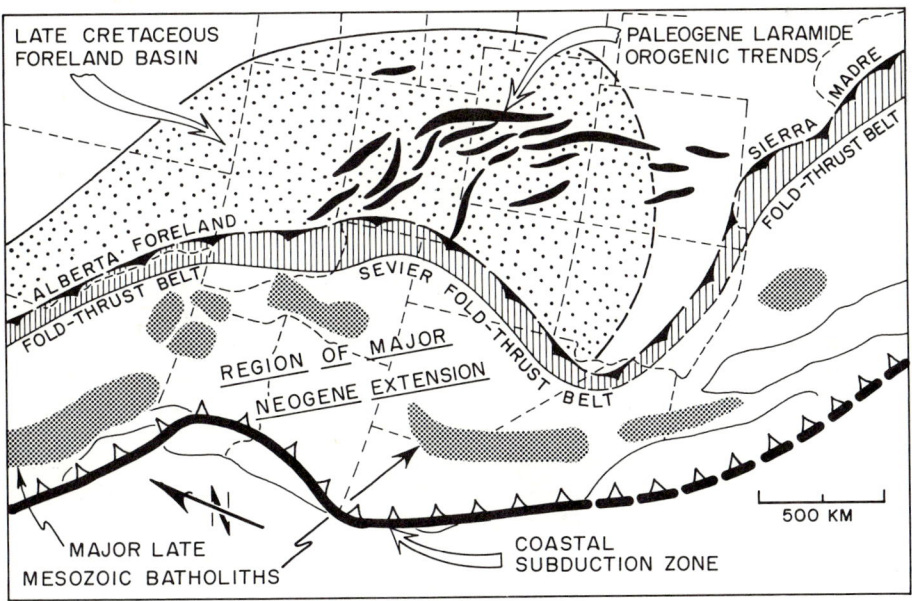

Figure 1. Sketch map of the Cordillera showing relation of Laramide orogenic trends to other major tectonic elements near the Mesozoic-Cenozoic boundary.

the coordinated diminution and migration of arc magmatism were jointly related to a shallowing of the angle of subduction of the consumed slab descending beneath the continent from the coastal subduction zone. For the region south of the Colorado Plateau (see Fig. 1), Coney and Reynolds (1977) have shown elegantly how the spotty Laramide magmatism swept inland to Colorado, and then back again toward the coast, between Late Cretaceous and mid-Tertiary time. They have ascribed this behavior to successive shallowing and then steepening of the angle of slab descent. Under such changing regimes of subduction, the descending slab reaches melting depth in the mantle at varying distances inland from the coastal subduction zone.

We here accept this hypothesis of changing slab dip to explain the Paleogene magmatic lull. As discussed elsewhere, we are able by kinematic analysis of plate motions to exclude the chief alternative hypothesis of coastal transform activity (Dickinson and Snyder, 1978; compare Coney, 1976, 1978).

SUBDUCTION MODES

The modern Andes constitute our best modern example of a continental-margin arc-trench system analogous to the pre-Neogene Cordillera (Hamilton, 1969). Recent work in the central Andes (Barazangi and Isacks, 1976; Megard and Philip, 1976) documents two contrasting modes of subduction (see Fig. 2): (1) a familiar mode, which we regard as Sevier-like, involving slab descent into the asthenosphere at a steep angle, and (2) an unfamiliar mode, which we regard as Laramide-like, in which the descending slab of oceanic lithosphere moves into the mantle at an angle shallow enough to maintain contact with the overriding plate of continental lithosphere.

Where the subducted slab dives steeply into the asthenosphere, a typical level of steady arc magmatism is stimulated at a normal distance from the trench. The usual type of standard arc orogen is thus formed in response to plate consumption. Strong crustal deformation reflected by seismicity is restricted to (1) the subduction zone near the trench, (2) the inclined seismic zone marking the upper tier of the descending slab in the mantle, (3) local subvolcanic sites near crustal magma chambers, and (4) an antithetic back-arc seismic belt along the so-called sub-Andean trend. The last-named feature is viewed here as analogous to the Sevier fold-thrust belt east of the Mesozoic arc trend, marked now by its roots in the Sierra Nevada batholith.

Where the subducted slab glides subhorizontally beneath the overriding plate of lithosphere, arc magmatism is generally suppressed. Evidently, the descending slab must actually penetrate asthenosphere to induce the magmatism. Moreover, the mantle seismic zone is truly inclined only near the trench; it forms a nearly horizontal zone slightly deeper than 100 km as it passes beneath the dormant arc massif. Crustal earthquakes are widespread across the full width of the arc massif, where reverse faults and block uplifts are common. Erosion has bitten deep into the prevolcanic terrane of the dormant arc and locally has exposed basement terranes on line with the trend of the active arc segments in which steep subduction is continuing (for example, see Audebaud and others, 1973).

Because of the existence of the Paleogene magmatic null in the Cordillera, we infer here that the shallow mode of subduction was directly responsible for the Laramide orogeny, as suggested previously by Coney (1976, 1978). We postulate that the dynamic effects of a subhorizontally subducted plate scraping along beneath the overriding continental plate were recorded by the crustal buckling and fracturing of basement rocks in the Laramide structures. Lowell (1974) also related Laramide deformation directly to shallow subduction, but appealed mainly to associated buoyant effects, rather than to the dynamic shear that seems to us to be the most likely linkage between the two. In either case, the Laramide problem reduces in gross outline to an analysis of the mechanical behavior of a surface slab of lithosphere subject to the influence of a subterranean slab sliding beneath it. Relative motion between the two plates was probably oriented along a northeast-trending line (Coney, 1976, 1978).

LARAMIDE EVENTS

The model proposed for Laramide tectonism can be tested by examining in detail the space-time patterns of magmatism and diastrophism for the Sevier-Laramide and adjoining segments of the Cordillera. We here present a preliminary assessment of the pertinent relationships using refinements of data we have presented elsewhere (Dickinson, 1976; Snyder and others, 1976). Inferred trends and migrations of arc magmatism at various times (Figs. 3 through 5) are based primarily on about 2,500 radiometric dates for Cenozoic igneous rocks in the Western United States (Noblett and others, 1977). On Figures 3 through 5, stippled areas are the magmatic loci (Snyder and others, 1976) to which most available radiometric data are confined. Some loci are separated by relatively amagmatic regions such as southernmost Nevada, the Colorado Plateau, and central Wyoming. Other loci are separated by young lava fields that mask the possible continuity of older igneous suites in the Pacific Northwest.

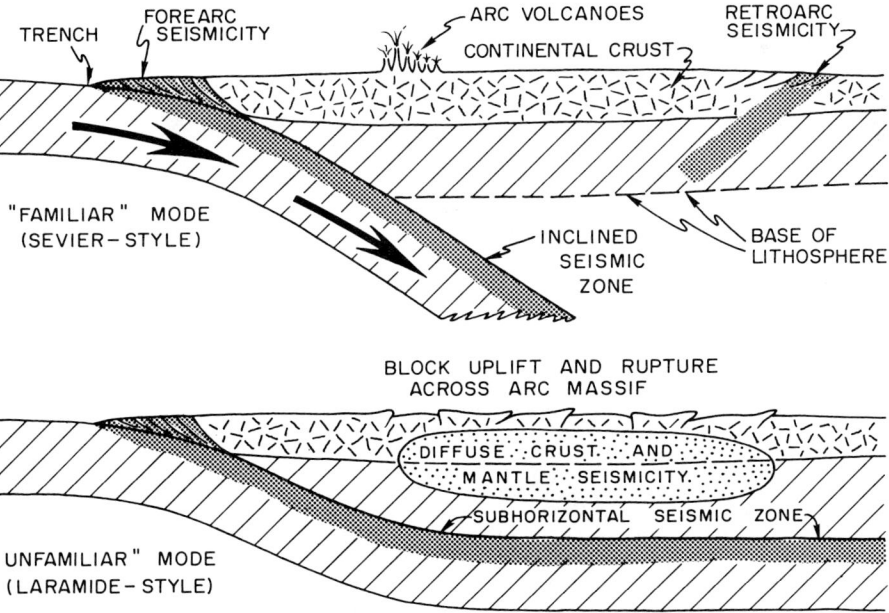

Figure 2. Diagram of two types of arc orogens displaying two modes of subduction, steep (top) and shallow (bottom) plate descent. Modified after Barazangi and Isacks (1976) and Megard and Philip (1976).

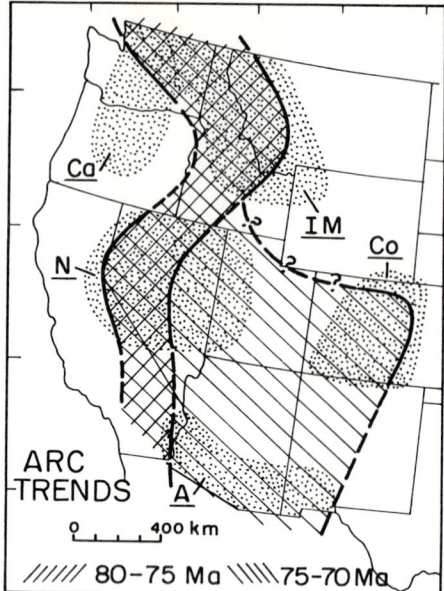

Figure 3. Areal distribution of arc-related igneous activity in the Western United States during latest Cretaceous time. Stippled magmatic loci (Snyder and others, 1976) include Ca, Cascades; IM, Idaho-Montana; N, Nevada; Co, Colorado; A, Arizona (Ma = million years before present).

Figure 4. Diagrammatic map showing Paleogene migration of arc volcanism in the Western United States (Ma = million years before present).

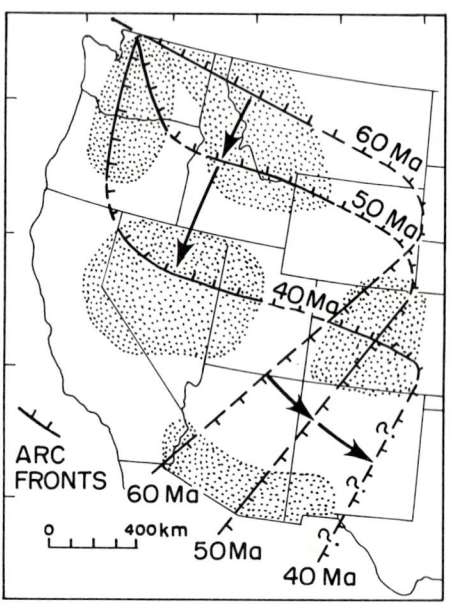

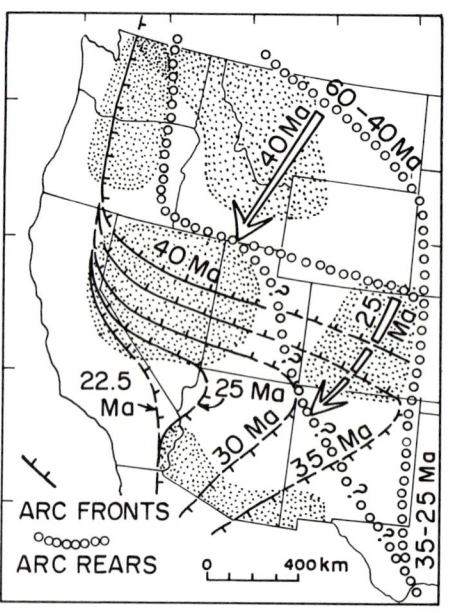

Figure 5. Diagrammatic map showing mid-Cenozoic migration of arc volcanism in the Western United States (Ma = million years before present).

Paleogene Magmatic Null

Figure 3 indicates the general eastward expansion of diffuse arc magmatism that continued at a low level of activity when the intense Late Cretaceous magmatism along the Sierran arc trend sputtered out in latest Cretaceous time. The pattern of activity is not well controlled for this time span, because most of the dated rocks are plutonic and many of their radiometric dates may reflect uplift and cooling rather than actual emplacement. Nevertheless, the trend of arc migration is clearly inland. We infer, therefore, that the angle of plate descent beneath the Cordillera changed from steep to shallow between 80 and 70 m.y. B.P.

The Paleogene magmatic null began to develop at about 70 m.y. B.P., probably near the Idaho-Nevada-Oregon common corner along the northern edge of the Nevada locus. Within 5 to 10 m.y., arc magmatism had been snuffed out within the Nevada locus and over much of the Pacific Northwest as well. Figure 4 depicts the changing shape and size of the magmatic gap in the western Cordillera as the null evolved through Paleogene time. On the diagram, the magmatic gap that existed at various times is marked by the series of regions that lie west of the successive hachured magmatic fronts. The latter are drawn schematically to connect the westernmost occurrences of arc igneous activity within the various magmatic loci.

In detail, the magmatic null was thus diachronous. North of the Nevada locus, the magmatic gap expanded from 70 to 60 m.y. B.P. (across the Cretaceous-Tertiary boundary), whereupon it began to contract and had essentially disappeared by 45 m.y. B.P. To the south, however, the magmatic gap continued to expand until about 40 m.y. B.P. before beginning to close. Presumably, the subducted slab (which controlled the position of the magmatic front) was to some extent flexing or breaking, and thus whipping about, in its subterranean position. The Colorado locus can be regarded as a sort of pivot point between the northern and southern regions of contrasting slab behavior. Intermittent Laramide volcanism was, therefore, apparently continuous within the Colorado locus throughout the period from 70 to 50 m.y. B.P. (Tweto, 1975).

Figure 5 depicts the manner in which arc volcanism was rekindled within the Nevada locus by the progressive southward sweep of the magmatic front (compare Armstrong and Higgins, 1973, Fig. 2; Stewart and others, 1977, Fig. 2). An analogous shift of arc magmatism carried arc activity back across the Arizona locus as well. We infer that the angle of plate descent beneath the Cordillera changed gradually from shallow to steep between 40 to 45 and 20 to 25 m.y. B.P. (compare Coney and Reynolds, 1977). The arc front thus swept forward as the dip of the subducted slab increased. Presumably, each successive position of the magmatic front was a register of some critical depth contour on the subducted slab. The due-east trend of the arc front as it swept down across the Nevada locus can thus be interpreted as the record of a flexure in the subducted slab. The flexure, which evidently propagated southward with time, lay between a steeply dipping slab to the north and a gently dipping slab to the south. The occurrence of successive age belts of mid-Cenozoic igneous rocks along due-east trends in the Great Basin has been interpreted in identical fashion by P. W. Lipman (in Stewart and others, 1977). The presence of a subducted slab thus having transient due-east strikes locally may well be the correct explanation for the indication of a nominally flat-dipping slab inferred earlier by Lipman and others (1971) from geochemical arguments based on the potassium content of arc-related Cenozoic igneous rocks.

By the beginning of Neogene time, the magmatic null and gap were past, and a continuous arc trend again extended parallel to the continental margin in the western Cordillera. This Neogene arc was the one later disrupted by evolution of the San Andreas transform (Dickinson and Snyder, 1978).

Laramide Orogenic Timing

Figure 6 illustrates the close spatial relationship between the Paleogene magmatic gap and the Laramide orogenic tract. Low-angle thrusting within the thin-skinned fold-thrust belt continued throughout Laramide time to the north and to the south in Canada and Mexico, but essentially ceased west of the main Laramide tract (Burchfiel and Davis, 1975). Wrench-style deformation suggestive of relative east-to-west translation of basement is present along the northern edge of the Laramide tract (Sales, 1968). This behavior can be interpreted as a tectonic transition between (1) continuing thin-skinned deformation to the north where crustal contraction was concentrated at the fold-thrust belt in Canada and northern Montana and (2) the Laramide region of thick-skinned buckling farther south where crustal contraction was distributed over a much wider orogenic transect in Wyoming and southern Montana.

Still farther south, a Paleogene lowland lying in the region of the present Colorado Plateau was a residual topographic trough trapped between vestigial Mesozoic highlands of the Sevier orogenic belt on the west and nascent Laramide highlands on the east (Hunt, 1956). The reduced vigor of Laramide deformation in the plateau region, as compared to its effects in the Rockies, remains enigmatic to us (see Coney, 1976); perhaps the presence of Paleozoic structures inherited from the Ancestral Rockies somehow enhanced response to Laramide deformation (Coney, 1978).

Strict contemporaneity between the development of the Paleogene magmatic null and the course of the Laramide deformation is difficult to demonstrate, but roughly coeval timing is evident. Laramide structures formed generally between 70 and 45 m.y. B.P. (Coney, 1972, Fig. 2), with the most intense deformation within the core of the Laramide tract in Wyoming and Colorado coming between 65 and 50 m.y. B.P. (Berg, 1962). The Paleogene magmatic gap was largest in the midst of the indicated Laramide time span, from 60 to 55 m.y. B.P., but was prominent throughout the whole period from 70 to 45 m.y. B.P. (see Figs. 4, 5). Neither the Laramide deformation nor the magmatic null predated about 75 m.y. B.P. anywhere. The Laramide deformation had definitely ended and the magmatic gap was everywhere rapidly contracting by 40 m.y. B.P. In California, activity on the proto–San Andreas fault, whose origin may have been related somehow to the same plate interactions, was restricted to some part of the period 70 to 55 m.y. B.P. (Dickinson and others, 1978).

From the genetic relationship that we infer here between the Laramide orogeny and the Paleogene magmatic null, we would expect comparable phases of Laramide deformation to be younger in the southern Rockies than in the northern Rockies. There are hints that such may be the case. For example, the post-Laramide Challis volcanic rocks of Idaho and Absaroka volcanic rocks of Wyoming were erupted in middle to late Eocene time (Prostka and others, 1977), whereas deformation may have continued through all or part of that period in Colorado where the post-Laramide volcanic rocks are of latest Eocene or earliest Oligocene age (Tweto, 1975).

Some evidence of Laramide-style deformation might also be expected in areas west of the classic Laramide tract but within the magmatic gap (see Fig. 6). The monoclines of the Colorado Plateau are a clear-cut example (Coney, 1976; Davis, this volume). Perhaps the episode of early to middle Eocene folding and high-angle

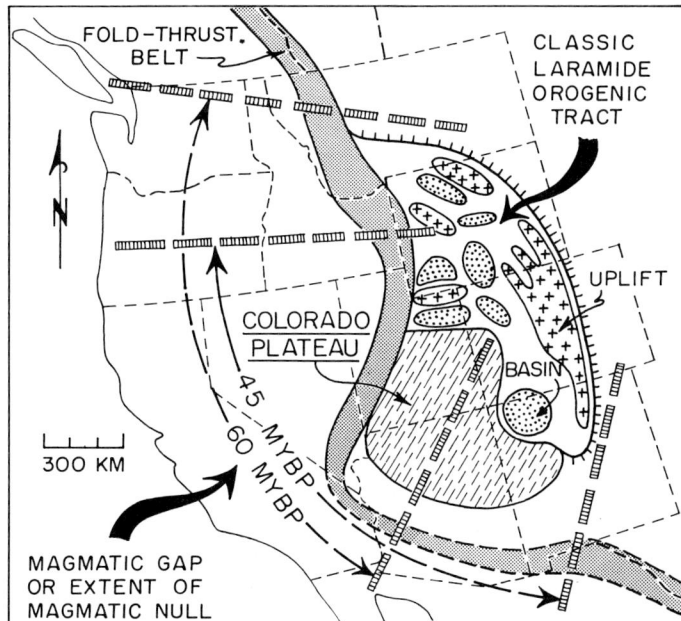

Figure 6. Diagrammatic map showing spatial relationship between Paleogene magmatic gap (after Figs. 4, 5) and area of classic Laramide orogeny.

faulting in the northern Cascades (Misch, 1966) could also have been a local effect of the Laramide tectonism. Important deformation may have occurred in the Great Basin during early Cenozoic time and could conceivably record Laramide events that are largely masked now throughout the Basin and Range province by younger volcanic cover and much more widespread late Cenozoic faulting.

CONCLUSIONS

Our chain of logic thus runs as follows:
1. The Laramide orogeny had unusual attributes that distinguish it from standard models for arc, collision, and transform orogens.
2. The classic Laramide orogeny in an eastern segment of the Cordillera occupied a time span coincident with the development of a magmatic null and cessation of thrusting in a paired western segment of the Cordillera.
3. Recent work in the central Andes shows that both cessation of volcanism and block deformation across the width of the arc massif are related to subhorizontal subduction of consumed oceanic plate beneath the continental plate.
4. Therefore, we ascribe classic Laramide deformation to the mechanical effects on the overriding plate of the overlapped underlying plate; that is, a regime in which the subducted plate scrapes horizontally beneath the surficial plate.

In his discussion of Rocky Mountain tectonics, Grose (1972) concluded that the Laramide orogeny was caused by "primary horizontal or gently inclined compressive stress . . . [that] . . . acted in the lower crust and/or mantle generally from west to east" in association with plate consumption far away along the coast. In effect, our inferences here suggest a specific plate-tectonics context for such postulated stresses.

The greatest weakness in our argument is the lack of evidence for wholesale disruption of the sub-Andean foreland basin, as occurred in Wyoming and Colorado

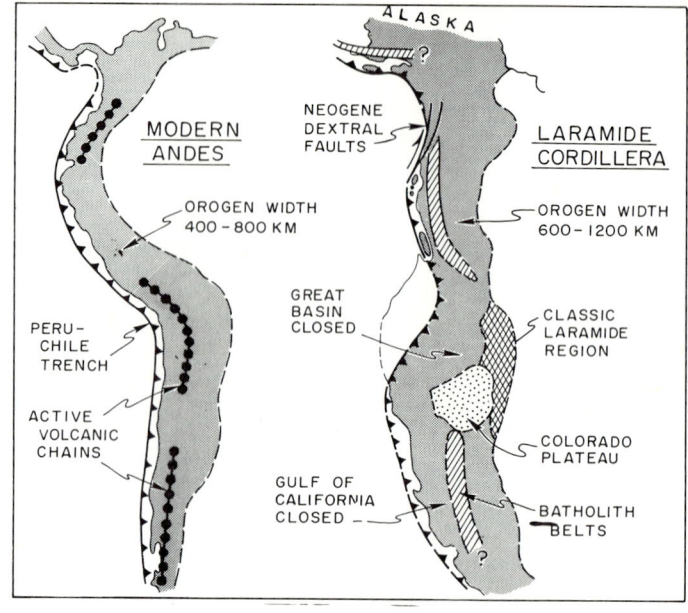

Figure 7. Diagrammatic maps at same scale comparing widths of Cordilleran orogen during Laramide time in North America and modern Andean orogen of South America. About 250 km of Neogene extension in Great Basin has been restored (see Hamilton and Myers, 1966).

for the Rocky Mountain Cretaceous foreland basin during the Paleogene Laramide orogeny. We are content, however, that our correlation of two general styles of deformation with two modes of subduction, shallow and steep, is valid. Furthermore, Figure 7 shows that the total inferred width of the Cordilleran orogen, as reconstructed for Laramide time by reversal of Neogene tectonic events, is comparable to the full width of the modern Andean system from trench to foreland. No matter how the reconstruction is accomplished for the San Andreas fault and the Basin and Range province, the widest parts of the Laramide-age Cordillera are somewhat wider than the widest parts of the Andean orogen today, but narrower segments of the Laramide-age Cordillera are not as wide as some segments of the Andean orogen. Moreover, the single Andean example simply cannot be taken as definitive of the maximum widths conceivable for all past continental-margin arc-trench systems. Our view of the Laramide orogen as a special variation of arc orogen thus remains tenable when overall dimensions are considered. Note also on Figure 7 that the characteristic lengths of alternating active arc segments and dormant magmatic gaps appear comparable for the Laramide Cordillera and the Quaternary Andes.

We close with the thought that the next step in the resolution of the classic Laramide problem must be an analysis of the mechanical behavior of a plate of lithosphere given the general sorts of boundary conditions inferred here.

ACKNOWLEDGMENTS

This work was supported by the Earth Sciences Section of the National Science Foundation with NSF Grants DES72-01728 and EAR76-22636. Our interest in the Laramide problem stems in part from discussions with P. J. Coney and L. T. Grose, and our thinking was aided at a critical stage by a set of term papers at Stanford University dealing with the sedimentary evolution of Laramide basins in Wyoming (by J. T. Bateson, D. C. Dawson, D. I. Fletcher, C. K. Keller,

D. S. Storm, and G. Yuan in W. R. Dickinson's class on sedimentary basins). Suggestions from P. C. Coney and W. B. Hamilton improved the text.

REFERENCES CITED

Armstrong, F. C. and Oriel, S. S., 1965, Idaho-Wyoming thrust belt: Am. Assoc. Petroleum Geologists Bull., v. 49, p. 1847–1866.

Armstrong, R. L., 1968, Sevier orogenic belt in Nevada and Utah: Geol. Soc. America Bull., v. 79, p. 429–458.

―――1974, Magmatism, orogenic timing, and orogenic diachronism in the Cordillera from Mexico to Canada: Nature, v. 247, p. 348–351.

Armstrong, R. L., and Higgins, R. E., 1973, K-Ar dating of the beginning of Tertiary volcanism in the Mojave Desert, California: Geol. Soc. America Bull., v. 84, p. 1095–1100.

Audebaud, Etienne, and others, 1973, Les traits géologiques essentiels des Andes centrales (Pérou-Bolivie): Rev. Géographie Phys. et Géologie Dynam., v. 15, p. 73–114.

Barazangi, M., and Isacks, B. L., 1976, Spatial distribution of earthquakes and subduction of the Nazca plate beneath South America: Geology, v. 4, p. 686–692.

Berg, R. R., 1962, Mountain flank thrusting in Rocky Mountain foreland, Wyoming and Colorado: Am. Assoc. Petroleum Geologists Bull., v. 46, p. 2019–2032.

Burchfiel, B. C., and Davis, G. A., 1975, Nature and controls of Cordilleran orogenesis, western United States; extensions of an earlier synthesis (Rodgers volume): Am. Jour. Sci., v. 275-A, p. 363–396.

Coney, P. J., 1972, Cordilleran tectonics and North America plate motion: Am. Jour. Sci., v. 272, p. 603–628.

―――1976, Plate tectonics and the Laramide orogeny, in Woodward, L. A., and Northrop, S. A., eds., Tectonics and mineral resources of southwestern North America: New Mexico Geol. Soc. Spec. Pub. No. 6, p. 5–10.

―――1978, Mesozoic-Cenozoic Cordilleran plate tectonics, in Smith, R. B., and Eaton, G. P., eds., Cenozoic tectonics and regional geophysics of western Cordillera: Geol. Soc. America Mem. 152 (in press).

Coney, P. J., and Reynolds, S. J., 1977, Cordilleran Benioff zones: Nature, v. 270, p. 403–406.

Cross, T. A., and Pilger, R. H., Jr., 1978, Constraints on absolute motion and plate interaction inferred from the Cenozoic igneous activity in the western United States: Am. Jour. Sci., v. 278 (in press).

Davis, G. H., 1978, Monocline fold pattern of the Colorado Plateau, in Matthews, Vincent, III, ed., Laramide folding associated with basement block faulting in the Western United States: Geol. Soc. America Mem. 151 (this volume).

Dewey, J. F., and Bird, J. M., 1970, Mountain belts and the new global tectonics: Jour. Geophys. Research, v. 75, p. 2625–2647.

Dickinson, W. R., 1976, Sedimentary basins developed during evolution of Mesozoic-Cenozoic arc-trench system in western North America: Canadian Jour. Earth Sci., v. 13, p. 1268–1287.

Dickinson, W. R., and Snyder, W. S., 1978, Geometry of triple junctions and subducted slabs related to San Andreas transform: Jour. Geophys. Research (in press).

Dickinson, W. R., Graham, S. A., and Ingersoll, R. V., 1978, Paleogene sediment dispersal and paleotectonics in northern California: Am. Assoc. Petroleum Geologists Bull. (in press).

Dorr, J. A., Jr., Spearing, D. R., and Steidtmann, J. R., 1977, Deformation and deposition between a foreland uplift and an impinging thrust belt, Hoback Basin, Wyoming: Geol. Soc. America Special Paper 177, 82 p.

Evitt, W. R., and Pierce, S. T., 1975, Early Tertiary ages from the coastal belt of the Franciscan Complex, northern California: Geology, v. 3, p. 433–436.

Grose, L. T., 1972, Tectonics, in Mallory, W. W., ed., Geologic atlas of the Rocky Mountain region: Denver, Colo., Rocky Mtn. Assoc. Geologists, p. 35–44.

Hamilton, W., 1969, Mesozoic California and the underflow of Pacific mantle: Geol. Soc. America Bull., v. 80, p. 2409–2430.

Hamilton, W. B., and Myers, W. B., 1966, Cenozoic tectonics of the western United States: Rev. Geophysics, v. 4, p. 509–549.

Harland, W. B., 1971, Tectonic transpression in Caledonian Spitzbergen: Geol. Mag., v. 108, p. 27–42.

Helmstaedt, Herwart, 1974, Overplating, a major factor in the tectonic evolution of the Colorado Plateau: EOS (Am. Geophys. Union Trans.), v. 55, p. 448.

Hunt, C. B., 1956, Cenozoic geology of the Colorado Plateau: U.S. Geol. Survey Prof. Paper 279, 99 p.

Lipman, P. W., Prostka, H. J., and Christiansen, R. L., 1971, Evolving subduction zones in the western United States, as interpreted from igneous rocks: Science, v. 174, p. 821–825.

Lowell, J. D., 1972, Spitzbergen Tertiary orogenic belt and the Spitzbergen fracture zone: Geol. Soc. America Bull., v. 83, p. 3091–3102.

——1974, Plate tectonics and foreland basement deformation: Geology, v. 2, p. 275–278.

Megard, F., and Philip, H., 1976, Plio-Quaternary tectono-magmatic zonation and plate tectonics in the central Andes: Earth and Planetary Sci. Letters., v. 33, p. 231–238.

Misch, Peter, 1966, Tectonic evolution of the northern Cascades of Washington State: Canadian Inst. Mining and Metallurgy Spec. Vol. 8, p. 101–148.

Noble, D. C., and McKee, E. H., 1977, Spatial distribution of earthquakes and subduction of the Nazca plate beneath South America: Comment: Geology, v. 5, p. 576–578.

Noblett, J. B., Snyder, W. S., Dickinson, W. R., Silberman, M. L., 1977, Age determinations of Cenozoic rocks in the Western United States: unpub. ms., 133 p. (computer listing available at cost through W. R. Dickinson, Geology Dept., Stanford Univ., Stanford, Calif. 94305).

Prostka, H. J., McIntyre, D. H., and Skipp, Betty, 1977, Interrelations of lower Tertiary volcanic fields in northwestern interior United States: Geol. Soc. America Abs. with Programs, v. 9, p. 755–756.

Sales, J. K., 1968, Cordilleran foreland deformation: Am. Assoc. Petroleum Geologists Bull., v. 52, p. 2016–2044.

Snyder, W. S., Dickinson, W. R., and Silberman, M. L., 1976, Tectonic implications of space-time patterns of Cenozoic magmatism in the Western United States: Earth and Planetary Sci. Letters, v. 32, p. 91–106.

Stewart, J. H., Moore, W. J., and Zietz, I., 1977, East-west patterns of Cenozoic igneous rocks, aeromagnetic anomalies, and mineral deposits, Nevada and Utah: Geol. Soc. America Bull., v. 88, p. 67–77.

Travers, W. B., 1972, A trench off central California in Late Eocene–early Oligocene time, *in* Shagan, R., and others, eds., Studies in Earth and space sciences: Geol. Soc. America Mem. 132, p. 173–182.

Tweto, Ogden, 1975, Laramide (late Cretaceous–early Tertiary) orogeny in the Southern Rocky Mountains, *in* Curtis, B. F., ed., Cenozoic history of the Southern Rocky Mountains: Geol. Soc. America Mem. 144, p. 1–44.

Wilcox, R. E., Harding, T. P., and Seely, D. R., 1973, Basic wrench tectonics: Am. Assoc. Petroleum Geologists Bull., v. 57, p. 74–96.

MANUSCRIPT RECEIVED BY THE SOCIETY DECEMBER 14, 1977
MANUSCRIPT ACCEPTED JANUARY 19, 1978

Index

Anisotropy
  effect on basement deformation, 348–349
  importance of, 3

Basement
  definition of, 9
  lateral expansion of, 131
  peneplaned surface on, 8, 105–106, 111
    lack of arching of, 10, 110
    lack of folding of, 10, 110, 122, 135
  structural response of, 9–21, 318
Basement blocks
  compressional folds associated with rotation of. See Folds, free configuration beneath drape folds. See Drape fold geometry, influence of forcing member
  possible decoupling from lower crust, 34
  rotation of, 8, 10, 20, 72, 105–106, 317, 322–331
    paleomagnetic evidence of, 117
  splintering along mountain fronts where ductile shales are absent, 27
Basins, shape of, 8, 113
Bedding-plane slip, 3, 23–24, 66, 71, 75, 96
Boundary condition
  at base of basement block, 20, 35, 323–324, 331
  differences in Sanford and Hafner models, 19
  in Owl Creek Mountains, 20
  in Uinta Mountains, 20
  in Wind River Mountains, 20

Calcite twin-lamellae, 82, 85, 88, 152, 255–257
Coefficient of sliding friction, 3, 5, 24

Depth of burial, 3, 5, 8–9, 105, 200
Displacement field, 2, 19
Drape fold geometry
  abrupt changes of strike, 25, 110–120, 141, 159, 220
  at corners of basement blocks, 25–26, 30, 82–89, 132, 136, 141, 153, 225
  development of blocks, 1–5, 23, 25, 28, 59, 119, 144–145, 189, 204
  difference in Mesozoic and Paleozoic sections, 26, 51–77, 117–119, 191, 246, 266
  influence of forcing member, 5, 25, 105–110, 112–113, 115–116, 121, 130, 134–135, 148, 173, 183, 185, 190, 203, 225–227
  influence of layered package, 22–31
  loss of throw along strike, 25, 88, 116, 134, 206
  in non-welded, non-thinning sections, 22–27
  reflecting underlying block configuration. See Drape fold geometry, influence of forcing member relationship between vertical displacement and lateral movement of basement blocks, 53–63
  steplike character of fold axis, 111
  in welded, non-thinning sections, 27–28
Drape folds
  bifurcation of, 144, 174, 185, 189, 218
  definition of, 2–6
  disharmonic structures associated with. See Drape subsidiary folds
  displacement limit for, 23
  displacement vectors of, 148–149
  displacement without faulting of, 5, 122, 144, 177, 203–204
  implied sense of shear in the layered rocks of, 64–66
  as an invalid model for the formation of monoclines, 277–278, 285–288
  kinematic analyses of, 51–72, 155
  mass balance problems in, 89–94, 149–155
  overturned strata in the steep limb of, 111, 131, 159, 161–162, 192, 208, 242
  specific examples of. See Localities
  three-dimensional movement patterns in, 140–156
  as a valid model for the formation of monoclines, 110, 139–155, 165–234
Drape subsidiary folds, 26, 75–76, 117–119

Elastic behavior of rocks in the crust, 339
Equations
  for analysis of buckling and draping, 302–306
  for functional relation between fault displacement and lateral displacement of layered rocks, 54–56
  for multilayers, 306–310
  for stability index, 345

Experimental studies
  at high confining pressures, 29, 72–74, 79–97
  of kaolinite and modeling clay over wooden blocks, 216
  philosophy of, 94–96
  sand-box type, 18–21, 120, 236, 264, 342. *See also* Sanford-type sand-box experiments
  simulating Colorado Plateau structures, 228–231
  simulating Uinta Mountain structure, 21

Fault angle
  bounding mountain blocks, 8, 127, 130
  effect leaving forcing member, 6, 84–89
  in high-confining-pressure experiments, 84–89
  in sand-box experiments, 18
Faulting through of drape folds, 23, 27, 122
Faults
  associated with folds, 251–254
  curving of, 10, 18, 86–89, 130, 318–324
  cylindrical, 319
  dying out in a fold, 108, 110, 189, 216, 244
  en echelon, 103, 105, 174
  flattening of, 18, 86–89
  lack of coincidence between natural and predicted, 349
  low-angle examples, 11
  multiple periods of movement on (Palisades fault), 241–246
  normal, 6, 10, 18–19, 23, 61, 251, 326–327
  occurrence of high-angle and low-angle, 8, 19–21
  potential from stress diagram, 17, 323–324
  sequence of formation
    in high-confining-pressure experiments, 89
    in sand-box experiments, 18–19, 343
  tear, 127–128, 148, 189, 209–214
  wrench, 15, 34, 127, 314, 356
Fold hinges
  fixed, 23, 209
  migration of, 23
  shattered, 23, 31, 189, 248–249
Folds
  buckle, 206, 218–219, 264, 278–282, 288–292
  compressional. *See* Folds, free
  domes, 111, 114–115, 166, 170, 177. *See also* Plutons causing domes in Black Hills
  drape. *See* Drape folds
  forced. *See* Drape folds
  free, 3, 6, 25, 31, 48
  geometry not typical of drape folds, 113, 132, 204, 206, 209, 227. *See also* Drape subsidiary folds; Folds, free
  monoclines, 110, 139–156, 165–311. *See also* Plateau-type uplifts
Forcing member, characteristics of, 5

Graben, 21, 121, 202
Gravity glide blocks, 113, 132, 136, 227

Hafner-type diagrams, 17–22, 314, 322
Horizontal compression, 10–15, 162–163, 193, 216, 228, 331
Horizontal movements
  compared to vertical uplift in Colorado Plateau, 226
  problems with, 10–15, 226

Kink bands, 253, 282–284

Layered rocks. *See also* Anisotropy
  behavior of, 22–31
    thick carbonate package, 22–27
    thick clastic package, 28–31
  bulk behavior of, 5–6
  detachment from basement, 22–27, 152–153, 197–214
  difference in lateral transport of Paleozoic and Mesozoic, 64–66
  thickening of, 190, 206, 246–247, 268
    at block corners, 26, 83, 88–94
  thinning of, 6, 26, 28–30, 83, 88–94, 120, 122, 149, 151–152, 189, 191, 203, 246, 268
  welding to forcing member, 5
Layer-parallel orientation of σ_1, 89, 131, 253, 260, 263–264, 274
Localities mentioned in text
  Beartooth Mountains, Montana-Wyoming, 8, 11–13, 15, 19, 23, 317, 326–328
  Belle Fourche anticline, South Dakota, 180
  Belleview dome and basin, Colorado, 111, 114–115
  Bighorn Basin, Wyoming, 8, 12, 17, 39, 43, 75
  Big Horn Mountains, Wyoming, 10, 12–13, 15, 21, 25, 125–155
  Black Hills monocline, Wyoming, 174–177, 186–190
  Black Hills uplift, Montana–South Dakota–Wyoming, 165–196
  Black Point monocline, Arizona, 217
  Blue Mountain block, Colorado, 108–110
  Brady structure, Wyoming, 44, 46
  Camp Creek monocline, Montana, 179
  Carter Lake anticline, Colorado, 111–112
  Cascade anticline, South Dakota, 185–186
  Casper Mountain, Wyoming, 6, 15, 26–28, 75, 153
  Chilson anticline, South Dakota, 183
  Circle Cliffs uplift, Utah, 225–226
  Circle Ridge anticline, Wyoming, 13
  Coconino Point monocline, Arizona, 217–218
  Colorado Plateau, Arizona–Colorado–New Mexico–Utah, 2, 215–237
  Comb Ridge monocline, Arizona, 217–218
  Cottonwood Canyon, Wyoming, 15, 139–156
  Cottonwood Creek anticline, South Dakota, 179
  Cow Springs monocline, Arizona, 217–218
  Crazy Jug monocline, Arizona, 217
  Dead Indian Hill, Wyoming, 31

Localities mentioned in text (continued)
  Defiance uplift, Arizona, 225–226
  Dinosaur National Monument, Colorado-Utah, 75, 197–214, 298
  Dowe Pass anticlines, Colorado, 120–121
  Dry Fork Ridge structure, Wyoming, 13, 64, 131
  Dudley anticline, South Dakota, 185
  East Defiance monocline, Arizona, 217–219, 235–271
  East Kaibab monocline, Arizona, 217–218
  Echo Cliffs monocline, Arizona, 217–218
  Echo Cliffs uplift, Arizona, 225–226
  Elk Basin structure, Wyoming, 13
  Elk Mountain, Wyoming, 15, 30, 157–163
  Fanny Peak monocline, Wyoming-South Dakota, 173–177, 186–190
  Freezeout Mountains, Wyoming, 5, 15, 30
  Front Range, Colorado, 9, 15, 20, 25, 30, 75, 101–124
  Grandview monocline, Arizona, 217–218
  Granite Mountains, Wyoming, 11
  Green River Basin, Wyoming, 13, 48
  Gros Ventre Mountains, Wyoming, 12–13, 25
  Hamilton Dome, Wyoming, 6
  Hanna Basin, Wyoming, 8, 12–13, 27, 48
  Hartville uplift, Wyoming, 174–175
  Horn area, Wyoming, 126, 132–136
  Island Park, Utah, 203, 205
  Kaibab uplift, Arizona, 225–226
  La Flemme anticline, Wyoming-South Dakota, 180
  Loveland anticline, Colorado, *See* Localities mentioned in text, Milner Mountain, Colorado
  Medicine Bow Mountains, Wyoming, 15, 157
  Milner Mountain, Colorado, 117–120
  Mitten Park, Utah, 203, 205–206
  Monument Hill monocline, Wyoming, 178–179
  Monument upwarp, Utah, 225–226
  Moorcroft terrace, Wyoming, 178
  Mule Creek terrace, Wyoming, 174–177
  Never Summer Range, Colorado, 324
  Newcastle ramp, Wyoming, 177
  Oil Butte–Pine Ridge anticline, South Dakota, 178
  Oraibi monocline, Arizona, 218
  Organ Rock monocline, Arizona, 218
  Owl Canyon block, Colorado, 107–108
  Owl Creek Mountains, Wyoming, 6, 10–11, 13, 150, 317, 325–326
  Palisades monocline, Arizona, 238–271, 275
  Pat O'Hara structure, Wyoming, 31
  Pedro terrace, Wyoming, 177
  Piney Creek area, Wyoming, 126–132
  Powder River Basin, Wyoming, 8, 12, 44–45, 75, 174
  Pryor Mountains, Montana, 13
  Rabbit Mountain, Colorado, 113, 116
  Rattlesnake Mountain, Wyoming, 6, 11, 13–14, 22, 27, 31, 53, 131, 161
  Red Lake monocline, Arizona, 217–218
  Red Rock fold, Colorado, 204–206, 209, 212
  San Rafael swell, Utah, 225–226
  Seminole Mountains, Wyoming, 15, 25, 27
  Split Mountain, Utah, 203–205, 209
  Stockade Beaver Creek fold, Wyoming, 187–188
  Teton Mountains, Wyoming, 13
  Thornton dome, Wyoming, 177
  Uinta Mountain graben, Colorado, 202, 204
  Uinta Mountains, Colorado-Utah, 4, 6, 8–9, 11, 13, 15, 20, 150, 197–214
  Uncompahgre uplift, Colorado, 6, 8, 10, 15, 28–30, 225–226
  West Defiance monocline, Arizona, 217–218
  Whitewood anticline, South Dakota, 180
  Williams Range thrust, Colorado, 315
  Wind River Basin, Wyoming, 8, 47, 75
  Wind River Mountains, Wyoming, 8–11, 48, 150, 317, 322–325
  Yampa fold, Colorado, 204, 206, 212, 298
  Zuni uplift, New Mexico, 225–226

Mathematical solutions. *See* Equations
Mohr circle diagram, 345, 349
Monoclinal flexuring, general model of, 276–284

Orogen types
  arc, 355–356
  collision, 356
  transform, 356
Orthogonal pattern of monoclines in Colorado Plateau, 220. *See also* Precambrian control of Laramide structures in Colorado Plateau
Overthrusting
  directions of, 12
  hypothesis for Piney Creek lobe, 129

Petrofabric study, 82–88, 255–257
Plateau-type uplifts, 8, 15, 19
  difficulty of producing with horizontal motions, 15
  non-vertical faults, 58–59
  vertical faults, 58
Plate tectonics, 314, 355–366
Plutons causing domes in Black Hills, 170
Precambrian control of Laramide structures in Colorado Plateau, 216–217, 225–227, 266

Ramp structures, 172–179, 187, 198, 204, 209, 214
Rate of deformation, 339
Remote sensing, 103, 105–106
Rocky Mountains foreland
  age of deformation in, 356–358. *See also* Timing of monocline formation in Colorado Plateau
  analogy with Andes, 358, 364
  areas of research needing further study, 34

Rocky Mountains foreland (continued)
  correlation with magmatic null to the west, 357, 359–361
  generalized geologic map of, 7
  hypotheses of tectonics, 328–330
  loading conditions, 330–333
  mechanical basis of deformation, 333
  statement of structural problem, 9
  trends of structural features, 8, 158, 162–163
  ultimate causes of deformation, 31–34, 331, 355–365
  vertical movements in, 15–21, 162–163, 230

Sanford-type sand-box experiments, 18
  application to basement, 18, 342
  application to layered rocks, 19
  misapplication to layered rocks, 120, 236, 264
Sedimentary rocks. See Layered rocks
Seismic interpretation of drape folds as faults, 40
Seismic migrated versus nonmigrated time sections, 43, 46
Seismic modeling, 39–42
Seismic pseudostructures, 40
Seismic ray-path diagram, 40–41
Seismic synthetic section, 40–41
Shear fracture
  as indicators of movement of material toward corners of drape fold, 154
  trajectories, 17, 323–324, 338–344

Shear stress, amount needed for slip, 3
Shortening
  of the basement, 11, 75, 136, 149, 230–231
  normal to fold axis, 206
State of stress
  calculations, 341
  stability index, 344–346
Stratigraphic package. See Layered rocks
Stress
  at corner of basement block, 30–31, 86–87, 119
  gravitationally induced, 131
  normal, 3, 5
  orientation. See Layer-parallel orientation of σ_1
  reorientation, 346–352
  trajectories in experimental drape folds, 82, 87, 89
  trajectories from theory, 323–324, 341–342
Subduction modes, 358–359

Terrace structures, 172, 174–179, 218
Thrust-block hypothesis and basement-block-uplift hypothesis comparison for Piney Creek area, Wyoming, 129–131
Thickening. See Layered rocks
Thinning. See Layered rocks
Timing of monocline formation in Colorado Plateau, 227–228

Upthrust problem 317–318

WITHDRAWN